VOLUME EIGHTY FIVE

Current Topics in MEMBRANES

Ion Channels and Calcium Signaling in the Microcirculation

CURRENT TOPICS IN MEMBRANES, VOLUME 85

Series Editor

IRENA LEVITAN
Division of Pulmonary and Critical Care Medicine
University of Illinois at Chicago
Chicago, IL, United States

VOLUME EIGHTY FIVE

Current Topics in MEMBRANES

Ion Channels and Calcium Signaling in the Microcirculation

Edited by

WILLIAM F. JACKSON

Department of Pharmacology and Toxicology, Michigan State University, East Lansing, MI, United States

ELSEVIER

ACADEMIC PRESS

An imprint of Elsevier

Academic Press is an imprint of Elsevier
50 Hampshire Street, 5th Floor, Cambridge, MA 02139, United States
525 B Street, Suite 1650, San Diego, CA 92101, United States
The Boulevard, Langford Lane, Kidlington, Oxford OX5 1GB, United Kingdom
125 London Wall, London, EC2Y 5AS, United Kingdom

First edition 2020

ISBN: 978-0-12-820090-2
ISSN: 1063-5823

For information on all Academic Press publications
visit our website at https://www.elsevier.com/books-and-journals

Publisher: Zoe Kruze
Acquisitions Editor: Ashlie Jackman
Editorial Project Manager: Shellie Bryant
Production Project Manager: Abdulla Sait
Cover Designer: Alan Studholme

Typeset by SPi Global, India

Contents

Contributors

Sher Ali
Department of Pharmacology, Center for Molecular and Cellular Signaling in the Cardiovascular System, University of Nevada, Reno, School of Medicine, Reno, NV, United States

Josy Augustine
Wellcome-Wolfson Institute for Experimental Medicine, Queen's University of Belfast, Belfast, United Kingdom

Peter Barabas
Wellcome-Wolfson Institute for Experimental Medicine, Queen's University of Belfast, Belfast, United Kingdom

Laura Chambers
Department of Pharmacology and Toxicology, Michigan State University, East Lansing, MI, United States

Yen-Lin Chen
Robert M. Berne Cardiovascular Research Center, University of Virginia-School of Medicine, Charlottesville, VA, United States

Tim M. Curtis
Wellcome-Wolfson Institute for Experimental Medicine, Queen's University of Belfast, Belfast, United Kingdom

Kim A. Dora
Department of Pharmacology, University of Oxford, Oxford, United Kingdom

Anne M. Dorrance
Department of Pharmacology and Toxicology, Michigan State University, East Lansing, MI, United States

Scott Earley
Department of Pharmacology, Center for Molecular and Cellular Signaling in the Cardiovascular System, University of Nevada, Reno, School of Medicine, Reno, NV, United States

Ibra S. Fancher
Division of Pulmonary, Critical Care, Sleep and Allergy, Department of Medicine, University of Illinois at Chicago, Chicago, IL, United States

José A. Fernández
Wellcome-Wolfson Institute for Experimental Medicine, Queen's University of Belfast, Belfast, United Kingdom

Daniela C.G. Garcia
Department of Physiology, School of Medicine, University of Maryland, Baltimore, MD, United States

Christopher J. Garland
Department of Pharmacology, University of Oxford, Oxford, United Kingdom

William F. Jackson
Department of Pharmacology and Toxicology, Michigan State University, East Lansing, MI, United States

Hamish A.L. Lemmey
Department of Pharmacology, University of Oxford, Oxford, United Kingdom

Irena Levitan
Division of Pulmonary, Critical Care, Sleep and Allergy, Department of Medicine, University of Illinois at Chicago, Chicago, IL, United States

Thomas A. Longden
Department of Physiology, School of Medicine, University of Maryland, Baltimore, MD, United States

Mary K. McGahon
Wellcome-Wolfson Institute for Experimental Medicine, Queen's University of Belfast, Belfast, United Kingdom

J. Graham McGeown
Wellcome-Wolfson Institute for Experimental Medicine, Queen's University of Belfast, Belfast, United Kingdom

Frederick C. Monson
West Chester University, West Chester, PA, United States

Maria Sancho
Robarts Research Institute and the Department of Physiology & Pharmacology, University of Western Ontario, London, ON, Canada

Swapnil K. Sonkusare
Robert M. Berne Cardiovascular Research Center; Department of Molecular Physiology and Biological Physics, University of Virginia-School of Medicine, Charlottesville, VA, United States

Pratish Thakore
Department of Pharmacology, Center for Molecular and Cellular Signaling in the Cardiovascular System, University of Nevada, Reno, School of Medicine, Reno, NV, United States

Nathan R. Tykocki
Michigan State University, East Lansing, MI, United States

Donald G. Welsh
Robarts Research Institute and the Department of Physiology & Pharmacology, University of Western Ontario, London, ON, Canada

Preface

Ion channels are transmembrane proteins that form water-filled channels through cell and organelle membranes. They provide a controlled pathway for diffusion of charged, inorganic ions to cross lipid bilayer membranes down their electrochemical gradients and are essential for all life functions of cells in our body. Despite the explosion in information about ion channel expression, structure, and function during the past 35 years, our understanding of the patterns of expression and function of ion channels in the membranes of cells that make up the microcirculation has lagged.

As outlined in the introduction to this volume (Chapter 1), the microcirculation really is the business end of the cardiovascular system. With few exceptions, every cell in our body depends on the microcirculation—without which they would suffocate, starve, and drown in their own waste products. The devasting consequences of ischemic stroke and myocardial infarction are but two graphic examples of the consequences of lack of blood flow to the microcirculation. Our understanding of the expression and function of ion channels in vascular smooth muscle and endothelial cells that make up the vessels in the microcirculation is challenging because of the difficulty in isolating cells from specific segments of microvascular beds, and also the heterogeneity of structure and function of the microcirculation in different organs and tissues. However, specific knowledge of the expression and function of ion channels and the Ca^{2+} and membrane potential signals that they generate is becoming apparent. This was the impetus for the present volume: to provide a glimpse at the ion channels and function of these channels in cells in several microvascular beds by leaders in the field who are actively trying to remedy the paucity of information on this important part of the cardiovascular system.

Chapter 1 provides an introduction and overview of the structure and function of the microcirculation and lays out the central role played by ion channels and Ca^{2+} signals in these processes. Chapter 2 addresses the major ion channels involved in the generation and maintenance of pressure-induced myogenic tone in peripheral arterioles. The regulation of inward rectifier channels by fluid shear stress in endothelial cells is the topic of Chapter 3. The important role played by two members of the transient receptor potential family of ion channels, TRPV4 and TRPA1, in endothelium-dependent vasodilation of resistance arteries and arterioles is addressed in Chapters 4 and 5, respectively. Vascular ion channel expression and function controlled by the mineralocorticoid receptor in health and disease is the focus of Chapter 6. Chapter 7 looks at the expression and function of ion channels and myogenic tone in retinal arterioles and their modulation during diseases such as diabetes. The regulation of microvascular smooth muscle and endothelial cell inwardly rectifying K^+ channels by membrane lipids is the topic of Chapter 8. The heterogeneity of expression and function of ion channels in brain capillary and arteriolar endothelial cells is covered in Chapter 9. Chapter 10 reviews the function of the urinary bladder and its vasculature and shows how inwardly rectifying K^+ channels depress myogenic reactivity of urinary bladder arterioles which appears important for their function in health and disease. Finally, Chapter 11 focuses on a review of the ion channels involved in communication between endothelial cells and smooth muscle cells and the negative feedback process in the regulation of arteriolar myogenic- and agonist-induced tone.

As will be apparent in the chapters that make up this volume, our understanding of the expression and function of ion channels and the Ca^{2+} signals that they generate in cells that comprise the microcirculation is on the rise. It will also become apparent that our journey to understand the control and function of the microcirculation remains incomplete, particularly with regard to the changes that lead to or are caused by the disease states that plague developed societies: hypertension, diabetes, and obesity. It is hoped that this volume will encourage more study of the body's microcirculations, which may provide us with new targets for preventive strategies and treatments to combat diseases that impact the microcirculation.

Acknowledgments

Many thanks to the authors of this volume for their contributions and for their assistance with the review process for each chapter. Special thanks to Dr. Irena Levitan, University of Illinois, Chicago who hatched the original idea for this volume in addition to contributing a chapter. Gratitude is also offered to Dr. Nathan Tykocki, Michigan State University, for Chapter 10 of this volume and for his assistance with the editing process.

William F. Jackson
Department of Pharmacology and Toxicology
Michigan State University
East Lansing, MI, United States

CHAPTER ONE

Introduction to ion channels and calcium signaling in the microcirculation

William F. Jackson*
Department of Pharmacology and Toxicology, Michigan State University, East Lansing, MI, United States
*Corresponding author: e-mail address: jacks783@msu.edu

Contents

Abstract

The microcirculation is the network of feed arteries, arterioles, capillaries and venules that supply and drain blood from every tissue and organ in the body. It is here that exchange of heat, oxygen, carbon dioxide, nutrients, hormones, water, cytokines, and immune cells takes place; essential functions necessary to maintenance of homeostasis throughout the life span. This chapter will outline the structure and function of each microvascular segment highlighting the critical roles played by ion channels in the microcirculation. Feed arteries upstream from the true microcirculation and arterioles within the microcirculation contribute to systemic vascular resistance and blood pressure control. They also control total blood flow to the downstream microcirculation with arterioles being responsible for distribution of blood flow within a tissue or organ dependent on the metabolic needs of the tissue. Terminal arterioles control blood flow and blood pressure to capillary units, the primary site of diffusional exchange between blood and tissues due to their large surface area. Venules collect blood from capillaries and are important sites for fluid exchange and immune cell trafficking. Ion channels in microvascular smooth muscle cells, endothelial cells and pericytes importantly contribute to all of these functions through generation of intracellular Ca^{2+} and membrane potential signals in these cells.

Current Topics in Membranes, Volume 85
ISSN 1063-5823
https://doi.org/10.1016/bs.ctm.2020.01.001

1. What is the microcirculation?

The microcirculation is the network of feed arteries, arterioles, capillaries and venules that supply and drain blood from every tissue and organ in the body (Fig. 1). The specific architecture of this system of microvessels (vessel numbers, lengths and orientation) differs dependent on the structure of the parenchyma that they perfuse. Despite structural differences, the constant feature of the microcirculation is that it is the business-end of the cardiovascular system. It is here that exchange of heat, oxygen, carbon dioxide, nutrients, hormones, water, cytokines, and immune cells takes place; essential functions necessary for maintenance of homeostasis throughout the life span (Durán, Sánchez, & Breslin, 2011; Renkin, 1984). Dysregulation of the microcirculation accompanies diseases such as diabetes (Edmonds, 2019; Frisbee, Lewis, & Wiseman, 2019; Strain & Paldanius, 2018), hypertension (Csiszar et al., 2017; Rizzoni et al., 2019; Santisteban & Iadecola, 2018; Ziegler, Abdel Rahman, Jurisch, & Kupatt, 2019) and obesity (Sorop et al., 2017) and may, in fact, be a causal factor in the development of some pathologies (see Chapter "Regulation of ion channels in the microcirculation by mineralocorticoid receptor activation" by Chambers and Dorrance). For example, some forms of dementia (van der Flier et al., 2018) and coronary disease in women (Sidik, Morrow, & Berry, 2020) may be caused by dysfunction of the microcirculation in these organs.

2. What are the components of the microcirculation?

2.1 Feed arteries and arterioles

Feed arteries and arterioles are the primary sites of hydraulic resistance to blood flow in the body (Segal, 2000; Zweifach & Lipowsky, 1984). Changes in their caliber, effected by contraction (vasoconstriction) or relaxation (vasodilation) of the VSMCs in their walls, strongly determines their resistance to blood flow, local and systemic blood pressure, and the distribution of flow between and within the body's tissues and organs (Segal, 2000; Zweifach & Lipowsky, 1984). As will be outlined in a number of chapters in this volume, ion channels and the Ca^{2+} signals that they generate are at the heart of VSMC contractile activity and maintenance and changes in VSMC contractile tone.

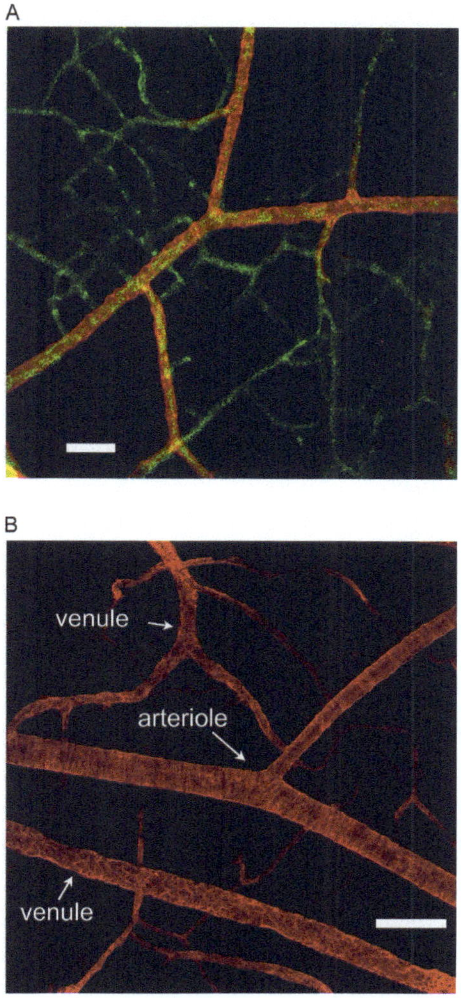

Fig. 1 Microvascular networks. (A) Terminal arterioles branch into capillaries. Methanol-fixed, hamster cheek pouch preparation labeled with a mouse monoclonal anti-α-smooth muscle actin antibody (Sigma, St. Louis, MO) to stain smooth muscle cells (red), and an anti-CD31 (PECAM-1) antibody (BD Pharmingen, San Diego, CA) to stain endothelial cells (green). Shown is a small fourth-order arteriole branching into terminal arterioles (red and yellow), which then branch into capillaries (green). Scale bar = 50 μm. (B) Smooth muscle cells in arterioles and pericytes in venules. Methanol-fixed hamster cheek pouch preparation labeled with mouse monoclonal anti-α-smooth muscle actin antibody (Sigma, St. Louis, MO) to stain smooth muscle cells (circumferential banding pattern on arterioles) and pericytes (web-like pattern on venules). Shown is a second-order arteriole with a third-order branch, parallel with a collecting venule (bottom). At the top of the image, post-capillary venules combine to form a small collecting venule. Scale bar = 100 μm. *From Jackson, W. F. (2012). Microcirculation. In J. A. Olson & E. N. Hill (Eds.),* Muscle *(pp. 1197–1206). Boston/Waltham: Academic Press with permission.*

Feed arteries are the small arteries immediately proximal to arterioles; they are the last arterial branch of the cardiovascular system before the true beginning of the microcirculation. Feed arteries have been included in this Introduction because these vessels, also called resistance arteries, contribute significantly to peripheral vascular resistance (30–40%) and because coordinated vasodilation/vasoconstriction of feed arteries when downstream arterioles dilate/constrict is essential for proper delivery of blood flow to the downstream microcirculation to meet the metabolic demands of active tissues and organs (Segal, 2000; Segal & Duling, 1986). This coordinated activity is enacted by conduction of electrical signals (membrane hyperpolarization or depolarization) along the endothelium and communication of these signals to overlying VSMCs via both homocellular (EC to EC) and heterocellular (EC to VSMC) gap junctions (Socha, Behringer, & Segal, 2012) (see Chapters "Endothelial TRPV4 channels and vasodilator reactivity" by Chen and Sonkusare; "Regulation of vascular tone by transient receptor potential ankyrin 1 channels" by Thakore et al.; "Ion channels in capillary endothelium" by Garcia and Longden; "Intrinsic regulation of microvascular tone by myoendothelial feedback circuits" by Lemmey et al. in this volume for more information), with ion channels being the source of the altered membrane potential and contributing to the membrane properties that allow cell-to-cell conduction of the ion channel-generated electrical signals (Socha et al., 2012) (see Chapters "Endothelial inwardly-rectifying K^+ channels as a key component of shear stress-induced mechanotransduction" by Fancher and Levitan; "Endothelial TRPV4 channels and vasodilator reactivity" by Chen and Sonkusare; "Regulation of vascular tone by transient receptor potential ankyrin 1 channels" by Thakore et al.; "K_{IR} channels in the microvasculature: Regulatory properties and the lipid-hemodynamic environment" by Sancho and Welsh; "Ion channels in capillary endothelium" by Garcia and Longden; "Intrinsic regulation of microvascular tone by myoendothelial feedback circuits" by Lemmey et al. in this volume).

Feed arteries and arterioles consist of an EC tube (with ECs oriented parallel to the long axis and in the direction of flow) and the basal lamina associated with the ECs, an internal elastic lamina, overlying VSMCs oriented perpendicular to the long axis of the vessels and adventitial connective tissue (Rhodin, 2014; Simionescu & Simionescu, 1984). There are periodic perforations in the internal elastic lamina that allow projections of EC and VSMC membranes to contact the other cell type forming myoendothelial projections (Fig. 2). These projections may contain gap junction connexin

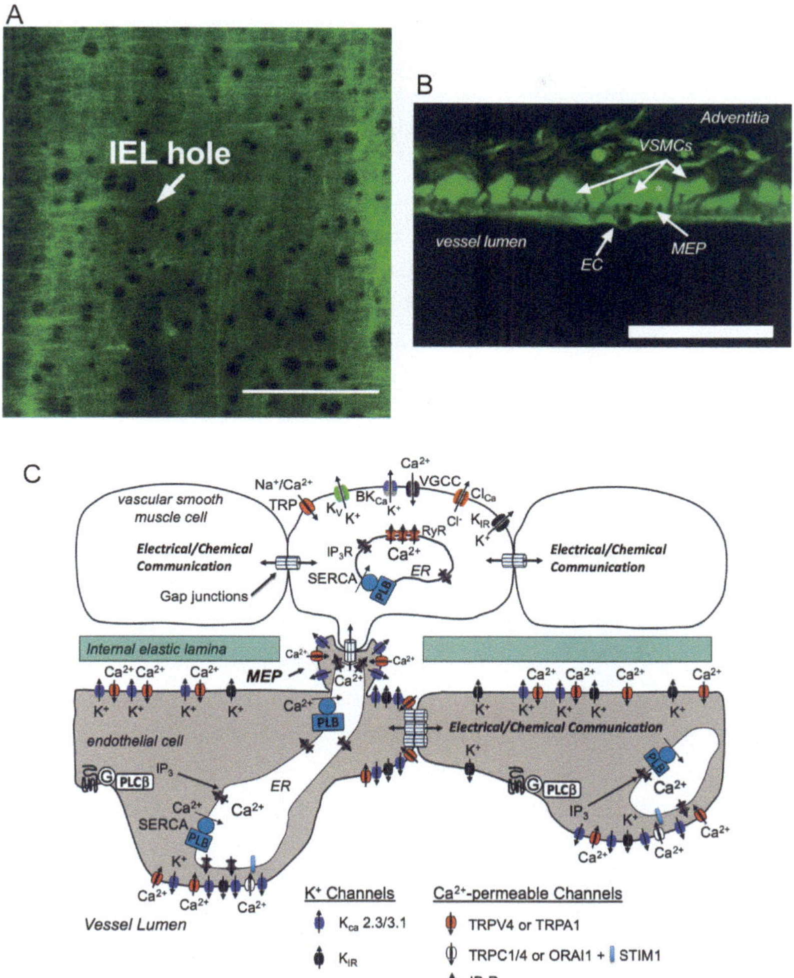

Fig. 2 Holes in the internal elastic lamina (IEL) allow communication between vascular smooth muscle cells (VSMCs) and endothelial cells (ECs). Panel (A) shows an en face z-projection (sum of 34 images, 0.58 μm), two-photon autofluorescence image (850 nm excitation, 500–550 emission) of the IEL in an excised, pressurized (80 cm H_2O) rat mesenteric artery. White arrow points to a hole in the IEL that allows for myoendothelial projections to pass through the IEL. The scale bar = 100 μm. Panel (B) shows a cross sectional image of one wall of an excised, pressurized (60 cm H_2O) hamster cremaster arteriole. The VSMCs and ECs were loaded with the cell permeant, calcein acetoxymethyl ester which fluoresces green (510–530 nm) when excited with two-photon illumination (725 nm). A longitudinal cross section of an EC is shown on the lumen side of the wall as indicated, with multiple VSMCs overlying the EC. A white arrow points to a myoendothelial projection (MEP) in one VSMC as shown. Note that the VSMC labeled * in the figure has multiple MEPs approaching the underlying EC.

(Continued)

proteins to form myoendothelial junctions allowing electrical and small molecule communication between ECs and VSMCs (Fig. 2) (Little, Beyer, & Duling, 1995a; Little, Xia, & Duling, 1995b; Rhodin, 2014; Sandow & Hill, 2000). See Chapters "Endothelial TRPV4 channels and vasodilator reactivity" by Chen and Sonkusare; "Regulation of vascular tone by transient receptor potential ankyrin 1 channels" by Thakore et al.; "Ion channels in capillary endothelium" by Garcia and Longden; "Intrinsic regulation of microvascular tone by myoendothelial feedback circuits" by Lemmey et al. in this volume for more information on myoendothelial communication.

What distinguishes feed arteries from downstream first-order arterioles is their proximity to the parenchymal cells that they perfuse, and the number of layers of VSM that comprise the media of these vessels. Feed arteries are usually external to the parenchyma and have at least two layers of VSMCs in their walls, whereas arterioles are embedded in the parenchyma that they perfuse and typically have only a single layer of VSMCs.

First order arterioles branch from feed arteries, and this branching process continues for 3–5 generations (dependent on the tissue), ending in terminal arterioles, the last vessels endowed with VSMCs prior to the beginning of a capillary bed (Fig. 1). Arterioles contribute significantly to vascular resistance, blood pressure and their control in the cardiovascular system (Pries & Secomb, 2011; Renkin, 1984; Zweifach & Lipowsky, 1984). For example, the arteriolar network in skeletal muscle contributes 50% of the vascular resistance of this tissue (Fronek & Zweifach, 1975). In addition to contributing to systemic vascular resistance and blood

Fig. 2—Cont'd The fibrous material above the VSMCs is the autofluorescence of adventitial connective tissue as indicated. The scale bar = 25 μm. Panel (C) is a schematic representation of a cross section of one wall of an arteriole (as in Panel B) showing a MEP passing through a hole in the IEL. Heterocellular gap junctions are present allowing electrical and chemical (Ca^{2+}, IP_3, etc.) communication between ECs and VSMCs. Also shown are homocellular (EC-EC and VSMC-VSMC) gap junctions that also allow electrical and chemical communication as shown. A few classes of ion channels expressed by arteriolar VSMCs and ECs are also shown. TRP, transient receptor potential channels; K_V, voltage-gated K^+ channels; BK_{Ca}, large-conductance Ca^{2+}-activated K^+ channels; VGCC, voltage-gated Ca^{2+} channels, Cl_{Ca}, Ca^{2+}-activated Cl^- channels; K_{IR}, inwardly rectifying K^+ channels; IP_3R, inositol 1,4,5 trisphosphate receptor; RyR, ryanodine receptor; PLC, phospholipase C; SERCA, smooth endoplasmic reticulum Ca^{2+} ATPase; PLB, phospholamban; ORAI1 and STIM1 are components of store-operated Ca^{2+} channels found in many cells. *Panel (C): Adapted from Jackson, W. F. (2016). Endothelial cell ion channel expression and function in arterioles and resistance arteries. In I. Levitan & A. M. Dopico (Eds.),* Vascular ion channels in physiology and disease *(pp. 3–36). Switzerland: Springer International Publishing.*

pressure, arteriolar vascular resistance also importantly determines the flow rate of blood to and within organs and tissues providing convective transfer of oxygen, nutrients, hormones, blood cells, etc., for exchange between blood and the parenchyma (Davis, Hill, & Kuo, 2011; Durán et al., 2011; Renkin, 1984). Arteriolar vascular resistance also determines the blood pressure transmitted to downstream segments of the microcirculation affecting microvascular fluid exchange (Davis et al., 2011; Durán et al., 2011; Renkin, 1984). As with the upstream feed arteries, ion channels in VSMCs and ECs, and the Ca^{2+} signals that they generate importantly participate in all of these arteriolar functions.

2.2 Capillaries

Capillaries (1 − 20) branch from terminal arterioles and are long (100–1000 μm) endothelial cell tubes (5–7 μm in diameter) surrounded by a basement membrane and lacking a layer of VSMCs (Fig. 1A) (Rhodin, 2014; Simionescu & Simionescu, 1984). In skeletal muscle, for example, terminal arterioles give rise to 10–20 capillaries forming a microvascular unit; changes in VSMC tone in the terminal arterioles produce coordinated changes in blood flow through the capillaries within a unit (Segal, 2005). Microvascular units are also found in other organs and tissues, however, the number of capillaries within a unit may vary. Terminal arterioles, by controlling the number of capillaries perfused, also determine the effective surface area for diffusional exchange between blood and the parenchyma in addition to controlling the blood flow rate and convective transfer to the capillary units (Renkin, 1984; Segal, 2005; Wiedeman, 1984). Despite the lack of VSMCs, changes in the diameter of capillaries and their hydraulic resistance to blood flow can be produced by pericytes that invest the walls of some capillaries (Hamilton, Attwell, & Hall, 2010; Krueger & Bechmann, 2010; Kutcher & Herman, 2009; Pallone, Zhang, & Rhinehart, 2003; Puro, 2007). Although not covered in this volume, like VSMCs in arterioles, ion channels and Ca^{2+} signals also importantly contribute to the function of pericytes (Berweck, Lepple-Wienhues, Stöss, & Wiederholt, 1994; Burnette & White, 2006; Cao, Lee-Kwon, Silldorff, & Pallone, 2005; Hashitani & Lang, 2016; Li & Puro, 2001; Puro, 2007; Quignard, Harley, Duhault, Vanhoutte, & Feletou, 2003; Rhinehart, Zhang, & Pallone, 2002; Wu, Kawamura, Li, & Puro, 2001; Zhang et al., 2008; Zhang, Lin, Cao, Khurana, & Pallone, 2010; Zhang, Rhinehart, & Pallone, 2002).

While capillaries are the major site of exchange of heat, respiratory gases and solutes between blood and tissues largely because of their vast surface area available for exchange, it is worthy to note that there is significant diffusional exchange of O_2 across arterioles (Davis et al., 2011; Duling & Berne, 1970) and diffusional exchange of O_2 and CO_2 between arterioles and venules with which they are often paired (Fig. 1) (Johnson, 2011). Arteriolar VSMC tone also appears to be modulated by diffusion of vasoactive substances from adjacent venules (Davis et al., 2011; Falcone & Bohlen, 1990; Hester, 1990, 1993, 1995; Hester & Hammer, 2002; McKay, Gardner, Boyd, & Hester, 1998; Saito, Eraslan, & Hester, 1993; Saito, Eraslan, Lockard, & Hester, 1994; Segal, 2005). Thus, the exchange function of the microcirculation is not limited to capillaries.

Tissue and organ blood flow in health is tightly coupled to the metabolic activity of the parenchyma to which the blood is supplied: increases in the metabolic activity of the parenchyma results in a proportional increase in blood flow to supply the O_2 and substrates (glucose, fatty acids, etc.) and remove waste (CO_2, H^+, etc.) that sustain parenchymal cell metabolic activity. Efficient increases in blood flow to capillaries supplying active parenchymal cells (functional hyperemia) requires communication of the increased activity in the parenchyma to the capillaries and importantly to upstream arterioles which are the major control point for blood flow to the capillaries (Segal, 2005). As with coordination of dilation of feed arteries when downstream arterioles dilate, at least in the brain, this appears to be accomplished by retrograde conduction of Ca^{2+} and electrical signals from the capillaries along the endothelium (via EC-EC gap junction) to the VSMCs in upstream arterioles via myoendothelial gap junctions (Longden et al., 2017) (see Chapter "Ion channels in capillary endothelium" by Garcia and Longden for more information.). The chapter by Garcia and Longden also outlines differences in the expression and function of ion channels in capillary vs arteriolar ECs.

2.3 Venules

Capillaries (two or more) converge to form post-capillary collecting venules. Post-capillary venules consist of an endothelial cell tube and basement membrane with no overlying VSMCs in their walls. Instead, post-capillary venules are covered by a layer of pericytes (Rhodin, 2014; Shepro & Morel, 1993; Simionescu & Simionescu, 1984). Vascular smooth muscle cells (1–2 layers) are found in the wall of higher-order venules (Altura, 1978; Rhodin, 2014). Venules are important sites for microvascular fluid exchange

(Durán et al., 2011; Renkin, 1984), adhesion and transmigration of leukocytes and lymphocytes (Kogan & von Andrian, 2011; Ley, 2011) and also contribute to vascular capacitance (Rothe, 2011). The hemodynamic resistance of the venular side of the microcirculation is less than that of the arteriolar side because there are generally a larger number of venules of a given size compared to arterioles (Wiedeman, 1984; Zweifach & Lipowsky, 1984). Nonetheless, changes in the diameter of venules and their hemodynamic resistance to blood flow can have a significant effect on capillary and post-capillary venular hydrostatic pressure affecting fluid exchange across these microvessels and blood flow through the microcirculation (Davis et al., 2011). Although the ion channels expressed in the venular side of the microcirculation and the Ca^{2+} signals that they generate have not been as well studied, it is well established that venular fluid permeability correlates with increases in endothelial Ca^{2+} (Yuan, 2000; Zhou & He, 2010). Histamine-induced (Chen et al., 2013) and VEGF-induced (Pocock, Foster, & Bates, 2004) increases in microvascular fluid permeability appear to involve Ca^{2+} influx through TRPC6 channels. Spontaneous vasomotion and basal myogenic tone of rat gastric venules involves Ca^{2+} influx through nifedipine-sensitive L-type voltage-gated Ca^{2+} channels, Cl^- efflux through Ca^{2+}-activated Cl^- channels and release of Ca^{2+} from intracellular stores through both IP_3 receptors and ryanodine receptors in venular SMCs (Mitsui & Hashitani, 2016). Myogenic tone of porcine retinal venules also depends on Ca^{2+} influx through nifedipine-sensitive L-type Ca^{2+} channels (Chen et al., 2018). However, studies in hamster cheek pouch venules have shown that basal tone is insensitive to nifedipine, but instead is inhibited by blockers of N-type voltage-gated Ca^{2+} channels (Kushiro et al., 2004) suggesting species or regional heterogeneity in ion channel expression and function in venular SMCs. Additional research would seem to be warranted on the role played by ion channels in venules. Nonetheless, it is apparent that like upstream capillary ECs and arteriolar and resistance artery ECs and VSMCs, ion channels in venular ECs and SMCs importantly contribute to the function of venules. Chapter "Excitability and contractility in arterioles and venules from the urinary bladder" by Tykocki and Monson in this volume will explore ion channel expression and function in bladder venules compared with upstream arterioles.

2.4 Lymphatics

In parallel with the blood microcirculation is the lymphatic microcirculation that drains fluid filtered across capillaries and venules and serves as a pathway for transport of antigen presenting cells and lymphocytes to lymphoid organs (Aspelund, Robciuc, Karaman, Makinen, & Alitalo, 2016; Zawieja, von der Weid, & Gashev, 2011). Although not covered in this volume, ion channels and Ca^{2+} signaling in lymphatic SMCs and ECs, also participate in the function of these microvessels (Zawieja et al., 2011).

3. Ion channels and Ca^{2+} signaling in the microcirculation

As will be detailed in the chapters of this volume, ion channels in the plasma and intracellular membranes of vascular smooth muscle cells (VSMCs) and endothelial cells (ECs) and the intracellular Ca^{2+} signals that ion channels generate importantly contribute to all microvascular functions and the maintenance of homeostasis. Ion channels provide a major source of intracellular Ca^{2+} signals in both VSMCs (Tykocki, Boerman, & Jackson, 2017) (See Chapters "Ion channels and the regulation of myogenic tone in peripheral arterioles" by Jackson; "Ion channels and myogenic activity in retinal arterioles" by Barabas et al.; "Intrinsic regulation of microvascular tone by myoendothelial feedback circuits" by Lemmey et al. in this volume) and ECs (Jackson, 2016; Thakore & Earley, 2019) (See Chapters "Endothelial TRPV4 channels and vasodilator reactivity" by Chen and Sonkusare; "Regulation of vascular tone by transient receptor potential ankyrin 1 channels" by Thakore et al.; "Ion channels in capillary endothelium" by Garcia and Longden; "Intrinsic regulation of microvascular tone by myoendothelial feedback circuits" by Lemmey et al. in this volume). Intracellular Ca^{2+}, in turn, is the trigger for contraction (changes in tone) of VSMCs (Bolton, Prestwich, Zholos, & Gordienko, 1999; Harnett & Biancani, 2003; Kim, Appel, Vetterkind, Gangopadhyay, & Morgan, 2008; Lee, Poburko, Kuo, Seow, & van Breemen, 2002) (See Chapters "Ion channels and the regulation of myogenic tone in peripheral arterioles" by Jackson; "Ion channels and myogenic activity in retinal arterioles" by Barabas et al. in this volume), controls production of endothelial autacoids (Edwards, Feletou, & Weston, 2010) (See Chapters "Endothelial TRPV4 channels and vasodilator reactivity" by Chen and Sonkusare; "Regulation of vascular tone by transient receptor potential ankyrin 1 channels" by Thakore et al.; "Intrinsic regulation of microvascular tone by myoendothelial feedback circuits" by Lemmey et al. in this volume), modulates endothelial permeability (Goddard & Iruela-Arispe, 2013; Komarova & Malik, 2010), contributes to volume regulation in ECs and VSMCs (Hoffmann, Lambert, & Pedersen, 2009) and participates in the control of gene expression and proliferation of both cell types (Guan, Wang, & Zhou, 2006; Jackson, 2005; McCabe & Young, 1994; Munaron, 2006; Nilius & Droogmans, 2001; Vallot et al., 2000; Wang et al., 2001). Ion channels also importantly contribute to the determination and control of membrane potential (Offner, 1991). In VSMCs, membrane potential controls the open state probability of voltage-gated Ca^{2+} channels, a major source of activator Ca^{2+} in these cells (Tykocki et al., 2017). Membrane potential also affects the electrochemical gradient, the "force" that governs the movement of ions through open membrane channels (Offner, 1991), impacting the function of all ion channels in both ECs and VSMCs (Jackson, 2000, 2016). There is also evidence that membrane potential may affect Ca^{2+} release from intracellular stores as well as the Ca^{2+} sensitivity of contraction in VSMCs (del Valle-Rodriguez, Lopez-Barneo, & Urena, 2003; Fernández-Tenorio et al., 2010, 2011; Ganitkevich & Isenberg, 1993; Kukuljan, Rojas, Catt, & Stojilkovic, 1994; Liu et al., 2009; Mahaut-Smith, Martinez-Pinna, & Gurung, 2008; Okada, Yanagisawa, & Taira, 1993; Urena, del Valle-Rodriguez, & Lopez-Barneo, 2007; Yamagishi, Yanagisawa, & Taira, 1992; Yamamura, Ohya, Muraki, & Imaizumi, 2012; Yanagisawa, Yamagishi, & Okada, 1993). Changes in membrane potential are also an important signal for cell-cell communication in the wall of blood vessels (Socha et al., 2012) (See Chapters "Endothelial TRPV4 channels and vasodilator reactivity" by Chen and Sonkusare; "Regulation of vascular tone by transient receptor potential ankyrin 1 channels" by Thakore et al.; "Ion channels in capillary endothelium" by Garcia and Longden; "Intrinsic regulation of microvascular tone by myoendothelial feedback circuits" by Lemmey et al. in this volume for details). Thus, ion channels and the Ca^{2+} signals that they generate importantly impact all aspects of microvascular function.

4. Ion channels and the microcirculation: Future directions

A common theme that will arise in many of the subsequent chapters of this volume is that of heterogeneity in the ion channels expressed, the Ca^{2+} signals that the ion channels generate, and the functions of these Ca^{2+} signals in VSMCs, ECs and pericytes in the microcirculation. However, we know little about the mechanisms responsible for this heterogeneity which should be a topic for future research. Nonetheless, a significant consequence of this heterogeneity is that we cannot necessarily extrapolate from information gleaned from one microvascular bed to another and particularly from one segment of the microcirculation to another. For example, as will be shown in chapter "Ion channels and the regulation of myogenic tone in peripheral arterioles" by Jackson, ion channel function, Ca^{2+} signaling and the regulation of in skeletal muscle feed artery VSMCs differs from that in downstream arteriolar VSMCs and chapter "Ion channels in capillary endothelium" by Garcia and Longden, will show that ion channel expression and function differs between capillary and arteriolar ECs in the brain. Chapters "Regulation of ion channels in the microcirculation by mineralocorticoid receptor activation" by Chambers and Dorrance; "K_{IR} channels in the microvasculature: Regulatory properties and the lipid-hemodynamic environment" by Sancho and Welsh will show that the function and regulation of the same ion channels may differ between VSMCs and ECs. Chapter "Ion channels and myogenic activity in retinal arterioles" by Barabas et al. will show that ryanodine receptors and the Ca^{2+} sparks that they generate serve a positive feedback role to promote contraction of retinal arteriolar VSMCs. In contrast, in skeletal muscle feed artery VSMCs (Westcott, Goodwin, Segal, & Jackson, 2012; Westcott & Jackson, 2011) and in brain resistance artery VSMCs (Nelson et al., 1995), ryanodine receptors and Ca^{2+} sparks participate in the negative feedback regulation of myogenic tone, while in skeletal muscle arterioles, ryanodine receptors are silent and do not generate Ca^{2+} sparks (Westcott et al., 2012; Westcott & Jackson, 2011) (Chapter "Ion channels and the regulation of myogenic tone in peripheral arterioles" by Jackson). Thus, each microvascular bed should be separately evaluated, which means that pattern of ion channel expression and their function in microvascular VSMCs and ECs from each segment of different vascular beds should be studied. This should provide ample research opportunities for the next generation of microcirculatory scientists.

References

Altura, B. M. (1978). Pharmacology of venular smooth muscle: New insights. *Microvascular Research*, *16*(1), 91–117.

Aspelund, A., Robciuc, M. R., Karaman, S., Makinen, T., & Alitalo, K. (2016). Lymphatic system in cardiovascular medicine. *Circulation Research*, *118*(3), 515–530. https://doi.org/10.1161/CIRCRESAHA.115.306544.

Berweck, S., Lepple-Wienhues, A., Stöss, M., & Wiederholt, M. (1994). Large conductance calcium-activated potassium channels in cultured retinal pericytes under normal and high-glucose conditions. *Pflügers Archiv*, *427*, 9–16.

Bolton, T. B., Prestwich, S. A., Zholos, A. V., & Gordienko, D. V. (1999). Excitation-contraction coupling in gastrointestinal and other smooth muscles. *Annual Review of Physiology*, *61*, 85–115.

Burnette, J. O., & White, R. E. (2006). PGI2 opens potassium channels in retinal pericytes by cyclic AMP-stimulated, cross-activation of PKG. *Experimental Eye Research*, *83*(6), 1359–1365. https://doi.org/10.1016/j.exer.2006.07.011.

Cao, C., Lee-Kwon, W., Silldorff, E. P., & Pallone, T. L. (2005). KATP channel conductance of descending vasa recta pericytes. *American Journal of Physiology. Renal Physiology*, *289*(6), F1235–F1245. https://doi.org/10.1152/ajprenal.00111.2005.

Chen, W., Oberwinkler, H., Werner, F., Gassner, B., Nakagawa, H., Feil, R., et al. (2013). Atrial natriuretic peptide-mediated inhibition of microcirculatory endothelial Ca^{2+} and permeability response to histamine involves cGMP-dependent protein kinase I and TRPC6 channels. *Arteriosclerosis, Thrombosis, and Vascular Biology*, *33*(9), 2121–2129. https://doi.org/10.1161/atvbaha.113.001974.

Chen, Y. L., Ren, Y., Xu, W., Rosa, R. H., Jr., Kuo, L., & Hein, T. W. (2018). Constriction of retinal venules to endothelin-1: Obligatory roles of ETA receptors, extracellular calcium entry, and Rho kinase. *Investigative Ophthalmology & Visual Science*, *59*(12), 5167–5175. https://doi.org/10.1167/iovs.18-25369.

Csiszar, A., Tarantini, S., Fulop, G. A., Kiss, T., Valcarcel-Ares, M. N., Galvan, V., et al. (2017). Hypertension impairs neurovascular coupling and promotes microvascular injury: Role in exacerbation of Alzheimer's disease. *GeroScience*, *39*(4), 359–372. https://doi.org/10.1007/s11357-017-9991-9.

Davis, M. J., Hill, M. A., & Kuo, L. (2011). Local regulation of microvascular perfusion. In R. L. Terjung (Ed.), *Comprehensive physiology* (pp. 161–284). John Wiley & Sons, Inc. https://doi.org/10.1002/cphy.cp020406

del Valle-Rodriguez, A., Lopez-Barneo, J., & Urena, J. (2003). Ca^{2+} channel-sarcoplasmic reticulum coupling: A mechanism of arterial myocyte contraction without Ca^{2+} influx. *The EMBO Journal*, *22*(17), 4337–4345. https://doi.org/10.1093/emboj/cdg432.

Duling, B. R., & Berne, R. M. (1970). Longitudinal gradients in periarteriolar oxygen tension. A possible mechanism for the participation of oxygen in local regulation of blood flow. *Circulation Research*, *27*(5), 669–678.

Durán, W. N., Sánchez, F. A., & Breslin, J. W. (2011). Microcirculatory exchange function. In R. L. Terjung (Ed.), *Comprehensive physiology* (pp. 81–123). John Wiley & Sons, Inc. https://doi.org/10.1002/cphy.cp020404.

Edmonds, M. (2019). Vascular disease in the lower limb in type 1 diabetes. *Cardiovascular Endocrinology & Metabolism*, *8*(1), 39–46. https://doi.org/10.1097/XCE.0000000000000168.

Edwards, G., Feletou, M., & Weston, A. H. (2010). Endothelium-derived hyperpolarising factors and associated pathways: A synopsis. *Pflügers Archiv*, *459*(6), 863–879. https://doi.org/10.1007/s00424-010-0817-1.

Falcone, J. C., & Bohlen, H. G. (1990). EDRF from rat intestine and skeletal muscle venules causes dilation of arterioles. *The American Journal of Physiology—Heart and Circulatory Physiology*, *258*, H1515–H1523.

Fernández-Tenorio, M., González-Rodríguez, P., Porras, C., Castellano, A., Moosmang, S., Hofmann, F., et al. (2010). Short communication: Genetic ablation of L-type Ca^{2+} channels abolishes depolarization-induced Ca^{2+} release in arterial smooth muscle. *Circulation Research*, *106*(7), 1285–1289. https://doi.org/10.1161/circresaha.109.213967.

Fernández-Tenorio, M., Porras-Gonzalez, C., Castellano, A., del Valle-Rodriguez, A., Lopez-Barneo, J., & Urena, J. (2011). Metabotropic regulation of RhoA/Rho-associated kinase by L-type Ca^{2+} channels. *Circulation Research*, *108*, 1348–1357. CIRCRESAHA.111.240127. https://doi.org/10.1161/circresaha.111.240127.

Frisbee, J. C., Lewis, M. T., & Wiseman, R. W. (2019). Skeletal muscle performance in metabolic disease: Microvascular or mitochondrial limitation or both? *Microcirculation*, *26*(5). e12517. https://doi.org/10.1111/micc.12517.

Fronek, K., & Zweifach, B. W. (1975). Microvascular pressure distribution in skeletal muscle and the effect of vasodilation. *The American Journal of Physiology*, *228*(3), 791–796.

Ganitkevich, V. Y., & Isenberg, G. (1993). Membrane potential modulates inositol 1,4,5-trisphosphate-mediated Ca^{2+} transients in guinea-pig coronary myocytes. *Journal of Physiology (London)*, *470*, 35–44.
Goddard, L. M., & Iruela-Arispe, M. L. (2013). Cellular and molecular regulation of vascular permeability. *Thrombosis and Haemostasis*, *109*(3), 407–415. https://doi.org/10.1160/th12-09-0678.
Guan, Y. Y., Wang, G. L., & Zhou, J. G. (2006). The ClC-3 Cl^- channel in cell volume regulation, proliferation and apoptosis in vascular smooth muscle cells. *Trends in Pharmacological Sciences*, *27*(6), 290–296.
Hamilton, N. B., Attwell, D., & Hall, C. N. (2010). Pericyte-mediated regulation of capillary diameter: A component of neurovascular coupling in health and disease. *Frontiers in Neuroenergetics*, *2*, 1–14. https://doi.org/10.3389/fnene.2010.00005.
Harnett, K. M., & Biancani, P. (2003). Calcium-dependent and calcium-independent contractions in smooth muscles. *The American Journal of Medicine*, *115*(Suppl. 3A), 24S–30S.
Hashitani, H., & Lang, R. J. (2016). Spontaneous activity in the microvasculature of visceral organs: Role of pericytes and voltage-dependent Ca(2+) channels. *The Journal of Physiology*, *594*(3), 555–565. https://doi.org/10.1113/JP271438.
Hester, R. L. (1990). Venular-arteriolar diffusion of adenosine in hamster cremaster microcirculation. *The American Journal of Physiology—Heart and Circulatory Physiology*, *258*, H1918–H1924.
Hester, R. L. (1993). Uptake of metabolites by postcapillary venules: Mechanism for the control of arteriolar diameter. *Microvascular Research*, *46*, 254–261.
Hester, R. L. (1995). Venular endothelium: Metabolic sensors? *News in Physiological Sciences*, *10*, 50–51.
Hester, R. L., & Hammer, L. W. (2002). Venular-arteriolar communication in the regulation of blood flow. *American Journal of Physiology. Regulatory, Integrative and Comparative Physiology*, *282*(5), R1280–R1285.
Hoffmann, E. K., Lambert, I. H., & Pedersen, S. F. (2009). Physiology of cell volume regulation in vertebrates. *Physiological Reviews*, *89*(1), 193–277. https://doi.org/10.1152/physrev.00037.2007.
Jackson, W. F. (2000). Ion channels and vascular tone. *Hypertension*, *35*(1 Pt. 2), 173–178.
Jackson, W. F. (2005). Potassium channels and proliferation of vascular smooth muscle cells. *Circulation Research*, *97*(12), 1211–1212.
Jackson, W. F. (2016). Endothelial cell ion channel expression and function in arterioles and resistance arteries. In I. Levitan & A. M. Dopico (Eds.), *Vascular ion channels in physiology and disease* (pp. 3–36). Switzerland: Springer International Publishing.
Johnson, P. C. (2011). Introduction: overview of the microcirculation. In D. M. Pollock (Ed.), *Comprehensive physiology* (pp. xi–xxiv). John Wiley & Sons, Inc. https://doi.org/10.1002/cphy.cp0204fm02
Kim, H. R., Appel, S., Vetterkind, S., Gangopadhyay, S. S., & Morgan, K. G. (2008). Smooth muscle signalling pathways in health and disease. *Journal of Cellular and Molecular Medicine*, *12*(6A), 2165–2180. https://doi.org/10.1111/j.1582-4934.2008.00552.x.
Kogan, A. N., & von Andrian, U. H. (2011). Lymphocyte trafficking. In R. L. Terjung (Ed.), *Comprehensive physiology* (pp. 449–482). John Wiley & Sons, Inc. https://doi.org/10.1002/cphy.cp020410
Komarova, Y., & Malik, A. B. (2010). Regulation of endothelial permeability via paracellular and transcellular transport pathways. *Annual Review of Physiology*, *72*, 463–493. https://doi.org/10.1146/annurev-physiol-021909-135833.
Krueger, M., & Bechmann, I. (2010). CNS pericytes: Concepts, misconceptions, and a way out. *Glia*, *58*(1), 1–10. https://doi.org/10.1002/glia.20898.

Kukuljan, M., Rojas, E., Catt, K. J., & Stojilkovic, S. S. (1994). Membrane potential regulates inositol 1,4,5-trisphosphate-controlled cytoplasmic Ca^{2+} oscillations in pituitary gonadotrophs. *The Journal of Biological Chemistry*, *269*(7), 4860–4865.
Kushiro, T., Watanabe, N., Takahashi, A., Koike, M., Saito, F., Otsuka, Y., et al. (2004). Different effects of L-type and L + N-type calcium channel blockers on hamster cheek pouch venules. *Journal of Cardiovascular Pharmacology*, *44*(6), 672–675. https://doi.org/10.1097/00005344-200412000-00008.
Kutcher, M. E., & Herman, I. M. (2009). The pericyte: Cellular regulator of microvascular blood flow. *Microvascular Research*, 77(3), 235–246. https://doi.org/10.1016/j.mvr.2009.01.007.
Lee, C. H., Poburko, D., Kuo, K. H., Seow, C. Y., & van Breemen, C. (2002). Ca(2+) oscillations, gradients, and homeostasis in vascular smooth muscle. *American Journal of Physiology. Heart and Circulatory Physiology*, *282*(5), H1571–H1583.
Ley, K. (2011). The microcirculation in inflammation. In R. L. Terjung (Ed.), *Comprehensive physiology* (pp. 387–488). John Wiley & Sons, Inc. https://doi.org/10.1002/cphy.cp020409
Li, Q., & Puro, D. G. (2001). Adenosine activates ATP-sensitive K(+) currents in pericytes of rat retinal microvessels: Role of A1 and A2a receptors. *Brain Research*, *907*(1–2), 93–99.
Little, T. L., Beyer, E. C., & Duling, B. R. (1995a). Connexin 43 and connexin 40 gap junctional proteins are present in arteriolar smooth muscle and endothelium in vivo. *American Journal of Physiology*, *268*, H729–H739.
Little, T. L., Xia, J., & Duling, B. R. (1995b). Dye tracers define differential endothelial and smooth muscle coupling patterns within the arteriolar wall. *Circulation Research*, *76*, 498–504.
Liu, Q. H., Zheng, Y. M., Korde, A. S., Yadav, V. R., Rathore, R., Wess, J., et al. (2009). Membrane depolarization causes a direct activation of G protein-coupled receptors leading to local Ca^{2+} release in smooth muscle. *Proceedings of the National Academy of Sciences of the United States of America*, *106*(27), 11418–11423, 0813307106 [pii]. https://doi.org/10.1073/pnas.0813307106.
Longden, T. A., Dabertrand, F., Koide, M., Gonzales, A. L., Tykocki, N. R., Brayden, J. E., et al. (2017). Capillary K+-sensing initiates retrograde hyperpolarization to increase local cerebral blood flow. *Nature Neuroscience*, *20*(5), 717–726. https://doi.org/10.1038/nn.4533.
Mahaut-Smith, M. P., Martinez-Pinna, J., & Gurung, I. S. (2008). A role for membrane potential in regulating GPCRs? *Trends in Pharmacological Sciences*, *29*(8), 421–429. S0165-6147(08)00145-4 [pii]https://doi.org/10.1016/j.tips.2008.05.007.
McCabe, R. D., & Young, D. B. (1994). Potassium inhibits cultured vascular smooth muscle cell proliferation. *American Journal of Hypertension*, 7, 346–350.
McKay, M. K., Gardner, A. L., Boyd, D., & Hester, R. L. (1998). Influence of venular prostaglandin release on arteriolar diameter during functional hyperemia. *Hypertension*, *31*(1 Pt. 2), 213–217.
Mitsui, R., & Hashitani, H. (2016). Mechanisms underlying spontaneous constrictions of postcapillary venules in the rat stomach. *Pflügers Archiv*, *468*(2), 279–291. https://doi.org/10.1007/s00424-015-1752-y.
Munaron, L. (2006). Intracellular calcium, endothelial cells and angiogenesis. *Recent Patents on Anti-Cancer Drug Discovery*, *1*(1), 105–119.
Nelson, M. T., Cheng, H., Rubart, M., Santana, L. F., Bonev, A. D., Knot, H. J., et al. (1995). Relaxation of arterial smooth muscle by calcium sparks. *Science*, *270*(5236), 633–637. https://doi.org/10.1126/science.270.5236.633.
Nilius, B., & Droogmans, G. (2001). Ion channels and their functional role in vascular endothelium. *Physiological Reviews*, *81*(4), 1415–1459.
Offner, F. F. (1991). Ion flow through membranes and the resting potential of cells. *The Journal of Membrane Biology*, *123*(2), 171–182. https://doi.org/10.1007/bf01998087.

Okada, Y., Yanagisawa, T., & Taira, N. (1993). BRL 38227 (levcromakalim)-induced hyperpolarization reduces the sensitivity to Ca^{2+} of contractile elements in canine coronary artery. *Naunyn-Schmiedeberg's Archives of Pharmacology*, *347*, 438–444.

Pallone, T. L., Zhang, Z., & Rhinehart, K. (2003). Physiology of the renal medullary microcirculation. *American Journal of Physiology. Renal Physiology*, *284*(2), F253–F266. https://doi.org/10.1152/ajprenal.00304.2002.

Pocock, T. M., Foster, R. R., & Bates, D. O. (2004). Evidence of a role for TRPC channels in VEGF-mediated increased vascular permeability in vivo. *American Journal of Physiology. Heart and Circulatory Physiology*, *286*(3), H1015–H1026. https://doi.org/10.1152/ajpheart.00826.2003.

Pries, A. R., & Secomb, T. W. (2011). Blood flow in microvascular networks. In R. L. Terjung (Ed.), *Comprehensive physiology* (pp. 3–36). John Wiley & Sons, Inc. https://doi.org/10.1002/cphy.cp020401.

Puro, D. G. (2007). Physiology and pathobiology of the pericyte-containing retinal microvasculature: New developments. *Microcirculation*, *14*(1), 1–10. https://doi.org/10.1080/10739680601072099.

Quignard, J. F., Harley, E. A., Duhault, J., Vanhoutte, P. M., & Feletou, M. (2003). K+ channels in cultured bovine retinal pericytes: Effects of beta-adrenergic stimulation. *Journal of Cardiovascular Pharmacology*, *42*(3), 379–388.

Renkin, E. M. (1984). Control of microcirculation and blood-tissue exchange. In E. M. Renkin & C. C. Michel (Eds.), *Handbook of physiology: Sec. 2, the cardiovascular system, vol. IV, microcirculation, part 2* (pp. 627–687). Bethesda: American Physiological Society.

Rhinehart, K., Zhang, Z., & Pallone, T. L. (2002). Ca(2+) signaling and membrane potential in descending vasa recta pericytes and endothelia. *American Journal of Physiology. Renal Physiology*, *283*(4), F852–F860.

Rhodin, J. A. G. (2014). Architecture of the vessel wall. In R. L. Terjung (Ed.), *Comprehensive physiology* (pp. 1–31). John Wiley & Sons, Inc. https://doi.org/10.1002/cphy.cp020201

Rizzoni, D., Rizzoni, M., Nardin, M., Chiarini, G., Agabiti-Rosei, C., Aggiusti, C., et al. (2019). Vascular aging and disease of the small vessels. *High Blood Pressure & Cardiovascular prevention: The Official Journal of the Italian Society of Hypertension*, *26*(3), 183–189. https://doi.org/10.1007/s40292-019-00320-w.

Rothe, C. F. (2011). Venous system: physiology of the capacitance vessels. In R. L. Terjung (Ed.), *Comprehensive physiology* (pp. 397–452). John Wiley & Sons, Inc. https://doi.org/10.1002/cphy.cp020313

Saito, Y., Eraslan, A., & Hester, R. L. (1993). Importance of venular flow in control of arteriolar diameter in hamster cremaster muscle. *The American Journal of Physiology—Heart and Circulatory Physiology*, *265*, H1294–H1300.

Saito, Y., Eraslan, A., Lockard, V., & Hester, R. L. (1994). Role of venular endothelium in control of arteriolar diameter during functional hyperemia. *The American Journal of Physiology—Heart and Circulatory Physiology*, *267*, H1227–H1231.

Sandow, S. L., & Hill, C. E. (2000). Incidence of myoendothelial gap junctions in the proximal and distal mesenteric arteries of the rat is suggestive of a role in endothelium-derived hyperpolarizing factor-mediated responses. *Circulation Research*, *86*(3), 341–346.

Santisteban, M. M., & Iadecola, C. (2018). Hypertension, dietary salt and cognitive impairment. *Journal of Cerebral Blood Flow and Metabolism*, *38*(12), 2112–2128. https://doi.org/10.1177/0271678X18803374.

Segal, S. S. (2000). Integration of blood flow control to skeletal muscle: Key role of feed arteries. *Acta Physiologica Scandinavica*, *168*(4), 511–518. https://doi.org/10.1046/j.1365-201x.2000.00703.x.

Segal, S. S. (2005). Regulation of blood flow in the microcirculation. *Microcirculation, 12*(1), 33–45. Q8452611821846W1 [pii]https://doi.org/10.1080/10739680590895028.

Segal, S. S., & Duling, B. R. (1986). Communication between feed arteries and microvessels in hamster striated-muscle—Segmental vascular-responses are functionally coordinated. *Circulation Research, 59*(3), 283–290.

Shepro, D., & Morel, N. M. L. (1993). Pericyte physiology. *The FASEB Journal,* 7, 1031–1038.

Sidik, N., Morrow, A., & Berry, C. (2020). Human microcirculation in ischemic heart disease. *Arteriosclerosis, Thrombosis, and Vascular Biology, 40*(1), 11–13. https://doi.org/10.1161/ATVBAHA.119.313579.

Simionescu, M., & Simionescu, N. (1984). Ultrastructure of the microvascular wall: Functional correlations. In E. M. Renkin & C. C. Michel (Eds.), *Handbook of physiology, sec 2: The cardiovascular system, vol. IV microcirculation. Part 2* (pp. 41–101). Bethesda, MD: The American Physiological Society.

Socha, M. J., Behringer, E. J., & Segal, S. S. (2012). Calcium and electrical signalling along endothelium of the resistance vasculature. *Basic & Clinical Pharmacology & Toxicology, 110*(1), 80–86. https://doi.org/10.1111/j.1742-7843.2011.00798.x.

Sorop, O., Olver, T. D., van de Wouw, J., Heinonen, I., van Duin, R. W., Duncker, D. J., et al. (2017). The microcirculation: A key player in obesity-associated cardiovascular disease. *Cardiovascular Research, 113*(9), 1035–1045. https://doi.org/10.1093/cvr/cvx093.

Strain, W. D., & Paldanius, P. M. (2018). Diabetes, cardiovascular disease and the microcirculation. *Cardiovascular Diabetology, 17*(1), 57. https://doi.org/10.1186/s12933-018-0703-2.

Thakore, P., & Earley, S. (2019). Transient receptor potential channels and endothelial cell calcium signaling. *Comprehensive Physiology, 9*(3), 1249–1277. https://doi.org/10.1002/cphy.c180034.

Tykocki, N. R., Boerman, E. M., & Jackson, W. F. (2017). Smooth muscle ion channels and regulation of vascular tone in resistance arteries and arterioles. *Comprehensive Physiology,* 7(2), 485–581. https://doi.org/10.1002/cphy.c160011.

Urena, J., del Valle-Rodriguez, A., & Lopez-Barneo, J. (2007). Metabotropic Ca^{2+} channel-induced calcium release in vascular smooth muscle. *Cell Calcium, 42*(4–5), 513–520. https://doi.org/10.1016/j.ceca.2007.04.010.

Vallot, O., Combettes, L., Jourdon, P., Inamo, J., Marty, I., Claret, M., et al. (2000). Intracellular Ca(2+) handling in vascular smooth muscle cells is affected by proliferation. *Arteriosclerosis, Thrombosis, and Vascular Biology, 20*(5), 1225–1235.

van der Flier, W. M., Skoog, I., Schneider, J. A., Pantoni, L., Mok, V., Chen, C. L. H., et al. (2018). Vascular cognitive impairment. *Nature Reviews. Disease Primers, 4,* 18003. https://doi.org/10.1038/nrdp.2018.3.

Wang, Y., Chen, J., Taylor, C. W., Hirata, Y., Hagiwara, H., Mikoshiba, K., et al. (2001). Crucial role of type 1, but not type 3, inositol 1,4,5-trisphosphate (IP(3)) receptors in IP(3)-induced Ca(2+) release, capacitative Ca(2+) entry, and proliferation of A7r5 vascular smooth muscle cells. *Circulation Research, 88*(2), 202–209.

Westcott, E. B., & Jackson, W. F. (2011). Heterogeneous function of ryanodine receptors, but not IP3 receptors, in hamster cremaster muscle feed arteries and arterioles. *American Journal of Physiology. Heart and Circulatory Physiology, 300*(5), H1616–H1630. https://doi.org/10.1152/ajpheart.00728.2010.

Westcott, E. B., Goodwin, E. L., Segal, S. S., & Jackson, W. F. (2012). Function and expression of ryanodine receptors and inositol 1,4,5-trisphosphate receptors in smooth muscle cells of murine feed arteries and arterioles. *The Journal of Physiology, 590*(8), 1849–1869. https://doi.org/10.1113/jphysiol.2011.222083.

Wiedeman, M. P. (1984). Architecture. In E. M. Renkin & C. C. Michel (Eds.), *Handbook of physiology: Section 2: The cardiovascular system, vol. IV, microcirculation, part1* (pp. 11–40). Bethesda: The American Physiological Society.

Wu, D. M., Kawamura, H., Li, Q., & Puro, D. G. (2001). Dopamine activates ATP-sensitive K+ currents in rat retinal pericytes. *Visual Neuroscience*, *18*(6), 935–940.
Yamagishi, T., Yanagisawa, T., & Taira, N. (1992). K^+ channel openers, cromakalim and Ki4032, inhibit agonist-induced Ca^{2+} release in canine coronary artery. *Naunyn-Schmiedeberg's Archives of Pharmacology*, *346*, 691–700.
Yamamura, H., Ohya, S., Muraki, K., & Imaizumi, Y. (2012). Involvement of inositol 1,4,5-trisphosphate formation in the voltage-dependent regulation of the Ca(2+) concentration in porcine coronary arterial smooth muscle cells. *The Journal of Pharmacology and Experimental Therapeutics*, *342*(2), 486–496. https://doi.org/10.1124/jpet.112.194233.
Yanagisawa, T., Yamagishi, T., & Okada, Y. (1993). Hyperpolarization induced by K^+ channel openers inhibits Ca^{2+} influx and Ca^{2+} release in coronary artery. *Cardiovascular Drugs and Therapy/Sponsored by the International Society of Cardiovascular Pharmacotherapy*, 7(Suppl. 3), 565–574.
Yuan, S. Y. (2000). Signal transduction pathways in enhanced microvascular permeability. *Microcirculation*, 7(6 Pt. 1), 395–403.
Zawieja, D. C., von der Weid, P.-Y., & Gashev, A. A. (2011). Microlymphatic biology. In R. L. Terjung (Ed.), *Comprehensive physiology* (pp. 125–158). John Wiley & Sons, Inc. https://doi.org/10.1002/cphy.cp020405.
Zhang, Z., Rhinehart, K., & Pallone, T. L. (2002). Membrane potential controls calcium entry into descending vasa recta pericytes. *American Journal of Physiology. Regulatory, Integrative and Comparative Physiology*, *283*(4), R949–R957.
Zhang, Q., Cao, C., Zhang, Z., Wier, W. G., Edwards, A., & Pallone, T. L. (2008). Membrane current oscillations in descending vasa recta pericytes. *American Journal of Physiology. Renal Physiology*, *294*(3), F656–F666. https://doi.org/10.1152/ajprenal.00493.2007.
Zhang, Z., Lin, H., Cao, C., Khurana, S., & Pallone, T. L. (2010). Voltage-gated divalent currents in descending vasa recta pericytes. *American Journal of Physiology. Renal Physiology*, *299*(4), F862–F871. https://doi.org/10.1152/ajprenal.00321.2010.
Zhou, X., & He, P. (2010). Endothelial $[Ca^{2+}]i$ and caveolin-1 antagonistically regulate eNOS activity and microvessel permeability in rat venules. *Cardiovascular Research*, *87*(2), 340–347. https://doi.org/10.1093/cvr/cvq006.
Ziegler, T., Abdel Rahman, F., Jurisch, V., & Kupatt, C. (2019). Atherosclerosis and the capillary network; pathophysiology and potential therapeutic strategies. *Cell*, *9*(1). https://doi.org/10.3390/cells9010050.
Zweifach, B. W., & Lipowsky, H. H. (1984). Pressure-flow relations in blood and lymph microcirculation. In E. M. Renkin & C. C. Michel (Eds.), *Handbook of physiology, section 2: The cardiovascular system, vol. IV, microcirculation. Part 1* (pp. 251–305). Bethesda, MD: The American Physiological Society.

CHAPTER TWO

Ion channels and the regulation of myogenic tone in peripheral arterioles

William F. Jackson*
Department of Pharmacology and Toxicology, Michigan State University, East Lansing, MI, United States
*Corresponding author: e-mail address: jacks783@msu.edu

Contents

Abstract

Myogenic tone is a hall-mark feature of arterioles in the microcirculation. This pressure-induced, contractile activation of vascular smooth muscle cells (VSMCs) in the wall of these microvessels importantly contributes to the regulation and maintenance of blood pressure; blood flow to and within organs and tissues; and capillary pressure and fluid balance. Ion channels play a central role in the genesis and maintenance of myogenic tone. Mechanosensitive ion channels such as TRPC6 may serve as one of the sensors of pressure-induced membrane stress/strain, and TRPC6 along with TRPM4 channels are responsible pressure-induced VSMC depolarization that may be bolstered by the activity of Ca^{2+}-activated Cl^- channels and inhibition of voltage-gated K^+ (K_V) channels, inwardly-rectifying K^+ (K_{IR}) channels and ATP-sensitive K^+ (K_{ATP}) channels. Membrane potential depolarization activates voltage-gated Ca^{2+} channels (VGCCs), with CaV1.2 channels playing a central role. Calcium entry through CaV1.2 channels, which is

Current Topics in Membranes, Volume 85
ISSN 1063-5823
https://doi.org/10.1016/bs.ctm.2020.01.002

amplified by Ca^{2+} release through IP_3 receptors in the form of Ca^{2+} waves in some arterioles, provides the major source of activator calcium responsible for arteriolar myogenic tone. Stabilizing negative-feedback comes from depolarization- and Ca^{2+}-induced activation of large-conductance Ca^{2+}-activated K^+ channels and depolarization-induced activation of K_V channels. Myogenic tone also is dampened by tonic activity of K_{IR} and K_{ATP} channels. While much has been learned about ion channel expression and function in myogenic tone, additional studies are required to fill in our knowledge gaps due to significant regional differences in ion channel expression and function and a lack of data specifically from VSMCs in arterioles.

1. Introduction

Arterioles in the peripheral microcirculation substantially contribute to total peripheral resistance and blood pressure (Pries & Secomb, 2011; Renkin, 1984; Zweifach & Lipowsky, 1984); control blood flow to and within organs and tissues; and modulate the pressure in capillaries and venules contributing to tissue fluid filtration/reabsorption. These arteriolar functions are effected by vascular smooth muscle cells (VSMCs) (Davis, Hill, & Kuo, 2011; Renkin, 1984) that wrap in a single circumferential layer around the endothelial cell "tube" that forms the basic structure of arterioles (Rhodin, 2014; Simionescu & Simionescu, 1984). For the purpose of this chapter, arterioles are defined as arterial vessels with a single layer of VSMCs and that are embedded within the parenchyma which they perfuse. Feed arteries (or resistance arteries), the last branch of the arterial system before arterioles, are external to the parenchyma and generally have more than one layer of VSMCs overlying the endothelium. Changes in the contractile activity of the VSMCs (contraction or relaxation) produce coordinated constriction or dilation of the arteriolar lumen resulting in changes in the hydraulic resistance to fluid flow and changes in the pressure drop along the length of these vessels (Zweifach & Lipowsky, 1984). A hall-mark feature of arterioles is pressure-induced myogenic tone, the steady-state contractile activity of vascular smooth muscle cells in the arteriolar wall that sets the baseline degree of contraction of the VSMCs at a given blood pressure (Davis et al., 2011; Davis & Hill, 1999; Hill, Zou, Potocnik, Meininger, & Davis, 2001; Johnson, 1980). Arterioles can either constrict or dilate in response to physiological or pathophysiological stimuli from this baseline level of tone (Davis et al., 2011; Renkin, 1984).

Vascular smooth muscle cells express a diverse array of ion channels that contribute to myogenic tone and its regulation (Tykocki, Boerman, &

Jackson, 2017). This includes more than five classes of K^+ channels (Tykocki et al., 2017), one or more classes of voltage-gated Ca^{2+} channels (Tykocki et al., 2017), several types of Cl^- channels (Bannister et al., 2012; Dam, Boedtkjer, Aalkjaer, & Matchkov, 2014; Dam, Boedtkjer, Nyvad, Aalkjaer, & Matchkov, 2014; Heinze et al., 2014; Hubner, Schroeder, & Ehmke, 2015; Leblanc et al., 2015; Matchkov, Boedtkjer, & Aalkjaer, 2015) and a variety of transient-receptor-potential (TRP) channels (Tykocki et al., 2017) expressed in the plasma membrane and two or more Ca^{2+} permeable channels in the endoplasmic reticulum (Tykocki et al., 2017) that can be involved in the generation and modulation of myogenic tone. The purpose of this chapter is to review the functional expression of a number of these channels in VSMCs that appear to importantly contribute to the regulation of myogenic tone in arterioles in the peripheral microcirculation. The reader is directed to a recent extensive review of ion channel expression and function in arterioles and resistance arteries for a more general treatment of this topic (Tykocki et al., 2017). Ion channel expression and function in endothelial cells will not be discussed and the reader is directed to recent reviews for information and references on that topic (Jackson, 2016b; Thakore & Earley, 2019).

2. Overview of myogenic tone in arterioles and resistance arteries

The precise mechanism by which blood pressure activates arteriolar VSMCs to generate myogenic tone remains in question, largely because there appear to be multiple mechanisms involved that depend on the species studied, tissue location and the experimental conditions under which this process is examined. When pressure is increased in a resistance artery or arteriole, this radial force will be transmitted to the wall of the arteriole increasing tangential wall stress (T) given by the law of Laplace: $T = Pr/\Delta$, where P = pressure in the lumen of the vessel, r is the lumen radius and Δ is the thickness of the wall. The increase in T will cause an initial passive dilation of the vessel resulting in "stretch" of the VSMCs, until the VSMCs can generate sufficient active tension to overcome the new pressure-related T. The increased T or the strain (stretch) that ensues then activates some sensing mechanism in the membrane of the VSMCs to activate the VSMCs to actively contract to overcome the pressure-induced T and develop myogenic tone. Studies in cerebral resistance arteries suggest the increased T (or the strain that ensues) is sensed, perhaps, by a mechanosensitive-G_q-coupled receptor (msGPCR) such as the angiotensin II receptor 1B (ATR1B) (Mederos, Storch, & Gudermann, 2016; Pires et al., 2017; Schleifenbaum et al., 2014), cysteinyl leukotriene 1 receptors (CysLT1)

(Mederos et al., 2016; Storch, Blodow, Gudermann, & Mederos, 2015), or P2Y6 purinergic receptors (Brayden, Li, & Tavares, 2013) as well as mechanosensitive ion channels such as TRPC6 (Welsh, Morielli, Nelson, & Brayden, 2002) (Fig. 1). Signaling mechanisms involving integrins and myogenic tone also have been proposed (Davis et al., 2001; Martinez-Lemus, Crow, Davis, & Meininger, 2005). The precise mechanism by which these receptors and ion channels are activated by the increased stress or strain is not yet known. However studies in other systems suggest that increased

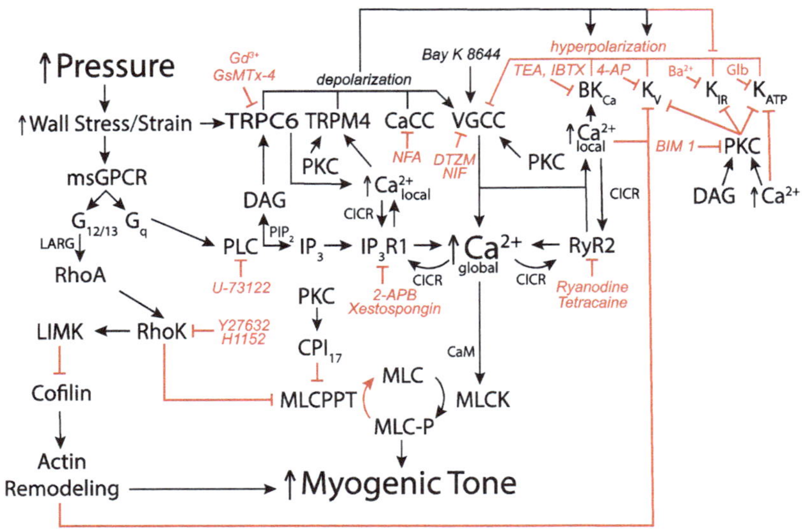

Fig. 1 Ion channels and signaling pathways in myogenic tone. Schematic diagram of some of the major ion channels and signaling pathways involved in myogenic tone in resistance arteries and arterioles. Black arrows show stimulation, increases or activation of signaling molecules, ion channels or enzymes that participate in myogenic tone. Red capped lines indicate inhibition, decreases or deactivation of signaling molecules, ion channels or enzymes involved in myogenic tone. Also shown are pharmacological agents that we have used to interrogate the ion channels and signaling pathways in arteriolar myogenic tone. Abbreviations: CaCC, Ca^{2+}-activated Cl^- channel; VGCC, voltage-gated Ca^{2+} channel; BK_{Ca}, large-conductance Ca^{2+}-activated K^+ channel; K_V, voltage-gated K^+ channel; K_{IR}, inwardly-rectifying K^+ channel; K_{ATP}, ATP-sensitive K^+ channel; msGPCR, mechanosensitive G-protein-coupled receptor; DAG, diacylglycerol; PKC, protein kinase C; NFA, niflumic acid; DTZM, diltiazem; NIF, nifedipine; TEA, tetraethylammonium; IBTX, iberiotoxin; 4-AP, 4-aminopyridine; GLIB, glibenclamide; BIM I, bisindolylmaleimide I; PLC, phospholipase C; PIP_2, phosphatidylinositol bisphosphate; IP_3, inositol, 1,4,5 trisphosphate; IP_3R1, IP_3 receptor 1; RyR, ryanodine receptor; CICR, Ca^{2+}-induced-Ca^{2+} release; LARG, guanine nucleotide exchange factor LARG; RhoA, small G-protein Rho; 2-APB, 2-aminoethoxydiphenyl borate; RhoK, Rho kinase; LIMK, LIM kinase; CPI_{17}, C-kinase potentiated protein phosphatase-1 inhibitor; MLCPPT, myosin light-chain phosphatase; MLC, myosin light-chain; MLCK, myosin light-chain kinase; See text for more details and references.

membrane stress/strain may be transmitted to membrane proteins to activate channels/receptors by: (1) transmission of force via connections with the extracellular matrix and the cytoskeleton, (2) changes in membrane curvature, (3) tension caused by membrane thinning, and/or membrane tension-induced conformational changes (Leiphart, Chen, Peredo, Loneker, & Janmey, 2019). Regardless of the specifics, it is proposed that activation of the msGPCR leads to activation of phospholipase $C_{\gamma 1}$ (PLCγ1), hydrolysis of phosphatidyl inositol 4,5-bisphosphate (PIP_2) and formation of inositol 1,4,5 trisphosphate (IP_3) and diacylglycerol (DAG) (Gonzales et al., 2014). The DAG produced by this process may contribute to the activation of TRPC6 channels (Inoue et al., 2009; Mederos y Schnitzler et al., 2008) although evidence against this idea has been presented (Anfinogenova, Brett, Walsh, Harraz, & Welsh, 2011) (Fig. 1). TRPC3 also can be activated by DAG produced during agonist-induced activation of G_q-protein-coupled receptors (Reading, Earley, Waldron, Welsh, & Brayden, 2005), but these channels have not been implicated in the signaling pathway of myogenic tone. Phosphatidyl choline-specific PLC also has been proposed to participate in myogenic tone of murine mesenteric resistance arteries (Mauban, Zacharia, Fairfax, & Wier, 2015).

The IP_3 formed from hydrolysis of PIP_2 then binds to IP_3 receptors (IP_3R) in the peripheral endoplasmic reticulum (ER). Coupled with the Ca^{2+} entry through TRPC6 channels, this results in release of Ca^{2+} from the ER through the IP_3R via Ca^{2+}-induced Ca^{2+} release. The Ca^{2+} released through the IP_3R then activates TRPM4 channels (Gonzales et al., 2014; Gonzales, Amberg, & Earley, 2010), a process that may involve protein kinase C which is activated by PLC-dependent production of DAG and Ca^{2+} (Earley, Straub, & Brayden, 2007). Along with the inward cation current through TRPC6 channels, the cation influx through TRPM4 channels results in membrane depolarization, activation of voltage-gated Ca^{2+} channels and Ca^{2+} influx (Gonzales et al., 2014; Gonzales, Garcia, Amberg, & Earley, 2010) (Fig. 1). Increased activity of PKC tethered to CaV1.2 by A-kinase anchoring proteins (AKAPs) may facilitate this process (Navedo, Amberg, Votaw, & Santana, 2005).

The local increase in Ca^{2+} from TRPC6, IP_3R and CaV1.2 may also activate Ca^{2+}-activated Cl^- channels (CaCC) contributing to pressure-induced membrane depolarization in some vessels (Bulley et al., 2012; Heinze et al., 2014), particularly small arterioles (Heinze et al., 2014) (Fig. 1). Membrane depolarization is also supported by inhibition of several K^+ channels (Fig. 1). The PLC-dependent production of DAG and the increase in local Ca^{2+} concentration activate PKCs. Increased PKC activity inhibits basal K_V channel (Aiello, Clement-Chomienne, Sontag, Walsh, & Cole, 1996; Clement-Chomienne, Walsh, & Cole, 1996; Hayabuchi, Standen, & Davies, 2001; Kidd, Bulley, & Jaggar, 2017; Ko et al., 2010) and K_{ATP} channel (Hayabuchi, Davies, & Standen, 2001; Hu, Huang, Jan, & Jan, 2003; Jiao, Garg, Yang, Elton, & Hu, 2008; Sampson, Davies, Barrett-Jolley, Standen, & Dart, 2007) activity, promoting membrane depolarization. The increase in intracellular Ca^{2+} associated with stretch-induced activation of VSMCs also inhibits K_V channel activity promoting membrane depolarization and increased myogenic tone (Cox & Petrou, 1999; Gelband, Ishikawa, Post, Keef, & Hume, 1993; Ishikawa, Hume, & Keef, 1993). Increased intracellular Ca^{2+} acting through the Ca^{2+}-activated phosphatase, calcineurin, also inhibits K_{ATP} activity (Wilson, Jabr, & Clapp, 2000). Vascular smooth muscle stretch also has been shown to inhibit inwardly-rectifying K^+ (K_{IR}) channels in vascular smooth by a mechanism involving PKC (Wu et al., 2007). Membrane depolarization induced by activation of TRPC6, TRPM4, VGCCs and CaCCs also will decrease the activity of open K_{IR} due to the voltage-dependent block of these channels by polyamines and Mg^{2+} (Tykocki et al., 2017). Thus, closure of open K^+ channels reduces outward current and enhances the pressure-induced membrane depolarization and myogenic tone.

The global increase in intracellular Ca^{2+} promotes Ca^{2+} binding to the calcium binding protein, calmodulin (CaM), and the Ca^{2+}-CaM complex activates myosin light-chain kinase (MLCK), resulting in phosphorylation of the 20 KD myosin light-chains (Cole & Welsh, 2011; Zou, Ratz, & Hill, 1995). Phosphorylated myosin light-chains then allow the formation of actin-myosin cross bridges, the cross-bridge cycle and smooth muscle contraction (increased myogenic tone) (Cole & Welsh, 2011; Zou et al., 1995) (Fig. 1). The myosin light-chains are dephosphorylated by myosin light-chain phosphatase (MLCPPT); the ratio of the activity of MLCK/MLCPPT determines the Ca^{2+} sensitivity of the contractile machinery (Cole & Welsh, 2011). Activation of G-proteins activates the small GTPase, RhoA, via guanine nucleotide exchange factors (GEFs) such as LARG which couples $G_{12/13}$ to activation of RhoA (Chennupati et al., 2019) (Fig. 1). RhoA activation then activates Rho Kinase that has several effects in VSMCs related to myogenic tone. First, activated Rho kinase phosphorylates and inhibits MLCPPT (Cole & Welsh, 2011). This results in Ca^{2+} sensitization of the contractile machinery as the ratio of the activity of MLCK/MYPPT increases. Second, activated Rho kinase also modulates the actin cytoskeleton through LIM kinase (LIMK)-mediated inhibition of

Cofilin (Loirand, Guerin, & Pacaud, 2006). Remodeling of cortical actin induced by this process inhibits K_V channels and promotes membrane depolarization and increased myogenic tone (Luykenaar, El-Rahman, Walsh, & Welsh, 2009). Finally, the RhoA/Rho kinase pathway has also been implicated in the modulation of TRP channel function (Li & Brayden, 2017) and voltage-gated Ca^{2+} channel activity (Guan, Baty, Zhang, Remedies, & Inscho, 2019) (Fig. 1).

In addition to modulating ion channel activity, activated PKC also can affect the Ca^{2+} sensitivity of the contractile machinery in VSMCs (Cole & Welsh, 2011). Activated PKC phosphorylates the protein CPI_{17} and phosphorylated CPI_{17} inhibits MLCPPT to promote Ca^{2+} sensitization of the contractile machinery supporting increased myogenic tone (Cole & Welsh, 2011).

3. Mechanisms of myogenic tone in peripheral arterioles: Role of mechanosensitive G-protein-coupled receptors and inositol 1,4,5 trisphosphate receptors

Myogenic tone in second-order cremaster arterioles, as a "model" peripheral microvessel, appears to follow the general pathway outlined above. Consistent with this mechanism, both hamster and mouse cremaster arterioles display asynchronous, IP_3-dependent Ca^{2+} waves that appear to contribute to global Ca^{2+} levels and maintenance of myogenic tone: Inhibition of PLCs with U73122 inhibits Ca^{2+} waves, decreases global intracellular Ca^{2+} levels and causes vasodilation in both hamster (Westcott & Jackson, 2011) (Fig. 2) and mouse (Westcott et al., 2012) cremaster arterioles. Identical results were obtained by blockade of IP_3R with either 2-APB or xestospongin D or depletion of internal Ca^{2+} stores with thapsigargin (Westcott et al., 2012; Westcott & Jackson, 2011). These findings support the general mechanisms shown in Fig. 1. Despite the asynchronous Ca^{2+} waves that are observed in pressurized hamster and mouse cremaster arterioles studied by pressure myography, vasomotion (oscillations in diameter), associated with the Ca^{2+} waves, is not usually observed (at least on a detectable scale). We think that this may mean that the contractile machinery "filters" these oscillations, responding instead to a time-averaged value of the global intracellular Ca^{2+} concentration. In contrast, when vasoconstriction is induced with a G-protein-coupled agonist, such as phenylephrine, synchronous Ca^{2+} waves are observed in the VSMCs that are associated with diameter oscillations in synchrony with the Ca^{2+} waves

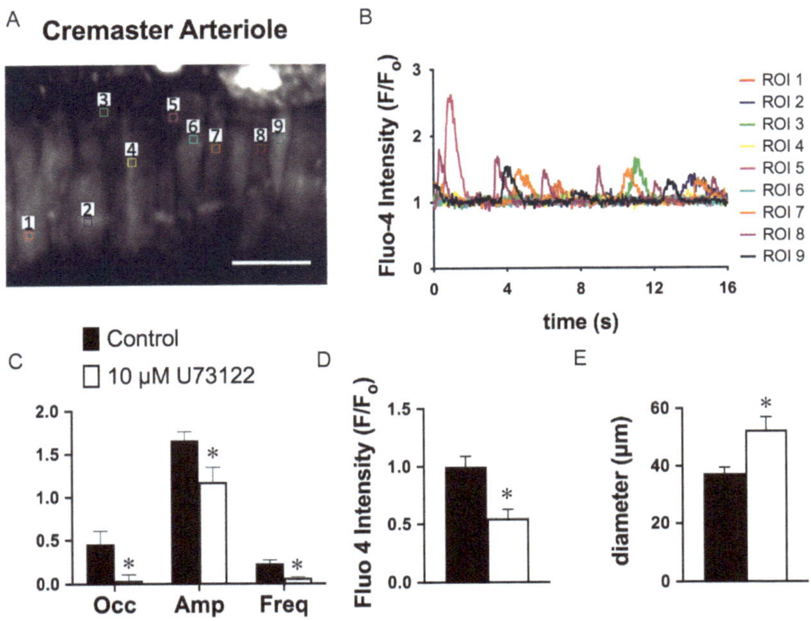

Fig. 2 Smooth muscle cells in cremaster arterioles display Ca^{2+} waves sensitive to phospholipase C inhibition. Panel (A): Image of a tangential, confocal slice through a hamster cremaster arteriole loaded with Fluo-4 and studied by pressure myography, along with nine regions of interest (ROI's) that were placed at the peak of Ca^{2+} waves as described (Jackson & Boerman, 2017). The scale bar equals 20 μm. Panel (B): Plot of relative fluorescence (F/F_o) for the 9 ROI's in panel (A) showing the amplitude and duration of Ca^{2+} events at these locations. Application of SparkAn software to these data with a threshold of 15% above baseline revealed 19 Ca^{2+} waves with mean amplitude = 1.52 ± 0.07 F/F_o (range = 1.2–2.6) during the 16 s recording period shown for this vessel. Similar results were obtained in four additional cremaster arterioles, confirming prior studies (Westcott & Jackson, 2011). Panel (C) shows inhibition of the occurrence (Occ—number of cells with waves), the amplitude (Amp) and the frequency (Hz) of Ca^{2+} waves by the PLC inhibitor, U-73122 ($n = 8$). Panel (D): In addition to inhibition of Ca^{2+} waves, U-73122 also caused a decrease in global Ca^{2+} levels measured as the average Fluo-4 fluorescence in projections of 500 images before and after treatment with U-73122 relative to Control fluorescence (F/F_o) (Westcott & Jackson, 2011) ($n = 5$). Panel (E): Associated with the U-73122-induced fall in global Ca^{2+} was a significant dilation (Westcott & Jackson, 2011) ($n = 4$). Similar results were obtained with the IP_3 receptor antagonists 2-APB and xestospongin or depletion of Ca^{2+} stores using thapsigargin in both hamster (Westcott & Jackson, 2011) and mouse (Westcott, Goodwin, Segal, & Jackson, 2012) second-order cremaster arterioles. These data support the hypothesis that PLC, IP_3, IP_3R and the resultant Ca^{2+} waves contribute to global Ca^{2+} levels and myogenic tone in these arterioles. * = $P < 0.05$ by paired *t*-tests. *Panel (A and B): Reproduced from Jackson, W. F., & Boerman, E. M. (2017). Regional heterogeneity in the mechanisms of myogenic tone in hamster arterioles.* American Journal of Physiology. Heart and Circulatory Physiology, 313*(3), H667–H675. doi:10.1152/ajpheart.00183.2017 with permission. Panels (C–E): Data from Westcott and Jackson Westcott, E. B., & Jackson, W. F. (2011). Heterogeneous function of ryanodine receptors, but not IP3 receptors, in hamster cremaster muscle feed arteries and arterioles.* American Journal of Physiology. Heart and Circulatory Physiology, 300*(5), H1616–H1630. doi:10.1152/ajpheart.00728.2010, used with permission.*

(Jackson, unpublished observations) as has been reported in mesenteric arteries (Mauban, Lamont, Balke, & Wier, 2001). The increase in Ca^{2+} sensitivity induced by phenylephrine acting at α_1-adrenoreceptors may partially explain why vasomotion is observed in the presence of an agonist rather than pressure alone. Additional research will be required to resolve this issue.

Where hamster cremaster arterioles deviate from the model shown in Fig. 1 is in the reliance on ATR1: the ATR1 antagonist, Losartan, which inhibits myogenic tone in the renal circulation (Mederos y Schnitzler et al., 2008) and effectively blocks Ang II-induced vasoconstriction in hamster cremaster arterioles (Jackson & Boerman, 2017), has no effect on basal myogenic tone in second-order hamster cremaster arterioles (Jackson & Boerman, 2017). In support of our findings with Losartan, Hong et al. (2016) have shown that the ATR1 inverse agonist, Candesartan (1 μM), has no effect on myogenic tone in first-order rat cremaster arterioles. However, they went on to demonstrate that a higher concentration (10 μM) of Candesartan does inhibit myogenic tone in these arterioles (Hong et al., 2016). Interpretation of these data are difficult because of the high concentration of Candesartan used, but they suggest that there may be regional or species differences in the details of the mechanisms underlying myogenic tone. Consistent with the notion of regional heterogeneity in mechanisms of myogenic tone, Brayden and colleagues have shown that 1 μM Candesartan inhibits myogenic tone of rat pial arteries, but has no significant effect on myogenic tone of downstream parenchymal arterioles, microvessels that appear to use P2Y6 receptors as the sensors responsible for myogenic tone (Brayden et al., 2013). The msGPCR responsible for activation of PLC in second-order cremaster arterioles remains to be established.

Hamster cheek pouch arterioles develop myogenic tone similar in magnitude to that in cremaster arterioles, however, PLC, IP_3 and IP_3R do not appear to be involved in this process (Jackson & Boerman, 2017). Cheek pouch arterioles do not display Ca^{2+} waves when studied under conditions identical to those in cremaster arterioles and effective inhibition of PLC with U73122 (Fig. 3) or of IP_3R with 2-APB has no effect on myogenic tone in these vessels (Jackson & Boerman, 2017). These data demonstrate that there are substantial regional differences in the mechanisms responsible for myogenic tone in peripheral arterioles, and in particular that msGPCR-PLC-IP_3-IP_3R-dependent mechanisms are not an essential component of the process in all vessels. They also open the possibility that the mechanisms may not be fixed, but able to vary as the conditions demand. The membrane signaling mechanism responsible for myogenic tone in hamster cheek pouch

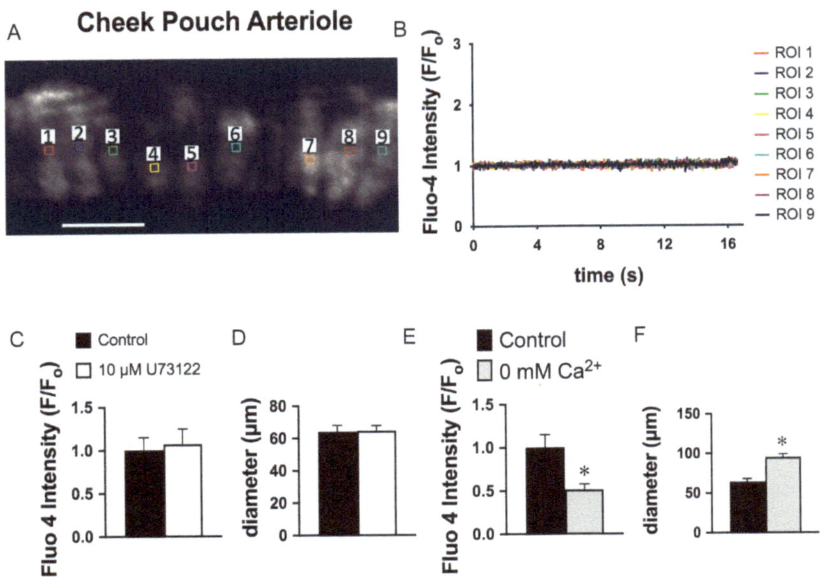

Fig. 3 Smooth muscle cells in cheek pouch arterioles do not display Ca^{2+} waves and myogenic tone is insensitive to phospholipase C inhibition. Panel (A): Image of a tangential, confocal slice through a hamster cheek pouch arteriole loaded with Fluo-4 and studied by pressure myography, along with nine regions of interest (ROI's) that were placed on the cells as described (Jackson & Boerman, 2017). The scale bar equals 20 μm. Panel (B): In the nine ROI's shown, no events with amplitude greater than 15% above baseline were detected during the 16 s recording period shown. Panel (C): U-73122 had no effect on global Ca^{2+} levels in hamster cheek pouch measured as the average Fluo-4 fluorescence from projections of 500 images as described (Jackson & Boerman, 2017) ($n=3$). Panel (D): U-73122 also had no significant effect on the diameter of cheek pouch arterioles (n = 3). Data identical to Panels (C and D) were obtained for the IP3R antagonist, 2-APB indicating that PLC and IP_3R do not contribute to global Ca^{2+} levels or resting myogenic tone in cheek pouch arterioles (Jackson & Boerman, 2017). Panels (E and F) show, as a positive control, that removal of extracellular Ca^{2+} decreases global Ca^{2+} levels and dilates cheek pouch arterioles (n = 3). * = $P < 0.05$ by paired *t*-tests. *Panel (A–D): Reproduced from Jackson, W. F., & Boerman, E. M. (2017). Regional heterogeneity in the mechanisms of myogenic tone in hamster arterioles.* American Journal of Physiology. Heart and Circulatory Physiology, 313*(3), H667–H675. doi:10.1152/ajpheart.00183.2017 with permission.*

arterioles remains to be established, but mechanosensitive ion channels, such as TRPC6 (Welsh et al., 2002) or mechanisms involving integrins (Davis et al., 2001; Martinez-Lemus et al., 2005) may be speculated.

In VSMCs of resistance arteries upstream from the microcirculation, Ca^{2+} waves that contribute to global Ca^{2+} levels and myogenic tone depend on both IP_3R and RyR (Fig. 4); inhibition of RyR reduces the occurrence

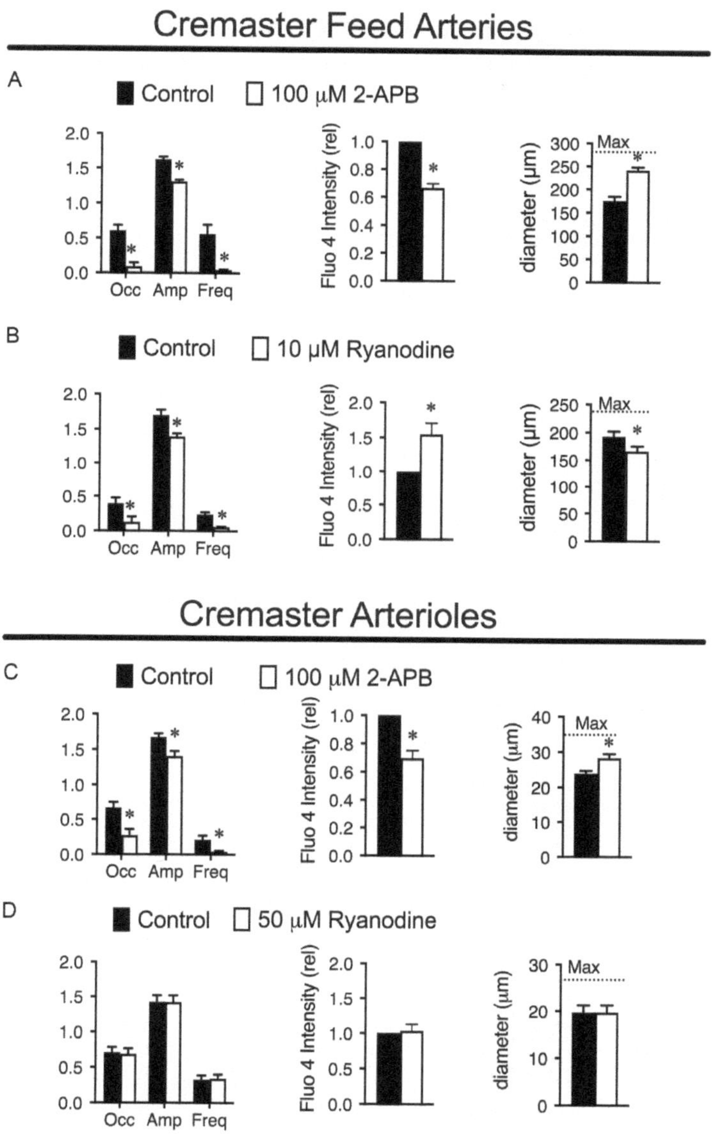

Fig. 4 Ryanodine inhibits Ca^{2+} waves in hamster cremaster feed arteries, but not in downstream arterioles. Data are means ± 95% confidence intervals for occurrence (Occ) of Ca^{2+} waves (panels A–D). All other data are means ± SE for amplitude (Amp, F/F_o), and frequency (Freq, Hz) of Ca^{2+} waves (panels A–D); global Fluo-4 intensity, an index of global intracellular Ca^{2+} (panels A–D); and diameter (panels A–D) in the absence (Control) or presence of the IP_3 receptor antagonist, 2-aminoethoxydiphenyl borate (2-APB, 100 μM; Panels (A) and (C), $n = 6$); or the ryanodine receptor antagonist, ryanodine (RYR, 10 μM; Panel (B) and 50 μM; Panel (D)). * = significantly

(Continued)

and frequency of Ca^{2+} waves (Westcott et al., 2012; Westcott & Jackson, 2011). However, in contrast to the effects of inhibition of IP_3R, which also inhibits Ca^{2+} waves, inhibition of RyR leads to an increase in global intracellular Ca^{2+} and an increase in myogenic tone (Fig. 4B) (Westcott et al., 2012; Westcott & Jackson, 2011). The increase in Ca^{2+} and tone with inhibition of RyR results from inhibition of RyR-mediated Ca^{2+} sparks and the loss of Ca^{2+} spark-mediated activation of BK_{Ca} channels (Westcott et al., 2012; Westcott & Jackson, 2011), as will be discussed below. In downstream cremaster arterioles, in contrast to effects of inhibition of IP_3R signaling (Fig. 4C) (Westcott et al., 2012; Westcott & Jackson, 2011), effective inhibition of RyR has no effect on Ca^{2+} waves or myogenic tone (Fig. 4D) (Westcott et al., 2012; Westcott & Jackson, 2011) in the arterioles. RyR appear to be silent in cremaster arterioles and do not contribute to Ca^{2+} waves or the regulation of arteriolar myogenic tone (Westcott et al., 2012; Westcott & Jackson, 2011). It should be noted that RyR are not silent in all arterioles. In retinal arterioles, RyR-dependent Ca^{2+} sparks contribute to contraction of VSMCs in this microcirculation (See "Ion channels and myogenic activity in retinal arterioles" in this volume). In cerebral parenchymal arterioles, RyR contribute to Ca^{2+} waves under resting conditions, but can generate Ca^{2+} sparks when extracellular pH is reduced leading to activation of BK_{Ca} channels in VSMCs from these arterioles (Dabertrand, Nelson, & Brayden, 2012). Thus, there appear to be significant regional differences in RyR function.

Fig. 4—Cont'd different from Control value, $P<0.05$, $n=6$ by paired *t*-tests. Dashed lines in panels (A–D) represent the maximum diameters of the vessels in 0 Ca^{2+} PSS. Panels (A) and (C): 2-APB inhibits Ca^{2+} waves in both feed arteries and downstream arterioles. Similar results were obtained with another IP_3R antagonist, xestospongin (Westcott & Jackson, 2011) or depletion of Ca^{2+} stores with thapsigargin (Westcott & Jackson, 2011). These findings demonstrate that Ca^{2+} waves are dependent on IP_3R in both cremaster feed arteries and arterioles. In feed arteries (Panel B), Ca^{2+} waves are also inhibited by ryanodine supporting a role for RyR in these resistance arteries. However, in downstream arterioles (Panel (D)), even 50 μM ryanodine has no effect on Ca^{2+} waves. See Westcott and Jackson (2011) and text for more information. *Data from Westcott, E. B., & Jackson, W. F. (2011). Heterogeneous function of ryanodine receptors, but not IP3 receptors, in hamster cremaster muscle feed arteries and arterioles.* American Journal of Physiology. Heart and Circulatory Physiology, 300*(5), H1616–H1630. doi:10.1152/ajpheart.00728. 2010 used with permission.*

4. VGCCs and myogenic tone in peripheral arterioles

There is considerable evidence, mostly from ex vivo studies of resistance arteries and arterioles studied by pressure myography, that CaV1.2, which form L-type VGCCs, provide a significant source of activator Ca^{2+} responsible for myogenic tone: in hamster cremaster and hamster cheek pouch arterioles maximal inhibition of CaV1.2 VGCC with diltiazem produces 86% dilation in cremaster arterioles vs 54% dilation in cheek pouch arterioles both studied by pressure myography under identical conditions (Jackson & Boerman, 2017). These data support a major role for CaV1.2 VGCC in myogenic tone, but also suggest that other sources of Ca^{2+} significantly contribute to the process, particularly in cheek pouch arterioles. Likely sources include TRP channels expressed in these VSMCs such as TRPC1, TRPC3, or TRPC6 (although this needs to be studied), or Ca^{2+} entry through transporters such as the Na^+/Ca^{2+} exchanger (Zhang et al., 2010). Measurement of membrane potential in pressurized cremaster and cheek pouch arterioles indicates that the latter VSMCs are slightly hyperpolarized (-33 mV vs -28 mV in cremaster arterioles) (Fig. 5A) that might account for lower activity of CaV1.2 in the cheek pouch. In addition, Ba^{2+} current density through nifedipine-sensitive CaV1.2 channels are smaller in VSMCs from cheek pouch than cremaster arterioles (Fig. 5B). Despite these differences, intracellular Ca^{2+} measured with Fura 2 (Fig. 5C) and myogenic tone (Fig. 5D) are similar in cremaster and cheek pouch arterioles.

In cremaster arterioles, we think that IP_3R serve to amplify Ca^{2+} signals through VGCC (Jackson & Boerman, 2018). We have shown that activation of VGCC by the CaV1.2 agonist Bay K8644 or by membrane depolarization increases the frequency and occurrence of Ca^{2+} waves and increases global myoplasmic Ca^{2+}, whereas removal of extracellular Ca^{2+} or deactivation of VGCC by reduced pressure or membrane hyperpolarization reduce the frequency and occurrence of Ca^{2+} waves and decrease global myoplasmic Ca^{2+} (Jackson & Boerman, 2018). These data suggest that IP_3R serve to amplify Ca^{2+} signals through CaV1.2 VGCC in cremaster arterioles. This process appears to be lacking in cheek pouch arterioles studied under identical conditions and may also help explain why CaV1.2 VGCC contribute less to myogenic tone in these microvessels (Jackson & Boerman, 2017).

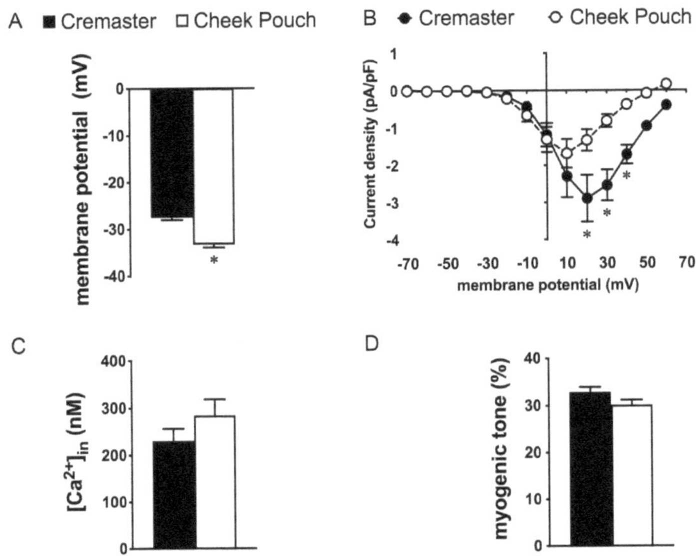

Fig. 5 Cheek pouch arteriolar VSMCs are slightly hyperpolarized and have smaller VGCC currents than cremaster VSMCs. Panel (A): Sharp electrode membrane potential recorded from VSMCs in cremaster and cheek pouch arterioles studied by pressure myography at 80 cm H_2O. See Burns, Cohen, and Jackson (2004) for detailed methods. Data for cremaster replotted from Burns et al. (2004). Under similar recording conditions, cheek pouch arterioles were more hyperpolarized -33 ± 1 mV ($n = 63$ cells in eight arterioles) than cremaster arterioles (-27 ± 1 mV) ($n = 34$ arterioles; $P < 0.05$ by *t*-test). Panel (B): Whole-cell Ba^{2+} current density (pA/pF) recorded in isolated VSMCs as described in (Cohen & Jackson, 2003) ($n = 5$ cells for each group; $* = P < 0.05$ by two-way ANOVA followed by Sidak test for comparison of means). Panel (C): Intracellular Ca^{2+} concentration measured with FURA 2 as described (Burns et al., 2004; Jackson, Boerman, Lange, Lundback, & Cohen, 2008; Westcott & Jackson, 2011). Data are replotted from Burns et al. (2004) for cremaster ($n = 7$) and Brekke, Jackson, and Segal (2006) for cheek pouch arterioles ($n = 7$) ($P > 0.05$ by *t*-test). Panel (D): Data are replotted from Jackson and Boerman (2017) for $n = 38$ arterioles for both cremaster and cheek pouch ($P > 0.05$ by *t*-test). Despite the apparent difference in membrane potential (panel A) and the activity of VGCCs (panel B) intracellular Ca^{2+} and resting myogenic tone are similar in these two classes of arterioles.

In vivo, the role played by CaV1.2 in arteriolar myogenic tone appears more variable. We (Jackson, 2012; Welsh, Jackson, & Segal, 1998) and others (Hill & Meininger, 1994) have shown that in pentobarbital-anesthetized rodents, inhibition of CaV1.2 channels with either diltiazem or nifedipine has no steady-state effect on arteriolar myogenic tone in hamster cheek pouch (Welsh et al., 1998), hamster cremaster (Jackson, 2012) (Fig. 6A) and rat cremaster arterioles (Hill & Meininger, 1994).

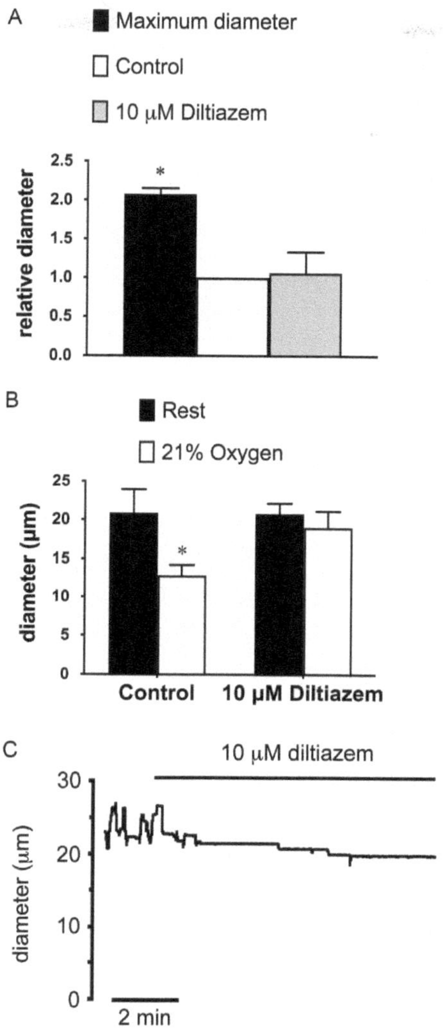

Fig. 6 Lack of effect of diltiazem on resting myogenic tone in hamster cremaster arterioles, in vivo. Panel (A): Data are relative diameters ± SE ($n=4$) of arterioles in superfused hamster cremaster muscles studied by intravital microscopy as described (Jackson, 1986, 1993a, 1993b) at rest (Control—open bar), after topical application of the endothelium-dependent dilator, methacholine (10 μM) (Maximum—black bar), or in the presence of 10 μM diltiazem (gray bar), as indicated. Note that the vessels developed myogenic tone (Control diameters less than Maximum), yet diltiazem no effect on vessel diameter. * = significantly different from Control, $P < 0.05$. Panel (B): Data are mean diameters ± SE (n = 4) of hamster cremaster muscle arterioles studied by intravital microscopy. Shown are arteriolar diameters at rest and in the presence of elevated superfusate O_2 tension, before (Control) and during superfusion with diltiazem, as indicated. As described previously (Jackson, 1986, 1993b, 1998), under control conditions,

(Continued)

This lack of effect of CaV1.2 VGCC blockers does not appear to be due to a lack of efficacy because the blockers inhibited oxygen-induced arteriolar constriction (Jackson, 2012; Welsh et al., 1998) (Fig. 6B) and vasomotion (Hill & Meininger, 1994; Jackson, 2012) (Fig. 6C) processes dependent on CaV1.2 VGCCs. The source of activator Ca^{2+} after blockade of CaV1.2 VGCC in vivo remains to be established. It is likely that some other VGCC such as CaV3.X T-type VGCC (Abd El-Rahman et al., 2013; Dam, Boedtkjer, Aalkjaer, & Matchkov, 2014; Gustafsson, Andreasen, Salomonsson, Jensen, & Holstein-Rathlou, 2001; Hansen, Jensen, Andreasen, & Skott, 2001; Morita, Cousins, Onoue, Ito, & Inoue, 1999) or other classes of VGCC can take over this important role and maintain arteriolar tone within a homeostatic range. In vivo, arterioles are exquisitely sensitive to the ambient PO_2: increased PO_2 results in profound arteriolar constriction of peripheral arterioles (Jackson, 2016a) (Fig. 6B). At rest, arterioles are exposed to much lower PO_2 values than the systemic value of 100 mmHg. In second-order cheek pouch and cremaster arterioles, we estimate that the resting PO_2 is about 30 mmHg, and not 100 mmHg as seen in large, systemic arteries, due to substantial pre-capillary diffusion of O_2 out of blood into the surrounding tissue (Jackson, 2016a). Experimentally, by superfusion of tissues with physiological salt solution equilibrated with elevated PO_2, we and others have shown that arteriolar tone can be substantially increased and that this is the result of membrane depolarization and activation of CaV1.2 VGCC due to O_2-dependent production of vasoconstrictors like leukotrienes and 20-HETE (Jackson, 2016a). It is worthy to note that most pressure myography studies of resistance arteries and arterioles are performed at PO_2's in excess of 100 mmHg (usually air saturated with a PO_2 about 150 mmHg). This theoretically could explain why ex vivo studies of resistance arteries and arterioles consistently show a major role for CaV1.2 VGCC, whereas in vivo studies, particularly where the PO_2 is controlled to much

Fig. 6—Cont'd elevated O_2 tension caused significant arteriolar constriction. However, this response was abolished by diltiazem. These data demonstrate the efficacy of diltiazem in this preparation. * = significantly different from Rest, $P < 0.05$. Panel (C): Shown is a digitized trace of arteriolar internal diameter in a hamster cremaster studied by intravital microscopy before and during superfusion with diltiazem as indicated. As shown, the arteriole displayed vasomotion, the oscillatory diameter shown at the beginning of the trace. This behavior was abolished when the muscle was superfused with diltiazem. These data also demonstrate the efficacy of diltiazem in this preparation. *Data are replotted from Jackson, W. F. (2012). Microcirculation. In J. A. Olson & E. N. Hill (Eds.),* Muscle *(pp. 1197–1206). Boston/Waltham: Academic Press.*

lower physiological levels, the contribution of CaV1.2 VGCC to resting tone appears less. However, despite the high PO_2, the degree of tone in pressure myograph systems with high PO_2 is quite similar to the level of tone observed in arterioles studied in vivo at much lower PO_2 values. If the ambient PO_2 were the sole explanation of the higher activity of VGCC ex vivo compared to in vivo, one would expect that myogenic tone would be substantially higher in the pressure myograph than is typically observed. In addition, we have previously shown that changes in PO_2 between 15 and 150 mmHg do not appear to have effects on VSMC membrane potential (Jackson, 2000a), outward K^+ currents (Jackson, 2000a), currents through VGCCs (Cohen & Jackson, 2003) or IP_3-mediated Ca^{2+} transients (Cohen & Jackson, 2003) in hamster cremaster VSMCs, in vitro. Therefore, why CaV1.2 VGCCs appear to be silent, in vivo, in cheek pouch and cremaster arterioles, but active in isolated vessels studied in the pressure myograph remains unknown.

4.1 Non-selective cation channels and myogenic tone in peripheral arterioles

As outlined in Fig. 1, cation influx through TRPC6 and TRPM4 channels appear to be an integral part of myogenic tone in cerebral resistance arteries (Gonzales et al., 2014). Cation channels inhibitable by the lanthanide Gd^{3+} (10 nM–1 μM) and the Chilean rose tarantula venom toxin GsMTx-4 (5 μM) also appear essential for the development of myogenic tone in both second-order hamster cremaster and cheek pouch arterioles (Jackson & Boerman, 2017). For example, Gd^{3+} inhibited $67 \pm 4\%$ of Ca^{2+}-dependent myogenic tone with a $\log IC_{50} = -7.3 \pm 0.1$ (~55 nM) in both cheek pouch and cremaster arterioles (Jackson & Boerman, 2017). While the precise identity of the cation channels that contribute to membrane depolarization and activation of VGCC in these vessels remain to be established, the strong inhibition of myogenic tone by Gd^{3+} and GsMTx-4 suggests that TRPC6 or other mechanosensitive ion channels inhibited by these substances also importantly contribute to myogenic tone in these hamster arterioles. A role for TRPM4 in myogenic tone of cremaster and cheek pouch arterioles has not been directly established. It is also not known if the apparent activity of TRP channels contributes to myogenic tone only through effects on membrane potential (increased activity will depolarize VSMCs and activate CaV1.2 VGCCs) or if Ca^{2+} influx through these channels contributes to activator Ca^{2+} in the VSMCs. Additional research will be required to answer these questions.

4.2 Ca^{2+} activated Cl^- channels and myogenic tone in peripheral arterioles

Pressure-induced membrane depolarization of arteriolar VSMCs may also be supported by CaCCs (Bulley et al., 2012; Heinze et al., 2014). Consistent with this hypothesis, a CaCC blocker, niflumic acid, dilates isolated cheek pouch arterioles studied by pressure myography (Fig. 7), where CaV1.2 VGCCs contribute to resting myogenic tone (Jackson & Boerman, 2017). We suspect that it is Ca^{2+} entry through CaV1.2 VGCC that activates the CaCCs in these vessels, because we have also found that niflumic acid dilates O_2-constricted cheek pouch arterioles, in vivo, in a concentration-dependent manner (Fig. 8A and B), and attenuates the O_2-induced VSMC depolarization that drives this constriction (Fig. 8C and D). However, niflumic acid had no significant effect on resting diameter, in vivo, a condition where CaV1.2 channels do not contribute to resting myogenic tone: arterioles were $28 \pm 3\,\mu m$ at rest and were $32 \pm 5\,\mu m$ in the presence of $100\,\mu M$ niflumic acid (maximum diameter $= 40 \pm 3\,\mu m$, $n = 4$, $P > 0.05$ by paired t-test). A role for CaCC in second-order cremaster arterioles has not been directly studied.

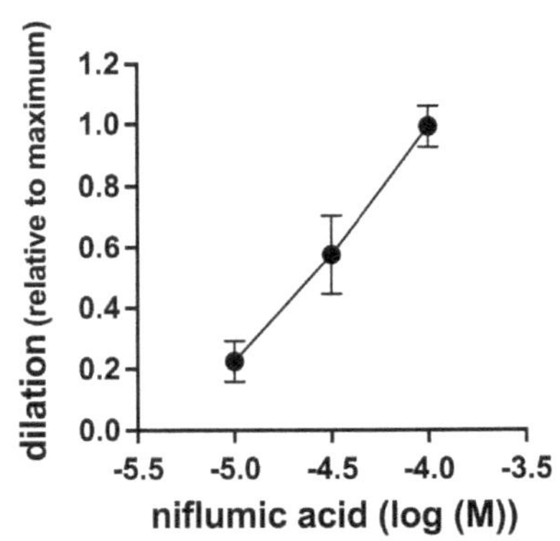

Fig. 7 Block of Ca^{2+}-activated Cl^- channels dilates hamster cheek pouch arterioles studied by pressure myography. Data are mean dilations relative to maximal diameter $\pm$ SE ($n = 3$) in the presence of 30–100 μM niflumic acid in isolated cheek pouch arterioles studied by pressure myography as in Jackson and Boerman (2017). Resting diameters were $66 \pm 3.2\,\mu m$ and maximum diameters in 0 mM Ca^{2+} were $91 \pm 8.9\,\mu m$. As shown, niflumic acid produced concentration-dependent dilation of cheek pouch arterioles studied by pressure myography. These data suggest that Ca^{2+}-activated Cl^- channels (CaCCs) contribute to pressure-induced myogenic tone in cheek pouch arterioles, in vitro.

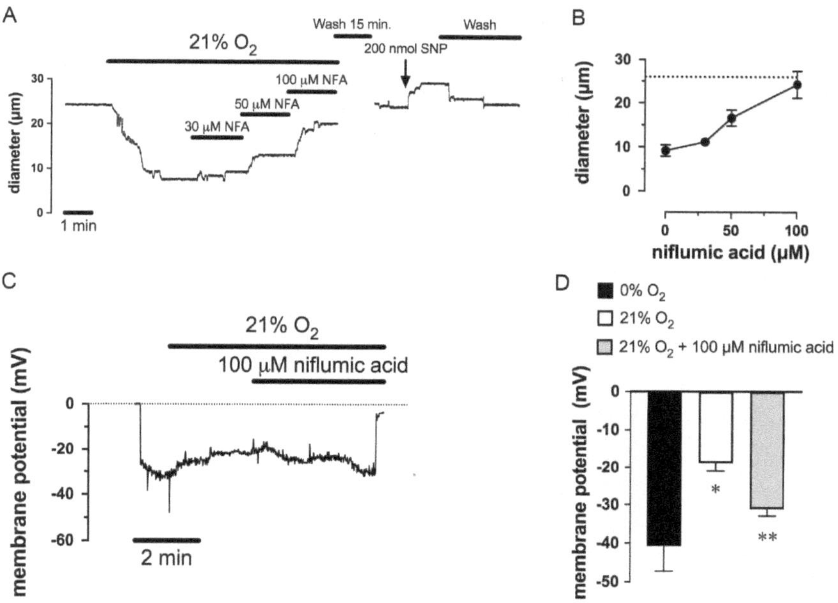

Fig. 8 Ca^{2+}-activated Cl^- channels (CaCC) contribute to O_2-induced constriction of hamster cheek pouch arterioles, in vivo. Panel (A): digitized diameter trace of a hamster cheek pouch arteriole studied by intravital microscopy as described (Jackson, 1993b). As indicated, superfusion of the preparation with solution equilibrated with 21% O_2 caused sustained vasoconstriction. Addition of niflumic acid (NFA) from 30 to 100 μM produced concentration-dependent arteriolar dilation. Arteriolar diameter recovered after washout of niflumic acid and removal of O_2 from the superfusate. Sodium nitroprusside (SNP) applied as a 200 nmol bolus dilated the arteriole demonstrating significant resting tone. Panel (B): Summary data for experiments as shown in panel (A). Data are mean diameters ± SE (n = 3) of O_2-constricted arterioles in the absence (0 μM) or presence (30–100 μM) of niflumic acid in the superfusate. Niflumic acid produced concentration-dependent dilation of O_2 constricted arterioles supporting a role for CaCC in O_2-induced constriction of hamster cheek pouch arterioles. Panel (C): Digitized trace of arteriolar VSMC membrane potential recorded with sharp electrodes as described (Welsh et al., 1998). From a resting membrane potential of about −33 mV, increasing superfusate O_2 content to 21% depolarized the cell to about −20 mV as we have previously described (Welsh et al., 1998). Addition of 100 μM niflumic acid repolarized the cell to about −30 mV. Panel (D): Summary data for experiments as in panel (C). Data are means ± SE ($n = 8$) under control conditions with 0% O_2 ($PO_2 = 15$ mmHg), with 21% O_2 in the superfusate and with 21% O_2 and 100 μM niflumic acid in the superfusate. Elevated O_2 tension depolarized the VSMCs in cheek pouch arterioles as reported (Welsh et al., 1998), and niflumic acid (inhibition of CaCC) lead to significant repolarization (* = $P < 0.05$ vs 0% O_2; ** = $P < 0.05$ vs 21% O_2 but not different from 0% O_2 by Tukey's test after ANOVA).

4.3 Lack of a role for PKC in arteriolar myogenic tone in peripheral arterioles

There is considerable evidence in resistance arteries for a role of PKC in myogenic tone, with studies in cerebral arteries (Earley et al., 2007; Osol, Laher, & Cipolla, 1991; Slish, Welsh, & Brayden, 2002; Yeon et al., 2002), coronary arteries (Miller, Dellsperger, & Gutterman, 1997), subcutaneous resistance arteries (Coats, Johnston, MacDonald, McMurray, & Hillier, 2001), small arteries from rat gracilis muscle (Massett, Ungvari, Csiszar, Kaley, & Koller, 2002), rabbit facial vein (Laher & Bevan, 1989), murine mesenteric resistance arteries (Mauban et al., 2015) and ferret aorta (Pawlowski & Morgan, 1992) showing that inhibition of PKC inhibits myogenic tone. In contrast, studies in second-order hamster cremaster and cheek pouch arterioles using the PKC inhibitor, bisindolylmaleimide I (BIM I), which effectively blocks phorbol ester-induced arteriolar constriction in these preparations, has no significant effect on myogenic tone (Jackson & Boerman, 2017). Similarly, in first-order rat cremaster arterioles (Hill, Falcone, & Meininger, 1990) and rat ophthalmic arteries (Ito, Jarajapu, Grant, & Knot, 2007) effective inhibition of PKC had no effect on myogenic tone. The cause of these regional differences in the role played by PKC has not been established. Nonetheless, these data support the hypothesis that there are significant regional- and perhaps species-dependent differences in the mechanisms of myogenic tone and specifically the role played by PKC.

4.4 Rho kinase and myogenic tone in peripheral arterioles

Rho kinase has been proposed to be involved in the regulation of Ca^{2+} entry into rat aortic and mesenteric VSMCs (Ghisdal, Vandenberg, & Morel, 2003). Rho kinase has also been proposed to affect VSMC membrane potential by affecting the activity of K_V channels (Luykenaar et al., 2009) and TRPM4 channels (Li & Brayden, 2017) in cerebral arteries, which in turn, affect VSMC Ca^{2+} signaling through voltage-dependent activation of VGCC. In second-order hamster cremaster and cheek pouch arterioles, we found that the Rho kinase inhibitors Y27632 (Jackson & Boerman, 2017) and H1152 (Fig. 9) inhibited >80% of Ca^{2+}-dependent myogenic tone (Jackson & Boerman, 2017). However, the loss of tone was associated with inhibition of IP_3R-dependent Ca^{2+} waves in cremaster arterioles and a substantial decrease in global myoplasmic Ca^{2+} in both cremaster (Fig. 9) and cheek pouch arterioles (Jackson & Boerman, 2017). These data support

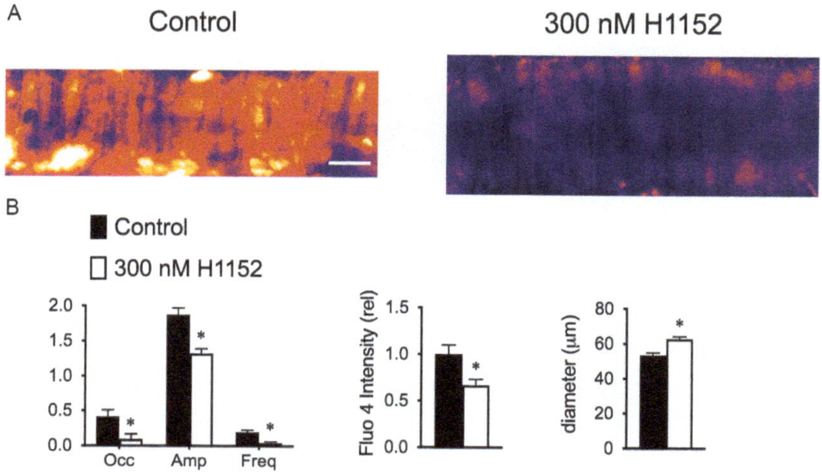

Fig. 9 Rho kinase antagonist inhibits Ca^{2+} signaling and myogenic tone in hamster cremaster arterioles. Panel (A): Shown are maximum intensity projections of 500 images of a hamster cremaster arteriole loaded with Fluo-4 before (Control) and in the presence of the Rho kinase inhibitor, H1152, showing the decrease in global Ca^{2+} produced by this inhibitor. The scale bar in the Control image is 20 µm and applies to the H1152 image as well. Panel (B): Summary data showing that H1152 inhibited Ca^{2+} waves, reduced global Ca^{2+} and dilated hamster cremaster arterioles studied by pressure myography (Jackson & Boerman, 2017). Data are the occurrence of Ca^{2+} waves (Occ; number of cells showing waves/total number of cells studied) ±95% confidence interval (n = 4), or means ± SE (n = 4) for the amplitude (Amp; F/F_o) or frequency (Freq; Hz) of Ca^{2+} waves; global Fluo-4 levels (an index of global Ca^{2+}) and arteriolar internal diameter. Identical results were obtained using the Rho kinase inhibitor, 10 µM Y-27632 (Jackson & Boerman, 2017). * = significantly different from Control value, $P < 0.05$ by paired *t*-tests. *Data are from Jackson, W. F., & Boerman, E. M. (2017). Regional heterogeneity in the mechanisms of myogenic tone in hamster arterioles.* American Journal of Physiology. Heart and Circulatory Physiology, 313*(3), H667–H675. doi:10.1152/ajpheart.00183.2017 with permission.*

a significant role for Rho kinase in arteriolar VSMC Ca^{2+} signaling, in addition to possible effects on Ca^{2+} sensitization. Studies in renal arterioles have shown that Y27632 and another Rho kinase inhibitor, RKI 1447, both inhibit myogenic tone and Ca^{2+} signaling through voltage-gated Ca^{2+} channels (Guan et al., 2019). In cerebral VSMCs, Y27632 produces a small (~15%) inhibition of currents through CaV1.2 VGCC by (Wu et al., 2009), although it is not known if this was a direct effect of the drug on the channel, or because Rho kinase modulates the function of these ion channels. Our findings that Y27632 and H1152 dilated cheek pouch arterioles (Jackson & Boerman, 2017), which do not appear to rely on a msGPCR-G-protein-PLC-dependent pathway to develop myogenic tone, suggest that either these drugs act off-target to inhibit Ca^{2+} influx through VGCCs, or that there is, perhaps, a Ca^{2+}-dependent pathway for activation of RhoA and Rho kinase that does not require activation of a GPCR.

4.5 Ion channels opposing myogenic tone in peripheral arterioles

Potassium channel activity, in general, opposes myogenic tone (Jackson, 2000b, 2005, 2017, 2018a; Jackson & Blair, 1998; Tykocki et al., 2017). Negative-feedback regulation of myogenic tone is achieved by the activity of large-conductance, Ca^{2+}-activated K^+ (BK_{Ca}) channels and K_V channels, essentially preventing vasospasm. The model of myogenic tone presented in Fig. 1 is inherently a positive feedback process. Pressure-induced "stretch" of VSMCs leading to activation of TRPC6 and TRPM4 channels, membrane depolarization and activation of CaV1.2 VGCC, if left unchecked, would strongly depolarize the plasma membrane of the VSMCs theoretically approaching the equilibrium potential for Ca^{2+} (~+120 mV) and leading to maximal activation of the contractile machinery in these cells. This does not happen because there is strong expression of both BK_{Ca} and K_V channels in the sarcolemma of the VSMCs that provides a negative-feedback signal to dampen and stabilize myogenic tone to maintain homeostasis (Jackson, 2000b, 2005, 2017, 2018a; Jackson & Blair, 1998; Tykocki et al., 2017).

The increase in intracellular Ca^{2+} (more on this topic below) and membrane depolarization associated with the myogenic response activate BK_{Ca} channels (Jackson, 2000b, 2005, 2017, 2018a; Jackson & Blair, 1998; Tykocki et al., 2017). The opening of only a few of these high conductance channels (~250 pS in cremaster arteriolar VSMCs; (Jackson & Blair, 1998)) leads to K^+ efflux from the VSMCs down the K^+ electrochemical gradient, limiting the pressure-induced depolarization of the VSMCs and preventing pressure-induced vasospasm (Jackson, 2000b, 2005, 2017, 2018a; Jackson & Blair, 1998; Tykocki et al., 2017). However, the source of Ca^{2+} that contributes to the activation of BK_{Ca} channels in peripheral arterioles appears to be different from what has been reported in upstream resistance arteries (Westcott et al., 2012; Westcott & Jackson, 2011). In cerebral (Nelson et al., 1995) and other resistance arteries (Westcott et al., 2012; Westcott & Jackson, 2011), Ca^{2+} released from groups of ryanodine receptors (RyR) into the subsarcolemmal space in the form of Ca^{2+} sparks appear to provide a major source of Ca^{2+} controlling the activity of BK_{Ca} channels. In these upstream resistance arteries, inhibition of ryanodine receptors with ryanodine or tetracaine, for example, eliminates Ca^{2+} sparks and increases myogenic tone to the same extent as blocking BK_{Ca} channels directly (Nelson et al., 1995; Westcott et al., 2012; Westcott & Jackson, 2011). In contrast, some arteriolar VSMCs do not display Ca^{2+} sparks at resting myogenic tone (Dabertrand et al., 2012; Westcott et al., 2012; Westcott & Jackson, 2011). In addition, inhibition of RyR with ryanodine or tetracaine which should inhibit RyR-dependent activation of BK_{Ca} channels, has no significant effect on myogenic tone implying that RyR are silent in these microvessels, as noted earlier in the chapter. For example, blockade of RyR in cremaster (Westcott et al., 2012; Westcott & Jackson, 2011) or cheek pouch arterioles (Fig. 10A and C) studied by pressure myography, has no significant effect on myogenic tone, whereas, direct block of BK_{Ca} channels produces the expected constriction (increase in myogenic tone) (Westcott et al., 2012; Westcott & Jackson, 2011) (Fig. 10B and C).

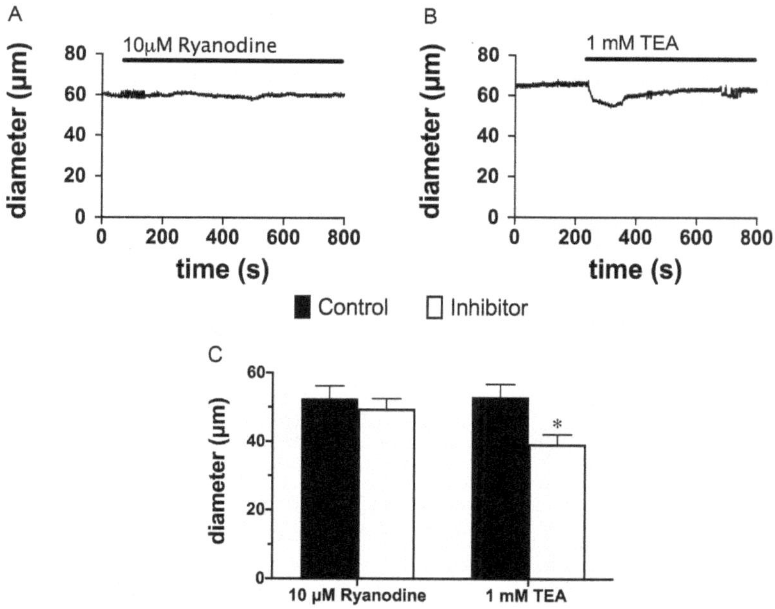

Fig. 10 Ryanodine receptors are silent, whereas BK_{Ca} channels are active and contribute to the negative-feedback regulation of myogenic tone in hamster cheek pouch arterioles, in vitro. Panels (A) and (B): Digitized diameter records demonstrating the lack of effect of ryanodine (panel A) but significant constriction produced by the BK_{Ca} channel blocker TEA (1 mM) in isolated cheek pouch arterioles studied by pressure myography as described (Jackson & Boerman, 2017). Panel (C): Summary data for experiments as in panels (A) and (B) showing that TEA but not ryanodine constricts hamster cheek pouch arterioles. Data are means ± SE (n = 15 arterioles). * = P < 0.05 by Sidak's multiple comparison method after two-way ANOVA.

Thus, BK_{Ca} channels do not appear to be controlled by RyR-related Ca^{2+} sparks in cremaster or cheek pouch arterioles. Instead, we think that Ca^{2+} entry through CaV1.2 VGCCs provides the major source of Ca^{2+} controlling the activity of BK_{Ca} channels. As has been reported in VSMCs from rabbit coronary arteries (Guia, Wan, Courtemanche, & Leblanc, 1999) and mouse mesenteric arteries (Suzuki, Yamamura, Ohya, & Imaizumi, 2013), BK_{Ca} channels are co-localized with VGCC in hamster cremaster arteriolar VSMCs (Jackson, 2018b). In vivo studies also support the hypothesis that Ca^{2+} entry through CaV1.2 VGCCs activate BK_{Ca} channels in cremaster arterioles (Jackson & Blair, 1998) and in cheek pouch arterioles (Fig. 11). The lack of role for RyR in

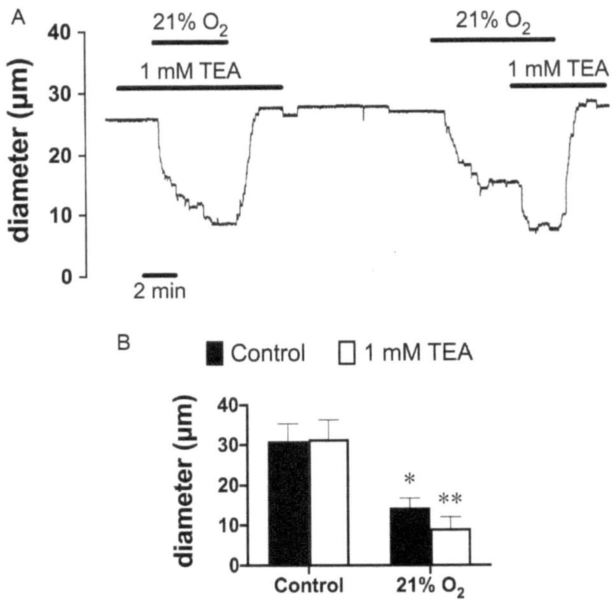

Fig. 11 Block of BK_{Ca} channels has no effect on resting diameter but potentiates O_2-induced arteriolar constriction, in vivo. Panel (A): Digitized diameter record of a hamster cheek pouch arteriole studied by intravital microscopy as in Jackson (1993b). At rest, 1 mM TEA had no effect on resting diameter but augmented constriction when the arterioles are exposed to elevated O_2 levels as shown. After washout of O_2 and TEA, elevated of O_2 levels constricted the arterioles, but to a larger diameter than in the presence of 1 mM TEA. Subsequent addition of TEA further constricted the arteriole as in the first trial of 21% O_2 + 1 mM TEA shown in the left side of panel (A). Panel (B): Summary data for experiments as in panel (A). Data are mean diameters ± SE ($n=5$). Under control conditions, TEA had no effect on arteriolar diameter, but in O_2 constricted arterioles, TEA caused further constriction confirming prior studies in hamster cremaster muscle that BK_{Ca} channels are active when VGCC are active (Jackson & Blair, 1998).

control of BK_{Ca} channels may be due to the level of expression of RyR isoforms and apparent lack of clustering of these channels in endoplasmic reticulum compared to RyR in upstream resistance arteries (Westcott et al., 2012).

The BK_{Ca} channels expressed by cremaster arteriolar VSMCs also have a substantially higher Ca^{2+} set point (the minimum Ca^{2+} concentration required for activity at physiological membrane potentials) (Jackson & Blair, 1998). This appears to be due to a lower expression of modulatory β-subunits (Yang et al., 2009) but could also be due to differences in spliced variant expression of the α-subunits that also can affect the response of these channels Ca^{2+} (Nourian et al., 2014) or differences in expression of other modulatory subunits (Evanson, Bannister, Leo, & Jaggar, 2014). The consequence of this high Ca^{2+} set point is that very high local levels of Ca^{2+} appear to be required to activate these channels in peripheral arteriolar VSMCs (>3 μM). We suspect that Ca^{2+} entering through VGCC adjacent to a BK_{Ca} channel, perhaps in the form of Ca^{2+} sparklets (Amberg, Navedo, Nieves-Cintron, Molkentin, & Santana, 2007), is sufficient to activate these channels, tuning the activity of the BK_{Ca} channels to match the activity of the VGCCs as previously proposed in coronary VSMCs (Guia et al., 1999) and mesenteric VSMCs (Suzuki et al., 2013). Calcium released, for example, through IP_3R in the form of Ca^{2+} waves in mouse and hamster feed arteries and downstream cremaster arteriolar VSMCs do not appear to activate BK_{Ca} channels because block of IP_3R reduces myogenic tone rather than increases tone as expected if BK_{Ca} channels were activated by the IP_3R-dependent Ca^{2+} waves (Westcott et al., 2012; Westcott & Jackson, 2011). In contrast, studies of cerebral arteries suggest that IP_3R are coupled to BK_{Ca} channels and contribute to the negative-feedback regulation of myogenic tone (Zhao et al., 2010). Why Ca^{2+} waves appear not to be coupled to BK_{Ca} channels in feed arteries and cremaster arterioles remains to be established. Most likely the Ca^{2+} waves do not raise subsarcolemmal Ca^{2+} to a level sufficient to activate arteriolar BK_{Ca} channels.

Studies of cerebral arteries and mesenteric arteries also have suggested a role for Ca^{2+} influx through CaV3.2 channels in controlling BK_{Ca} channel activity (Abd El-Rahman et al., 2013; Harraz et al., 2014, 2015). However, studies in a murine skeletal muscle feed artery have failed to replicate their findings (Mullan, Pettis, & Jackson, 2017) suggesting that there may be regional differences in coupling between CaV3.2 and BK_{Ca} channels.

Voltage-gated K^+ channels also contribute to the negative-feedback regulation of myogenic tone in peripheral arterioles in addition to BK_{Ca} channels (see Jackson, 2018a; Tykocki et al., 2017 for reviews). Studies have shown that VSMCs from resistance arteries and arterioles display a diverse array of K_V channels that include members of the K_V 1, 2, 3, 4, 7 and 11 families (Tykocki et al., 2017). These channels are activated by membrane depolarization and we showed that block of members of this large family of K^+ channels with 4-aminopyridine (4-AP) substantially reduces whole-cell outward K^+ currents at physiological membrane potentials (−90 to 0 mV) in cremaster arteriolar VSMCs and depolarizes isolated VSMCs (Jackson, Huebner, & Rusch, 1997) (Fig. 12A–D). In pressure myography experiments of hamster cremaster arterioles, 3 mM 4-AP constricts arterioles by about 20% (Jackson, 2018a) (Fig. 12E). Similar effects of 4-AP are seen in pressurized hamster cheek pouch arterioles (Fig. 12F). Thus, as was shown in cerebral resistance arteries (Knot & Nelson, 1995) and elsewhere (Tykocki et al., 2017), these data provide compelling evidence that in addition to BK_{Ca} channels, K_V channels also importantly contribute to the negative-feedback regulation of myogenic tone in peripheral arterioles. Although Ca^{2+} and PKC-dependent inhibition of K_V may dampen this negative-feedback, there appears to be sufficient K_V channel activity so that these channels still importantly contribute to the negative-feedback of VSMC membrane potential and myogenic tone.

Additional K^+ channels also contribute to the regulation of membrane potential of arteriolar VSMCs and hence myogenic tone. In mouse skeletal muscle resistance arteries and downstream arterioles, 100 μM Ba^{2+}, a selective blocker of inwardly-rectifying K^+ (K_{IR}) channels (Quayle, McCarron, Brayden, & Nelson, 1993), constricts these vessels by about 20% (Fig. 13A and B). These data suggest that K_{IR} channels are open under resting conditions and contribute to the regulation of resting membrane potential and myogenic tone. In first order hamster cremaster arterioles, 100 μM Ba^{2+} was reported to produce VSMC depolarization (4.4 ± 1.5 mV; $P=0.035$, $n=6$ by paired t-test) and a small ($10.4 \pm 6.5\%$; $P=0.17$, $n=6$ by paired t-test), but not statistically significant constriction (Burns et al., 2004). However, reexamination of these data indicated that one arteriole lost tone during the experiment. Elimination of this vessel from the analysis revealed that 100 μM Ba^{2+} constricted the remaining arterioles by $15.3 \pm 4.6\%$

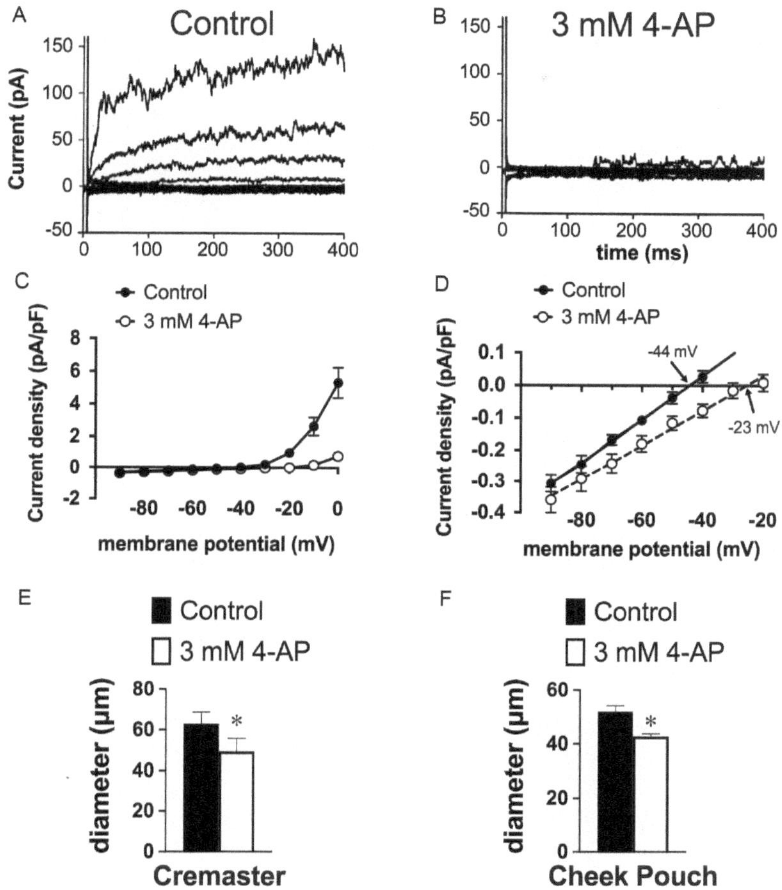

Fig. 12 K_V channels contribute to the negative-feedback regulation of arteriolar myogenic tone. Panel (A): Digitized whole-cell currents recorded from an enzymatically isolated rat cremaster VSMC using the perforated-patch technique (Jackson et al., 1997). Shown are outward currents elicited by 400 ms voltage pulses between −90 and 0 mV in 10 mV increments recorded under Control conditions. Panel (B): Whole-cell currents as in panel (A), but in the presence of 3 mM 4-AP to block K_V channels. Note that the majority of outward current was inhibited by 4-AP. Panel (C): Summary data for experiments as in panels (A and B). Data are mean current densities (pA/pF) ± SE (n = 8 cells) showing substantial inhibition of currents by 4-AP. Panel (D): Expanded scale of mean current densities for data shown in panel (C) showing that 4-AP shifted the whole-cell reversal potential (membrane potential where current density = 0), a measure of membrane potential, from −44 ± 2.7 to −23 ± 4.4 mV ($n = 8$, $P < 0.05$ by paired *t*-test) (Jackson et al., 1997). Panels (E) and (F): In addition to inhibition of K_V currents and membrane depolarization (panels A–E), 4-AP (3 mM) constricted hamster cremaster (panel E) and cheek pouch (panel F) arterioles studied by pressure myography as in Jackson and Boerman (2017). Data are mean diameters ± SE ($n = 8$ for cremaster and $n = 4$ for cheek pouch arterioles), * = $P < 0.05$ by paired *t*-tests. *Panels (A–D) are replotted from Jackson, W. F., Huebner, J. M., & Rusch, N. J. (1997). Enzymatic isolation and characterization of single vascular smooth muscle cells from cremasteric arterioles.* Microcirculation, *4(1), 35–50. doi:10.3109/10739689709148316 with permission.*

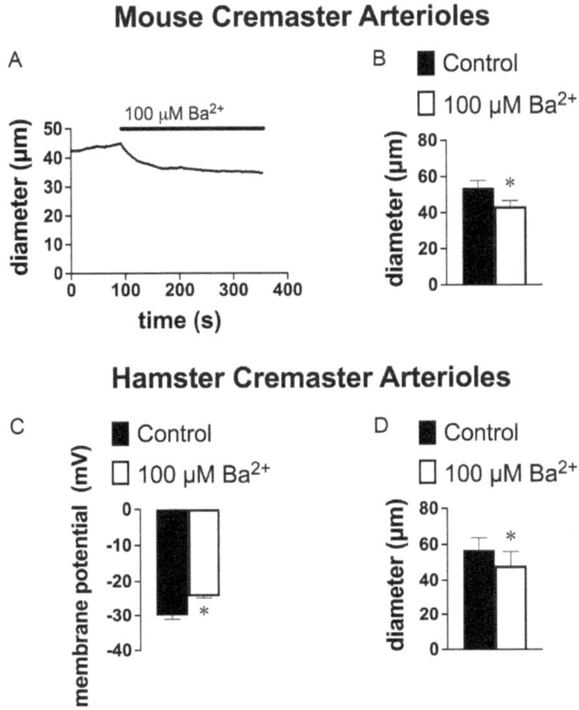

Fig. 13 K_{IR} channels are open under resting conditions and contribute to regulation of myogenic tone in cremaster arterioles. Panel (A): digitized diameter recording from a C57BL6 mouse cremaster arteriole studied by pressure myography as in Westcott et al. (2012). Addition of the K_{IR} blocker, Ba^{2+} (100 μM) constricted the arteriole about 20%. Similar results were obtained in mouse skeletal muscle feed arteries (Hayoz, Pettis, Bradley, Segal, & Jackson, 2017). Panel (B): Summary data for experiments as in panel (A). Data are mean diameters ± SE ($n=5$), $* = P < 0.05$ by paired *t*-test. Panels (C and D): Data are replotted from data in Burns et al. (2004) with one observation from a vessel that lost tone during the experiment removed as noted in the text from the data shown in Panels C and D. Data are mean membrane potentials or diameters ± SE (n = 5), $* = P < 0.05$ by paired *t*-test.

($n=5$, $P=0.03$ by paired *t*-test) (Fig. 13C and D), consistent with data from mouse cremaster arterioles (Fig. 13A and B). Barium also constricts murine skeletal muscle resistance arteries in pressure myography experiments by a similar amount (Hayoz et al., 2017). As with K_V channels, despite the potential for inhibition of the activity of K_{IR} channels by PKC and membrane depolarization, there are sufficient K_{IR} channels open at rest that they contribute to membrane potential and dampen pressure-induced membrane depolarization and myogenic tone.

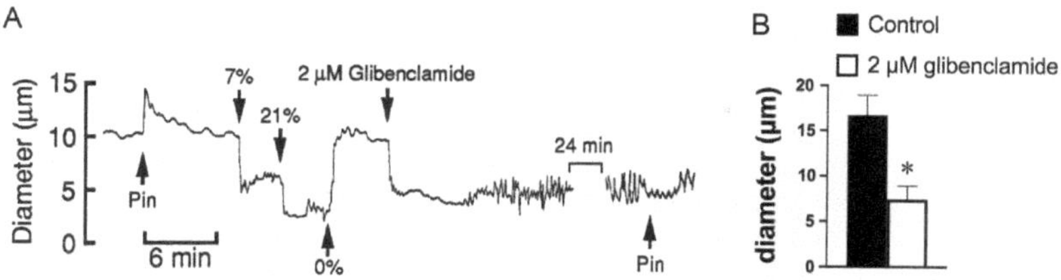

Fig. 14 Glibenclamide constricts hamster cremaster arterioles, in vivo. Panel (A): Digitized diameter record of a small third-order arteriole in a superfused hamster cremaster preparation. Starting at the top left of the diameter recording, topical application of 2.3 nmol of the K_{ATP} agonist, pinacidil, dilated the arteriole indicating that there are closed, recruitable K_{ATP} channels in this preparation. Elevation of the oxygen in the superfusate to 7% (PO_2 = 49 mmHg) and 21% (PO_2 = 150 mmHg) produced arteriolar constriction to demonstrate the reactivity of this arteriole. After return to the control condition (0% O_2, PO_2 = 15 mmHg) and diameter recovery, 2 μM glibenclamide was added to the superfusate, as indicated. This produced substantial constriction of the arteriole that was maintained as long as glibenclamide was present. The arteriole was then challenged again with pinacidil and no dilation was observed demonstrating the efficacy of glibenclamide for block of K_{ATP} channels. Panel (B): Summary data from five similar preparations from Jackson (1993a) showing that glibenclamide significantly constricts hamster cremaster arterioles, in vivo. The data are means ± SE. * = $P < 0.05$ by paired *t*-test. *Data shown are from Jackson, W. F. (1998). Potassium channels and regulation of the microcirculation.* Microcirculation, *5(2–3), 85–90. doi:10.1111/j.1549-8719.1998.tb00057.x with permission.*

In vivo studies in hamster cheek pouch and cremaster muscle suggest that ATP-sensitive K^+ (K_{ATP}) channels also are active at resting myogenic tone and blunt the pressure-induced VSMC activity (Jackson, 1993a): the K_{ATP} antagonist, glibenclamide, significantly constricts arterioles in these preparations supporting the hypothesis that these channels can be open at rest and contribute to the determination of membrane potential and myogenic tone (Jackson, 1993a) (Fig. 14). In addition, patch clamp studies of enzymatically isolated hamster cremaster arteriolar VSMCs suggest that K_{ATP} channels contribute to resting membrane potential within the physiological range (Jackson et al., 1997) (Fig. 15). However, ex vivo, in pressure myography experiments, glibenclamide did not produce significant membrane depolarization or constriction of hamster cremaster arterioles (Burns et al., 2004). The mechanism by which K_{ATP} channels are open in vivo and when cells are isolated from these arterioles but are silent when the arterioles are studied by pressure myography, remains to be established. Nonetheless, these channels may contribute to the regulation of resting membrane potential, particularly in vivo, and also dampen pressure-induced membrane depolarization and myogenic tone.

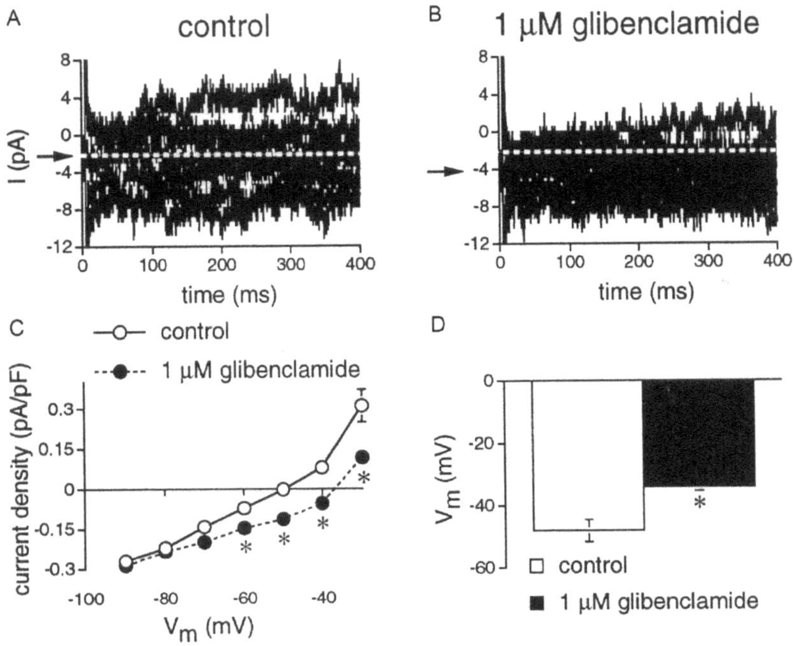

Fig. 15 K_{ATP} channels are open at rest in isolated cremaster VSMCs. Panel (A) shows a family of whole-cell currents in a cremaster VSMC studied by the perforated-patch technique as described (Jackson et al., 1997) in response to 400ms voltage pulses from −90 to −30mV in 10mV increments from a holding potential of −60mV. The arrow next to the current axis and the white dotted line indicate the holding current at −60mV. Panel (B): As in panel (A) in the presence of the K_{ATP} channel antagonist, glibenclamide (1 μM). The arrow next to current axis is the holding current at −60mV; the dotted white line is the Control holding current. Glibenclamide made the holding current more negative and inhibited currents between −60 and −30mV. Panel (C): Summary data showing that glibenclamide significantly inhibited currents between −60 and −30mV and reduced the slope of the I–V relationship in this range, resulting in a shift of the whole-cell reversal potential from about −50mV to about −35mV, consistent with membrane depolarization. Data are mean current densities ± SE ($n=6$ cells), $* = P < 0.05$ by Tukey's test after two-way ANOVA. Panel (D): Summary data as mean membrane potentials ± SE (n = 6 cells) recorded in current clamp as described (Jackson et al., 1997); $* = P < 0.05$ by paired *t*-test. Consistent with the changes in the I–V relationship shown in panel (C), glibenclamide significantly depolarized cremaster VSMCs. These data support the hypothesis that there are open K_{ATP} channels at rest in these VSMCs that contribute to resting membrane potential and the regulation of myogenic tone. *Data are replotted from Jackson, W. F. (1998). Potassium channels and regulation of the microcirculation.* Microcirculation, *5(2–3), 85–90. doi:10.1111/j.1549-8719.1998.tb00057.x; Jackson, W. F., Huebner, J. M., & Rusch, N. J. (1997). Enzymatic isolation and characterization of single vascular smooth muscle cells from cremasteric arterioles.* Microcirculation, *4(1), 35–50. doi:10.3109/10739689709148316 with permission.*

5. Summary and perspectives

Ion channels contribute substantially to myogenic tone of peripheral arterioles as outlined in this chapter. They provide the major source of activator Ca^{2+} and also play a dominant role in setting and modulating membrane potential that, in turn, controls the open-state probability of VGCCs, BK_{Ca} channels, K_V channels and K_{IR} channels. While much has been learned about ion channel expression and function in VSMCs (Tykocki et al., 2017), few of these studies have extended to VSMCs from arterioles in the microcirculation. The established heterogeneity in the mechanisms regulating myogenic tone between upstream arteries and downstream arterioles (Brayden et al., 2013; Cipolla et al., 2014; Westcott et al., 2012; Westcott & Jackson, 2011) and between arterioles in different regions of the body (Jackson & Boerman, 2017; Kur, Bankhead, Scholfield, Curtis, & McGeown, 2013) imply that extrapolations cannot be made. Direct measurement of ion channel expression and function in arteriolar VSMCs from the specific organ/tissue of interest must be made. With the explosion in methods for single cell Omics (Chen et al., 2019), this should be possible, but will require modification of existing methods or development of new approaches to isolate specific segments of arterioles in the microvascular bed to be studied.

Three areas warrant further research. First, why are CaV1.2 relatively silent at rest in arterioles, in vivo? Second, what is the genesis of the regional differences in ion channel expression and mechanisms of myogenic tone and its regulation? Third, what is the physiological significance of the redundancy in function of ion channels in VSMCs? All of these questions will require a better understanding of the pattern of expression of all ion channels, their binding partners and the signaling pathways that control these channels along with details of their physiological functions in the regulation of membrane potential and Ca^{2+} signaling in the context of the cells and vessels in which they are expressed. This will require application of isolated VSMC approaches (single cell Omics; patch clamp electrophysiology; Ca^{2+} imaging, etc.), the use of isolated arterioles studied by pressure myography (arteriolar tone, VSMC membrane potential, local and global Ca^{2+} imaging, etc.), and also intravital imaging (arteriolar tone, membrane potential and Ca^{2+} imaging) of the same arterioles as studied ex vivo and from which the VSMCs were isolated to close the loop. Knock-out models while useful, may not be helpful here except perhaps as controls for single cell electrophysiology and expression studies, because deletion of a channel, for example, may disrupt signaling complexes associated with those channels and have significant off-target effects; knock-in experiments using ion channels with non-functional pores might be a better approach to exclude off-target effects of disrupting signaling complexes and other ion channel functions. Selective pharmacological blockade of specific ion channels is also a viable approach because of the paired experimental design that can be applied provided that such agents exist or can be developed through high-throughput screening approaches that are now available.

Acknowledgment

This research was supported by National Heart, Lung and Blood Institute grants HL-32469 and PO1-HL-070687.

Disclaimers

The content is solely the responsibility of the author and does not necessarily represent the official views of the National Institutes of Health.

Disclosures

No conflicts of interest, financial or otherwise, are declared by the author.

References

Abd El-Rahman, R. R., Harraz, O. F., Brett, S. E., Anfinogenova, Y., Mufti, R. E., Goldman, D., et al. (2013). Identification of L- and T-type Ca2+ channels in rat cerebral arteries: Role in myogenic tone development. *American Journal of Physiology. Heart and Circulatory Physiology*, *304*(1), H58–H71. https://doi.org/10.1152/ajpheart.00476.2012.

Aiello, E. A., Clement-Chomienne, O., Sontag, D. P., Walsh, M. P., & Cole, W. C. (1996). Protein kinase C inhibits delayed rectifier K+ current in rabbit vascular smooth muscle cells. *The American Journal of Physiology*, *271*(1 Pt. 2), H109–H119. https://doi.org/10.1152/ajpheart.1996.271.1.H109.

Amberg, G. C., Navedo, M. F., Nieves-Cintron, M., Molkentin, J. D., & Santana, L. F. (2007). Calcium sparklets regulate local and global calcium in murine arterial smooth muscle. *The Journal of Physiology*, *579*(Pt. 1), 187–201. https://doi.org/10.1113/jphysiol.2006.124420.

Anfinogenova, Y., Brett, S. E., Walsh, M. P., Harraz, O. F., & Welsh, D. G. (2011). Do TRPC-like currents and G protein-coupled receptors interact to facilitate myogenic tone development? *American Journal of Physiology. Heart and Circulatory Physiology*, *301*(4), H1378–H1388. https://doi.org/10.1152/ajpheart.00460.2011.

Bannister, J. P., Bulley, S., Narayanan, D., Thomas-Gatewood, C., Luzny, P., Pachuau, J., et al. (2012). Transcriptional upregulation of α2δ-1 elevates arterial smooth muscle cell voltage-dependent Ca2+ channel surface expression and cerebrovascular constriction in genetic hypertension. *Hypertension*, *60*(4), 1006–1015. https://doi.org/10.1161/HYPERTENSIONAHA.112.199661.

Brayden, J. E., Li, Y., & Tavares, M. J. (2013). Purinergic receptors regulate myogenic tone in cerebral parenchymal arterioles. *Journal of Cerebral Blood Flow and Metabolism, 33*(2), 293–299. https://doi.org/10.1038/jcbfm.2012.169.

Brekke, J. F., Jackson, W. F., & Segal, S. S. (2006). Arteriolar smooth muscle Ca^{2+} dynamics during blood flow control in hamster cheek pouch. *Journal of Applied Physiology (Bethesda, MD: 1985), 101*(1), 307–315. https://doi.org/10.1152/japplphysiol.01634.2005.

Bulley, S., Neeb, Z. P., Burris, S. K., Bannister, J. P., Thomas-Gatewood, C. M., Jangsangthong, W., et al. (2012). TMEM16A/ANO1 channels contribute to the myogenic response in cerebral arteries. *Circulation Research, 111*(8), 1027–1036. https://doi.org/10.1161/CIRCRESAHA.112.277145.

Burns, W. R., Cohen, K. D., & Jackson, W. F. (2004). K^+-induced dilation of hamster cremasteric arterioles involves both the Na^+/K^+-ATPase and inward-rectifier K^+ channels. *Microcirculation, 11*(3), 279–293. https://doi.org/10.1080/10739680490425985.

Chen, W., Li, S., Kulkarni, A. S., Huang, L., Cao, J., Qian, K., et al. (2019). Single cell omics: From assay design to biomedical application. *Biotechnology Journal, 15*, e1900262. https://doi.org/10.1002/biot.201900262.

Chennupati, R., Wirth, A., Favre, J., Li, R., Bonnavion, R., Jin, Y. J., et al. (2019). Myogenic vasoconstriction requires G12/G13 and LARG to maintain local and systemic vascular resistance. *eLife, 8*, e49374. https://doi.org/10.7554/eLife.49374.

Cipolla, M. J., Sweet, J., Chan, S. L., Tavares, M. J., Gokina, N., & Brayden, J. E. (2014). Increased pressure-induced tone in rat parenchymal arterioles vs. middle cerebral arteries: Role of ion channels and calcium sensitivity. *Journal of Applied Physiology (Bethesda, MD: 1985), 117*(1), 53–59. https://doi.org/10.1152/japplphysiol.00253.2014.

Clement-Chomienne, O., Walsh, M. P., & Cole, W. C. (1996). Angiotensin II activation of protein kinase C decreases delayed rectifier K+ current in rabbit vascular myocytes. *The Journal of Physiology, 495*(Pt. 3), 689–700. https://doi.org/10.1113/jphysiol.1996.sp021626.

Coats, P., Johnston, F., MacDonald, J., McMurray, J. J., & Hillier, C. (2001). Signalling mechanisms underlying the myogenic response in human subcutaneous resistance arteries. *Cardiovascular Research, 49*(4), 828–837. https://doi.org/10.1016/s0008-6363(00)00314-x.

Cohen, K. D., & Jackson, W. F. (2003). Hypoxia inhibits contraction but not calcium channel currents or changes in intracellular calcium in arteriolar muscle cells. *Microcirculation, 10*(2), 133–141. https://doi.org/10.1038/sj/mn.7800178.

Cole, W. C., & Welsh, D. G. (2011). Role of myosin light chain kinase and myosin light chain phosphatase in the resistance arterial myogenic response to intravascular pressure. *Archives of Biochemistry and Biophysics, 510*(2), 160–173. https://doi.org/10.1016/j.abb.2011.02.024.

Cox, R. H., & Petrou, S. (1999). Ca(2+) influx inhibits voltage-dependent and augments Ca(2+)-dependent K(+) currents in arterial myocytes. *The American Journal of Physiology, 277*(1), C51–C63. https://doi.org/10.1152/ajpcell.1999.277.1.C51.

Dabertrand, F., Nelson, M. T., & Brayden, J. E. (2012). Acidosis dilates brain parenchymal arterioles by conversion of calcium waves to sparks to activate BK channels. *Circulation Research, 110*(2), 285–294. https://doi.org/10.1161/CIRCRESAHA.111.258145.

Dam, V. S., Boedtkjer, D. M., Aalkjaer, C., & Matchkov, V. (2014). The bestrophin- and TMEM16A-associated Ca(2+)-activated Cl(−) channels in vascular smooth muscles. *Channels (Austin, Texas), 8*(4), 361–369. https://doi.org/10.4161/chan.29531.

Dam, V. S., Boedtkjer, D. M., Nyvad, J., Aalkjaer, C., & Matchkov, V. (2014). TMEM16A knockdown abrogates two different Ca(2+)-activated Cl (−) currents and contractility of smooth muscle in rat mesenteric small arteries. *Pflügers Archiv, 466*(7), 1391–1409. https://doi.org/10.1007/s00424-013-1382-1.

Davis, M. J., & Hill, M. A. (1999). Signaling mechanisms underlying the vascular myogenic response. *Physiological Reviews, 79*(2), 387–423. https://doi.org/10.1152/physrev.1999.79.2.387.

Davis, M. J., Hill, M. A., & Kuo, L. (2011). Local regulation of microvascular perfusion. In R. L. Terjung (Ed.), *Comprehensive physiology*, (pp. 161–284). John Wiley & Sons, Inc. https://doi.org/10.1002/cphy.cp020406

Davis, M. J., Wu, X., Nurkiewicz, T. R., Kawasaki, J., Davis, G. E., Hill, M. A., et al. (2001). Integrins and mechanotransduction of the vascular myogenic response. *American Journal of Physiology. Heart and Circulatory Physiology*, *280*(4), H1427–H1433. https://doi.org/10.1152/ajpheart.2001.280.4.H1427.

Earley, S., Straub, S. V., & Brayden, J. E. (2007). Protein kinase C regulates vascular myogenic tone through activation of TRPM4. *American Journal of Physiology. Heart and Circulatory Physiology*, *292*(6), H2613–H2622. https://doi.org/10.1152/ajpheart.01286.2006.

Evanson, K., Bannister, J. P., Leo, M. D., & Jaggar, J. H. (2014). LRRC26 is a functional BK channel auxiliary gamma subunit in arterial smooth muscle cells. *Circulation Research*, *115*, 423–431. https://doi.org/10.1161/CIRCRESAHA.115.303407.

Gelband, C. H., Ishikawa, T., Post, J. M., Keef, K. D., & Hume, J. R. (1993). Intracellular divalent cations block smooth muscle K^+ channels. *Circulation Research*, *73*(1), 24–34. https://doi.org/10.1161/01.res.73.1.24.

Ghisdal, P., Vandenberg, G., & Morel, N. (2003). Rho-dependent kinase is involved in agonist-activated calcium entry in rat arteries. *The Journal of Physiology*, *551*(Pt. 3), 855–867. https://doi.org/10.1113/jphysiol.2003.047050.

Gonzales, A. L., Amberg, G. C., & Earley, S. (2010). Ca2+ release from the sarcoplasmic reticulum is required for sustained TRPM4 activity in cerebral artery smooth muscle cells. *American Journal of Physiology. Cell Physiology*, *299*(2), C279–C288. https://doi.org/10.1152/ajpcell.00550.2009.

Gonzales, A. L., Garcia, Z. I., Amberg, G. C., & Earley, S. (2010). Pharmacological inhibition of TRPM4 hyperpolarizes vascular smooth muscle. *American Journal of Physiology. Cell Physiology*, *299*(5), C1195–C1202. https://doi.org/10.1152/ajpcell.00269.2010.

Gonzales, A. L., Yang, Y., Sullivan, M. N., Sanders, L., Dabertrand, F., Hill-Eubanks, D. C., et al. (2014). A PLCgamma1-dependent, force-sensitive signaling network in the myogenic constriction of cerebral arteries. *Science Signaling*, 7(327), ra49. https://doi.org/10.1126/scisignal.2004732.

Guan, Z., Baty, J. J., Zhang, S., Remedies, C. E., & Inscho, E. W. (2019). Rho kinase inhibitors reduce voltage-dependent Ca(2+) channel signaling in aortic and renal microvascular smooth muscle cells. *American Journal of Physiology. Renal Physiology*, *317*(5), F1132–F1141. https://doi.org/10.1152/ajprenal.00212.2018.

Guia, A., Wan, X., Courtemanche, M., & Leblanc, N. (1999). Local Ca2+ entry through L-type Ca2+ channels activates Ca2+-dependent K+ channels in rabbit coronary myocytes. *Circulation Research*, *84*(9), 1032–1042. https://doi.org/10.1161/01.res.84.9.1032.

Gustafsson, F., Andreasen, D., Salomonsson, M., Jensen, B. L., & Holstein-Rathlou, N. (2001). Conducted vasoconstriction in rat mesenteric arterioles: Role for dihydropyridine-insensitive Ca(2+) channels. *American Journal of Physiology. Heart and Circulatory Physiology*, *280*(2), H582–H590. https://doi.org/10.1152/ajpheart.2001.280.2.H582.

Hansen, P. B., Jensen, B. L., Andreasen, D., & Skott, O. (2001). Differential expression of T- and L-type voltage-dependent calcium channels in renal resistance vessels. *Circulation Research*, *89*(7), 630–638. https://doi.org/10.1161/hh1901.097126.

Harraz, O. F., Abd El-Rahman, R. R., Bigdely-Shamloo, K., Wilson, S. M., Brett, S. E., Romero, M., et al. (2014). Ca(V)3.2 channels and the induction of negative feedback in cerebral arteries. *Circulation Research*, *115*(7), 650–661. https://doi.org/10.1161/CIRCRESAHA.114.304056.

Harraz, O. F., Brett, S. E., Zechariah, A., Romero, M., Puglisi, J. L., Wilson, S. M., et al. (2015). Genetic ablation of CaV3.2 channels enhances the arterial myogenic response by modulating the RyR-BKCa axis. *Arteriosclerosis, Thrombosis, and Vascular Biology, 35*(8), 1843–1851. https://doi.org/10.1161/ATVBAHA.115.305736.

Hayabuchi, Y., Davies, N. W., & Standen, N. B. (2001). Angiotensin II inhibits rat arterial KATP channels by inhibiting steady-state protein kinase A activity and activating protein kinase Ce. *The Journal of Physiology, 530*(Pt. 2), 193–205. https://doi.org/10.1111/j.1469-7793.2001.0193l.x.

Hayabuchi, Y., Standen, N. B., & Davies, N. W. (2001). Angiotensin II inhibits and alters kinetics of voltage-gated K(+) channels of rat arterial smooth muscle. *American Journal of Physiology. Heart and Circulatory Physiology, 281*(6), H2480–H2489. https://doi.org/10.1152/ajpheart.2001.281.6.H2480.

Hayoz, S., Pettis, J., Bradley, V., Segal, S. S., & Jackson, W. F. (2017). Increased amplitude of inward rectifier K+ currents with advanced age in smooth muscle cells of murine superior epigastric arteries. *American Journal of Physiology. Heart and Circulatory Physiology, 312*(6), H1203–H1214. https://doi.org/10.1152/ajpheart.00679.2016.

Heinze, C., Seniuk, A., Sokolov, M. V., Huebner, A. K., Klementowicz, A. E., Szijarto, I. A., et al. (2014). Disruption of vascular Ca2+-activated chloride currents lowers blood pressure. *The Journal of Clinical Investigation, 124*(2), 675–686. https://doi.org/10.1172/JCI70025.

Hill, M. A., Falcone, J. C., & Meininger, G. A. (1990). Evidence for protein kinase C involvement in arteriolar myogenic reactivity. *The American Journal of Physiology, 259*(5 Pt. 2), H1586–H1594. https://doi.org/10.1152/ajpheart.1990.259.5.H1586.

Hill, M. A., & Meininger, G. A. (1994). Calcium entry and myogenic phenomena in skeletal muscle arterioles. *The American Journal of Physiology, 267*(3 Pt. 2), H1085–H1092. https://doi.org/10.1152/ajpheart.1994.267.3.H1085.

Hill, M. A., Zou, H., Potocnik, S. J., Meininger, G. A., & Davis, M. J. (2001). Invited review: Arteriolar smooth muscle mechanotransduction: Ca(2+) signaling pathways underlying myogenic reactivity. *Journal of Applied Physiology (Bethesda, MD: 1985), 91*(2), 973–983. https://doi.org/10.1152/jappl.2001.91.2.973.

Hong, K., Zhao, G., Hong, Z., Sun, Z., Yang, Y., Clifford, P. S., et al. (2016). Mechanical activation of angiotensin II type 1 receptors causes actin remodelling and myogenic responsiveness in skeletal muscle arterioles. *The Journal of Physiology, 594*(23), 7027–7047. https://doi.org/10.1113/JP272834.

Hu, K., Huang, C. S., Jan, Y. N., & Jan, L. Y. (2003). ATP-sensitive potassium channel traffic regulation by adenosine and protein kinase C. *Neuron, 38*(3), 417–432. https://doi.org/10.1016/s0896-6273(03)00256-3.

Hubner, C. A., Schroeder, B. C., & Ehmke, H. (2015). Regulation of vascular tone and arterial blood pressure: Role of chloride transport in vascular smooth muscle. *Pflügers Archiv, 467*(3), 605–614. https://doi.org/10.1007/s00424-014-1684-y.

Inoue, R., Jensen, L. J., Jian, Z., Shi, J., Hai, L., Lurie, A. I., et al. (2009). Synergistic activation of vascular TRPC6 channel by receptor and mechanical stimulation via phospholipase C/diacylglycerol and phospholipase A2/omega-hydroxylase/20-HETE pathways. *Circulation Research, 104*(12), 1399–1409. https://doi.org/10.1161/CIRCRESAHA.108.193227.

Ishikawa, T., Hume, J. R., & Keef, K. D. (1993). Modulation of K^+ and Ca^{2+} channels by histamine H_1-receptor stimulation in rabbit coronary artery cells. *The Journal of Physiology, 468*, 379–400. https://doi.org/10.1113/jphysiol.1993.sp019777.

Ito, I., Jarajapu, Y. P., Grant, M. B., & Knot, H. J. (2007). Characteristics of myogenic tone in the rat ophthalmic artery. *American Journal of Physiology. Heart and Circulatory Physiology, 292*(1), H360–H368. https://doi.org/10.1152/ajpheart.00630.2006.

Jackson, W. F. (1986). Prostaglandins do not mediate arteriolar oxygen reactivity. *The American Journal of Physiology*, *250*, H1102–H1108.

Jackson, W. F. (1993a). Arteriolar tone is determined by activity of ATP-sensitive potassium channels. *American Journal of Physiology*, *265*(5), H1797–H1803.

Jackson, W. F. (1993b). Regional differences in mechanism of action of oxygen on hamster arterioles. *The American Journal of Physiology*, *265*(2 Pt. 2), H599–H603. https://doi.org/10.1152/ajpheart.1993.265.2.H599.

Jackson, W. F. (1998). Potassium channels and regulation of the microcirculation. *Microcirculation*, *5*(2–3), 85–90. https://doi.org/10.1111/j.1549-8719.1998.tb00057.x.

Jackson, W. F. (2000a). Hypoxia does not activate ATP-sensitive K+ channels in arteriolar muscle cells. *Microcirculation*, *7*(2), 137–145.

Jackson, W. F. (2000b). Ion channels and vascular tone. *Hypertension*, *35*(1 Pt. 2), 173–178. https://doi.org/10.1161/01.hyp.35.1.173.

Jackson, W. F. (2005). Potassium channels in the peripheral microcirculation. *Microcirculation*, *12*(1), 113–127. https://doi.org/10.1080/10739680590896072.

Jackson, W. F. (2012). Microcirculation. In J. A. Olson & E. N. Hill (Eds.), *Muscle* (pp. 1197–1206). Boston/Waltham: Academic Press.

Jackson, W. F. (2016a). Arteriolar oxygen reactivity: Where is the sensor and what is the mechanism of action? *The Journal of Physiology*, *594*(18), 5055–5077. https://doi.org/10.1113/JP270192.

Jackson, W. F. (2016b). Endothelial cell ion channel expression and function in arterioles and resistance arteries. In I. Levitan & A. M. Dopico (Eds.), *Vascular ion channels in physiology and disease* (pp. 3–36). Switzerland: Springer International Publishing.

Jackson, W. F. (2017). Potassium channels in regulation of vascular smooth muscle contraction and growth. *Advances in Pharmacology*, *78*, 89–144. https://doi.org/10.1016/bs.apha.2016.07.001.

Jackson, W. F. (2018a). KV channels and the regulation of vascular smooth muscle tone. *Microcirculation*, *25*(1). https://doi.org/10.1111/micc.12421.

Jackson, W. F. (2018b). Methods for the isolation of and study of Ca^{2+} signaling in arteriolar smooth muscle cells. In M. Trebak & S. Earley (Eds.), *Signal transduction and smooth muscle* (1st ed., pp. 49–64). Boca Raton: CRC Press.

Jackson, W. F., & Blair, K. L. (1998). Characterization and function of Ca(2+)-activated K+ channels in arteriolar muscle cells. *The American Journal of Physiology*, *274*(1), H27–H34. https://doi.org/10.1152/ajpheart.1998.274.1.H27.

Jackson, W. F., & Boerman, E. M. (2017). Regional heterogeneity in the mechanisms of myogenic tone in hamster arterioles. *American Journal of Physiology. Heart and Circulatory Physiology*, *313*(3), H667–H675. https://doi.org/10.1152/ajpheart.00183.;2017.

Jackson, W. F., & Boerman, E. M. (2018). Voltage-gated Ca(2+) channel activity modulates smooth muscle cell calcium waves in hamster cremaster arterioles. *American Journal of Physiology. Heart and Circulatory Physiology*, *315*(4), H871–H878. https://doi.org/10.1152/ajpheart.00292.2018.

Jackson, W. F., Boerman, E. M., Lange, E. J., Lundback, S. S., & Cohen, K. D. (2008). Smooth muscle alpha1D-adrenoceptors mediate phenylephrine-induced vasoconstriction and increases in endothelial cell Ca^{2+} in hamster cremaster arterioles. *British Journal of Pharmacology*, *155*(4), 514–524. https://doi.org/10.1038/bjp.2008.276.

Jackson, W. F., Huebner, J. M., & Rusch, N. J. (1997). Enzymatic isolation and characterization of single vascular smooth muscle cells from cremasteric arterioles. *Microcirculation*, *4*(1), 35–50. https://doi.org/10.3109/10739689709148316.

Jiao, J., Garg, V., Yang, B., Elton, T. S., & Hu, K. (2008). Protein kinase C-epsilon induces caveolin-dependent internalization of vascular adenosine 5′-triphosphate-sensitive K+ channels. *Hypertension*, *52*(3), 499–506. https://doi.org/10.1161/HYPERTENSION AHA.108.110817.

Johnson, P. C. (1980). The myogenic response. In D. F. Bohr, A. P. Somlyo, & H. V. Sparks (Eds.), *Handbook of physiology: Section 2, the cardiovascular system, volume II, vascular smooth muscle* (pp. 409–442). Bethesda, MD: American Physiological Society.

Kidd, M. W., Bulley, S., & Jaggar, J. H. (2017). Angiotensin II reduces the surface abundance of KV 1.5 channels in arterial myocytes to stimulate vasoconstriction. *The Journal of Physiology*, *595*(5), 1607–1618. https://doi.org/10.1113/JP272893.

Knot, H. J., & Nelson, M. T. (1995). Regulation of membrane potential and diameter by voltage-dependent K+ channels in rabbit myogenic cerebral arteries. *The American Journal of Physiology*, *269*(1 Pt. 2), H348–H355. https://doi.org/10.1152/ajpheart.1995.269.1.H348.

Ko, E. A., Park, W. S., Firth, A. L., Hong, D. H., Choi, S. W., Heo, H. J., et al. (2010). Increased sensitivity of serotonin on the voltage-dependent K+ channels in mesenteric arterial smooth muscle cells of OLETF rats. *Progress in Biophysics and Molecular Biology*, *103*(1), 88–94. https://doi.org/10.1016/j.pbiomolbio.2010.02.003.

Kur, J., Bankhead, P., Scholfield, C. N., Curtis, T. M., & McGeown, J. G. (2013). Ca(2+) sparks promote myogenic tone in retinal arterioles. *British Journal of Pharmacology*, *168*(7), 1675–1686. https://doi.org/10.1111/bph.12044.

Laher, I., & Bevan, J. A. (1989). Staurosporine, a protein kinase C inhibitor, attenuates Ca2+-dependent stretch-induced vascular tone. *Biochemical and Biophysical Research Communications*, *158*(1), 58–62. https://doi.org/10.1016/s0006-291x(89)80176-7.

Leblanc, N., Forrest, A. S., Ayon, R. J., Wiwchar, M., Angermann, J. E., Pritchard, H. A., et al. (2015). Molecular and functional significance of Ca(2+)-activated Cl(−) channels in pulmonary arterial smooth muscle. *Pulmonary Circulation*, *5*(2), 244–268. https://doi.org/10.1086/680189.

Leiphart, R. J., Chen, D., Peredo, A. P., Loneker, A. E., & Janmey, P. A. (2019). Mechanosensing at cellular interfaces. *Langmuir*, *35*(23), 7509–7519. https://doi.org/10.1021/acs.langmuir.8b02841.

Li, Y., & Brayden, J. E. (2017). Rho kinase activity governs arteriolar myogenic depolarization. *Journal of Cerebral Blood Flow and Metabolism*, *37*(1), 140–152. https://doi.org/10.1177/0271678X15621069.

Loirand, G., Guerin, P., & Pacaud, P. (2006). Rho kinases in cardiovascular physiology and pathophysiology. *Circulation Research*, *98*(3), 322–334. https://doi.org/10.1161/01.RES.0000201960.04223.3c.

Luykenaar, K. D., El-Rahman, R. A., Walsh, M. P., & Welsh, D. G. (2009). Rho-kinase-mediated suppression of KDR current in cerebral arteries requires an intact actin cytoskeleton. *American Journal of Physiology. Heart and Circulatory Physiology*, *296*(4), H917–H926. https://doi.org/10.1152/ajpheart.01206.2008.

Martinez-Lemus, L. A., Crow, T., Davis, M. J., & Meininger, G. A. (2005). Alphavbeta3- and alpha5beta1-integrin blockade inhibits myogenic constriction of skeletal muscle resistance arterioles. *American Journal of Physiology. Heart and Circulatory Physiology*, *289*(1), H322–H329. https://doi.org/10.1152/ajpheart.00923.2003.

Massett, M. P., Ungvari, Z., Csiszar, A., Kaley, G., & Koller, A. (2002). Different roles of PKC and MAP kinases in arteriolar constrictions to pressure and agonists. *American Journal of Physiology. Heart and Circulatory Physiology*, *283*(6), H2282–H2287. https://doi.org/10.1152/ajpheart.00544.2002.

Matchkov, V. V., Boedtkjer, D. M., & Aalkjaer, C. (2015). The role of Ca(2+) activated Cl(−) channels in blood pressure control. *Current Opinion in Pharmacology*, *21*, 127–137. https://doi.org/10.1016/j.coph.2015.02.003.

Mauban, J. R., Lamont, C., Balke, C. W., & Wier, W. G. (2001). Adrenergic stimulation of rat resistance arteries affects Ca(2+) sparks, Ca(2+) waves, and Ca(2+) oscillations. *American Journal of Physiology. Heart and Circulatory Physiology*, *280*(5), H2399–H2405.

Mauban, J. R., Zacharia, J., Fairfax, S., & Wier, W. G. (2015). PC-PLC/sphingomyelin synthase activity plays a central role in the development of myogenic tone in murine resistance arteries. *American Journal of Physiology. Heart and Circulatory Physiology*, *308*(12), H1517–H1524. https://doi.org/10.1152/ajpheart.00594.2014.

Mederos, Y. S. M., Storch, U., & Gudermann, T. (2016). Mechanosensitive Gq/11 protein-coupled receptors mediate myogenic vasoconstriction. *Microcirculation*, *23*(8), 621–625. https://doi.org/10.1111/micc.12293.

Mederos y Schnitzler, M., Storch, U., Meibers, S., Nurwakagari, P., Breit, A., Essin, K., et al. (2008). Gq-coupled receptors as mechanosensors mediating myogenic vasoconstriction. *The EMBO Journal*, *27*(23), 3092–3103. https://doi.org/10.1038/emboj.2008.233.

Miller, F. J., Jr., Dellsperger, K. C., & Gutterman, D. D. (1997). Myogenic constriction of human coronary arterioles. *The American Journal of Physiology*, *273*(1 Pt. 2), H257–H264. https://doi.org/10.1152/ajpheart.1997.273.1.H257.

Morita, H., Cousins, H., Onoue, H., Ito, Y., & Inoue, R. (1999). Predominant distribution of nifedipine-insensitive, high voltage-activated Ca2+ channels in the terminal mesenteric artery of guinea pig. *Circulation Research*, *85*(7), 596–605. https://doi.org/10.1161/01.res.85.7.596.

Mullan, B., Pettis, J., & Jackson, W. F. (2017). T-type voltage-gated Ca(2+) channels do not contribute to the negative feedback regulation of myogenic tone in murine superior epigastric arteries. *Pharmacology Research & Perspectives*, *5*(3), e00320. https://doi.org/10.1002/prp2.320.

Navedo, M. F., Amberg, G. C., Votaw, V. S., & Santana, L. F. (2005). Constitutively active L-type Ca2+ channels. *Proceedings of the National Academy of Sciences of the United States of America*, *102*(31), 11112–11117. https://doi.org/10.1073/pnas.0500360102.

Nelson, M. T., Cheng, H., Rubart, M., Santana, L. F., Bonev, A. D., Knot, H. J., et al. (1995). Relaxation of arterial smooth muscle by calcium sparks. *Science*, *270*(5236), 633–637. https://doi.org/10.1126/science.270.5236.633.

Nourian, Z., Li, M., Leo, M. D., Jaggar, J. H., Braun, A. P., & Hill, M. A. (2014). Large conductance Ca2+-activated K+ channel (BKCa) alpha-subunit splice variants in resistance arteries from rat cerebral and skeletal muscle vasculature. *PLoS One*, *9*(6), e98863. https://doi.org/10.1371/journal.pone.0098863.

Osol, G., Laher, I., & Cipolla, M. (1991). Protein kinase C modulates basal myogenic tone in resistance arteries from the cerebral circulation. *Circulation Research*, *68*(2), 359–367. https://doi.org/10.1161/01.res.68.2.359.

Pawlowski, J., & Morgan, K. G. (1992). Mechanisms of intrinsic tone in ferret vascular smooth muscle. *The Journal of Physiology*, *448*, 121–132. https://doi.org/10.1113/jphysiol.1992.sp019032.

Pires, P. W., Ko, E. A., Pritchard, H. A. T., Rudokas, M., Yamasaki, E., & Earley, S. (2017). The angiotensin II receptor type 1b is the primary sensor of intraluminal pressure in cerebral artery smooth muscle cells. *The Journal of Physiology*, *595*(14), 4735–4753. https://doi.org/10.1113/JP274310.

Pries, A. R., & Secomb, T. W. (2011). Blood flow in microvascular networks. In R. L. Terjung (Ed.), *Comprehensive physiology*, (pp. 3–36). John Wiley & Sons, Inc. https://doi.org/10.1002/cphy.cp020401

Quayle, J. M., McCarron, J. G., Brayden, J. E., & Nelson, M. T. (1993). Inward rectifier K^+ currents in smooth muscle cells from rat resistance-sized cerebral arteries. *The American Journal of Physiology*, *265*(5 Pt. 1), C1363–C1370. https://doi.org/10.1152/ajpcell.1993.265.5.C1363.

Reading, S. A., Earley, S., Waldron, B. J., Welsh, D. G., & Brayden, J. E. (2005). TRPC3 mediates pyrimidine receptor-induced depolarization of cerebral arteries. *American Journal of Physiology. Heart and Circulatory Physiology*, *288*(5), H2055–H2061.

Renkin, E. M. (1984). Control of microcirculation and blood-tissue exchange. In E. M. Renkin & C. C. Michel (Eds.), *Handbook of physiology: Section 2, the cardiovascular system, vol. IV, microcirculation, part 2* (pp. 627–687). Bethesda: American Physiological Society.

Rhodin, J. A. G. (2014). Architecture of the vessel wall. In R. L. Terjung (Ed.), *Comprehensive physiology*, (pp. 1–31). John Wiley & Sons, Inc. https://doi.org/10.1002/cphy.cp020201

Sampson, L. J., Davies, L. M., Barrett-Jolley, R., Standen, N. B., & Dart, C. (2007). Angiotensin II-activated protein kinase C targets caveolae to inhibit aortic ATP-sensitive potassium channels. *Cardiovascular Research, 76*(1), 61–70. https://doi.org/10.1016/j.cardiores.2007.05.020.

Schleifenbaum, J., Kassmann, M., Szijarto, I. A., Hercule, H. C., Tano, J. Y., Weinert, S., et al. (2014). Stretch-activation of angiotensin II type 1a receptors contributes to the myogenic response of mouse mesenteric and renal arteries. *Circulation Research, 115*(2), 263–272. https://doi.org/10.1161/CIRCRESAHA.115.302882.

Simionescu, M., & Simionescu, N. (1984). Ultrastructure of the microvascular wall: Functional correlations. In E. M. Renkin & C. C. Michel (Eds.), *Handbook of physiology, section 2: The cardiovascular system, vol. IV, microcirculation. Part 2,* (pp. 41–101). Bethesda, MD: The American Physiological Society.

Slish, D. F., Welsh, D. G., & Brayden, J. E. (2002). Diacylglycerol and protein kinase C activate cation channels involved in myogenic tone. *American Journal of Physiology. Heart and Circulatory Physiology, 283*(6), H2196–H2201. https://doi.org/10.1152/ajpheart.00605.2002.

Storch, U., Blodow, S., Gudermann, T., & Mederos, Y. S. M. (2015). Cysteinyl leukotriene 1 receptors as novel mechanosensors mediating myogenic tone together with angiotensin II type 1 receptors-brief report. *Arteriosclerosis, Thrombosis, and Vascular Biology, 35*(1), 121–126. https://doi.org/10.1161/ATVBAHA.114.304844.

Suzuki, Y., Yamamura, H., Ohya, S., & Imaizumi, Y. (2013). Caveolin-1 facilitates the direct coupling between large conductance Ca2+-activated K+ (BKCa) and Cav1.2 Ca2+ channels and their clustering to regulate membrane excitability in vascular myocytes. *The Journal of Biological Chemistry, 288*(51), 36750–36761. https://doi.org/10.1074/jbc.M113.511485.

Thakore, P., & Earley, S. (2019). Transient receptor potential channels and endothelial cell calcium signaling. *Comprehensive Physiology, 9*(3), 1249–1277. https://doi.org/10.1002/cphy.c180034.

Tykocki, N. R., Boerman, E. M., & Jackson, W. F. (2017). Smooth muscle ion channels and regulation of vascular tone in resistance arteries and arterioles. *Comprehensive Physiology,* 7(2), 485–581. https://doi.org/10.1002/cphy.c160011.

Welsh, D. G., Jackson, W. F., & Segal, S. S. (1998). Oxygen induces electromechanical coupling in arteriolar smooth muscle cells: A role for L-type Ca^{2+} channels. *The American Journal of Physiology, 274*(6), H2018–H2024. https://doi.org/10.1152/ajpheart.1998.274.6.H2018.

Welsh, D. G., Morielli, A. D., Nelson, M. T., & Brayden, J. E. (2002). Transient receptor potential channels regulate myogenic tone of resistance arteries. *Circulation Research, 90*(3), 248–250. https://doi.org/10.1161/hh0302.105662.

Westcott, E. B., Goodwin, E. L., Segal, S. S., & Jackson, W. F. (2012). Function and expression of ryanodine receptors and inositol 1,4,5-trisphosphate receptors in smooth muscle cells of murine feed arteries and arterioles. *The Journal of Physiology, 590*(8), 1849–1869. https://doi.org/10.1113/jphysiol.2011.222083.

Westcott, E. B., & Jackson, W. F. (2011). Heterogeneous function of ryanodine receptors, but not IP3 receptors, in hamster cremaster muscle feed arteries and arterioles. *American Journal of Physiology. Heart and Circulatory Physiology, 300*(5), H1616–H1630. https://doi.org/10.1152/ajpheart.00728.2010.

Wilson, A. J., Jabr, R. I., & Clapp, L. H. (2000). Calcium modulation of vascular smooth muscle ATP-sensitive K(+) channels: Role of protein phosphatase-2B. *Circulation Research*, *87*(11), 1019–1025. https://doi.org/10.1161/01.res.87.11.1019.

Wu, B. N., Chen, M. L., Dai, Z. K., Lin, Y. L., Yeh, J. L., Wu, J. R., et al. (2009). Inhibition of voltage-gated L-type calcium channels by labedipinedilol-A involves protein kinase C in rat cerebrovascular smooth muscle cells. *Vascular Pharmacology*, *51*(2–3), 65–71. https://doi.org/10.1016/j.vph.2009.03.001.

Wu, B. N., Luykenaar, K. D., Brayden, J. E., Giles, W. R., Corteling, R. L., Wiehler, W. B., et al. (2007). Hyposmotic challenge inhibits inward rectifying K+ channels in cerebral arterial smooth muscle cells. *American Journal of Physiology. Heart and Circulatory Physiology*, *292*(2), H1085–H1094. https://doi.org/10.1152/ajpheart.00926.2006.

Yang, Y., Murphy, T. V., Ella, S. R., Grayson, T. H., Haddock, R., Hwang, Y. T., et al. (2009). Heterogeneity in function of small artery smooth muscle BKCa: Involvement of the beta1-subunit. *The Journal of Physiology*, *587*(Pt. 12), 3025–3044. https://doi.org/10.1113/jphysiol.2009.169920.

Yeon, D. S., Kim, J. S., Ahn, D. S., Kwon, S. C., Kang, B. S., Morgan, K. G., et al. (2002). Role of protein kinase C- or RhoA-induced Ca2+ sensitization in stretch-induced myogenic tone. *Cardiovascular Research*, *53*(2), 431–438. https://doi.org/10.1016/S0008-6363(01)00496-5.

Zhang, J., Ren, C., Chen, L., Navedo, M. F., Antos, L. K., Kinsey, S. P., et al. (2010). Knockout of Na+/Ca2+ exchanger in smooth muscle attenuates vasoconstriction and L-type Ca2+ channel current and lowers blood pressure. *American Journal of Physiology. Heart and Circulatory Physiology*, *298*(5), H1472–H1483. https://doi.org/10.1152/ajpheart.00964.2009.

Zhao, G., Neeb, Z. P., Leo, M. D., Pachuau, J., Adebiyi, A., Ouyang, K., et al. (2010). Type 1 IP3 receptors activate BKCa channels via local molecular coupling in arterial smooth muscle cells. *The Journal of General Physiology*, *136*(3), 283–291. https://doi.org/10.1085/jgp.201010453.

Zou, H., Ratz, P. H., & Hill, M. A. (1995). Role of myosin phosphorylation and [Ca2+]i in myogenic reactivity and arteriolar tone. *The American Journal of Physiology*, *269*(5 Pt. 2), H1590–H1596, doi: 10.1152y1995.269.5.H1590.

Zweifach, B. W., & Lipowsky, H. H. (1984). Pressure-flow relations in blood and lymph microcirculation. In E. M. Renkin & C. C. Michel (Eds.), *Handbook of physiology, section 2: The cardiovascular system, vol. IV, microcirculation. Part 1* (pp. 251–305). Bethesda, MD: The American Physiological Society.

CHAPTER THREE

Endothelial inwardly-rectifying K^+ channels as a key component of shear stress-induced mechanotransduction

Ibra S. Fancher*, Irena Levitan
Division of Pulmonary, Critical Care, Sleep and Allergy, Department of Medicine, University of Illinois at Chicago, Chicago, IL, United States
*Corresponding author: e-mail address: ifancher@uic.edu

Contents

Abstract

It has been recognized for decades that fluid shear stress plays a major role in vascular function. Acting on the endothelium shear stress induces vasorelaxation of resistance arteries and plays a major role in the propensity of the major arteries to atherosclerosis. Many elements of shear-induced signaling have been identified yet we are just

Current Topics in Membranes, Volume 85
ISSN 1063-5823
https://doi.org/10.1016/bs.ctm.2020.02.002

beginning to decipher the roles that mechanosensitive ion channels may play in the signaling pathways initiated by shear stress. Endothelial inwardly-rectifying K^+ channels were identified as potential primary mechanosensors in the late 1980s yet until our recent works, highlighted in the forthcoming chapter, the functional effect of a shear-activated K^+ current was completely unknown. In this chapter, we present the physiological effects of shear stress in arteries in health and disease and highlight the most prevalent of today's investigated mechanosensitive ion channels. Ultimately, we focus on Kir2.1 channels and discuss in detail our findings regarding the downstream signaling events that are induced by shear-activated endothelial Kir2.1 channels. Most importantly, we examine our findings regarding hypercholesterolemia-induced inhibition of Kir channel shear-sensitivity and the impact on endothelial function in the context of flow (shear)-mediated vasodilation and atherosclerosis.

1. Introduction

It has been recognized for decades that fluid shear stress, a frictional force on the vessel wall generated by the blood flow, plays a major role in vascular function. Acting on the endothelium, the monolayer inner lining of the vessels, shear stress induces vasorelaxation of resistance arteries and plays a major role in the propensity of the major arteries to atherosclerosis (Davies, Barbeee, Lal, Robotewskyj, & Griem, 1995; Davies, Civelek, Fang, & Fleming, 2013; Davies, Spaan, & Krams, 2005; Ku, Giddens, Zarins, & Glagov, 1985; Zhou, Li, & Chien, 2014). Specifically, numerous studies showed that atherosclerotic lesions form in the regions of branches and bifurcations that are exposed to non-unidirectional disturbed or oscillatory flow whereas straight segments of the arteries exposed to unidirectional laminar flow are relatively atheroresistant (Chatzizisis et al., 2007; Davies et al., 2013; Zhou et al., 2014). The causative role of shear stress in atherosclerosis development was further confirmed by introducing flow disturbance into a straight carotid artery artificially by partial ligation: arteries that underwent partial ligation developed massive atherosclerosis whereas arteries that were left intact in the same mouse had virtually none (Nam et al., 2009). Notably, endothelial flow sensation can occur acutely (flow-induced vasodilation) and in a chronic setting (e.g., vessel remodeling). Therefore, understanding the mechanisms of endothelial flow sensation and mechanotransduction is one of the central questions in endothelial and vascular biology.

Endothelial inwardly-rectifying K^+ (Kir) channels emerged very early as potential primary mechanosensors because of their virtually immediate response to flow, which occurs within a second (Olesen, Clapham, & Davies, 1988). However, their actual role in flow-induced mechanotransduction remained unknown until very recently when our studies demonstrated that endothelial Kir are required for flow-induced activation of Akt 1 and production of nitric oxide (NO) (Ahn et al., 2017), a major flow-sensitive pathway contributing to vasorelaxation and anti-inflammatory effects of shear stress. In this review, we will summarize the early studies that led to the identification of Kir channels as a potential flow sensor and present and discuss evidence from our studies demonstrating that endothelial Kir are essential for flow-induced NO production and vasorelaxation. It is important to emphasize, however, that while we clearly showed that endothelial Kir channels are critical for flow-induced vasorelaxation, there is no evidence that they serve as primary mechanosensors and this distinction will be further discussed in the review.

Our studies over more than a decade also focused on the sensitivity of endothelial Kir channels to membrane cholesterol and plasma hypercholesterolemia (Levitan, 2009; Rosenhouse-Dantsker, Mehta, & Levitan, 2012). Briefly, we showed that endothelial Kir are suppressed by the elevation of membrane cholesterol (Romanenko, Rothblat, & Levitan, 2002), by exposing the cells to proatherogenic lipoproteins and by plasma hypercholesterolemia in vivo in pigs and in ApoE$^{-/-}$ mice (Fancher et al., 2018; Fang et al., 2006). We propose, therefore, suppression of endothelial Kir channels by plasma hypercholesterolemia is a major contributor to endothelial dysfunction under these conditions. In this review, we will summarize the current understanding about the mechanisms of cholesterol-induced suppression of Kir channels and discuss evidence that it contributes to the impairment of vasorelaxation in hypercholesterolemia.

2. Arterial hemodynamic environments and general principles of endothelial flow sensation

2.1 Role of shear stress in conduit arteries

Shear stress (tau) is defined as a stress that is applied parallel or tangential to a face of a material, as opposed to a normal stress which is applied perpendicularly. This tangential friction force induces deformation of endothelial cells during the systolic and diastolic cardiac phases. The magnitude of the shear stress can be estimated by Poiseuille's law: shear stress (tau) is proportional to the flow rate (Q) and blood viscosity (μ) and inversely proportional to the third power of the internal radius (R) (see Fig. 1).

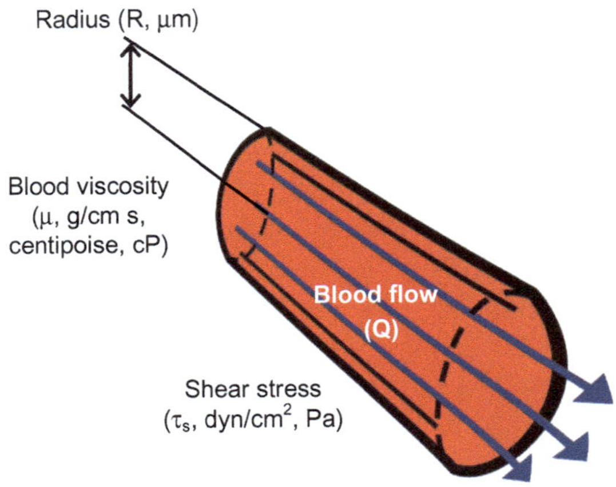

Fig. 1 Components of fluid shear stress. The figure illustrates the Poiseuille's law by depicting of a typical laminar flow with velocity Q in a blood vessel with radius R.

$$\text{Poiseuille's law } (\tau_s) = \frac{4\mu Q}{\pi R^3}$$

Both, the magnitude and the directionality of shear stress vary significantly within the same artery depending on the geometry of the vessel and local bends, branches and bifurcations. In the long and unbranched sections of the arteries, flow increases and decrease its velocity during the cardiac cycle but stays overall laminar and unidirectional. These regions are generally resistant to the development of atherosclerosis. In contrast, at bifurcations and branching points, the flow patterns become very complex with flow separation giving rise to flow reversals that create oscillatory or recirculating flows and strong shear stress gradients. These patterns are typically designated as "disturbed flow." Notably, shear stress recirculation also correlates with low shear stress, the patterns typically observed on the outside of the bifurcations or at the inner surface of bends, such as aortic arch (Fig. 2). These are exactly the sites with the highest propensity for lesion development (Chatzizisis et al., 2007; Davies et al., 2013; Ku et al., 1985; Zhou et al., 2014). Furthermore, it is well established that unidirectional flow induces increased expression of Kruppel-like factors 2 and 4, transcription factors that induce multiple anti-inflammatory pathways (Davies et al., 2013; Zhou et al., 2014). More recently, it was also established that flow plays a major role in the regulation of a novel master regulator of vascular homeostasis, a complex of Yes-associated protein (YAP) and transcriptional coactivator with PDZ-binding motif (TAZ) with unidirectional flow inhibiting and disturbed flow activating YAP/TAZ pathway

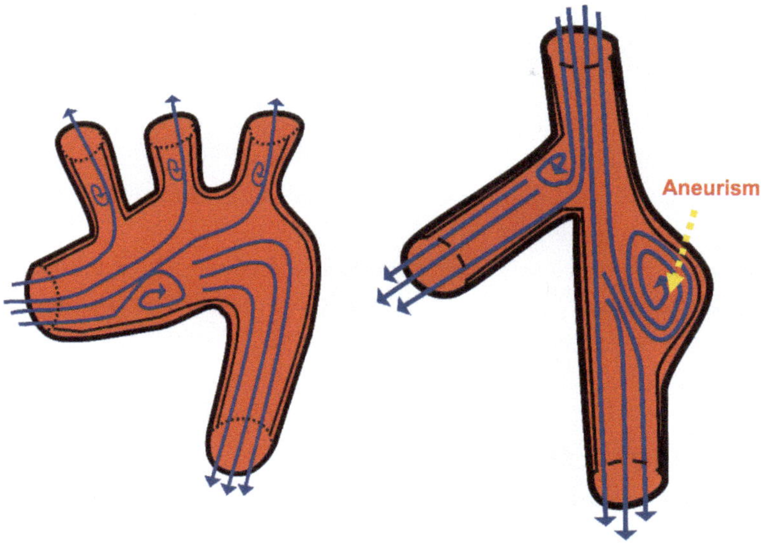

Fig. 2 Patterns of shear stress at aortic arch and bifurcations with an aneurism showing regions that are atherosusceptible (turbulence; low shear stress) and atheroprotective regions (high shear stress).

(Wang et al., 2016). Flow-induced inhibition of YAP/TAZ was shown to be a potent suppressor of the formation of atherosclerotic lesions (Wang et al., 2016).

2.2 Role of shear stress in resistance arteries

The role of shear stress in resistance arteries is multifaceted and well-established as an important component of acute arterial responses such as flow-mediated vasodilation (Henrion et al., 1997; Liu et al., 2003) and more chronic phenomenon such as vascular development/angiogenesis and remodeling (Chouinard-Pelletier, Jahnsen, & Jones, 2013; Tulis, Unthank, & Prewitt, 1998; Wragg et al., 2014). As evidence of this, genetic disruption or pharmacological inhibition of elements critical to cellular mechanotransduction pathways induced by shear stress drastically impairs the appropriate physiological response. With regard to the acute dilatory response to shear stress and the control of tone in the resistance vasculature, parallel pathways—one driven by the production of nitric oxide (NO), the other driven by multiple endothelial derived hyperpolarizing factors (EDHFs)—differentially contribute to the active flow-induced dilation in resistance arteries in series. In our studies and findings from others, first order arteries (~150–200 μm in diameter at basal tone) branching from conduit vessels in which NO is the dominant vasoactive dilator in response to shear stress, exhibit approximately half of the dilatory contributions from NO and EDHFs (Ahn et al., 2017; Miura et al., 2001; Watanabe, Yashiro, Mizuno, & Ohhashi, 2005). Specifically, in our studies we showed that flow-sensitive endothelial Kir channels are upstream of the NO production while Ca^{2+}-sensitive K^+ channels are likely responsible for NO-independent signaling (Ahn et al., 2017).

This extraordinary correlation between the hemodynamic environment and lesion formation in conduit arteries and a major role that shear stress plays in vasorelaxation of resistance arteries and control of vascular tone led to decades of the search for endothelial flow sensors.

2.3 Principles of flow sensation

Many studies proposed various mechanisms to serve as primary putative flow sensors, including flow-sensitive ion channels, intercellular adhesion molecules, integrins, cytoskeleton, glycocalyx, cilia and more but overall most of these mechanisms can be segregated into three general principles: (i) generation of tension at the cell-cell or cell-substrate interfaces; (ii) direct displacement of structures protruding into the lumen at the apical membrane and (iii) generation of tension and/or deformation of the membrane bilayer and/or in the cytoskeleton. These mechanisms are depicted in Fig. 3.

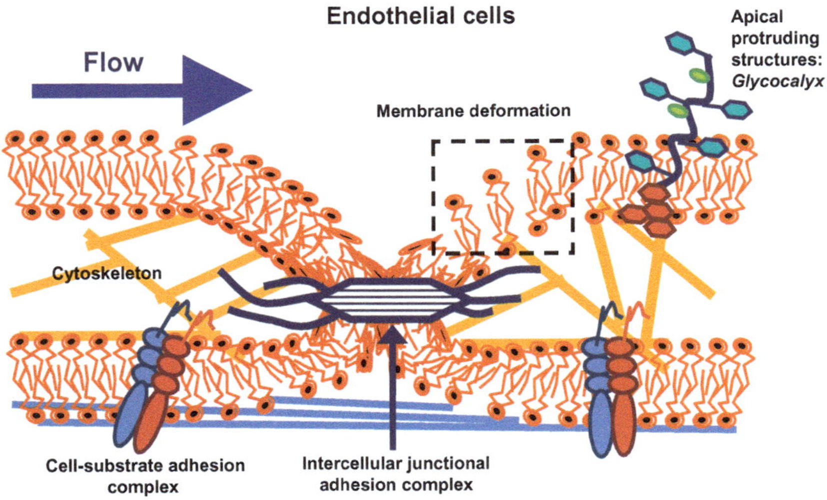

Fig. 3 Mechanisms of flow sensation showing intercellular and cell-substrate adhesion complexes, apical membrane protruding structures and membrane deformation. Cytoskeleton interconnects between all the elements.

We only briefly describe these general principles below and the reader is referred to numerous excellent reviews for a more detailed description and analysis of these pathways (e.g., Hahn & Schwartz, 2009; Zhou et al., 2014).

2.3.1 Adhesion mechanosensitive complex

Application of shear stress is expected to generate tension at the cell-cell junctions and it has been proposed that this tension can be sensed by the intercellular adhesion molecules that anchor between two cells. Specifically, it was proposed that tension can be generated on the junctional proteins, the immunoglobulin family receptor platelet endothelial cell adhesion molecule (PECAM1) and/or a major protein of endothelial adherent junctions VE-Cadherin. Indeed, earlier studies demonstrated that a lack of either PECAM1 or VE-Cadherin abrogate flow-induced endothelial alignment, a hallmark of endothelial response to flow and the reconstitution of this complex together with a Vascular Endothelial Growth Factor (VEGF) receptor in non-endothelial cells confers flow sensitivity (Tzima et al., 2005). Later studies established that flow-induced tension indeed develops on PECAM1 but not on VE-Cadherin and that force is transmitted to the activation of VEGFR receptors via activation of a Src family kinase (Baeyens, Bandyopadhyay, Coon, Yun, & Schwartz, 2016; Conway et al., 2013). This initiates one of the major flow-induced signaling pathways: activation of Phospho-Inositide 3 Kinase (PI3K), generation of PIP3, activation of Akt1 and the activation endothelial NO synthase (eNOS) to generate NO (Baeyens et al., 2016; Zhou et al., 2014). However, Conway et al. also found that force generation on PECAM1 is not a primary event but depends on the integrity of the cytoskeleton and specifically vimentin (Conway et al., 2013). As described in detail in the next section of this review, our recent study demonstrated that endothelial flow-sensitive Kir channels are essential for flow-induced eNOS activation, NO production and NO-dependent vasorelaxation in intact resistance arteries isolated from mesenteric adipose of mice and in subcutaneous adipose of humans (Ahn et al., 2017). It is not clear yet, however, whether Kir channels are upstream or downstream from VEGFR2/PI3K activation. These adhesion complexes are shown as a schematic in Fig. 3.

2.3.2 Apical membrane protruding structures

Another common principle for primary mechanosensation is displacement of structures protruding into the vessel lumen on the apical membranes of endothelial cells. Indeed, two major apical structures have been identified to contribute to flow-induced endothelial response: (i) glycocalyx, a layer of macromolecules with glycosaminoglycan (GAG) chains, such as heparan sulfate (HS), covalently attached to a core protein, that coat the apical surface of endothelial cells (see Fig. 3) (Reitsma, Slaaf, Vink, van Zandvoort, & oude Egbrink, 2007; Tarbell & Ebong, 2008; Tarbell, Shi, Dunn, & Jo, 2014) and (ii) primary cilia, a microtubule-based organelle that protrudes from the surface of most mammalian cells (Praetorius, 2015). The relationships and the relative contributions of these mechanisms to different flow-induced pathways are still poorly understood.

2.3.3 Membrane tension and cytoskeletal deformation

It is also expected that mechanical signal generated by the flow results in an increase in membrane tension, possibly changing membrane curvature and/or tethering to the cytoskeletal structures. It was proposed that these perturbations of the membrane and/or cytoskeleton lead to activation of mechanosensitive ion channels or other membrane proteins that are sensitive to membrane stretch or connected to different cytoskeletal structures (Zhou et al., 2014). Currently, there is little direct evidence to support these hypotheses. In the next section of this review, we will discuss what is known about the flow sensitivity of Kir channels, as well as other endothelial channels that were found to be involved in shear-stress induced endothelial responses.

3. Evidence for flow-sensitivity of endothelial ion channels

3.1 Inwardly-rectifying K^+ channels

The initial discovery by Olesen et al. that endothelial ion channel activity could be affected by mechanical shear revealed that a K^+ current (originally designated $I^{K.s}$ and now known to be conducted by Kir channels) increased in response to shear stress applied across bovine aortic endothelial cells (BAECs) cultured in a capillary tube (Olesen et al., 1988). In this study, flow was delivered through a capillary tube at varying rates to administer different levels of shear across cells at the opening of the tube to be accessed by a patch electrode for current detection. Use of the capillary tubes mimics the enclosed arterial system so that known levels of shear stress could be accurately calculated and delivered. Notably, the shear stress stimulus evoked a selective K^+ current that is significantly different from previously identified nonselective cation stretch-activated channels (Nilius & Droogmans, 2001).

Flow activation of the current was initiated within seconds and reversed rapidly upon cessation of shear stress (Olesen et al., 1988). The flow-sensitive K^+ channels have been identified as inwardly-rectifying K^+ channels (Kir 2), belonging to a class of strong rectifiers: K^+ channels characterized by strong voltage dependence of the conductance pathway producing small outward currents in the physiological ranges of the membrane potential and generally expressed in numerous cell types (Hibino et al., 2010; Kubo et al., 2005). Kir channels were also reported to be expressed in several subtypes of endothelial cells, including aortic endothelium (Fang et al., 2005; Romanenko et al., 2002), microvascular cerebral and mesenteric endothelium (Crane, Walker, Dora, & Garland, 2003; Millar, Wang, Brown, Barrand, & Hladky, 2008) and hamster cremaster arterial ECs (Jackson, 2005).However, their role in endothelial flow-sensation until recently was not well understood.

Since the revolutionary finding of Olesen et al. (1988), further studies from our group demonstrated shear stress induces activation of Kir channels in both human and primary endothelial cells (Ahn et al., 2017; Fancher et al., 2018; Fang et al., 2006). An example of shear stress activation of Kir currents in primary freshly isolated mesenteric endothelial cells is shown in Fig. 4A. The recordings were performed in the same day post-digestion without exposure to culture conditions.

To investigate the shear-sensitivity of Kir channels, we developed a modification of a parallel plate flow chamber, shown in Fig. 4, that allows electrophysiological recordings under well-defined shear stress (Levitan, Helmke, & Davies, 2000). As mentioned above, applying perfusion in an open chamber or delivering a stream of a solution directly onto the cells from above or from a side, methods that have been tried in other studies (e.g., Hoger, Ilyin, Forsyth, & Hoger, 2002), do not generate a well-defined flow pattern and poorly reproduce the shear force generated by blood flow in arteries (Levitan et al., 2000). Seeding cells into capillaries, as was done by Olesen et al. (1988) is a very valid method but extremely difficult and cumbersome to obtain stable recordings. Thus, we developed a chamber with narrow open slits in the top plate of the chamber that allow a recoding pipette to access cells seeded within the chamber but are also narrow enough to prevent the overflow due to surface tension forces. The principles and the design are described in detail in our early study (Levitan et al., 2000) and more recently in a video publication by the Journal of Visualized Experiments (Fancher & Levitan, 2019). Briefly, the fluid-chamber adhesion results in an "enclosed" chamber to which we gravity perfuse

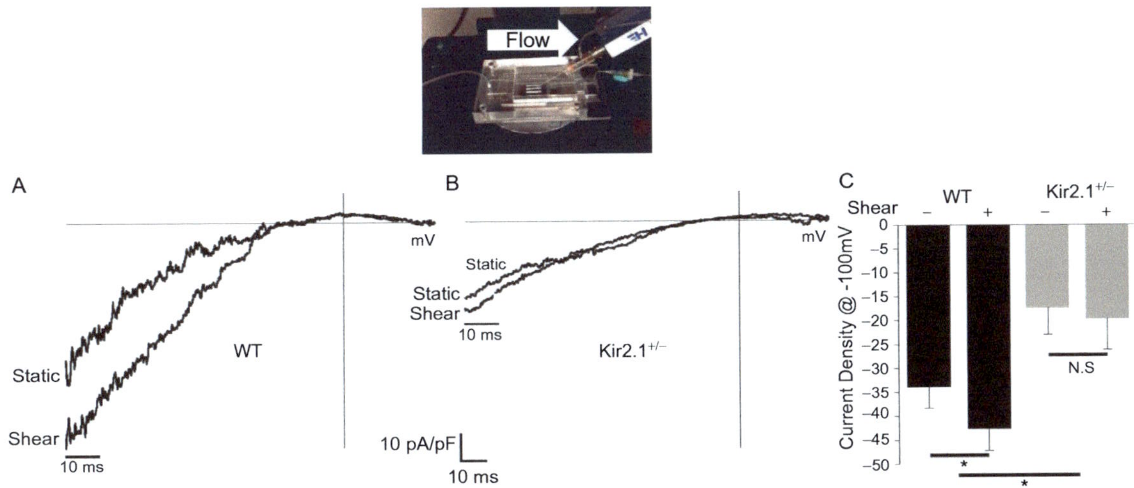

Fig. 4 Endothelial Kir2.1 underlies the shear-sensitive K+ current in freshly isolated mesenteric endothelial cells. Representative recordings of endothelial Kir currents from endothelial cells isolated from (A) WT and (B) Kir2.1+/− mouse mesenteric arteries. (C) Group data reveals that Kir2.1 deficiency results in the loss of shear-activated Kir currents. The unique parallel plate flow chamber we use to assess shear-activated Kir channel currents in endothelial cells seeded in the chamber is shown above the figure. *Adapted with permissions from Barbera, N., Ayee, M. A. A., Akpa, B. S., & Levitan, I. (2018). Molecular dynamics simulations of Kir2.2 interactions with an ensemble of cholesterol molecules. Biophysical Journal, 115(7), 1264–1280.*

electrophysiological bath solutions at varying reservoir heights to deliver the desired and well-defined flow and, therefore, shear stress across the endothelial cells by calculating known parameters as a function of shear in a parallel chamber:

$$\tau = \frac{6Q\mu}{wh^2}$$

where Q is flow rate in mL/s, μ is viscosity in g/cm s (0.009 g/cm s for bath solution), w is the chamber width in cm (2.2 cm), and h is the chamber height in cm (0.1 cm). This unique device allows for the real time investigation of shear-induced ion channel activation in both cells from culture and freshly isolated cell preparations. In both preparations, we have identified robust activation of Kir currents in the physiological range of shear stress (Ahn et al., 2017; Fancher et al., 2018; Fang et al., 2006).

First, we extended our studies to human aortic endothelial cells (HAECs) and demonstrated that similarly to BAECs, shear stress induced small but robust activation of Kir current (Fang et al., 2006). As expected, flow activation of Kir channels was accompanied by membrane hyperpolarization (Fang et al., 2006). To determine if this effect was localized to ECs in conduit arteries, we next tested the Kir flow sensitivity in the endothelium of mouse mesenteric resistance arteries (Ahn et al., 2017). First, we tested the expression of different Kir2 isoforms and the sensitivity of Kir currents to shear stress in freshly isolated mouse mesenteric endothelial cells. As noted above, the reason to focus on the Kir2 family of the inward rectifiers is the biophysical profile of strong rectification, typical for this sub-family (Kubo et al., 2005). In addition, we already established that endothelial Kir current in HAECs is underlined by the two members of Kir2 family, Kir2.1 and Kir2.2 (Fang et al., 2005). In terms of Kir2 expression, we found that the dominant isoform in mesenteric endothelial cells is Kir2.1 with very low, almost negligible expressions of the other Kir2 isoforms (Ahn et al., 2017). We also found a robust response of the Kir current to flow, as shown below in Fig. 4A. Notably, in this work, testing if Kir channels were sensitive to shear stress was performed in native cells never exposed to culture conditions and tested the same day post digestion.

To confirm the molecular identity of the flow-sensitive Kir channel, we used a Kir2.1-deficient mouse model. Specifically, we used Kir2.1$^{+/-}$ model because a full deletion of Kir2.1 is lethal hours within birth (Zaritsky, Eckman, Wellman, Nelson, & Schwarz, 2000). In contrast, we found that Kir2.1$^{+/-}$ is fully viable and reproduces well. First, we established that as expected the expression of Kir2.1 is significantly decreased in Kir2.1$^{+/-}$ as compared to WT litter mates both on the mRNA and protein levels. We then showed that Kir currents in mesenteric endothelial cells isolated from the arteries of Kir2.1$^{+/-}$ are significantly reduced and virtually lack flow sensitivity (Fig. 4B). These data unequivocally demonstrated that endothelial Kir2.1 channels are the flow-sensitive Kir channel in mesenteric endothelium.

In addition to Kir, there are several other ion channels that are implicate in shear stress mechanotransduction, particularly TRPV4 and Piezo, both non-selective cation channels. In this review, we briefly discuss evidence for the direct sensitivity of these channels to fluid shear stress. The reader is referred to a recent excellent review by Gerhold and Schwartz (2016) (Gerhold & Schwartz, 2016) for more general information about the role of mechanosensitive ion channels in endothelial flow response.

3.2 Transient receptor potential channel Vallinoid family type 4 (TRPV4)

A member of the transient receptor potential family, TRPV4, was originally identified as a channel gated by hypotonic stress which was interpreted as sensitivity to membrane tension and thus classified as a mechanosensitive ion channel (Liedtke et al., 2000; Strotmann, Harteneck, Nunnenmacher, Schultz, & Plant, 2000). Evidence for the shear stress sensitivity of TRPV4 channels was shown using fura2, a Ca^{2+} sensitive dye, in cultured endothelial cells whereby increases in flow resulted in elevated influx of Ca^{2+}, an effect that was inhibited by blocking TRPV4 with ruthenium red or by downregulating TRPV4 expression (Mendoza et al., 2010). However, with regard to hypotonic swelling, two arguments against TRPV4 as being directly mechanically activated are presented: First, hypotonic swelling has been shown to induce arachidonic acid production, a potent activator of TRPV4, which may result in the observed increases in channel activity under hypoosmotic conditions, in which case activation of TRPV4 by osmotic stress would be indirect (Vriens et al., 2004). Furthermore, our recent study demonstrated that osmotic stress actually does not increase membrane tension in endothelial cells, as assessed by measuring tether force of nanotubes pulled from the membrane by Atomic Force cantilever (Ayee, LeMaster, Teng, Lee, & Levitan, 2018). In light of this, it is important to clarify that the contributions of TRPV4 in shear stress-mediated endothelial signaling is clear yet whether or not the channel itself is directly activated by shear, and not a secondary shear-induced mediator, remains unknown.

Furthermore, several studies demonstrated that TRPV4-induced increase in intracellular Ca^{2+} results in the activation of Ca^{2+}-sensitive K^+ channels, specifically the small- (SK) and intermediate (IK) K^+ channels, which in turn contributes to flow-induced vasodilation (Brahler et al., 2009; Sonkusare et al., 2012).

3.3 Piezo channels

Piezo channels have been identified as stretch-sensitive, directly activated by applying negative pressure to the membrane via a recording patch pipette (Coste et al., 2010, 2012). Piezo was also shown to be activated by touch, poking the cells with a blunt probe (Ranade et al., 2014). The first evidence that suggested that Piezo channels are sensitive to shear stress arose from experiments using a perfusion tube positioned very near the cell for recording shear-induced currents (Ranade et al., 2014). Increasing pressure in the tube corresponded to elevated wall shear stress within the tube and induced increases in whole-cell Piezo current. However, while the mechanically activated increase in current is clear, this method of applying fluid-flow across a cell from a tube above the recording cell does not equate to fluid shear stress generated by flow but likely mechanical deformation of the cell sufficient to induce Piezo activation. Flow-sensitivity of Piezo1 was also demonstrated by Li et al. showing significant increase in current in response to increased rate of perfusion of the recording bath but again, it is not clear whether well-defined shear was applied and thus, it is not certain that the mechanical stimulus that activated the channels in this system was shear stress (Li et al., 2014). A similar approach was applied to test the effect of shear stress on single channel activity of Piezo 1 (Rode et al., 2017). Specifically, a patch pipette with an attached membrane patch was maneuvered to the opening of a large perfusion pipette through which the flow was applied. There are clearly several concerns about this approach. First, a cell from which recording is made needs to be detached from the substrate which is very different from the physiological shear stress in the blood vessels. Second, for the single channel recording, a small membrane patch is excised from a cell and typically due to the negative pressure in the recording pipette, this membrane patch is drawn into the pipette and is actually not really exposed to the shear outside of the pipette. Thus, which clearly Piezo channels respond to a mechanical stimulus, the physical nature of the mechanical disturbance remains uncertain. In contrast, while the methods described for electrophysiological recordings introduce potential artifacts with regard specifically to shear-stress activation of Piezo channels, the Piezo-dependent, shear-induced Ca^{2+}-measurement experiments reveal a much clearer picture. In these experiments, single cell Ca^{2+} influx was assessed in the presence and absence of Piezo using chambers designed for the delivery of controlled flow and shear force and it was clearly shown that Piezo channels are critical for shear stress-induced Ca^{2+} response (Li et al., 2014; Ranade, Qiu, et al., 2014; Rode et al., 2017).

4. The role of Kir channels in endothelial responses to flow

When the shear-activated K^+ current was originally discovered in endothelial cells, it was assumed that the shear-induced increase in endothelial Kir channel activity would result in endothelial hyperpolarization, thereby promoting Ca^{2+} influx and downstream signaling cascades. However, until our recent studies, the effects on what truly occurs upon shear-activation of Kir channels was speculative.

4.1 Role of endothelial Kir in flow-induced vasodilation and NO release

Our recent studies determined that genetic deficiency of Kir2.1, which we identified as the major shear-sensitive Kir channel expressed in mouse mesenteric endothelial cells, plays a key role in flow-induced vasodilation (FIV) of mesenteric resistance arteries (Ahn et al., 2017). As described briefly above, FIV of resistance arteries is a major mechanism to regulate vascular resistance and blood flow in the circulation (Kuo, Chilian, & Davis, 1991; Spier, Delp, Stallone, Dominguez 2nd, & Muller-Delp, 2007). Specifically, we showed that mesenteric arteries excised from $Kir2.1^{+/-}$ mice show a reduction in FIV as compared to WT control mesenteric resistance arteries (Ahn et al., 2017). Furthermore, in contrast to WT arteries, which were sensitive to the endothelial nitric oxide synthase inhibitor N(gamma)-nitro-L-arginine methyl ester (LNAME), FIV in arteries from $Kir2.1^{+/-}$ mice were unaffected by LNAME, indicating impaired flow-induced NO production (Fig. 5A). Indeed, this was confirmed using the NO specific dye, diaminorhodamine-4M, to detect flow-induced NO production in WT vs $Kir2.1^{+/-}$ mesenteric arteries whereby significantly less NO was detected in arteries of $Kir2.1^{+/-}$ mice (Fig. 5B). Remarkably, FIV was restored when WT Kir2.1 channels were overexpressed in mesenteric endothelium ex vivo using adenoviral transduction

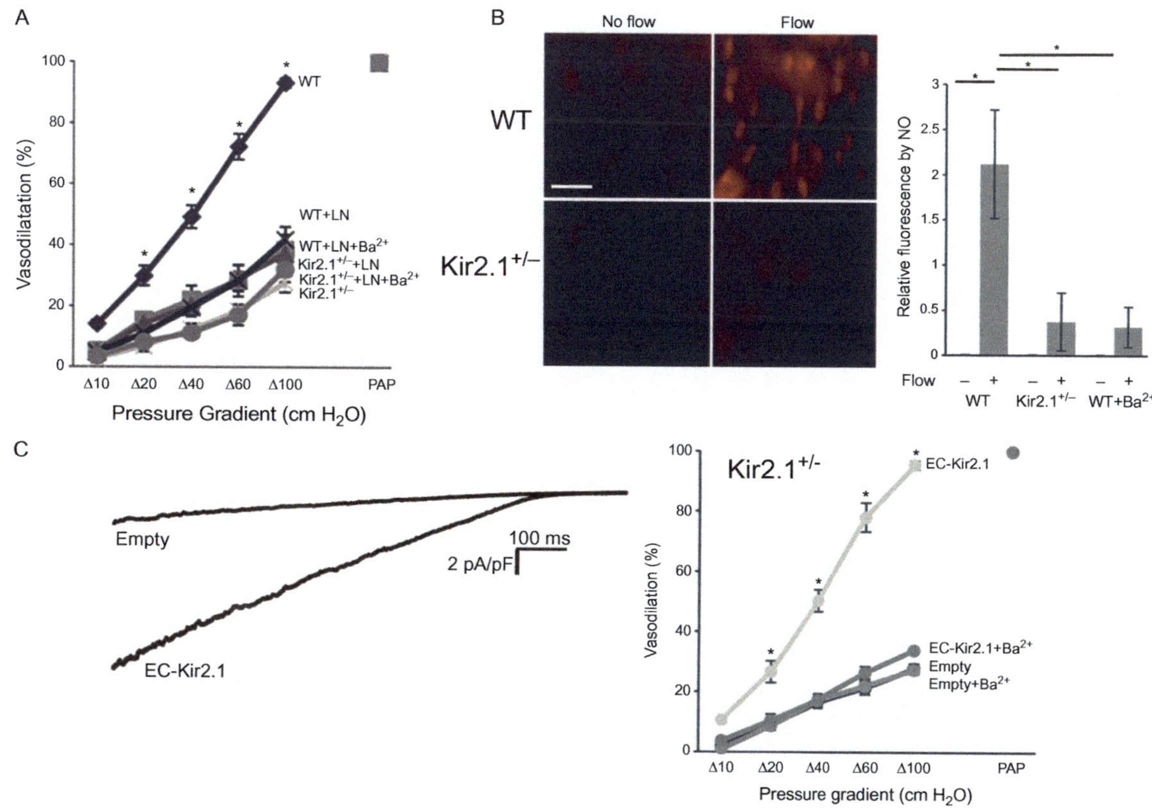

Fig. 5 Kir2.1 channels are critical to flow-induced vasodilation and flow-induced NO production. (A) FIV was measured in isolated and pressurized mesenteric arteries from WT and Kir2.1$^{+/-}$ mice and the effects of LNAME (100 μM) and Ba2+ (30 μM) were assessed. (B) Representative images and group data showing flow-induced NO production in isolated and pressurized mesenteric arteries from WT and Kir2.1$^{+/-}$ mice using the NO-specific dye, DAR4M. (C, left) HAECs were transduced with empty or adenovirus containing endothelium-specific (VE-Cadherin promoter) driven expression of WT Kir2.1 (EC-Kir2.1) and currents were recorded to confirm functional expression of the channel in endothelium. (C, right) FIV was measured in isolated and pressurized mouse mesenteric arteries from Kir2.1$^{+/-}$ mice following transduction with empty or EC-Kir2.1 adenovirus. *Adapted with permissions from Barbera, N., Ayee, M. A. A., Akpa, B. S., & Levitan, I. (2018). Molecular dynamics simulations of Kir2.2 interactions with an ensemble of cholesterol molecules. Biophysical Journal, 115(7), 1264–1280.*

(Fig. 5C). The NO-independent component of FIV of mesenteric arteries were completely unaffected by the loss of Kir2.1 and was instead mediated by Ca^{2+}-sensitive K^+ channels (Ahn et al., 2017). As described above, flow activation of Ca^{2+}-sensitive K^+ channels is expected to be mediated by the activation of flow-sensitive cation channels, TRPV4 and/or piezo1 (Gerhold & Schwartz, 2016). Our findings supported an essential role for shear-activation of endothelial Kir2.1 channels in flow-induced NO production and NO-dependent component of the FIV and suggest that this occurs independently of EDHF pathway mediated by cation and Ca^{2+}-sensitive K^+ channels. Interestingly, Harraz et al. showed that activation of specific Gq-protein coupled receptors caused simultaneous TRPV4 activation and Kir2.1 channel suppression in brain capillary endothelial cells (Harraz, Longden, Hill-Eubanks, & Nelson, 2018). This effect was determined to be due to hydrolysis of phosphatidylinositol 4,5-bishphosphate (PIP2), an inhibitor of TRPV4 channels and essential membrane component required for Kir channel activity, and adds to the complexity of vascular endothelial signaling given a unique stimulus. Furthermore, a recent study showed that PIP2 is required for endothelial Kir2.1 flow-sensitivity as depleting PIP2 from the cell abolished the response of Kir to flow (Sancho et al., 2019). In this context, it is not clear whether PIP2 may be involved in the flow sensitivity of both TRPV4 and Kir2.1: if flow induces an increase in membrane PIP2 content, that would be consistent with playing a role in flow sensitivity of Kir2.1 but not TRPV4 and vice versa. Clearly, more studies are needed to clarify these questions.

Our discovery that Kir 2.1 is required for the flow-induced eNOS activation raises the question about the mechanistic link that couple Kir channels to eNOS. We addressed this question in vitro using a cone-and-plate rheometer, a well-established device used for investigating endothelial responses to physiological levels of shear (Ahn et al., 2017; Wu et al., 2015), to expose primary mesenteric endothelial cells in culture to shear stress. By following this technique with Western blot, we further identified that flow-induced Kir activation resulted in Akt phosphorylation/activation, a well-established signaling event upstream of shear stress-induced eNOS activation (Dimmeler et al., 1999; Fisslthaler, Dimmeler, Hermann, Busse, & Fleming, 2000). Shear-induced activation of Akt, and eNOS was completely absent in Kir2.1$^{+/-}$ ECs. Most importantly, these signaling events could be fully restored in vitro in mesenteric endothelial cells of Kir2.1$^{+/-}$ mice by re-expression of the WT Kir2.1 channel, establishing for the first time the critical role of endothelial Kir2.1 channels in the major Akt/eNOS/NO shear stress mechanotransduction signaling cascade.

Furthermore, our findings indicate that Kir channels are responsible for much more than simply generation of an electrical driving force favorable for Ca^{2+} entry into ECs and, in contrast, are critical to Akt activation, a pathway that is considered to be Ca^{2+} independent. While major advancements regarding Kir channels and downstream signaling are presented here, how shear-induced Kir channel activation results in Akt activation represents the crux of our ongoing studies.

It is important to note that the role of Kir in the response of the endothelium to flow may be different in distinct vascular beds. A recent and eloquent study by Sancho et al. identified a very interesting dichotomy for Kir channels in the cerebral vasculature with regard to the mechanosensitivity of Kir channels; in cerebral vascular endothelium, Kir channels were activated by low levels of laminar shear (0.1 dynes/cm^2) applied across cells in a custom flow chamber, while in cerebral vascular smooth muscle, this level of shear stress did not influence Kir currents (Sancho et al., 2019). The authors attribute this distinction in shear-stress sensitivity between Kir in cerebral endothelium and Kir in smooth muscle to different cellular lipid mediators, highlighting an essential role for endothelial phosphatidylinositol 4,5-bisphosphate (PIP2) in shear-stress sensitivity of Kir channels. Interestingly, removal of cholesterol from the membrane of cerebral vascular endothelial cells using methyl-β-cylcodextrin had no effect on Kir activity in response to shear stress, a treatment we have shown to increase Kir currents in virtually all other endothelial or other cells we have tested with this approach (Levitan, Singh, & Rosenhouse-Dantsker, 2014). Together, these findings support the notion that Kir channels in distinct vascular beds may play unique roles in regulating blood flow in response to hyperemic stimuli. The findings by Sancho et al. also highlight the importance of the very different cellular pools of lipids, proteins, and other regulatory elements in cells that can regulate Kir channel function in order to integrate the appropriate physiological response.

4.2 Role of endothelial Kir in agonist-induced vasodilation

In addition to shear stress-induced mechanotransduction, the role of Kir channels in receptor-mediated, agonist-induced endothelium-dependent vasodilation has been explored by others. A distinguished role for endothelial Kir in receptor-mediated vasodilation is a "booster" or amplifier of hyperpolarization induced by other K^+ channels (Sonkusare, Dalsgaard, Bonev, & Nelson, 2016). This phenomenon is well studied in vascular smooth muscles, where it was shown using a combination of modeling and experimental approaches that the negative slope conductance characteristic of Kir2 channels results activation of these channels in response to membrane hyperpolarization, thus further amplifying the hyperpolarization response (Jantzi et al., 2006; Smith et al., 2008). Furthermore, the sensitivity of Kir channels to the extracellular K^+, another well studied mechanism, provides an additional mechanism for their potential role of "boosters." Briefly, since an increase in extracellular K^+ increases the activity of the channels, it was demonstrated in several systems that K^+ efflux generated by other K+ channels, primarily Ca^{2+}-sensitive K^+ channels may be sufficient to activate Kir channels and thus amplify the hyperpolarization (Longden & Nelson, 2015). Sonkusare et al. recently demonstrated that a similar phenomenon is observed in endothelial cells of third order resistance arteries (Sonkusare et al., 2016). Specifically, they showed that activation of muscarinic receptors or TRPV4 induced Ca^{2+} influx and activation of SK/IK channels whereby increases in membrane hyperpolarization and extracellular K+, both of which increase the activity of Kir channels, subsequently activates endothelial Kir2.1 to boost the endothelium-dependent dilation. However, how exactly Kir channel activation enhances the SK/IK driven dilatory effect, whether it be through NO as we identified in FIV or via an undetermined EDHF mechanism, remains unclear. In contrast, however, blocking Kir channels with Ba^{2+} did not affect conducted vasodilation induced by hyperpolarizing factors in mesenteric arteries suggesting the booster effect of Kir can be tissue specific (Takano, Dora, Spitaler, & Garland, 2004). These mechanisms and the possible interpretations of the contradictory results are described in detail in a recent excellent review (Jackson, 2017). Notably, our studies demonstrated a critical role of Kir channels in FIV not only in mesenteric arteries, as described above, but also in cerebral arteries (Boriushkin, Fancher, & Levitan, 2019).

4.3 Role of endothelial Kir in developmental angiogenesis

We recently showed that Kir2.1 channels also play a significant role in developmental retinal angiogenesis (Boriushkin et al., 2019). Specifically, we showed that Kir2.1-deficient mice have increased vessel density in a developing retina, an effect that we found to be due to deficient vessel pruning and remodeling. Notably, while we did not demonstrate directly a link between deficient pruning of retinal vessels in Kir2.1-deficient mice and shear stress, several earlier studies demonstrated that shear stress play a major role in the regression/pruning of blood vessels (Franco et al., 2015;

Lucitti et al., 2007), which is a normal process in developmental angiogenesis (Isogai, Lawson, Torrealday, Horiguchi, & Weinstein, 2003). It is tempting to propose that endothelial Kir2.1 channels may couple between shear stress and vascular remodeling.

5. Cholesterol suppression of Kir channels in endothelial function

5.1 Mechanisms of cholesterol-induced suppression of Kr2 channels

Cholesterol is a major lipid component of plasma membranes in all mammalian cells and it is well-known to have a major impact on the physical properties of lipid bilayer: an increase in membrane cholesterol decreases membrane fluidity, increases lipid order and enhances the formation of ordered lipid domains (Simons & Ikonen, 1997, 2000; Yeagle, 1985, 1991). It is also well-known that changes in the level of membrane cholesterol regulate numerous types of ion channels with the predominant effect being suppression of the channel gating (Levitan, Fang, Rosenhouse-Dantsker, & Romanenko, 2010; Rosenhouse-Dantsker, Mehta, & Levitan, 2012).

Sensitivity of endothelial Kir channels to cholesterol was discovered in our early studies demonstrating not only that endothelial Kir is significantly suppressed by the elevation of membrane cholesterol but also that this effect is stereospecific: while cholesterol has a strong inhibitory effect, its chiral analogue, epicholesterol, has an opposite effect (Romanenko et al., 2002). This was surprising because the dominant theory at that time was that cholesterol regulates ion channels via changes the physical properties of the bilayer. This was clearly not the case for endothelial Kir channels leading us to hypothesize that cholesterol regulates Kir2 channels via specific sterol-protein interactions. Later studies fully confirmed this early conclusion. Using purified bacterial homologue of Kir2, KirBac1.1, we showed that KirBac1.1 is suppressed by cholesterol in an acellular environment of lipid vesicles that contained no other proteins but the channel (Singh, Rosenhouse-Dantsker, Nichols, Enkvetchakul, & Levitan, 2009) and that cholesterol-channel binding is required for the inhibitory effect (Singh, Shentu, Enkvetchakul, & Levitan, 2011). Similarly, D'Avanzo, Cheng, Doyle, and Nichols (2010) demonstrated that purified Kir2.1 is also sensitive to membrane cholesterol in the acellular environment and that this effect is also stereospecific (D'Avanzo et al., 2010).

The next major goal was to identify the binding site/s of cholesterol to Kir channels, which is a pre-requisite to understanding the molecular nature of the inhibitory effect. Our first and most straightforward hypothesis was that Kir channels interact with cholesterol via the residues on its protein-lipid bilayer interface. We found, however, that this is not the case and that mutating these residues have no effect on the sensitivity of Kir2.1 to cholesterol (Epshtein et al., 2009). Instead, we found specific residues in the C-terminus domain of Kir2.1 are critical for the cholesterol-induced inhibitory effect (Epshtein et al., 2009). Interestingly, these residues overlapped partially with the residues known to interact with a regulatory phospholipid $PI(4,5)P_2$ required for the activation of the Kir2 channels (Rohacs, Chen, Prestwich, & Logothetis, 1999; Zhang, He, Yan, Mirshahi, & Logothetis, 1999). In terms of identifying the binding site, however, energetic considerations made it very unlikely that these residues in the cytoplasmic domain of the channels actually constitute the binding. Instead, we believe that these residues participate in coupling of cholesterol binding to the gating mechanism, even though the molecular details of this process is not clear (Rosenhouse-Dantsker et al., 2012; Rosenhouse-Dantsker, Logothetis, & Levitan, 2011). To identify the binding site of cholesterol to Kir2 channels, we took an unbiased approach based on docking analysis and Molecular dynamics simulations of the cholesterol molecule to the entire channel protein (Barbera, Ayee, Akpa, & Levitan, 2018; Rosenhouse-Dantsker, Noskov, Durdagi, Logothetis, & Levitan, 2013). This was in contrast to a more standard approach of identifying the known cholesterol recognition motifs, such as CRAC and CARC (Di Scala et al., 2017; Fantini & Barrantes, 2013; Picazo-Juarez et al., 2011). Using the unbiased approach, we identified novel non-annular binding cholesterol binding sites in Kir2.1 channels providing the major insights into the mechanism of cholesterol-Kir2 interactions (Rosenhouse-Dantsker et al., 2013). A similar cholesterol-binding site was identified by other investigators using a slightly different approach (Fürst, Nichols, Lamoureux, & D'Avanzo, 2014). Notably, cholesterol binding sites of Kir2.1 channels have no homology and do not include the previously identified cholesterol recognition motifs. In this context, it is important to note that caution should be exercised in assuming that CRAC/CARC motifs are necessarily involved in cholesterol interactions with membrane proteins. Indeed, a thorough analysis of this question by Rosenhouse-Dantsker (2017) revealed that only a portion of the solved structures of cholesterol-protein interactions contain these motifs. Most recently, we explored cholesterol-Kir2 interactions in a larger set of simulations that allow following the diffusion of cholesterol from the bilayer to its binding sites and analyze the dynamics of these interactions. This study revealed that Kir2 channels continuously interact with multiple cholesterol molecules simultaneously that form a range of short-term and long-term interactions and the probabilities of these interactions increase with the increase of the level of cholesterol in the membrane, see Fig. 6, (Barbera et al., 2018). It is also interesting to note that in contrast to most ligands, cholesterol position within the channel binding site is highly dynamic and flexible with the cholesterol molecule exploring the space of the non-annular hydrophobic pockets of the channel protein (Ayee et al., 2018).

5.2 Cholesterol-induced suppression of endothelial Kir2 channels plays a major role in the loss of flow-induced vasodilation in hypercholesterolemia

Naturally, the major question is what is the relevance of cholesterol-induced suppression of endothelial Kir channels to endothelial dysfunction in vivo under hypercholesterolemic conditions? Our earlier study showed that endothelial Kir channels are suppressed by both, exposure to pro-atherogenic very-low density lipoproteins (VLDL) and by diet-induced hypercholesterolemia in a porcine model of atherosclerosis (Fang et al., 2006; Mohler Iii et al., 2007). More recently, we provided compelling evidence that cholesterol-induced suppression of endothelial Kir2.1 channels plays a major role in the impairment of FIV of resistance arteries in a mouse model of genetic hypercholesterolemia, ApoE$^{-/-}$ mice (Fancher et al., 2018). Specifically, we found that while resistance arteries isolated from ApoE$^{-/-}$ have significantly impaired vasodilation and flow-induced NO response compared to the wild type controls, arteries isolated from the double knockout mice, Kir2.1$^{+/-}$ApoE$^{-/-}$, were not different from Kir2.1$^{+/-}$ alone (Fig. 7A) indicating that hypercholesterolemia does not have an additional negative effect on FIV and flow-induced NO response in the absence of Kir2.1 channels. Strikingly, both, FIV and NO response in resistance arteries of ApoE$^{-/-}$ mice could be fully rescued by overexpressing Kir2.1 channels specifically in the endothelial cells using endothelial-specific adenoviral Kir2.1 construct (Fig. 7B, C) (Fancher et al., 2018). These observations indicate that hypercholesterolemia-induced impairment of FIV in ApoE$^{-/-}$ mice can be attributed to cholesterol-dependent suppression of endothelial Kir2.1 channels.

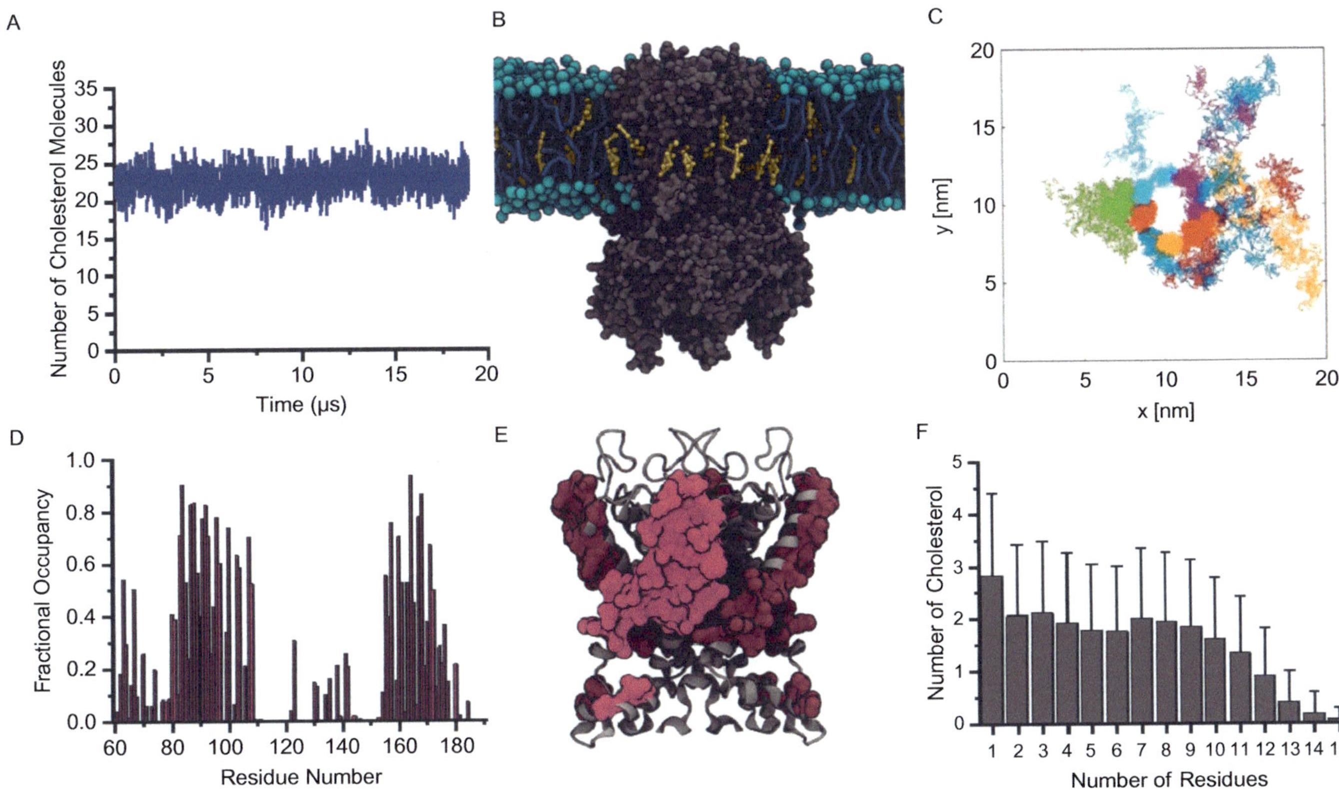

Fig. 6 (A) The average number of cholesterol molecules interacting with channel protein. (B) A representative snapshot of the simulation, showing multiple cholesterol interacting with the channel. (C) Top-down view of the channel in the membrane, with representative 5 μs trajectories of individual cholesterol molecules (D) The average fractional occupancy for individual residues in the transmembrane domain. (E) Residues in light purple and dark purple represent the residues interacting with cholesterol on different subunits. (F) Average number of cholesterol molecules interacting with the specified number of residues. *Adapted with permissions from Barbera, N., Ayee, M. A. A., Akpa, B. S., & Levitan, I. (2018). Molecular dynamics simulations of Kir2.2 interactions with an ensemble of cholesterol molecules.* Biophysical Journal, 115*(7), 1264–1280.*

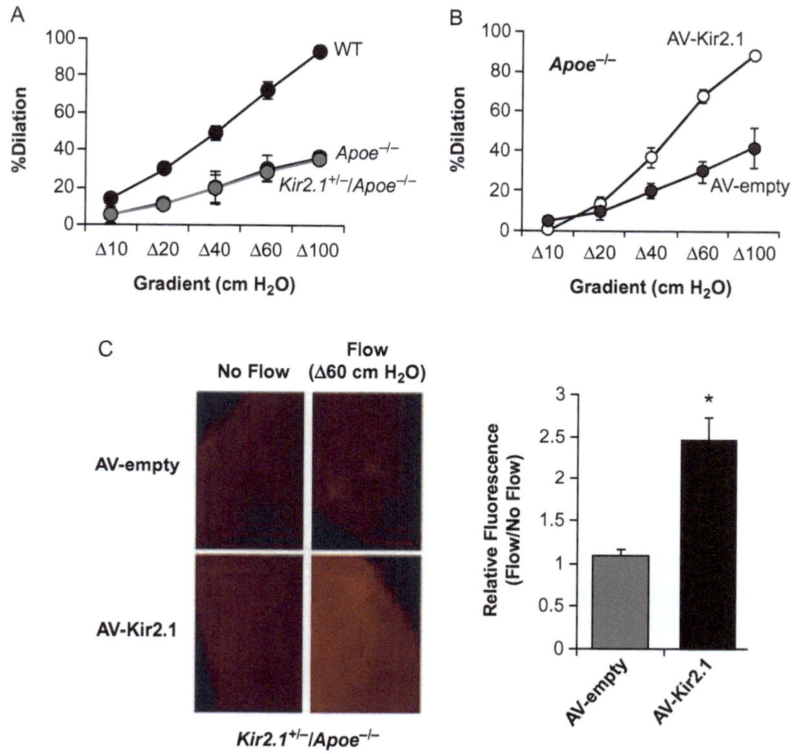

Fig. 7 Hypercholesterolemia-induced endothelial dysfunction is mediated by cholesterol suppression of Kir2.1 channels and subsequent prevention of NO production. (A) FIV was measured ex vivo in pressurized mesenteric arteries from WT, ApoE−/−, or Kir2.1+/−/ApoE−/− mice. Exposing arteries from ApoE−/− mice to adenovirus for endothelial specific over-expression of WT Kir2.1 (AV-Kir2.1) restored (B) FIV and (C) NO production as compared to arteries that were exposed to the empty control adenovirus (AV-empty). *Figures were adapted with permissions from the* Journal of the American Heart Association.

Furthermore, we also found that Kir2.1$^{+/-}$ApoE$^{-/-}$ double knockout mice developed a more severe atherosclerotic burden, as compared with ApoE$^{-/-}$ mice (Fancher et al., 2018). Notably, the exacerbating effect in Kir2.1$^{+/-}$ApoE$^{-/-}$ mice was most pronounced in the areas that are normally athero-resistant due to the athero-protective shear stress profile. In addition, the endothelial dysfunction and Kir channel impairment occurred well before lesion formation, indicating a potential role for Kir channel suppression in the early stages of atherogenesis. We suggest, therefore, that endothelial Kir2.1 plays a major role in FIV of resistance arteries and also in conferring endothelial athero-resistant phenotype. Clearly, further studies are needed to explore this possibility.

6. Conclusions, open questions and future directions

In summary, in this review we present the most current state of knowledge about the role of endothelial Kir channels in vascular function under normal and hypercholesterolemic conditions. A major recent breakthrough is our discovery that shear stress-induced activation of endothelial Kir channels is essential for flow-induced production of NO, identifying it as a key process in endothelial signaling. It is also intriguing that the effect of Kir on eNOS activation is mediated via activation/phosphorylation of Akt1 rather via an increase in Ca^{2+} influx, which is traditionally considered to be the downstream event of the opening of K^{+} channels and membrane hyperpolarization. Moreover, identifying shear-stress induced activation of Kir channels upstream of flow-induced activation of Akt1 has even broader potential impact that its role in eNOS activation. It is not clear yet, however, what is the mechanism that couples Kir channels and Akt1 activation. One intriguing hypothesis is that it might be mediated by a possible impact of membrane potential on the charged regulatory lipids from the phosphoinositide family, which are known to be involved in Akt1 activation. Another open question is what is the mechanism of shear stress activation of the channels, are the channels constitute primary flow sensors or are they effectors downstream of another sensor. If these channels are primary sensors, then most likely their activation is mediated by local membrane tension/deformation but there is no direct evidence that this is the case or even that shear stress really results in an increase in membrane tension. Clearly, more studies are needed to fully elucidate these mechanisms.

In terms of the role of cholesterol in the regulation of Kir channels, we briefly summarize the work done in our lab during the last 20 years to elucidate the mechanisms of specific and direct cholesterol-Kir interactions. We have to acknowledge, however, that while we made big strides to understand these mechanisms including identifying specific dynamic cholesterol binding sites within the hydrophobic cavities of the channel protein, we still just do not know how cholesterol binding leads to a decrease in channel activity or their ability to open and/or stay open. Most recently and significantly, we found that hypercholesterolemia-induced impairment of endothelial function in hypercholesterolemic mouse model, particularly flow-induced eNOS activation and endothelium-dependent FIV, can be attributed to cholesterol-induced suppression of Kir function. A full recovery of the hypercholesterolemia-induced loss by the over-expression of the channels suggests that it might be developed as a potential therapeutic target. Determining whether Kir channels are also involved in hypercholesterolemia-induced endothelial dysfunction in vivo is the next frontier.

Acknowledgments

This work was supported by NIH R01s HL073965 (I.L.), HL141120 (I.L.), and HL083298 (I.L.). The authors would like to thank Gregory Kowalsky for generating cartoon images and schematics in figures for this chapter.

References

Ahn, S. J., Fancher, I. S., Bian, J. T., Zhang, C. X., Schwab, S., Gaffin, R., et al. (2017). Inwardly rectifying K+ channels are major contributors to flow-induced vasodilatation in resistance arteries. *The Journal of Physiology, 595*(7), 2339–2364.

Ayee, M. A. A., LeMaster, E., Teng, T., Lee, J., & Levitan, I. (2018). Hypotonic challenge of endothelial cells increases membrane stiffness with no effect on tether force. *Biophysical Journal, 114*(4), 929–938.

Baeyens, N., Bandyopadhyay, C., Coon, B. G., Yun, S., & Schwartz, M. A. (2016). Endothelial fluid shear stress sensing in vascular health and disease. *The Journal of Clinical Investigation, 126*(3), 821–828.

Barbera, N., Ayee, M. A. A., Akpa, B. S., & Levitan, I. (2018). Molecular dynamics simulations of Kir2.2 interactions with an ensemble of cholesterol molecules. *Biophysical Journal, 115*(7), 1264–1280.

Boriushkin, E., Fancher, I. S., & Levitan, I. (2019). Shear-stress sensitive inwardly-rectifying K(+) channels regulate developmental retinal angiogenesis by vessel regression. *Cellular Physiology and Biochemistry, 52*(6), 1569–1583.

Brahler, S., Kaistha, A., Schmidt, V. J., WÃlfle, S. E., Busch, C., Kaistha, B. P., et al. (2009). Genetic deficit of SK3 and IK1 channels disrupts the endothelium-derived hyperpolarizing factor vasodilator pathway and causes hypertension. *Circulation, 119*(17), 2323–2332.

Chatzizisis, Y. S., Coskun, A. U., Jonas, M., Edelman, E. R., Feldman, C. L., & Stone, P. H. (2007). Role of endothelial shear stress in the natural history of coronary atherosclerosis and vascular remodeling: Molecular, cellular, and vascular behavior. *Journal of the American College of Cardiology, 49*(25), 2379–2393.

Chouinard-Pelletier, G., Jahnsen, E. D., & Jones, E. A. (2013). Increased shear stress inhibits angiogenesis in veins and not arteries during vascular development. *Angiogenesis, 16*(1), 71–83.

Conway, D. E., Breckenridge, M. T., Hinde, E., Gratton, E., Chen, C. S., & Schwartz, M. A. (2013). Fluid shear stress on endothelial cells modulates mechanical tension across VE-cadherin and PECAM-1. *Current Biology, 23*(11), 1024–1030.

Coste, B., Mathur, J., Schmidt, M., Earley, T. J., Ranade, S., Petrus, M. J., et al. (2010). Piezo1 and Piezo2 are essential components of distinct mechanically activated cation channels. *Science (New York, NY), 330*(6000), 55–60.

Coste, B., Xiao, B., Santos, J. S., Syeda, R., Grandl, J., Spencer, K. S., et al. (2012). Piezo proteins are pore-forming subunits of mechanically activated channels. *Nature, 483*(7388), 176–181.

Crane, G. J., Walker, S. D., Dora, K. A., & Garland, C. J. (2003). Evidence for a differential cellular distribution of inward rectifier K channels in the rat isolated mesenteric artery. *Journal of Vascular Research, 40*(2), 159.

D'Avanzo, N., Cheng, W. W. L., Doyle, D. A., & Nichols, C. G. (2010). Direct and specific activation of human inward rectifier K+ channels by membrane phosphatidylinositol 4,5-bisphosphate. *Journal of Biological Chemistry*, *285*(48), 37129–37132.

Davies, P. F., Barbeee, K. A., Lal, R., Robotewskyj, A., & Griem, M. L. (1995). Hemodynamics and atherogenesis. *Annals of the New York Academy of Sciences*, *748*, 86–103.

Davies, P. F., Civelek, M., Fang, Y., & Fleming, I. (2013). The atherosusceptible endothelium: Endothelial phenotypes in complex haemodynamic shear stress regions in vivo. *Cardiovascular Research*, *99*(2), 315–327.

Davies, P. F., Spaan, J. A., & Krams, R. (2005). Shear stress biology of the endothelium. *Annals of Biomedical Engineering*, *V33*(12), 1714.

Di Scala, C., Baier, C. J., Evans, L. S., Williamson, P. T. F., Fantini, J., & Barrantes, F. J. (2017). Relevance of CARC and CRAC cholesterol-recognition motifs in the nicotinic acetylcholine receptor and other membrane-bound receptors. *Current Topics in Membranes*, *80*, 3–23.

Dimmeler, S., Fleming, I., Fisslthaler, B., Hermann, C., Busse, R., & Zeiher, A. M. (1999). Activation of nitric oxide synthase in endothelial cells by Akt-dependent phosphorylation. *Nature*, *399*(6736), 601.

Epshtein, Y., Chopra, A., Rosenhouse-Dantsker, A., Kowalsky, G., Logothetis, D. E., & Levitan, I. (2009). Identification of a C-terminus domain critical for the sensitivity of Kir2.1 channels to cholesterol. *PNAS*, *106*, 8055–8060.

Fancher, I. S., Ahn, S. J., Adamos, C., Osborn, C., Oh, M. J., Fang, Y., et al. (2018). Hypercholesterolemia-induced loss of flow-induced vasodilation and lesion formation in apolipoprotein E-deficient mice critically depend on inwardly rectifying K(+) channels. *Journal of the American Heart Association*, 7(5), e007430. pii: e007430. doi: 007410.001161/JAHA.007117.007430.

Fancher, I. S., & Levitan, I. (2019). Electrophysiological recordings of single-cell ion currents under well-defined shear stress. *Journal of Visualized Experiments*, *150*, e59776.

Fang, Y., Mohler, E. R., III, Hsieh, E., Osman, H., Hashemi, S. M., Davies, P. F., et al. (2006). Hypercholesterolemia suppresses inwardly rectifying K+ channels in aortic endothelium in vitro and in vivo. *Circulation Research*, *98*(8), 1064–1071.

Fang, Y., Schram, G., Romanenko, V. G., Shi, C., Conti, L., Vandenberg, C. A., et al. (2005). Functional expression of Kir2.x in human aortic endothelial cells: The dominant role of Kir2.2. *American Journal of Physiology Cell Physiology*, *289*(5), C1134–C1144.

Fantini, J., & Barrantes, F. J. (2013). How cholesterol interacts with membrane proteins: An exploration of cholesterol-binding sites including CRAC, CARC and tilted domains. *Frontiers in Physiology*, *4*, 31.

Fisslthaler, B., Dimmeler, S., Hermann, C., Busse, R., & Fleming, I. (2000). Phosphorylation and activation of the endothelial nitric oxide synthase by fluid shear stress. *Acta Physiologica Scandinavica*, *168*(1), 81–88.

Franco, C. A., Jones, M. L., Bernabeu, M. O., Geudens, I., Mathivet, T., Rosa, A., et al. (2015). Dynamic endothelial cell rearrangements drive developmental vessel regression. *PLoS Biology*, *13*(4), e1002125.

Fürst, O., Nichols, C. G., Lamoureux, G., & D'Avanzo, N. (2014). Identification of a cholesterol-binding pocket in inward rectifier k(+) (kir) channels. *Biophysical Journal*, *107*, 2786–2796.

Gerhold, K. A., & Schwartz, M. A. (2016). Ion channels in endothelial responses to fluid shear stress. *Physiology (Bethesda)*, *31*(5), 359–369.

Hahn, C., & Schwartz, M. A. (2009). Mechanotransduction in vascular physiology and atherogenesis. *Nature Reviews Molecular Cell Biology*, *10*(1), 53–62.

Harraz, O. F., Longden, T. A., Hill-Eubanks, D., & Nelson, M. T. (2018). PIP2 depletion promotes TRPV4 channel activity in mouse brain capillary endothelial cells. *eLife*, *7*, e38689.

Henrion, D., Terzi, F., Matrougui, K., Duriez, M., Boulanger, C. M., Colucci-Guyon, E., et al. (1997). Impaired flow-induced dilation in mesenteric resistance arteries from mice lacking vimentin. *The Journal of Clinical Investigation*, *100*(11), 2909–2914.

Hibino, H., Inanobe, A., Furutani, K., Murakami, S., Findlay, I., & Kurachi, Y. (2010). Inwardly rectifying potassium channels: Their structure, function, and physiological roles. *Physiological Reviews*, *90*(1), 291–366.

Hoger, J. H., Ilyin, V. I., Forsyth, S., & Hoger, A. (2002). Shear stress regulates the endothelial Kir2.1 ion channel. *PNAS*, *99*(11), 7780–7785.

Isogai, S., Lawson, N. D., Torrealday, S., Horiguchi, M., & Weinstein, B. M. (2003). Angiogenic network formation in the developing vertebrate trunk. *Development*, *130*(21), 5281–5290.

Jackson, W. F. (2005). Potassium channels in the peripheral microcirculation. *Microcirculation*, *12*, 113–127.

Jackson, W. F. (2017). Boosting the signal: Endothelial inward rectifier K(+) channels. *Microcirculation (New York, N.Y.: 1994)*, *24*(3), 24. https://doi.org/10.1111/micc.12319.

Jantzi, M. C., Brett, S. E., Jackson, W. F., Corteling, R., Vigmond, E. J., & Welsh, D. G. (2006). Inward rectifying potassium channels facilitate cell-to-cell communication in hamster retractor muscle feed arteries. *American Journal of Physiology-Heart and Circulatory Physiology*, *291*(3), H1319–H1328.

Ku, D. N., Giddens, D. P., Zarins, C. K., & Glagov, S. (1985). Pulsatile flow and atherosclerosis in the human carotid bifurcation. Positive correlation between plaque location and low oscillating shear stress. *Arteriosclerosis*, *5*(3), 293–302.

Kubo, Y., Adelman, J. P., Clapham, D. E., Jan, L. Y., Karschin, A., Kurachi, Y., et al. (2005). International Union of Pharmacology. LIV. Nomenclature and molecular relationships of inwardly rectifying potassium channels. *Pharmacological Reviews*, *57*(4), 509–526.

Kuo, L., Chilian, W. M., & Davis, M. J. (1991). Interaction of pressure- and flow-induced responses in porcine coronary resistance vessels. *The American Journal of Physiology*, *261*(6 Pt. 2), H1706–H1715.

Levitan, I. (2009). Cholesterol and Kir channels. *IUBMB Life*, *61*, 781–790.

Levitan, I., Fang, Y., Rosenhouse-Dantsker, A., & Romanenko, V. (2010). Cholesterol and ion channels. In J. R. Harris (Ed.), *Cholesterol binding and cholesterol transport proteins: Structure and function in health and disease* (pp. 509–549). Springer Science. *51*.

Levitan, I., Helmke, B. P., & Davies, P. F. (2000). A chamber to permit invasive manipulation of adherent cells in laminar flow with minimal disturbance of the flow field. *Annals of Biomedical Engineering*, *28*, 1184–1193.

Levitan, I., Singh, D. K., & Rosenhouse-Dantsker, A. (2014). Cholesterol binding to ion channels. *Frontiers in Physiology*, *5*, 65.

Li, J., Hou, B., Tumova, S., Muraki, K., Bruns, A., Ludlow, M. J., et al. (2014). Piezo1 integration of vascular architecture with physiological force. *Nature*, *515*(7526), 279–282.

Liedtke, W., Choe, Y., Martí-Renom, M. A., Bell, A. M., Denis, C. S., Sali, A., et al. (2000). Vanilloid receptor-related osmotically activated channel (VR-OAC), a candidate vertebrate osmoreceptor. *Cell*, *103*(3), 525–535.

Liu, Y., Zhao, H., Li, H., Kalyanaraman, B., Nicolosi, A. C., & Gutterman, D. D. (2003). Mitochondrial sources of H_2O_2 generation play a key role in flow-mediated dilation in human coronary resistance arteries. *Circulation Research*, *93*(6), 573–580.

Longden, T. A., & Nelson, M. T. (2015). Vascular inward rectifier K+ channels as external K+ sensors in the control of cerebral blood flow. *Microcirculation (New York, N.Y.: 1994)*, *22*(3), 183–196.

Lucitti, J. L., Jones, E. A., Huang, C., Chen, J., Fraser, S. E., & Dickinson, M. E. (2007). Vascular remodeling of the mouse yolk sac requires hemodynamic force. *Development*, *134*(18), 3317–3326.

Mendoza, S. A., Fang, J., Gutterman, D. D., Wilcox, D. A., Bubolz, A. H., Li, R., et al. (2010). TRPV4-mediated endothelial Ca2+ influx and vasodilation in response to shear stress. *American Journal of Physiology Heart and Circulatory Physiology*, *298*(2), H466–H476.

Millar, I. D., Wang, S., Brown, P. D., Barrand, M. A., & Hladky, S. B. (2008). Kv1 and Kir2 potassium channels are expressed in rat brain endothelial cells. *Pflügers Archiv*, *456*(2), 379–391.

Miura, H., Wachtel, R. E., Liu, Y., Loberiza, F. R., Jr., Saito, T., Miura, M., et al. (2001). Flow-induced dilation of human coronary arterioles: Important role of Ca2+-activated K+ channels. *Circulation*, *103*(15), 1992–1998.

Mohler Iii, E. R., Fang, Y., Gusic Shaffer, R., Moore, J., Wilensky, R. L., Parmacek, M., et al. (2007). Hypercholesterolemia suppresses Kir channels in porcine bone marrow progenitor cells in vivo. *Biochemical and Biophysical Research Communications*, *358*(1), 317–324.

Nam, D., Ni, C. W., Rezvan, A., Suo, J., Budzyn, K., Llanos, A., et al. (2009). Partial carotid ligation is a model of acutely induced disturbed flow, leading to rapid endothelial dysfunction and atherosclerosis. *American Journal of Physiology Heart and Circulatory Physiology*, *297*(4), H1535–H1543.

Nilius, B., & Droogmans, G. (2001). Ion channels and their functional role in vascular endothelium. *Physiological Reviews*, *81*, 1415–1459.

Olesen, S.-P., Clapham, D. E., & Davies, P. F. (1988). Hemodynamic shear stress activates a K+ current in vascular endothelial cells. *Nature*, *331*(January 14), 168–170.

Picazo-Juarez, G., Romero-Suarez, S., Nieto-Posadas, A., Llorente, I., Jara-Oseguera, A., Briggs, M., et al. (2011). Identification of a binding motif in the S5 helix that confers cholesterol-sensitivity to TRPV1. *Journal of Biological Chemistry*, *286*(28), 24966–24976.

Praetorius, H. A. (2015). The primary cilium as sensor of fluid flow: New building blocks to the model. A review in the theme: Cell signaling: Proteins, pathways and mechanisms. *American Journal of Physiology-Cell Physiology*, *308*(3), C198–C208.

Ranade, S. S., Qiu, Z., Woo, S. H., Hur, S. S., Murthy, S. E., Cahalan, S. M., et al. (2014). Piezo1, a mechanically activated ion channel, is required for vascular development in mice. *Proceedings of the National Academy of Sciences of the United States of America*, *111*(28), 10347–10352.

Ranade, S. S., Woo, S.-H., Dubin, A. E., Moshourab, R. A., Wetzel, C., Petrus, M., et al. (2014). Piezo2 is the major transducer of mechanical forces for touch sensation in mice. *Nature*, *516*(7529), 121–125.

Reitsma, S., Slaaf, D. W., Vink, H., van Zandvoort, M. A., & oude Egbrink, M. G. (2007). The endothelial glycocalyx: Composition, functions, and visualization. *Pflügers Archiv*, *454*(3), 345–359.

Rode, B., Shi, J., Endesh, N., Drinkhill, M. J., Webster, P. J., Lotteau, S. J., et al. (2017). Piezo1 channels sense whole body physical activity to reset cardiovascular homeostasis and enhance performance. *Nature Communications*, *8*(1), 350.

Rohacs, T., Chen, J., Prestwich, G. D., & Logothetis, D. E. (1999). Distinct specificities of inwardly rectifying K+ channels for phosphoinositides. *The Journal of Biological Chemistry*, *274*(51), 36065–36072.

Romanenko, V. G., Rothblat, G. H., & Levitan, I. (2002). Modulation of endothelial inward rectifier K+ current by optical isomers of cholesterol. *Biophysical Journal*, *83*, 3211–3222.

Rosenhouse-Dantsker, A. (2017). Insights into the molecular requirements for cholesterol binding to ion channels. *Current Topics in Membranes*, *80*, 187–208.

Rosenhouse-Dantsker, A., Logothetis, D. E., & Levitan, I. (2011). Cholesterol sensitivity of KIR2.1 is controlled by a belt of residues around the cytosolic pore. *Biophysical Journal*, *100*(2), 381.

Rosenhouse-Dantsker, A., Mehta, D., & Levitan, I. (2012). Regulation of ion channels by membrane lipids. *Comprehensive Phsyiology*, *2*, 31–68.

Rosenhouse-Dantsker, A., Noskov, S., Durdagi, S., Logothetis, D. E., & Levitan, I. (2013). Identification of novel cholesterol-binding regions in Kir2 channels. *Journal of Biological Chemistry*, *288*(43), 31154–31164.

Rosenhouse-Dantsker, A., Noskov, S., Han, H., Adney, S. K., Tang, Q.-Y., RodrÃguez-Menchaca, A. A., et al. (2012). Distant cytosolic residues mediate a two-way molecular switch that controls the modulation of inwardly rectifying potassium (Kir) channels by cholesterol and phosphatidylinositol 4,5-bisphosphate (PI(4,5)P2). *Journal of Biological Chemistry*, *287*(48), 40266–40278.

Sancho, M., Fabris, S., Hald, B. O., Brett, S. E., Sandow, S. L., Poepping, T. L., et al. (2019). Membrane lipid-KIR2.x channel interactions enable hemodynamic sensing in cerebral arteries. *Arteriosclerosis, Thrombosis, and Vascular Biology*, *39*(6), 1072–1087.

Simons, K., & Ikonen, E. (1997). Functional rafts in cell membranes. *Nature*, *387*, 569–572.

Simons, K., & Ikonen, E. (2000). How cells handle cholesterol. *Science*, *290*, 1721–1726.

Singh, D. K., Rosenhouse-Dantsker, A., Nichols, C. G., Enkvetchakul, D., & Levitan, I. (2009). Direct regulation of prokaryotic Kir channel by cholesterol. *Journal of Biological Chemistry*, *284*(44), 30727–30736.

Singh, D. K., Shentu, T.-P., Enkvetchakul, D., & Levitan, I. (2011). Cholesterol regulates prokaryotic Kir channel by direct binding to channel protein. *Biochimica et Biophysica Acta (BBA)—Biomembranes*, *1808*(10), 2527.

Smith, P. D., Brett, S. E., Luykenaar, K. D., Sandow, S. L., Marrelli, S. P., Vigmond, E. J., et al. (2008). KIR channels function as electrical amplifiers in rat vascular smooth muscle. *The Journal of Physiology*, *586*(4), 1147–1160.

Sonkusare, S. K., Bonev, A. D., Ledoux, J., Liedtke, W., Kotlikoff, M. I., Heppner, T. J., et al. (2012). Elementary Ca^{2+} signals through endothelial TRPV4 channels regulate vascular function. *Science*, *336*(6081), 597–601.

Sonkusare, S. K., Dalsgaard, T., Bonev, A. D., & Nelson, M. T. (2016). Inward rectifier potassium (Kir2.1) channels as end-stage boosters of endothelium-dependent vasodilators. *The Journal of Physiology*, *594*(12), 3271–3285.

Spier, S. A., Delp, M. D., Stallone, J. N., Dominguez, J. M., 2nd, & Muller-Delp, J. M. (2007). Exercise training enhances flow-induced vasodilation in skeletal muscle resistance arteries of aged rats: Role of PGI2 and nitric oxide. *American Journal of Physiology Heart and Circulatory Physiology*, *292*(6), H3119–H3127.

Strotmann, R., Harteneck, C., Nunnenmacher, K., Schultz, G., & Plant, T. D. (2000). OTRPC4, a nonselective cation channel that confers sensitivity to extracellular osmolarity. *Nature Cell Biology*, *2*(10), 695–702.

Takano, H., Dora, K. A., Spitaler, M. M., & Garland, C. J. (2004). Spreading dilatation in rat mesenteric arteries associated with calcium-independent endothelial cell hyperpolarization. *The Journal of Physiology*, *556*(Pt. 3), 887–903.

Tarbell, J. M., & Ebong, E. E. (2008). The endothelial glycocalyx: A mechano-sensor and -transducer. *Science Signaling*, *1*(40) pt8.

Tarbell, J. M., Shi, Z.-D., Dunn, J., & Jo, H. (2014). Fluid mechanics, arterial disease, and gene expression. *Annual Review of Fluid Mechanics*, *46*, 591–614.

Tulis, D. A., Unthank, J. L., & Prewitt, R. L. (1998). Flow-induced arterial remodeling in rat mesenteric vasculature. *The American Journal of Physiology*, *274*(3), H874–H882.

Tzima, E., Irani-Tehrani, M., Kiosses, W. B., Dejana, E., Schultz, D. A., Engelhardt, B., et al. (2005). A mechanosensory complex that mediates the endothelial cell response to fluid shear stress. *Nature*, *437*(7057), 426–431.

Vriens, J., Watanabe, H., Janssens, A., Droogmans, G., Voets, T., & Nilius, B. (2004). Cell swelling, heat, and chemical agonists use distinct pathways for the activation of the cation channel TRPV4. *Proceedings of the National Academy of Sciences of the United States of America*, *101*(1), 396–401.

Wang, K. C., Yeh, Y. T., Nguyen, P., Limqueco, E., Lopez, J., Thorossian, S., et al. (2016). Flow-dependent YAP/TAZ activities regulate endothelial phenotypes and atherosclerosis. *Proceedings of the National Academy of Sciences of the United States of America, 113*(41), 11525–11530.

Watanabe, S., Yashiro, Y., Mizuno, R., & Ohhashi, T. (2005). Involvement of NO and EDHF in flow-induced vasodilation in isolated hamster cremasteric arterioles. *Journal of Vascular Research, 42*(2), 137–147.

Wragg, J. W., Durant, S., McGettrick, H. M., Sample, K. M., Egginton, S., & Bicknell, R. (2014). Shear stress regulated gene expression and angiogenesis in vascular endothelium. *Microcirculation, 21*(4), 290–300.

Wu, C., Huang, R.-T., Kuo, C.-H., Kumar, S., Kim, C. W., Lin, Y.-C., et al. (2015). Mechanosensitive PPAP2B regulates endothelial responses to atherorelevant hemodynamic forces. *Circulation Research, 117*(4), e41–e53.

Yeagle, P. L. (1985). Cholesterol and the cell membrane. *Biochimica et Biophysica Acta, 822*, 267–287.

Yeagle, P. L. (1991). Modulation of membrane function by cholesterol. *Biochimie, 73*, 1303.

Zaritsky, J. J., Eckman, D. M., Wellman, G. C., Nelson, M. T., & Schwarz, T. L. (2000). Targeted disruption of Kir2.1 and Kir2.2 genes reveals the essential role of the inwardly rectifying K + current in K +-mediated vasodilation. *Circulation Research, 87*(2), 160–166.

Zhang, H., He, C., Yan, X., Mirshahi, T., & Logothetis, D. E. (1999). Activation of inwardly rectifying K+ channels by distinct Ptdlns(4,5)P2 interactions. *Nature Cell Biology, 1*, 183–188.

Zhou, J., Li, Y. S., & Chien, S. (2014). Shear stress-initiated signaling and its regulation of endothelial function. *Arteriosclerosis, Thrombosis, and Vascular Biology, 34*(10), 2191–2198.

CHAPTER FOUR

Endothelial TRPV4 channels and vasodilator reactivity

Yen-Lin Chen[a], Swapnil K. Sonkusare[a,b,*]

[a]Robert M. Berne Cardiovascular Research Center, University of Virginia-School of Medicine, Charlottesville, VA, United States

[b]Department of Molecular Physiology and Biological Physics, University of Virginia-School of Medicine, Charlottesville, VA, United States

[*]Corresponding author: e-mail address: sks2n@virginia.edu

Contents

Abstract

Transient receptor potential vanilloid 4 (TRPV4) ion channels on the endothelial cell membrane are widely regarded as a crucial Ca^{2+} influx pathway that promotes endothelium-dependent vasodilation. The downstream vasodilatory targets of endothelial TRPV4

Current Topics in Membranes, Volume 85
ISSN 1063-5823
https://doi.org/10.1016/bs.ctm.2020.01.007

channels vary among different vascular beds, potentially contributing to endothelial cell heterogeneity. Although numerous studies have examined the role of endothelial TRPV4 channels using specific pharmacological tools over the past decade, their physiological significance remains unclear, mainly due to a lack of endothelium-specific knockouts. Moreover, the loss of endothelium-dependent vasodilation is a significant contributor to vascular dysfunction in cardiovascular disease. The activity of endothelial TRPV4 channels is impaired in cardiovascular disease; therefore, strategies targeting the mechanisms that reduce endothelial TRPV4 channel activity may restore vascular function and provide therapeutic benefit. In this chapter, we discuss endothelial TRPV4 channel-dependent signaling mechanisms, the heterogeneity in endogenous activators and targets of endothelial TRPV4 channels, and the role of endothelial TRPV4 channels in the pathogenesis of cardiovascular diseases. We also discuss potentially interesting future research directions that may provide novel insights into the physiological and pathological roles of endothelial TRPV4 channels.

1. Introduction

Transient receptor potential (TRP) channels have emerged as important regulators of arterial diameter. Mammalian TRP channels are divided into six subfamilies based on their amino acid sequence homology—canonical (TRPC), vanilloid (TRPV), melastatin (TRPM), ankyrin (TRPA), mucolipin (TRPML), and polycystin (TRPP) (Earley & Brayden, 2015). Almost all the TRP channels are permeable to Ca^{2+} and influence Ca^{2+} signaling mechanisms in vascular endothelial cells (ECs) and smooth muscle cells (SMCs) (Earley & Brayden, 2015). Several TRP channels have been shown to regulate endothelial Ca^{2+}, although direct recordings of Ca^{2+} influx through the channels have only been shown for TRPA1, TRPV3, and TRPV4 channels. Studies over the past decade firmly establish a crucial role for TRPV4 channels in controlling vascular reactivity (Ottolini, Hong, & Sonkusare, 2019). In particular, EC (Sonkusare et al., 2012, 2014) and SMC (Earley, Heppner, Nelson, & Brayden, 2005; Tajada et al., 2017) TRPV4 channels are widely regarded as promoters of vasodilation, a phenomenon that is directly associated with a lowering of vascular resistance. However, global TRPV4 channel knockout mice do not show a distinct vascular phenotype under resting conditions, possibly due to compensatory changes or the absence of the channel from multiple cell types (Hong, Cope, Marziano, & Sonkusare, 2016b; Zhang et al., 2009). In this regard, functional TRPV4 channels are also known to be expressed in astrocytes, neurons, ependymal cells of the choroid plexus, kidney epithelium, keratinocytes, adipocytes, inner ear, bladder urothelium, cardiac myocytes and fibroblast, skeletal myocytes, and immune cells (Jones et al., 2019;

Shibasaki, Ikenaka, Tamalu, Tominaga, & Ishizaki, 2014; White et al., 2016). Thus, although numerous studies support a functional role for endothelial TRPV4 channels in causing vasodilation, a definitive answer to whether endothelial TRPV4 channel-mediated vasodilation is physiologically relevant awaits the development of endothelium-specific TRPV4 knockout animal models.

Numerous studies have provided crucial insights into the biophysical properties of TRPV4 channel. TRPV4 channel has been described to have a permeability ratio P_{Ca}/P_{Na} of ~6, and single-channel conductance of ~90 pS (Clapham, Montell, Schultz, & Julius, 2003). A single TRPV4 channel consists of four subunits, each with six transmembrane segments (S1–S6). Among the six transmembrane segments, S5, S6, and the interconnecting loop comprise the central cation-permeable pore, and S3 and S4 appear to form an agonist-binding pocket that influences channel gating (Nilius & Voets, 2013). The cytoplasmic amino- and carboxy-terminals of TRPV4 channels contain various functionally important domains, including an ankyrin repeat domain, a feature found in all channels of TRPC and TRPV subfamilies, a proline-rich region at the amino terminus, and a TRP box at the carboxy-terminal that is a binding site for calmodulin, actin, and tubulin. These functional domains enable protein–protein interactions that regulate channel assembly, channel activity and channel interaction with the cytoskeleton (Harteneck & Schultz, 2007; Nilius & Voets, 2013; White et al., 2016). Recently illustrated single-particle cryo-electron microscopy (cryo-EM) structure of *Xenopus tropicalis* TRPV4 channel under near-atomic resolution revealed unique interactions of transmembrane segments with the pore-forming domain. Moreover, the inner helices formed an intracellular gate in the ion-conduction pore but lacked an extracellular gate in the selectivity filter that was seen in other TRPV family members, possibly explaining the less selective nature of the channel compared to other TRPV channels. Anomalous X-ray diffraction analyses also identified a single ion-binding site in an unusually wide selectivity filter (Deng et al., 2018). Thus, unique structural features of TRPV4 channels may underlie their distinct gating mechanisms and selectivity features when compared to other ion channels of TRP family.

2. Methodologies for studying endothelial TRPV4 channel activity

Different methods have been used to study the activity of TRPV4 channels in native ECs. Direct studies of TRPV4 channel activity include the measurements of ionic currents through single TRPV4 channels or whole-cell TRPV4 channel in response to TRPV4 channel activators, and Ca^{2+} influx signals through individual TRPV4 channels on EC membranes. As an indirect way of estimating TRPV4 channel activity, numerous studies have also recorded whole-cell increases in EC Ca^{2+} in response to TRPV4 channel activation. The Ca^{2+} conductance of TRPV4 channels is approximately 100 times higher than that of L-type Ca^{2+} channel (Mercado et al., 2014). The large Ca^{2+} conductance of the channel makes it possible to record Ca^{2+} influx through individual TRPV4 channels using spinning disk confocal microscopy or total internal reflection fluorescence imaging (Sonkusare et al., 2012; Sullivan, Francis, Pitts, Taylor, & Earley, 2012). This technique has been termed "optical patch clamp" to indicate the optical measurements of single-channel activity (Demuro & Parker, 2005; Mercado et al., 2014; Sonkusare et al., 2014). The real-time recordings of Ca^{2+} influx, as opposed to whole-cell patch clamp, allow the use of intact tissue under physiological conditions, where cell–cell contacts and intercellular communications are maintained. Optical patch clamp also allows the determination of the spatial location of functional TRPV4 channels across the EC membrane. Ca^{2+} influx signals through endothelial TRPV4 channels were termed "TRPV4 sparklets" to imply their unique kinetic properties and single-channel behavior (Sonkusare et al., 2012). While TRPV4 channels are known to conduct cations other than Ca^{2+}, TRPV4 channel-mediated functional effects have mostly been attributed to Ca^{2+} influx through the channel (Bagher et al., 2012; Marziano et al., 2017; Mendoza et al., 2010; Sonkusare et al., 2012; Wu et al., 2009; Yin et al., 2008). Depending on the methodology used, TRPV4 channels have also been shown to increase whole-cell Ca^{2+} levels, an effect that is absent in $TRPV4^{-/-}$ mice (White et al., 2016). Whole-cell Ca^{2+} is an integral of individual Ca^{2+} signals. Therefore, TRPV4 channel-induced increases in whole-cell Ca^{2+} may represent an additive or synergistic interaction among Ca^{2+} signals mediated by TRPV4 channels and other ion channels at the cell membrane or endoplasmic reticulum membrane. The measurements of TRPV4 sparklets were performed in the presence of a sarco-endoplasmic reticulum ATPase inhibitor or phospholipase C (PLC) inhibitor to eliminate the interference from intracellular Ca^{2+} release signals (Sonkusare et al., 2012). In ECs, activation of TRPV4 channels induces intermediate (IK) and small (SK) conductance Ca^{2+}-sensitive K^+ channel currents, which can interfere with the measurement of ionic currents through TRPV4 channels. As an alternative, TRPV4 channel currents can be recorded in the presence of non-selective TRPV4 channel antagonist ruthenium red

(Sonkusare et al., 2012). Ruthenium red blocks TRPV4 channel at negative voltages, thereby inhibiting Ca^{2+} influx, and unblocks it at positive voltages so that outward K^+ currents through the channel can be recorded at positive voltages without interference from Ca^{2+}-activated IK/SK currents.

3. Signaling pathways for endothelial TRPV4 channel-mediated vasodilation

ECs are uniquely situated to sense the mechanical and humoral stimuli and convert them into vasodilation by increasing intracellular Ca^{2+}. In resistance-sized arteries, ECs send out projections to SMCs, called myoendothelial projections (MEPs), which electrically connect the two cell types via myoendothelial gap junctions (MEGJs). An increase in EC Ca^{2+} activates one or more of the following targets: (1) endothelial nitric oxide synthase (eNOS) resulting in the formation of NO. NO diffuses to SMC layer, stimulates cyclic GMP (cGMP)-protein kinase G (PKG) signaling in SMC, which activates myosin light chain phosphatase (MLCP), resulting in a decrease in contractile force and causing vasodilation (Davis, Hill, & Kuo, 2008); (2) IK and SK channels, thereby causing EC membrane hyperpolarization that is transmitted to SMCs via MEGJs, leading to deactivation of SMC voltage-dependent Ca^{2+} channels (VDCCs) and vasodilation; and (3) release of endothelium-derived hyperpolarizing factors, including but not limited to epoxyeicosatrienoic acids (EETs) and hydrogen peroxide (H_2O_2) (Ellinsworth et al., 2016), which hyperpolarize SMCs and cause vasodilation. Pressure myography studies show that TRPV4-IK/SK channel linkage underlies endothelium-dependent vasodilation in systemic resistance arteries (Fig. 1) (Bagher et al., 2012; Sonkusare et al., 2012). Pressure myography studies in pulmonary resistance arteries show that TRPV4 channel-eNOS linkage is the major endothelial vasodilatory pathway in this vascular bed (Fig. 1) (Marziano et al., 2017). Thus, TRPV4 channels appear to couple with differential downstream targets in different vascular beds, although the precise mechanisms for specific channel-target coupling remain unknown. Localization of TRPV4 channels and IK/SK channels at MEPs appears to be responsible for the selective TRPV4-IK/SK channel coupling in systemic resistance arteries (Sonkusare et al., 2014). However, the mechanisms underlying (1) the lack of TRPV4-eNOS coupling in systemic arteries; (2) preferential TRPV4-eNOS coupling in pulmonary arteries, and (3) the lack of TRPV4-IK/SK channel coupling in pulmonary arteries remain elusive.

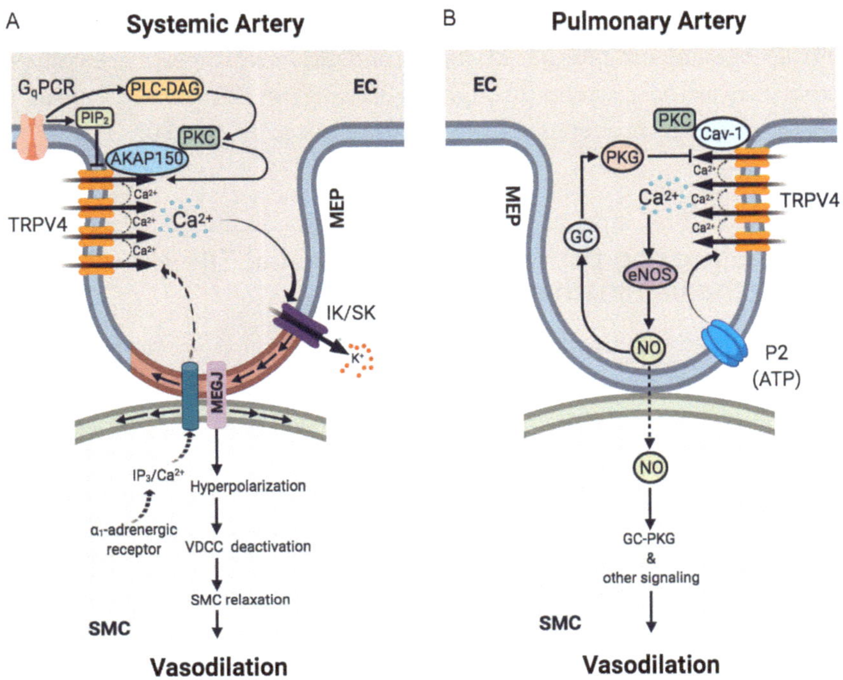

Fig. 1 TRPV4 channel-dependent Ca^{2+} signaling at endothelial projections to smooth muscle cells or myoendothelial projections (MEPs) regulates vasodilation in (A) systemic and (B) pulmonary arteries. (A) In systemic artery, Gq protein-coupled receptor (G_qPCR) stimulation activates phospholipase C (PLC)-diacylglycerol (DAG)-protein kinase C (PKC) signaling. PKC anchoring by AKAP150 enhances the activity and coupled openings of TRPV4 channels. In capillaries, G_qPCR signaling can lower the levels of PIP_2, thereby increasing TRPV4 channel activity. Ca^{2+} influx through TRPV4 channels selectively activates IK and SK channels at MEPs, thereby hyperpolarizing endothelial cell (EC) and smooth muscle cell (SMC) membranes. SMC membrane hyperpolarization leads to deactivation of voltage-dependent Ca^{2+} channels (VDCCs), thus relaxing SMCs and causing vasodilation. In response to SMC α_1-adrenergic stimulation, IP_3/Ca^{2+} from SMCs can cross MEGJs to enter EC layer to limit vasoconstriction via endothelial mechanisms. (B) In pulmonary artery, Ca^{2+} influx through TRPV4 channels selectively activates endothelial nitric oxide synthase (eNOS) to cause NO release and vasodilation. P2 purinergic receptor agonist ATP is an endogenous activator of the TRPV4-eNOS signaling. Endothelial caveolin-1 (Cav-1) enhances endothelial TRPV4 channel activity. Endothelium-derived NO subsequently causes the relaxation of SMCs leading to vasodilation via SMC guanylyl cyclase (GC)—protein kinase G (PKG)-dependent and -independent signaling mechanisms. NO also elicits activation of endothelial GC-PKG signaling, which lowers the coupled openings of TRPV4 channels, thereby limiting Ca^{2+} influx through TRPV4 channels. AKAP150, a-kinase anchoring protein 150; MEP, MEGJ, myoendothelial gap junction; IK/SK, intermediate/small-conductance Ca^{2+}-activated K^+ channels; SMC, smooth muscle cell; IP_3, inositol trisphosphate.

4. Protein regulators of endothelial TRPV4 channel activity

Several proteins are known to influence the activity of endothelial TRPV4 channels, including other ion channels, scaffolding proteins, and cytosolic signaling molecules. A recent study demonstrated the formation of a heteromeric TRPV4-TRPC1-TRPP2 (TRP polycystic 2) channel complex in primary rat mesenteric artery endothelial cells. Flow-induced shear stress is known to induce endothelium-dependent vasodilation that is associated with an increase in endothelial Ca^{2+} (Davies, 1995, 2009; Falcone, Kuo, & Meininger, 1993). Du et al. showed that TRPV4-TRPC1-TRPP2 complex is essential for flow-induced increase in intracellular Ca^{2+} (Du et al., 2014). TRPV4-TRPC1 channel complex was also shown to be responsible for increasing intracellular Ca^{2+} levels in human umbilical vein endothelial cells (Ma et al., 2010). Thus, the formation of heteromultimers with other cation channels is a potentially important mechanism for controlling TRPV4 channel activity. Other proteins that have been shown to alter TRPV4 channel activity include calmodulin, protein kinases, and structural proteins, particularly A-kinase anchoring protein 150 (AKAP150) and caveolin-1 (Cav-1) (Fig. 1).

TRPV4 channels have a calmodulin-binding site on the C-terminal, which enables the biphasic regulation of channel activity by Ca^{2+}. Lower concentrations of intracellular Ca^{2+} potentiate the activity of the channels, whereas higher concentrations inhibit channel activity (Strotmann, Schultz, & Plant, 2003). Similarly, studies in expression systems provided the first evidence that TRPV4 channel activity is potentiated by protein kinase A (PKA) or protein kinase C (PKC)-dependent channel phosphorylation (Cao et al., 2018; Fan, Zhang, & McNaughton, 2009). PKA and PKC appear to act on distinct sites on TRPV4 channels to potentiate its activity. Notably, AKAP150, which can anchor PKC and PKA, was found to promote TRPV4 channel phosphorylation by PKC and PKA in expression systems. In this regard, AKAP150 was localized at MEPs in native endothelium, where it enhanced PKC-dependent activation of TRPV4 channels (Sonkusare et al., 2014). Moreover, the classical endothelial muscarinic receptor agonist acetylcholine activated endothelial AKAP150-PKC-TRPV4 channel signaling at MEPs to cause vasodilation. TRPV4 channels at MEPs open in a coupled manner, indicating simultaneous openings of multiple channels in a cluster. Coupled openings of TRPV4 channels amplify the Ca^{2+} influx through a small number of channels, making it possible for one or two open channels to have a functional effect. Coupled openings of TRPV4 channels required the presence of AKAP150, as indicated by the loss of coupling in the arteries from AKAP150$^{-/-}$ mice (Sonkusare et al., 2014). Interestingly, the coupled openings of TRPV4 channels were not dependent on PKC, indicating that AKAP150 can enhance TRPV4 channel activity in PKC-dependent and -independent manner. Whether the PKC-independent component of AKAP150-regulation of TRPV4 channel activity involves PKA and/or channel regulation via direct binding of AKAP150 to TRPV4 channels remains unknown.

Cav-1 is an essential structural protein of the membrane invaginations called caveolae, which serve to localize key signaling elements (Fridolfsson, Roth, Insel, & Patel, 2014; Minshall, Sessa, Stan, Anderson, & Malik, 2003). There is no direct evidence supporting the regulation of TRPV4 channel activity by Cav-1, although Cav-1 has been shown to associate with TRPV4 channels in cultured ECs. Cav-1 co-immunoprecipitated with TRPV4 channels in cultured ECs, and endothelium-dependent vasodilation in response to TRPV4 channel activator was decreased in conduit arteries from Cav-1$^{-/-}$ mice (Saliez et al., 2008). A recent study showed that endothelial Cav-1 anchoring of PKC increases TRPV4 channel activity and promotes endothelium-dependent vasodilation in pulmonary resistance arteries (Marziano, Hong, Cope, Ottolini, & Sonkusare, 2019). The uterine arteries from rats showed a higher contribution of endothelial TRPV4 channels in controlling vasodilation during pregnancy, an effect that was attributed to a higher expression of TRPV4 channels and their association with caveolae (Senadheera et al., 2013). Thus, multiple studies suggest a potential regulation of TRPV4 channel activity by Cav-1, although further studies are needed to understand the precise linkages in endothelial Cav-1-TRPV4 channel signaling.

Protein kinase G (PKG), a cGMP-dependent protein kinase, has also been shown to regulate endothelial TRPV4 channel activity. PKG inhibited flow-induced TRPV4-TRPC1 channel-dependent Ca^{2+} influx in ECs, and flow-mediated vasodilation (Ma et al., 2010). Also, elevated intravascular pressure in pulmonary microvessels induced endothelial TRPV4 channel-dependent Ca^{2+} influx that increased vascular permeability, but concomitantly activated a negative-feedback mechanism involving cGMP-dependent attenuation of the TRPV4 channel activity (Yin et al., 2008). Recently, Marziano et al. demonstrated that TRPV4 channel-dependent release of

NO leads to vasodilation of pulmonary resistance arteries and activation of endothelial guanylyl cyclase (GC)-PKG-dependent negative-feedback loop that inhibits the coupled openings of TRPV4 channels (Marziano et al., 2017), thereby limiting Ca^{2+} influx through the channel. Whether PKG also requires anchoring by a scaffolding protein to modulate TRPV4 channel activity has not been tested.

5. Physiological stimuli that activate endothelial TRPV4 channels

Recent studies in cell type-specific knockout mice have immensely improved our understanding of the physiological roles of TRPV4 channels from individual cell types. Endothelium-specific TRPV4 knockout mouse has not been generated; thus, the precise physiological functions of endothelial TRPV4 channels remain unclear. Using highly selective and potent pharmacological tools (Table 1) and global knockout mice, numerous physiological stimuli have been shown to alter the activity of endothelial TRPV4 channels (Fig. 2) (Caires et al., 2017; Darby, Grace, Baratchi, & McIntyre, 2016; Meng, Zhao, Xie, Han, & Ji, 2018), as discussed below.

5.1 Mechanical stimuli (flow/shear stress/intravascular pressure)

Nikolaev et al. recently reported that TRPV4 channels are insensitive to mechanical stretch (Nikolaev et al., 2019). While the direct mechanosensing properties of TRPV4 channels remain in doubt, it is clear that mechano-activated signaling pathways can increase the activity of TRPV4 channels. Many studies, using genetic and pharmacological approaches, provide evidence that TRPV4 channels play an essential role in shear stress-induced activation of endothelial Ca^{2+} signaling. Shear stress-induced vasodilation of rat carotid artery and arteria gracilis was inhibited by non-selective TRPV4 channel inhibitor ruthenium red (Kohler et al., 2006). Also, shear stress- and flow/reperfusion-induced vasodilation was significantly attenuated in carotid arteries from TRPV4$^{-/-}$ mice (Hartmannsgruber et al., 2007). Moreover, increased luminal flow dilated carotid artery via EET-endothelium derived hyperpolarization (EDH) pathway, a response that was blunted in the arteries from TRPV4$^{-/-}$ mice. Interestingly, endothelial TRPV4 channels translocated from perinuclear area to the cell membrane in response to increased shear stress (Loot et al., 2008). In mouse mesenteric arteries, shear stress elicited a rapid Ca^{2+} influx through

Table 1 Pharmacological agonists and antagonists of TRPV4 channels.

Agonists	Species	Potency (EC_{50})	References
GSK1016790A	Rat, mouse, human	~10nM (Dunn, Hill-Eubanks, Liedtke, & Nelson, 2013; Thorneloe et al., 2008)	Marziano et al. (2017), Phuong et al. (2017), Sonkusare et al. (2012), Sonkusare et al. (2014), Thorneloe et al. (2012), Zhang, Papadopoulos, and Hamel (2013), Zhang et al. (2018)
RN1747	Mouse	~700nM (Vincent et al., 2009)	Marziano et al. (2017)
4α-phorbol 12,13-didecanoate	Mouse, rat	~400nM (Watanabe, Davis, et al., 2002)	Hartmannsgruber et al. (2007), Kohler et al. (2006), Marrelli, O'Neil, Brown, and Bryan (2007), Vriens et al. (2005), Watanabe, Vriens, et al. (2002)
Antagonists	**Species**	**Potency (IC_{50})**	**References**
GSK2193874	Human, canine, mouse, rat	2–100nM (Thorneloe et al., 2012)	Hong et al. (2018), Huh et al. (2012), Marziano et al. (2017), Thorneloe et al. (2012)
HC067047	Mouse, bovine	50–120nM (Everaerts et al., 2010)	Monaghan et al. (2015), Sonkusare et al. (2014)
RN1734	Rat	2–6μM (Vincent et al., 2009)	Naik and Walker (2018), Seki et al. (2017)
Ruthenium red	Mouse, rat	<200nM (Vincent et al., 2009; Watanabe, Davis, et al., 2002)	Kohler et al. (2006), Loot et al. (2008), Sonkusare et al. (2012), Watanabe, Vriens, et al. (2002)

endothelial TRPV4 channels and caused vasodilation via eNOS and EDH-dependent mechanisms (Mendoza et al., 2010).

Mechanical stretch was also shown to activate endothelial TRPV4 channels to induce capillary cell reorientation via phosphatidylinositol 3-kinase-dependent stimulation of integrin-integrin signaling (Thodeti et al., 2009). Additionally, Bagher et al. demonstrated that a decrease in intravascular pressure increases Ca^{2+} influx through endothelial TRPV4 channels at MEPs in cremaster arterioles (Bagher et al., 2012). Thus, evidence in the literature clearly supports the concept that endothelial TRPV4 channels are activated by mechanical stimuli, including shear stress, and changes in intravascular pressure.

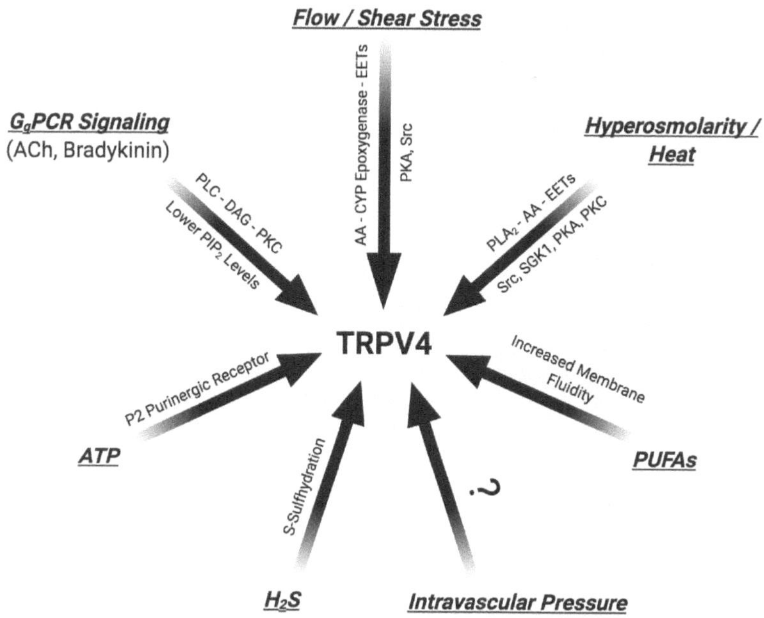

Fig. 2 Physiological stimuli and cellular regulators of endothelial TRPV4 channel activity. TRPV4, transient receptor potential vanilloid 4; G_qPCR, Gq protein-coupled receptors; ATP, adenosine triphosphate; CYP, cytochrome P450; Src, Src family kinases; PKA, protein kinase A; PKC, protein kinase C; SGK-1, serine/threonine kinase-1; EET, epoxyeicosatrienoic acid; PLA_2, phospholipase A2; PLC, phospholipase C; DAG, diacylglycerol; PUFA, ω-polyunsaturated fatty acids; H_2S, hydrogen sulfide; ACh, acetylcholine.

5.2 Changes in osmolarity and temperature

Early studies of TRPV4 channels revealed that the channel could be activated by increases in temperature or by changes in osmolarity that alter the cell size. Strotmann, Harteneck, Nunnenmacher, Schultz, and Plant (2000) demonstrated a rapid increase in TRPV4 channel activity and intracellular Ca^{2+} concentration with small decrease in extracellular osmolarity. Similarly, Vriens et al. showed that hypotonic extracellular solution activates phospholipase A2 (PLA_2)-dependent formation of arachidonic acid and EETs, which are purported TRPV4 channel activators (Vriens et al., 2004). Intriguingly, a heat ramp to ~40 °C increased the inward current through TRPV4 channels and intracellular Ca^{2+} concentration in freshly isolated mouse aorta ECs (Watanabe, Vriens, et al., 2002) supporting the idea that hyperthermia may activate endothelial TRPV4 channels to cause vasodilation and increase blood flow. Whether temperature- or osmolarity-induced activation of endothelial TRPV4 channels contributes to vasodilation has not been elucidated. Studies in other cell types suggest that PKA,

PKC, Src family kinases, and serine/threonine kinase-1 sensitize TRPV4 channels to activation by heat or mechanical stimuli (Fig. 2) (Darby et al., 2016). However, TRPV4 channel sensitization by PKA or Src family kinases has not been verified in ECs.

5.3 G-protein-coupled receptor (GPCR) activation

GPCR agonists are one of the best-known endogenous activators of endothelial TRPV4 channels. Gq protein-coupled receptor (G_qPCR) signaling has been associated with endothelial TRPV4 channel activation via two distinct mechanisms: (1) PLC-diacylglycerol (DAG)-PKC-dependent activation of TRPV4 channels in resistance arteries (Sonkusare et al., 2014); and (2) depletion of a negative regulator of TRPV4 channel activity, phosphatidylinositol 4,5-bisphosphate (PIP_2), in capillary endothelium (Harraz, Longden, Hill-Eubanks, & Nelson, 2018). Sonkusare et al. showed that the classical Gq-coupled muscarinic receptor agonist, acetylcholine, activated TRPV4 channels at MEPs via PLC-DAG-PKC signaling (Sonkusare et al., 2012, 2014). Bradykinin, an inflammatory mediator that acts on Gq-coupled bradykinin receptors, similarly stimulated the activity of TRPV4 channels at MEPs (Sonkusare et al., 2014). Anchoring of PKC by AKAP150 at MEPs was found to be essential for acetylcholine- or bradykinin-induced TRPV4 channel activation. Interestingly, Marziano et al. recently showed that endogenous purinergic receptor agonist adenosine triphosphate (ATP) dilates pulmonary resistance arteries via activation of endothelial TRPV4-eNOS-NO signaling (Marziano et al., 2017). This effect was found to be dependent on P2 purinergic receptors (Marziano et al., 2017). Pulmonary endothelium is known to express $P2Y_1$ and $P2Y_2$ receptor subtypes, both Gq-coupled receptors (Zemskov, Lucas, Verin, & Umapathy, 2011). Thus, it is conceivable that ATP activates TRPV4 channels in pulmonary endothelium via activation of one or both of these receptors.

Patch clamp studies in freshly isolated mouse brain capillary ECs indicated that G_qPCR activation increases TRPV4 channel currents mainly via PLC-mediated hydrolysis of plasma membrane PIP_2, a known inhibitor of TRPV4 channel activity (Harraz et al., 2018). It can be speculated that G_qPCR agonists activate endothelial TRPV4 channels via distinct signaling mechanisms in arterial and capillary ECs. Capillaries are characterized by the absence of SMCs, and therefore, MEPs. Thus, the signaling microdomains that enable G_qPCR-TRPV4 channel signaling in arterial endothelium could be absent from capillary endothelium. The absence of IK/SK channel currents in capillary endothelium further supports the differences in ion channel signaling networks between arterial and capillary endothelium (Longden et al., 2017).

Studies in expression systems show that TRPV4 channels can also be phosphorylated and activated by PKA, a phenomenon that is potentiated by AKAP150 (Cao et al., 2018; Fan et al., 2009). Whether PKA can activate endothelial TRPV4 channels is not known. PKA is activated by Gs-coupled receptor agonists, resulting in adenyl cyclase (AC)-cAMP-PKA signaling. Two main Gs-coupled receptors that are expressed on EC membrane include histamine and prostacyclin receptors (Alfranca, Iniguez, Fresno, & Redondo, 2006; Hekimian, Cote, Van Sande, & Boeynaems, 1992; Heltianu, Simionescu, & Simionescu, 1982). Whether endothelial histamine or prostacyclin receptors can activate TRPV4 channels is not known.

6. Cellular regulators of endothelial TRPV4 channel activity

6.1 Inositol 1,4,5-trisphosphate (IP_3)

A recent study by Takahashi et al. reported an IP_3 binding domain on TRPV4 channels and showed that binding of IP_3 to this domain increases the activity of TRPV4 channels (Takahashi et al., 2014). To this effect, Hong et al. recently showed that IP_3 generated in SMCs by the activation of SMC α_1-adrenergic receptors or thromboxane A2 receptors could travel to ECs via MEGJs, and activate endothelial TRPV4 channels. While the activation of α_1-adrenergic receptors or thromboxane A2 receptors causes vasoconstriction, the IP_3-dependent activation of endothelial TRPV4 channels serves as a negative-feedback mechanism that limits the vasoconstriction (Hong et al., 2018). Garland et al. suggested that increased intracellular Ca^{2+} from SMCs can cross MEGJs and directly activate endothelial IK and SK channels, thereby limiting SMC contraction. It was also proposed that the type of SMC stimulus and the extent of stimulation may determine the negative regulatory mechanism that is activated (Hong et al., 2018). The activity of IP_3Rs is known to be regulated by intracellular Ca^{2+}; indeed, Ca^{2+} is essentially a ligand for IP_3Rs. In this context, Heathcote et al. recently demonstrated that Ca^{2+} influx through endothelial TRPV4 channels activates Ca^{2+}-induced Ca^{2+} release through IP_3R ion channels, thus amplifying the Ca^{2+} response that underlies a large-scale endothelial communication system (Heathcote et al., 2019).

6.2 Hydrogen sulfide (H_2S)

H_2S, a gaseous mediator produced by ECs, was shown to cause vasodilation via endothelial TRPV4 channel activation in phenylephrine-constricted mesenteric arteries (Naik, Osmond, Walker, & Kanagy, 2016). H_2S also increased the activity of TRPV4 channel-dependent Ca^{2+} signals in ECs of pressurized mesenteric arteries. Interestingly, TRPV4 channel-dependent Ca^{2+} signaling resulted in the activation of endothelial big conductance, Ca^{2+}-activated K^+ (BK) channels. TRPV4 channel activator increased the BK channel currents in isolated ECs, and BK channel inhibitor abolished H_2S-induced vasodilation (Naik et al., 2016). While SMC TRPV4 channels have been linked with BK channel activation in the past (Earley et al., 2005), this was the first report on TRPV4-BK channel signaling in ECs. The mechanistic determinants of preferential TRPV4-BK or TRPV4-IK/SK channel signaling linkages will be an exciting topic for future investigations.

6.3 Metabolites of arachidonic acid

Mechanical stimuli are known to activate cytochrome P450 (CYP) epoxygenases and EET formation in ECs. EETs released in response to mechanical stimuli can regulate intracellular Ca^{2+} levels and vascular resistance. Increased flow/shear stress elicited vasodilation via CYP epoxygenase-dependent activation of endothelial TRPV4 channels in mouse carotid arteries (Loot et al., 2008). Marrelli et al. showed that PLA_2, via its main product, arachidonic acid (AA), increases endothelial TRPV4 channel activity and endothelial Ca^{2+} levels in cerebral arteries (Marrelli et al., 2007; Watanabe et al., 2003). Vriens et al. further showed that CYP epoxygenase metabolizes AA to biologically active EETs that increase the activity of endothelial TRPV4 channels (Vriens et al., 2005). Intriguingly, in human coronary arteries, the administration of AA caused EC membrane hyperpolarization and direct activation of TRPV4 channel-dependent Ca^{2+} entry (Cao et al., 2018; Zheng et al., 2013). This phenomenon also required basal phosphorylation of endothelial TRPV4 channels by PKA. Thus, the CYP epoxygenase-AA-EETs signaling axis appears to play an essential role in increasing the activity of endothelial TRPV4 channels and endothelium-dependent vasodilation (Fig. 2). It is important to note that not all EETs can activate TRPV4 channels. While 5,6-EET, 8,9-EET, and 11,12-EET activated TRPV4 channels in expression systems and in ECs, 14,15-EET had no effect on TRPV4 channel activity (Sonkusare et al., 2014; Thorneloe et al., 2012; Vriens et al., 2005; Watanabe et al., 2003; Zheng et al., 2013).

6.4 ω-3 Polyunsaturated fatty acids (PUFAs)

PUFAs from fish oil have been shown to have a protective effect on vascular function. Caires et al. recently demonstrated that ω-3 fatty acid metabolites (eicosapentaenoic acid, EPA; 17,18-epoxyeicosatetraenoic acid, 17,18-EEQ) contribute to the regulation of endothelial TRPV4 channel activity in *C. elegans* and provided evidence that PUFAs enhance TRPV4 channel activity in human microvascular endothelial cells by modifying the membrane fluidity and bending stiffness without altering channel expression or trafficking (Caires et al., 2017). These results reinforce the concept that lipid metabolites are crucial modulators of endothelial TRPV4 channel activity.

7. Heterogeneity in physiological regulators and downstream targets of endothelial TRPV4 channels

TRPV4 channel-dependent endothelial Ca^{2+} signaling mechanisms have been elucidated in several vascular beds. There appears to be considerable heterogeneity among different vascular beds regarding the molecular mechanisms that regulate TRPV4 channel activity. For example, the activity of endothelial TRPV4 channels is inhibited by a NO-GC-PKG-dependent negative-feedback pathway in pulmonary arteries. Still, such a mechanism is not observed in mesenteric arteries (Marziano et al., 2017). AKAP150 increases the activity of TRPV4 channels at MEPs in mesenteric arteries (Sonkusare et al., 2014); however, AKAP150 is not present in pulmonary arteries (Marziano et al., 2017). Physiological activators of TRPV4 channels also appear to be different in different vascular beds. For example, ATP can activate endothelial TRPV4 channels in pulmonary arteries but not in mesenteric arteries (Marziano et al., 2017) (Fig. 1). Moreover, acetylcholine and bradykinin activated endothelial TRPV4 channels in mesenteric arteries but not in pulmonary arteries. Additionally, endothelial TRPV4 channels appear to couple with distinct downstream signaling targets in different vascular beds. In systemic resistance arteries, TRPV4 channels exclusively couple with IK/SK channels (Bagher et al., 2012; Sonkusare et al., 2012), whereas in pulmonary resistance arteries, they couple predominantly with eNOS (Marziano et al., 2017). The molecular mechanisms that are responsible for the heterogeneity in regulatory mechanisms and downstream targets of endothelial TRPV4 channels remain unclear, and further research in this area may provide exciting insights on how differential TRPV4 channel regulation contributes to endothelial cell heterogeneity. Table 2 summarizes TRPV4 channel-dependent vasodilatory mechanisms in different vascular beds.

Table 2 Summary of endothelial TRPV4 channel-dependent vasodilatory mechanisms in different vascular beds.

Vascular bed	Species	Endothelial TRPV4 channel-dependent mechanism	References
Cerebral arteries	Rat, mouse, human	✓ PLA_2 activation increases endothelial Ca^{2+} influx through TRPV4 channels ✓ acetylcholine-dependent vasodilation via TRPV4 channels ✓ G_qPCR signaling activates endothelial TRPV4 channels over Kir2.1 channels	Harraz et al. (2018), Hatano, Suzuki, Itoh, and Muraki (2013), Marrelli et al. (2007), Zhang et al. (2013)
Retinal artery	Bovine	✓ TRPV4 channel control intracellular Ca^{2+} concentration in cultured retinal microvascular ECs	Monaghan et al. (2015)
Carotid artery	Rat, mouse	✓ TRPV4 channel activation triggers NO- and EDH-dependent vasodilation ✓ Dynamic TRPV4/SK channel complexes in caveolae of ECs promote flow-mediated vasodilation ✓ Flow-induced endothelium-dependent vasodilation via activation of cytochrome P450 epoxygenase-EETs-TRPV4 channel signaling	Goedicke-Fritz et al. (2015), Kohler et al. (2006), Loot et al. (2008)
Arteria gracilis	Rat	✓ TRPV4 channel activation triggers NO- and EDH-dependent vasodilation ✓ Acetylcholine activates TRPV4 channel-mediated Ca^{2+} signals ✓ TRPV4-SK/IK channel signaling ✓ Co-localization of Cav-1, TRPV4, and BK channels promotes vasodilation	Kohler et al. (2006), Naik and Walker (2018)
Microvascular ECs	Human	✓ Caveolar association of Cav-1, TRPV4 and SK channels in microvascular ECs ✓ Shear stress induces a partial de novo colocalization of IK channels with Cav-1 and TRPV4 channels ✓ PUFAs enhance TRPV4 channel activity by modifying cell membrane fluidity and bending stiffness	Caires et al. (2017), Goedicke-Fritz et al. (2015)
Feed artery	Human	✓ TRPV4 channels attenuate α1-adrenergic vasoconstriction	Gifford et al. (2014)

Coronary artery	Human	✓ Flow-induced activation of endothelial TRPV4 channels and mitochondrial reactive oxygen species production ✓ Arachidonic acid-activation of TRPV4 channels via PKA-mediated phosphorylation	Bubolz et al. (2012), Cao et al. (2018), Zheng et al. (2013)
Pulmonary artery	Rat, mouse	✓ Endothelium-dependent vasodilation mainly through TRPV4-eNOS signaling ✓ NO release activates endothelial GC-PKG-dependent negative-feedback pathway that limits TRPV4 channel activity ✓ ATP dilates pulmonary arteries mainly via TRPV4-eNOS signaling ✓ Endothelial Cav-1-PKC-TRPV4 channel signaling promotes vasodilation	Marziano et al. (2017), Marziano et al. (2019), Sukumaran et al. (2013)
Mesenteric artery	Mouse, rat	✓ EDH-dependent dilation in resistance arteries, EDH- and NO-dependent dilation in conduit arteries ✓ 11,12-EET induces TRPV4 channel activation and vasodilation ✓ Shear stress dilates via NO- and EDH-dependent mechanisms ✓ PKCα-TRPV4 channel signaling mediates vasodilation to acetylcholine ✓ Cooperative TRPV4 channel openings increase endothelial Ca^{2+} influx and cause vasodilation through IK and SK channels ✓ TRPV4 channels colocalize with Cav-1 ✓ PKC-AKAP150-TRPV4 channel signaling at MEPs underlies vasodilation ✓ Endothelial TRPV4-C1-P2 complex mediates flow-induced Ca^{2+} entry	Adapala et al. (2011), Du et al. (2014), Earley et al. (2009), Ma et al. (2013), Mendoza et al. (2010), Rath et al. (2012), Seki et al. (2017), Sonkusare et al. (2012)
Uterine radial artery	Rat	✓ The number of endothelial TRPV4 channels and caveolae is increased in pregnancy	Senadheera et al. (2013)
Bladder vessel	Rat, mouse	✓ Endothelial TRPV4 channels are expressed	Gevaert et al. (2007)
Cremaster artery	Rat	✓ Endothelial TRPV4 channels increase Ca^{2+} signal frequency at MEPs in response to low intraluminal pressure	Bagher et al. (2012)

8. Alterations in endothelial TRPV4 channel activity under non-physiological or disease conditions

Brachyolmia, a heterogeneous group of skeletal dysplasias, was the first human disease to be associated with abnormal TRPV4 channel activity (Rock et al., 2008). Several autosomal-dominant skeletal dysplasias and neuromuscular disorders have since been associated with altered TRPV4 channel activity (Dai et al., 2010; White et al., 2016). Recent studies in animal models and patients demonstrate a vital role for altered endothelial TRPV4 channel activity in the pathogenesis of cardiovascular diseases. Depending on the disease condition and vascular bed under consideration, the activity of endothelial TRPV4 channels is increased or decreased in vascular disorders.

8.1 Endothelial TRPV4 channel activity in pulmonary vascular disease

Pulmonary hypertension (PH) is characterized by elevated pulmonary arterial pressure (>25 mmHg), pulmonary vasoconstriction, and pulmonary vascular remodeling. Abnormal endothelial TRPV4 channel activity has been implicated in the pathogenesis of PH. Marziano et al. demonstrated that TRPV4 channels are a crucial pathway for endothelial Ca^{2+} influx and vasodilation in small pulmonary arteries that control pulmonary vascular resistance (Marziano et al., 2017). In PH, attenuated Ca^{2+} influx through endothelial TRPV4 channels resulted in impaired vasodilation of pulmonary resistance arteries and an increase in pulmonary arterial pressure. On the contrary, studies by Suresh et al. in a rat model of pulmonary arterial hypertension (PAH) showed that Ca^{2+} influx through TRPV4 channels is increased in cultured microvascular endothelial cells (MVECs) in PAH (Suresh et al., 2018). Moreover, studies in cultured ECs showed that the increase in TRPV4 channel activity was due to an elevated mitochondrial production of hydrogen peroxide in PAH. The differences in the results in small pulmonary arteries and MVECs could be attributed to differential regulation of TRPV4 channel activity in ECs from small arteries vs capillaries or differences in conditions (culture conditions vs native cells). Nevertheless, these results show that PH is associated with abnormal Ca^{2+} influx through endothelial TRPV4 channels. Therefore, the mechanisms that drive the abnormal TRPV4 channel activity in PH could potentially be targeted for therapeutic benefit.

Endothelial TRPV4 channels are critical controllers of lung capillary permeability, and elevated TRPV4 channel activity is associated with increased capillary permeability. In heart failure patients and animal models, TRPV4 channel expression in pulmonary vasculature was increased. Moreover, administering TRPV4 channel inhibitor (GSK2193874) prevented the increased vascular permeability and resultant pulmonary edema in an experimental model of elevated pulmonary venous pressure, and in a mouse model of myocardial infarction (Thorneloe et al., 2012). On the contrary, TRPV4 channel activator (GSK1016790A) decreased right ventricular and systemic pressures and led to a pulmonary circulatory collapse. In another study, the TRPV4 channel activator stimulated IK channels and eNOS, increased vascular permeability, and caused lung hemorrhage (Pankey, Zsombok, Lasker, & Kadowitz, 2014; Wandall-Frostholm et al., 2015). A study in isolated mouse lungs showed CYP epoxygenase-EET-dependent modulation of Ca^{2+} entry through TRPV4 channels in response to high pulmonary vascular pressure (Jian, King, Al-Mehdi, Liedtke, & Townsley, 2008). Yin et al. showed that pressure-induced Ca^{2+} entry through endothelial TRPV4 channels not only increased lung vascular permeability but also concomitantly activated a NO-cGMP-dependent negative-feedback loop, preserving the pulmonary vascular barrier (Yin et al., 2008). These studies prove that endothelial TRPV4 channels play a central role in the regulation of capillary permeability, particularly in the lung, and that higher TRPV4 channel activity increases capillary permeability.

8.2 Endothelial TRPV4 channel activity in hypertension

Systemic hypertension is characterized by elevated systolic/diastolic blood pressure (>130/80 mmHg). Endothelial dysfunction and increased contractility of resistance arteries are hallmarks of hypertension. In some of the earliest studies on the role of TRPV4 channels in cardiovascular disease, Earley et al. found that NOS inhibitor-induced hypertension was augmented in global $TRPV4^{-/-}$ mice. The authors proposed that endothelial TRPV4 channel-dependent vasodilation opposes hypertensive effects of the NOS inhibitor (Earley et al., 2009). Furthermore, in an angiotensin II-infused mouse model of hypertension, endothelial TRPV4 channel activity and acetylcholine stimulation of TRPV4 channels were impaired exclusively at MEPs. This abnormality in TRPV4 channel activity was attributed to lower AKAP150 levels in hypertension, which resulted in a loss of coupled openings of TRPV4 channel and acetylcholine-PKC activation of the channel, thereby lowering Ca^{2+} influx at MEPs (Sonkusare et al., 2014). High-fat diet-induced obesity was also associated with a decrease in TRPV4 channel activity at MEPs (Hong, Cope, Marziano, & Sonkusare, 2016a). It is not known whether the mechanisms for reduced TRPV4 channel activity are similar between angiotensin II-induced hypertension and obesity. Neither angiotensin II-induced hypertension nor obesity was associated with a loss of function of IK/SK channels (Hong et al., 2016a; Sonkusare et al., 2014). However, in mesenteric arteries from stroke-prone spontaneously hypertensive rats, the activity of endothelial TRPV4 and SK channels and EDH-mediated vasodilation were attenuated (Seki et al., 2017).

In a mouse model of salt-induced hypertension, hypercontractility of aorta was linked to TRPV4 channel-mediated increase in endothelial Ca^{2+} and subsequent activation of PLA_2, which increased the formation of prostaglandin $F_{2\alpha}$ ($PGF_{2\alpha}$) by endothelial COX2 (Zhang et al., 2018). Intriguingly, a study by Gao et al. in salt-induced hypertensive rats found enhanced TRPV4 channel expression in mesenteric arteries and higher TRPV4 channel-mediated vasodilation. These results suggested that a compensatory TRPV4 channel activation may prevent impaired vasodilation in salt-induced hypertension (Gao, Sui, Garavito, Worden, & Wang, 2009).

8.3 Endothelial TRPV4 channel activity in other vascular disorders

Endothelial TRPV4 channels also played a vital role in mediating vasodilation of cerebral arteries (Zhang et al., 2013). The activity of endothelial TRPV4 channels in cerebral arteries was decreased in a rat model of cerebrovascular pathology in Alzheimer's disease (Zhang et al., 2013). TRPV4 channel-mediated vasodilation was restored by superoxide dismutase or catalase in this model (Zhang et al., 2013), indicating that redox signaling impairs endothelial TRPV4 channel activity.

Endothelial TRPV4-SK channel signaling was impaired in streptozotocin-induced diabetic rats, possibly contributing to endothelial dysfunction in this model of type 1 diabetes (Ma et al., 2013). Moreover, the expression and activity of TRPV4 channels in retinal vascular ECs were attenuated following the exposure to high glucose (Monaghan et al., 2015), suggesting that hyperglycemia-induced TRPV4 channel dysfunction may contribute to diabetic retinopathy. Rath et al. showed that TRPV4 channels promote NO signaling and contribute to hypoxia-induced vasodilation of systemic arteries under ischemic conditions (Rath et al., 2012). Furthermore, TRPV4 channel inhibition restored high temperature-induced decline in vasoconstriction to phenylephrine in human skeletal muscle feed artery, suggesting that increased body temperature recruits endothelial TRPV4 channels to elicit a sympatholysis-like inhibition of α_1 adrenergic receptor-mediated vasoconstriction (Gifford et al., 2014). These results provide further evidence that TRPV4 channels are important for endothelial homeostasis, and impaired TRPV4 channel activity contributes to endothelial dysfunction in vascular disorders.

9. Summary

TRPV4 channels are a major Ca^{2+} influx pathway that controls endothelial Ca^{2+} signaling mechanisms and promotes endothelium-dependent vasodilation. There appears to be substantial heterogeneity in physiological regulators and downstream targets of TRPV4 channels. Many physiological stimuli (shear stress, stretch, osmolarity, temperature, GPCR signaling, H_2S, lipid metabolites) can activate endothelial TRPV4 channels to increase Ca^{2+} influx in ECs. Among all TRP channels, TRPV4 channels have some of the best (specific and potent) pharmacological activators and inhibitors available for reliably studying the channel activity (Table 1). Patch clamp electrophysiology has traditionally been used to study the ionic currents through TRPV4 channels in native ECs. Recent studies provide evidence that TRPV4 channel activity can be studied by optically recording Ca^{2+} influx through individual TRPV4 channels in the intact endothelium under physiological conditions. Multiple studies suggest that endothelial TRPV4 channel activity is impaired in vascular disorders, including systemic hypertension, diabetes, Alzheimer's disease, obesity, and pulmonary hypertension. The activity of TRPV4 channels in capillary endothelium appears to be increased in diseases characterized by elevated capillary permeability, particularly in the lung.

10. Perspectives and future directions

The role of endothelial TRPV4 channels in endothelium-dependent vasodilation is well-established. However, the physiological significance of TRPV4 channel-mediated vasodilation remains elusive. Global $TRPV4^{-/-}$ mice lack a distinct vascular phenotype, which could be attributed to compensatory upregulation of another Ca^{2+} entry pathway. Future studies in endothelium-specific TRPV4 knockout mice are needed for a more definitive assessment of the physiological roles of endothelial TRPV4 channels. Moreover, endothelium-specific knockout mice for proteins that control TRPV4 channel activity (AKAP150 or Cav-1, for example) will provide further insights into the physiological importance of endothelial TRPV4 channel regulation. If endothelium-dependent vasodilation lowers vascular resistance and blood pressure, then endothelium-specific knockout mice will be expected to show a higher resting blood pressure. The use of endothelium-specific knockout mice may also lead to discoveries on physiological stimuli that activate endothelial TRPV4 channels. The loss of endothelial function is a significant problem in almost all cardiovascular disorders. Understanding the mechanisms that lead to TRPV4 channel dysfunction may result in novel therapeutic strategies to restore endothelial Ca^{2+} signaling and vasodilation in disease conditions. A variety of physiological stimuli, regulators, and downstream targets couple with TRPV4 channels in different vascular beds. Understanding the molecular mechanisms for the differential regulation and targets of TRPV4 channel activity may provide crucial insights into molecular mechanisms that underlie endothelial cell heterogeneity. Different vascular beds are exposed to distinct physiological environments (intravascular pressure, flow, the concentration of dissolved oxygen, humoral mediators). Whether the unique physiological conditions experienced by ECs in a vascular bed determine the vasodilatory signaling linkages in that vascular bed remains unexplored. The majority of signaling elements for TRPV4 channel-dependent vasodilation are known to be concentrated at MEPs. Understanding the mechanisms for translocation of these signaling elements to MEPs will be an exciting topic for future investigations. Finally, substantial sex differences in TRPV4 channel activity have been shown in SMCs. Therefore, potential sex differences in endothelial TRPV4 channel activity and their physiological relevance need further examination.

Acknowledgment

This work was supported by grants from the NIH (HL142808, HL146914, and HL147555) to S.K.S.

References

Adapala, R. K., Talasila, P. K., Bratz, I. N., Zhang, D. X., Suzuki, M., Meszaros, J. G., et al. (2011). PKCalpha mediates acetylcholine-induced activation of TRPV4-dependent calcium influx in endothelial cells. *American Journal of Physiology. Heart and Circulatory Physiology*, *301*(3), H757–H765. https://doi.org/10.1152/ajpheart.00142.2011.

Alfranca, A., Iniguez, M. A., Fresno, M., & Redondo, J. M. (2006). Prostanoid signal transduction and gene expression in the endothelium: Role in cardiovascular diseases. *Cardiovascular Research*, *70*(3), 446–456. https://doi.org/10.1016/j.cardiores.2005.12.020.

Bagher, P., Beleznai, T., Kansui, Y., Mitchell, R., Garland, C. J., & Dora, K. A. (2012). Low intravascular pressure activates endothelial cell TRPV4 channels, local Ca^{2+} events, and IKCa channels, reducing arteriolar tone. *Proceedings of the National Academy of Sciences of the United States of America*, *109*(44), 18174–18179. https://doi.org/10.1073/pnas.1211946109.

Bubolz, A. H., Mendoza, S. A., Zheng, X., Zinkevich, N. S., Li, R., Gutterman, D. D., et al. (2012). Activation of endothelial TRPV4 channels mediates flow-induced dilation in human coronary arterioles: Role of Ca^{2+} entry and mitochondrial ROS signaling. *American Journal of Physiology. Heart and Circulatory Physiology*, *302*(3), H634–H642. https://doi.org/10.1152/ajpheart.00717.2011.

Caires, R., Sierra-Valdez, F. J., Millet, J. R. M., Herwig, J. D., Roan, E., Vasquez, V., et al. (2017). Omega-3 fatty acids modulate TRPV4 function through plasma membrane remodeling. *Cell Reports*, *21*(1), 246–258. https://doi.org/10.1016/j.celrep.2017.09.029.

Cao, S., Anishkin, A., Zinkevich, N. S., Nishijima, Y., Korishettar, A., Wang, Z., et al. (2018). Transient receptor potential vanilloid 4 (TRPV4) activation by arachidonic acid requires protein kinase A-mediated phosphorylation. *The Journal of Biological Chemistry*, *293*(14), 5307–5322. https://doi.org/10.1074/jbc.M117.811075.

Clapham, D. E., Montell, C., Schultz, G., & Julius, D. (2003). International union of pharmacology. XLIII. Compendium of voltage-gated ion channels: Transient receptor potential channels. *Pharmacological Reviews*, *55*(4), 591–596. https://doi.org/10.1124/pr.55.4.6.

Dai, J., Cho, T. J., Unger, S., Lausch, E., Nishimura, G., Kim, O. H., et al. (2010). TRPV4-pathy, a novel channelopathy affecting diverse systems. *Journal of Human Genetics*, *55*(7), 400–402. https://doi.org/10.1038/jhg.2010.37.

Darby, W. G., Grace, M. S., Baratchi, S., & McIntyre, P. (2016). Modulation of TRPV4 by diverse mechanisms. *The International Journal of Biochemistry & Cell Biology*, *78*, 217–228. https://doi.org/10.1016/j.biocel.2016.07.012.

Davies, P. F. (1995). Flow-mediated endothelial mechanotransduction. *Physiological Reviews*, *75*(3), 519–560. https://doi.org/10.1152/physrev.1995.75.3.519.

Davies, P. F. (2009). Hemodynamic shear stress and the endothelium in cardiovascular pathophysiology. *Nature Clinical Practice. Cardiovascular Medicine*, *6*(1), 16–26. https://doi.org/10.1038/ncpcardio1397.

Davis, M. J., Hill, M. A., & Kuo, L. (2008). Local regulation of microvascular perfusion. In R. F. Tuma, W. N. Duran, & K. Ley (Eds.), *Handbook of physiology: The cardiovascular system, microcirculation (section 2)* (2nd ed., pp. 161–284). San Diego, CA: Academic Press.

Demuro, A., & Parker, I. (2005). "Optical patch-clamping": Single-channel recording by imaging Ca^{2+} flux through individual muscle acetylcholine receptor channels. *The Journal of General Physiology*, *126*(3), 179–192. https://doi.org/10.1085/jgp.200509331.

Deng, Z., Paknejad, N., Maksaev, G., Sala-Rabanal, M., Nichols, C. G., Hite, R. K., et al. (2018). Cryo-EM and X-ray structures of TRPV4 reveal insight into ion permeation and gating mechanisms. *Nature Structural & Molecular Biology*, *25*, 252–260. https://doi.org/10.1038/s41594-018-0037-5.

Du, J., Ma, X., Shen, B., Huang, Y., Birnbaumer, L., & Yao, X. (2014). TRPV4, TRPC1, and TRPP2 assemble to form a flow-sensitive heteromeric channel. *The FASEB Journal*, *28*(11), 4677–4685. https://doi.org/10.1096/fj.14-251652.

Dunn, K. M., Hill-Eubanks, D. C., Liedtke, W. B., & Nelson, M. T. (2013). TRPV4 channels stimulate Ca^{2+}-induced Ca^{2+} release in astrocytic endfeet and amplify neurovascular coupling responses. *Proceedings of the National Academy of Sciences of the United States of America*, *110*(15), 6157–6162. https://doi.org/10.1073/pnas.1216514110.

Earley, S., & Brayden, J. E. (2015). Transient receptor potential channels in the vasculature. *Physiological Reviews, 95*(2), 645–690. https://doi.org/10.1152/physrev.00026.2014.
Earley, S., Heppner, T. J., Nelson, M. T., & Brayden, J. E. (2005). TRPV4 forms a novel Ca^{2+} signaling complex with ryanodine receptors and BKCa channels. *Circulation Research, 97*(12), 1270–1279. https://doi.org/10.1161/01.RES.0000194321.60300.d6.
Earley, S., Pauyo, T., Drapp, R., Tavares, M. J., Liedtke, W., & Brayden, J. E. (2009). TRPV4-dependent dilation of peripheral resistance arteries influences arterial pressure. *American Journal of Physiology. Heart and Circulatory Physiology, 297*(3), H1096–H1102. https://doi.org/10.1152/ajpheart.00241.2009.
Ellinsworth, D. C., Sandow, S. L., Shukla, N., Liu, Y., Jeremy, J. Y., & Gutterman, D. D. (2016). Endothelium-derived hyperpolarization and coronary vasodilation: Diverse and integrated roles of epoxyeicosatrienoic acids, hydrogen peroxide, and gap junctions. *Microcirculation, 23*(1), 15–32. https://doi.org/10.1111/micc.12255.
Everaerts, W., Zhen, X., Ghosh, D., Vriens, J., Gevaert, T., Gilbert, J. P., et al. (2010). Inhibition of the cation channel TRPV4 improves bladder function in mice and rats with cyclophosphamide-induced cystitis. *Proceedings of the National Academy of Sciences of the United States of America, 107*(44), 19084–19089. https://doi.org/10.1073/pnas.1005333107.
Falcone, J. C., Kuo, L., & Meininger, G. A. (1993). Endothelial cell calcium increases during flow-induced dilation in isolated arterioles. *The American Journal of Physiology, 264*(2 Pt. 2), H653–H659. https://doi.org/10.1152/ajpheart.1993.264.2.H653.
Fan, H. C., Zhang, X., & McNaughton, P. A. (2009). Activation of the TRPV4 ion channel is enhanced by phosphorylation. *The Journal of Biological Chemistry, 284*(41), 27884–27891. https://doi.org/10.1074/jbc.M109.028803.
Fridolfsson, H. N., Roth, D. M., Insel, P. A., & Patel, H. H. (2014). Regulation of intracellular signaling and function by caveolin. *The FASEB Journal, 28*(9), 3823–3831. https://doi.org/10.1096/fj.14-252320.
Gao, F., Sui, D., Garavito, R. M., Worden, R. M., & Wang, D. H. (2009). Salt intake augments hypotensive effects of transient receptor potential vanilloid 4: Functional significance and implication. *Hypertension, 53*(2), 228–235. https://doi.org/10.1161/HYPERTENSIONAHA.108.117499.
Gevaert, T., Vriens, J., Segal, A., Everaerts, W., Roskams, T., Talavera, K., et al. (2007). Deletion of the transient receptor potential cation channel TRPV4 impairs murine bladder voiding. *The Journal of Clinical Investigation, 117*(11), 3453–3462. https://doi.org/10.1172/JCI31766.
Gifford, J. R., Ives, S. J., Park, S. Y., Andtbacka, R. H., Hyngstrom, J. R., Mueller, M. T., et al. (2014). Alpha1- and alpha2-adrenergic responsiveness in human skeletal muscle feed arteries: The role of TRPV ion channels in heat-induced sympatholysis. *American Journal of Physiology. Heart and Circulatory Physiology, 307*(9), H1288–H1297. https://doi.org/10.1152/ajpheart.00068.2014.
Goedicke-Fritz, S., Kaistha, A., Kacik, M., Markert, S., Hofmeister, A., Busch, C., et al. (2015). Evidence for functional and dynamic microcompartmentation of Cav-1/TRPV4/K(Ca) in caveolae of endothelial cells. *European Journal of Cell Biology, 94*(7–9), 391–400. https://doi.org/10.1016/j.ejcb.2015.06.002.
Harraz, O. F., Longden, T. A., Hill-Eubanks, D., & Nelson, M. T. (2018). PIP2 depletion promotes TRPV4 channel activity in mouse brain capillary endothelial cells. *eLife, 7*, e38689. https://doi.org/10.7554/eLife.38689.
Harteneck, C., & Schultz, G. (2007). TRPV4 and TRPM3 as volume-regulated cation channels. In W. Liedtke & S. Heller (Eds.), *TRP ion channel function in sensory transduction and cellular signaling cascades*. Boca Raton, FL: CRC Press/Taylor & Francis.

Hartmannsgruber, V., Heyken, W. T., Kacik, M., Kaistha, A., Grgic, I., Harteneck, C., et al. (2007). Arterial response to shear stress critically depends on endothelial TRPV4 expression. *PLoS One*, *2*(9), e827. https://doi.org/10.1371/journal.pone.0000827.

Hatano, N., Suzuki, H., Itoh, Y., & Muraki, K. (2013). TRPV4 partially participates in proliferation of human brain capillary endothelial cells. *Life Sciences*, *92*(4–5), 317–324. https://doi.org/10.1016/j.lfs.2013.01.002.

Heathcote, H. R., Lee, M. D., Zhang, X., Saunter, C. D., Wilson, C., & McCarron, J. G. (2019). Endothelial TRPV4 channels modulate vascular tone by Ca^{2+}-induced Ca^{2+} release at inositol 1,4,5-trisphosphate receptors. *British Journal of Pharmacology*, *176*(17), 3297–3317. https://doi.org/10.1111/bph.14762.

Hekimian, G., Cote, S., Van Sande, J., & Boeynaems, J. M. (1992). H2 receptor-mediated responses of aortic endothelial cells to histamine. *The American Journal of Physiology*, *262*(1 Pt. 2), H220–H224. https://doi.org/10.1152/ajpheart.1992.262.1.H220.

Heltianu, C., Simionescu, M., & Simionescu, N. (1982). Histamine receptors of the microvascular endothelium revealed in situ with a histamine-ferritin conjugate: Characteristic high-affinity binding sites in venules. *The Journal of Cell Biology*, *93*(2), 357–364. https://doi.org/10.1083/jcb.93.2.357.

Hong, K., Cope, E. L., DeLalio, L. J., Marziano, C., Isakson, B. E., & Sonkusare, S. K. (2018). TRPV4 (transient receptor potential vanilloid 4) channel-dependent negative feedback mechanism regulates Gq protein-coupled receptor-induced vasoconstriction. *Arteriosclerosis, Thrombosis, and Vascular Biology*, *38*(3), 542–554. https://doi.org/10.1161/ATVBAHA.117.310038.

Hong, K., Cope, E. L., Marziano, C., & Sonkusare, S. K. (2016a). Loss of local Ca^{2+} signaling networks in the endothelium in diet induced obesity. *The FASEB Journal*, *30*(1), 1.

Hong, K., Cope, E. L., Marziano, C., & Sonkusare, S. K. (2016b). Smooth muscle α1-adrenergic receptor activation initiates a negative feedback mechanism that involves Ca^{2+} influx through endothelial TRPV4 channels. *The FASEB Journal*, *31*(1), 1.

Huh, D., Leslie, D. C., Matthews, B. D., Fraser, J. P., Jurek, S., Hamilton, G. A., et al. (2012). A human disease model of drug toxicity-induced pulmonary edema in a lung-on-a-chip microdevice. *Science Translational Medicine*, *4*(159). https://doi.org/10.1126/scitranslmed.3004249. 159ra147.

Jian, M. Y., King, J. A., Al-Mehdi, A. B., Liedtke, W., & Townsley, M. I. (2008). High vascular pressure-induced lung injury requires P450 epoxygenase-dependent activation of TRPV4. *American Journal of Respiratory Cell and Molecular Biology*, *38*(4), 386–392. https://doi.org/10.1165/rcmb.2007-0192OC.

Jones, J. L., Peana, D., Veteto, A. B., Lambert, M. D., Nourian, Z., Karasseva, N. G., et al. (2019). TRPV4 increases cardiomyocyte calcium cycling and contractility yet contributes to damage in the aged heart following hypoosmotic stress. *Cardiovascular Research*, *115*(1), 46–56. https://doi.org/10.1093/cvr/cvy156.

Kohler, R., Heyken, W. T., Heinau, P., Schubert, R., Si, H., Kacik, M., et al. (2006). Evidence for a functional role of endothelial transient receptor potential V4 in shear stress-induced vasodilatation. *Arteriosclerosis, Thrombosis, and Vascular Biology*, *26*(7), 1495–1502. https://doi.org/10.1161/01.ATV.0000225698.36212.6a.

Longden, T. A., Dabertrand, F., Koide, M., Gonzales, A. L., Tykocki, N. R., Brayden, J. E., et al. (2017). Capillary K+-sensing initiates retrograde hyperpolarization to increase local cerebral blood flow. *Nature Neuroscience*, *20*(5), 717–726. https://doi.org/10.1038/nn.4533.

Loot, A. E., Popp, R., Fisslthaler, B., Vriens, J., Nilius, B., & Fleming, I. (2008). Role of cytochrome P450-dependent transient receptor potential V4 activation in flow-induced vasodilatation. *Cardiovascular Research*, *80*(3), 445–452. https://doi.org/10.1093/cvr/cvn207.

Ma, X., Du, J., Zhang, P., Deng, J., Liu, J., Lam, F. F., et al. (2013). Functional role of TRPV4-KCa2.3 signaling in vascular endothelial cells in normal and streptozotocin-induced diabetic rats. *Hypertension*, *62*(1), 134–139. https://doi.org/10.1161/HYPERTENSIONAHA.113.01500.

Ma, X., Qiu, S., Luo, J., Ma, Y., Ngai, C. Y., Shen, B., et al. (2010). Functional role of vanilloid transient receptor potential 4-canonical transient receptor potential 1 complex in flow-induced Ca^{2+} influx. *Arteriosclerosis, Thrombosis, and Vascular Biology*, *30*(4), 851–858. https://doi.org/10.1161/ATVBAHA.109.196584.

Marrelli, S. P., O'Neil, R. G., Brown, R. C., & Bryan, R. M., Jr. (2007). PLA2 and TRPV4 channels regulate endothelial calcium in cerebral arteries. *American Journal of Physiology. Heart and Circulatory Physiology*, *292*(3), H1390–H1397. https://doi.org/10.1152/ajpheart.01006.2006.

Marziano, C., Hong, K., Cope, E. L., Kotlikoff, M. I., Isakson, B. E., & Sonkusare, S. K. (2017). Nitric oxide-dependent feedback loop regulates transient receptor potential vanilloid 4 (TRPV4) channel cooperativity and endothelial function in small pulmonary arteries. *Journal of the American Heart Association*, *6*(12). e007157. https://doi.org/10.1161/JAHA.117.007157.

Marziano, C., Hong, K., Cope, E. L., Ottolini, M., & Sonkusare, S. K. (2019). Peroxynitrite impairs caveolin-1-TRPV4 channel signaling in pulmonary hypertension. *The FASEB Journal*, *33*(1), 1.

Mendoza, S. A., Fang, J., Gutterman, D. D., Wilcox, D. A., Bubolz, A. H., Li, R., et al. (2010). TRPV4-mediated endothelial Ca^{2+} influx and vasodilation in response to shear stress. *American Journal of Physiology. Heart and Circulatory Physiology*, *298*(2), H466–H476. https://doi.org/10.1152/ajpheart.00854.2009.

Meng, G., Zhao, S., Xie, L., Han, Y., & Ji, Y. (2018). Protein S-sulfhydration by hydrogen sulfide in cardiovascular system. *British Journal of Pharmacology*, *175*(8), 1146–1156. https://doi.org/10.1111/bph.13825.

Mercado, J., Baylie, R., Navedo, M. F., Yuan, C., Scott, J. D., Nelson, M. T., et al. (2014). Local control of TRPV4 channels by AKAP150-targeted PKC in arterial smooth muscle. *The Journal of General Physiology*, *143*(5), 559–575. https://doi.org/10.1085/jgp.201311050.

Minshall, R. D., Sessa, W. C., Stan, R. V., Anderson, R. G., & Malik, A. B. (2003). Caveolin regulation of endothelial function. *American Journal of Physiology. Lung Cellular and Molecular Physiology*, *285*(6), L1179–L1183. https://doi.org/10.1152/ajplung.00242.2003.

Monaghan, K., McNaughten, J., McGahon, M. K., Kelly, C., Kyle, D., Yong, P. H., et al. (2015). Hyperglycemia and diabetes downregulate the functional expression of TRPV4 channels in retinal microvascular endothelium. *PLoS One*, *10*(6). e0128359. https://doi.org/10.1371/journal.pone.0128359.

Naik, J. S., Osmond, J. M., Walker, B. R., & Kanagy, N. L. (2016). Hydrogen sulfide-induced vasodilation mediated by endothelial TRPV4 channels. *American Journal of Physiology. Heart and Circulatory Physiology*, *311*(6), H1437–H1444. https://doi.org/10.1152/ajpheart.00465.2016.

Naik, J. S., & Walker, B. R. (2018). Endothelial-dependent dilation following chronic hypoxia involves TRPV4-mediated activation of endothelial BK channels. *Pflügers Archiv*, *470*(4), 633–648. https://doi.org/10.1007/s00424-018-2112-5.

Nikolaev, Y. A., Cox, C. D., Ridone, P., Rohde, P. R., Cordero-Morales, J. F., Vasquez, V., et al. (2019). Mammalian TRP ion channels are insensitive to membrane stretch. *Journal of Cell Science*, *132*(23). https://doi.org/10.1242/jcs.238360.

Nilius, B., & Voets, T. (2013). The puzzle of TRPV4 channelopathies. *EMBO Reports*, *14*, 152–163. https://doi.org/10.1038/embor.2012.219.

Ottolini, M., Hong, K., & Sonkusare, S. K. (2019). Calcium signals that determine vascular resistance. *Wiley Interdisciplinary Reviews. Systems Biology and Medicine, 11*, e1448. https://doi.org/10.1002/wsbm.1448.

Pankey, E. A., Zsombok, A., Lasker, G. F., & Kadowitz, P. J. (2014). Analysis of responses to the TRPV4 agonist GSK1016790A in the pulmonary vascular bed of the intact-chest rat. *American Journal of Physiology. Heart and Circulatory Physiology, 306*(1), H33–H40. https://doi.org/10.1152/ajpheart.00303.2013.

Phuong, T. T. T., Redmon, S. N., Yarishkin, O., Winter, J. M., Li, D. Y., & Krizaj, D. (2017). Calcium influx through TRPV4 channels modulates the adherens contacts between retinal microvascular endothelial cells. *The Journal of Physiology, 595*(22), 6869–6885. https://doi.org/10.1113/JP275052.

Rath, G., Saliez, J., Behets, G., Romero-Perez, M., Leon-Gomez, E., Bouzin, C., et al. (2012). Vascular hypoxic preconditioning relies on TRPV4-dependent calcium influx and proper intercellular gap junctions communication. *Arteriosclerosis, Thrombosis, and Vascular Biology, 32*(9), 2241–2249. https://doi.org/10.1161/ATVBAHA.112.252783.

Rock, M. J., Prenen, J., Funari, V. A., Funari, T. L., Merriman, B., Nelson, S. F., et al. (2008). Gain-of-function mutations in TRPV4 cause autosomal dominant brachyolmia. *Nature Genetics, 40*(8), 999–1003. https://doi.org/10.1038/ng.166.

Saliez, J., Bouzin, C., Rath, G., Ghisdal, P., Desjardins, F., Rezzani, R., et al. (2008). Role of caveolar compartmentation in endothelium-derived hyperpolarizing factor-mediated relaxation: Ca^{2+} signals and gap junction function are regulated by caveolin in endothelial cells. *Circulation, 117*(8), 1065–1074. https://doi.org/10.1161/CIRCULATIONAHA.107.731679.

Seki, T., Goto, K., Kiyohara, K., Kansui, Y., Murakami, N., Haga, Y., et al. (2017). Downregulation of endothelial transient receptor potential vanilloid type 4 channel and small-conductance of Ca^{2+}-activated K+ channels underpins impaired endothelium-dependent hyperpolarization in hypertension. *Hypertension, 69*(1), 143–153. https://doi.org/10.1161/HYPERTENSIONAHA.116.07110.

Senadheera, S., Bertrand, P. P., Grayson, T. H., Leader, L., Murphy, T. V., & Sandow, S. L. (2013). Pregnancy-induced remodelling and enhanced endothelium-derived hyperpolarization-type vasodilator activity in rat uterine radial artery: Transient receptor potential vanilloid type 4 channels, caveolae and myoendothelial gap junctions. *Journal of Anatomy, 223*(6), 677–686. https://doi.org/10.1111/joa.12127.

Shibasaki, K., Ikenaka, K., Tamalu, F., Tominaga, M., & Ishizaki, Y. (2014). A novel subtype of astrocytes expressing TRPV4 (transient receptor potential vanilloid 4) regulates neuronal excitability via release of gliotransmitters. *The Journal of Biological Chemistry, 289*(21), 14470–14480. https://doi.org/10.1074/jbc.M114.557132.

Sonkusare, S. K., Bonev, A. D., Ledoux, J., Liedtke, W., Kotlikoff, M. I., Heppner, T. J., et al. (2012). Elementary Ca^{2+} signals through endothelial TRPV4 channels regulate vascular function. *Science, 336*(6081), 597–601. https://doi.org/10.1126/science.1216283.

Sonkusare, S. K., Dalsgaard, T., Bonev, A. D., Hill-Eubanks, D. C., Kotlikoff, M. I., Scott, J. D., et al. (2014). AKAP150-dependent cooperative TRPV4 channel gating is central to endothelium-dependent vasodilation and is disrupted in hypertension. *Science Signaling*, 7(333), ra66. https://doi.org/10.1126/scisignal.2005052.

Strotmann, R., Harteneck, C., Nunnenmacher, K., Schultz, G., & Plant, T. D. (2000). OTRPC4, a nonselective cation channel that confers sensitivity to extracellular osmolarity. *Nature Cell Biology, 2*(10), 695–702. https://doi.org/10.1038/35036318.

Strotmann, R., Schultz, G., & Plant, T. D. (2003). Ca^{2+}-dependent potentiation of the nonselective cation channel TRPV4 is mediated by a C-terminal calmodulin binding site. *The Journal of Biological Chemistry, 278*(29), 26541–26549. https://doi.org/10.1074/jbc.M302590200.

Sukumaran, S. V., Singh, T. U., Parida, S., Narasimha Reddy Ch, E., Thangamalai, R., Kandasamy, K., et al. (2013). TRPV4 channel activation leads to endothelium-dependent relaxation mediated by nitric oxide and endothelium-derived hyperpolarizing factor in rat pulmonary artery. *Pharmacological Research, 78*, 18–27. https://doi.org/10.1016/j.phrs.2013.09.005.

Sullivan, M. N., Francis, M., Pitts, N. L., Taylor, M. S., & Earley, S. (2012). Optical recording reveals novel properties of GSK1016790A-induced vanilloid transient receptor potential channel TRPV4 activity in primary human endothelial cells. *Molecular Pharmacology, 82*(3), 464–472. https://doi.org/10.1124/mol.112.078584.

Suresh, K., Servinsky, L., Jiang, H., Bigham, Z., Yun, X., Kliment, C., et al. (2018). Reactive oxygen species induced Ca^{2+} influx via TRPV4 and microvascular endothelial dysfunction in the SU5416/hypoxia model of pulmonary arterial hypertension. *American Journal of Physiology. Lung Cellular and Molecular Physiology, 314*(5), L893–L907. https://doi.org/10.1152/ajplung.00430.2017.

Tajada, S., Moreno, C. M., O'Dwyer, S., Woods, S., Sato, D., Navedo, M. F., et al. (2017). Distance constraints on activation of TRPV4 channels by AKAP150-bound PKCalpha in arterial myocytes. *The Journal of General Physiology, 149*(6), 639–659. https://doi.org/10.1085/jgp.201611709.

Takahashi, N., Hamada-Nakahara, S., Itoh, Y., Takemura, K., Shimada, A., Ueda, Y., et al. (2014). TRPV4 channel activity is modulated by direct interaction of the ankyrin domain to PI(4,5)P(2). *Nature Communications, 5*, 4994. https://doi.org/10.1038/ncomms5994.

Thodeti, C. K., Matthews, B., Ravi, A., Mammoto, A., Ghosh, K., Bracha, A. L., et al. (2009). TRPV4 channels mediate cyclic strain-induced endothelial cell reorientation through integrin-to-integrin signaling. *Circulation Research, 104*(9), 1123–1130. https://doi.org/10.1161/CIRCRESAHA.108.192930.

Thorneloe, K. S., Cheung, M., Bao, W., Alsaid, H., Lenhard, S., Jian, M. Y., et al. (2012). An orally active TRPV4 channel blocker prevents and resolves pulmonary edema induced by heart failure. *Science Translational Medicine, 4*(159), 159ra148. https://doi.org/10.1126/scitranslmed.3004276.

Thorneloe, K. S., Sulpizio, A. C., Lin, Z., Figueroa, D. J., Clouse, A. K., McCafferty, G. P., et al. (2008). N-((1S)-1-{[4-((2S)-2-{[(2,4-dichlorophenyl)sulfonyl]amino}-3-hydroxypropanoyl)-1-piperazinyl]carbonyl}-3-methylbutyl)-1-benzothiophene-2-carboxamide (GSK1016790A), a novel and potent transient receptor potential vanilloid 4 channel agonist induces urinary bladder contraction and hyperactivity: Part I. *The Journal of Pharmacology and Experimental Therapeutics, 326*(2), 432–442. https://doi.org/10.1124/jpet.108.139295.

Vincent, F., Acevedo, A., Nguyen, M. T., Dourado, M., DeFalco, J., Gustafson, A., et al. (2009). Identification and characterization of novel TRPV4 modulators. *Biochemical and Biophysical Research Communications, 389*(3), 490–494. https://doi.org/10.1016/j.bbrc.2009.09.007.

Vriens, J., Owsianik, G., Fisslthaler, B., Suzuki, M., Janssens, A., Voets, T., et al. (2005). Modulation of the Ca^{2+} permeable cation channel TRPV4 by cytochrome P450 epoxygenases in vascular endothelium. *Circulation Research, 97*(9), 908–915. https://doi.org/10.1161/01.RES.0000187474.47805.30.

Vriens, J., Watanabe, H., Janssens, A., Droogmans, G., Voets, T., & Nilius, B. (2004). Cell swelling, heat, and chemical agonists use distinct pathways for the activation of the cation channel TRPV4. *Proceedings of the National Academy of Sciences of the United States of America, 101*(1), 396–401. https://doi.org/10.1073/pnas.0303329101.

Wandall-Frostholm, C., Dalsgaard, T., Bajoriunas, V., Olivan-Viguera, A., Sadda, V., Beck, L., et al. (2015). Genetic deficit of KCa3.1 channels protects against pulmonary circulatory collapse induced by TRPV4 channel activation. *British Journal of Pharmacology, 172*(18), 4493–4505. https://doi.org/10.1111/bph.13234.

Watanabe, H., Davis, J. B., Smart, D., Jerman, J. C., Smith, G. D., Hayes, P., et al. (2002). Activation of TRPV4 channels (hVRL-2/mTRP12) by phorbol derivatives. *The Journal of Biological Chemistry*, *277*(16), 13569–13577. https://doi.org/10.1074/jbc.M200062200.

Watanabe, H., Vriens, J., Prenen, J., Droogmans, G., Voets, T., & Nilius, B. (2003). Anandamide and arachidonic acid use epoxyeicosatrienoic acids to activate TRPV4 channels. *Nature*, *424*(6947), 434–438. https://doi.org/10.1038/nature01807.

Watanabe, H., Vriens, J., Suh, S. H., Benham, C. D., Droogmans, G., & Nilius, B. (2002). Heat-evoked activation of TRPV4 channels in a HEK293 cell expression system and in native mouse aorta endothelial cells. *The Journal of Biological Chemistry*, *277*(49), 47044–47051. https://doi.org/10.1074/jbc.M208277200.

White, J. P., Cibelli, M., Urban, L., Nilius, B., McGeown, J. G., & Nagy, I. (2016). TRPV4: Molecular conductor of a diverse orchestra. *Physiological Reviews*, *96*(3), 911–973. https://doi.org/10.1152/physrev.00016.2015.

Wu, S., Jian, M. Y., Xu, Y. C., Zhou, C., Al-Mehdi, A. B., Liedtke, W., et al. (2009). Ca^{2+} entry via alpha1G and TRPV4 channels differentially regulates surface expression of P-selectin and barrier integrity in pulmonary capillary endothelium. *American Journal of Physiology. Lung Cellular and Molecular Physiology*, *297*(4), L650–L657. https://doi.org/10.1152/ajplung.00015.2009.

Yin, J., Hoffmann, J., Kaestle, S. M., Neye, N., Wang, L., Baeurle, J., et al. (2008). Negative-feedback loop attenuates hydrostatic lung edema via a cGMP-dependent regulation of transient receptor potential vanilloid 4. *Circulation Research*, *102*(8), 966–974. https://doi.org/10.1161/CIRCRESAHA.107.168724.

Zemskov, E., Lucas, R., Verin, A. D., & Umapathy, N. S. (2011). P2Y receptors as regulators of lung endothelial barrier integrity. *Journal of Cardiovascular Disease Research*, *2*(1), 14–22. https://doi.org/10.4103/0975-3583.78582.

Zhang, D. X., Mendoza, S. A., Bubolz, A. H., Mizuno, A., Ge, Z. D., Li, R., et al. (2009). Transient receptor potential vanilloid type 4-deficient mice exhibit impaired endothelium-dependent relaxation induced by acetylcholine in vitro and in vivo. *Hypertension*, *53*(3), 532–538. https://doi.org/10.1161/HYPERTENSIONAHA.108.127100.

Zhang, L., Papadopoulos, P., & Hamel, E. (2013). Endothelial TRPV4 channels mediate dilation of cerebral arteries: Impairment and recovery in cerebrovascular pathologies related to Alzheimer's disease. *British Journal of Pharmacology*, *170*(3), 661–670. https://doi.org/10.1111/bph.12315.

Zhang, P., Sun, C., Li, H., Tang, C., Kan, H., Yang, Z., et al. (2018). TRPV4 (transient receptor potential vanilloid 4) mediates endothelium-dependent contractions in the aortas of hypertensive mice. *Hypertension*, *71*(1), 134–142. https://doi.org/10.1161/HYPERTENSIONAHA.117.09767.

Zheng, X., Zinkevich, N. S., Gebremedhin, D., Gauthier, K. M., Nishijima, Y., Fang, J., et al. (2013). Arachidonic acid-induced dilation in human coronary arterioles: Convergence of signaling mechanisms on endothelial TRPV4-mediated Ca^{2+} entry. *Journal of the American Heart Association*, *2*(3), e000080. https://doi.org/10.1161/JAHA.113.000080.

CHAPTER FIVE

Regulation of vascular tone by transient receptor potential ankyrin 1 channels

Pratish Thakore, Sher Ali, Scott Earley*
Department of Pharmacology, Center for Molecular and Cellular Signaling in the Cardiovascular System, University of Nevada, Reno, School of Medicine, Reno, NV, United States
*Corresponding author: e-mail address: searley@med.unr.edu

Contents

Abstract

The Ca^{2+}-permeable, non-selective cation channel, TRPA1 (transient receptor potential ankyrin 1), is the sole member of the ankyrin TRP subfamily. TRPA1 channels are expressed on the plasma membrane of neurons as well as non-neuronal cell types, such as vascular endothelial cells. TRPA1 is activated by electrophilic compounds, including dietary molecules such as allyl isothiocyanate, a derivative of mustard.

Current Topics in Membranes, Volume 85
ISSN 1063-5823
https://doi.org/10.1016/bs.ctm.2020.01.009

Endogenously, the channel is thought to be activated by reactive oxygen species and their metabolites, such as 4-hydroxynonenal (4-HNE). In the context of the vasculature, activation of TRPA1 channels results in a vasodilatory response mediated by two distinct mechanisms. In the first instance, TRPA1 is expressed in sensory nerves of the vasculature and, upon activation, mediates release of the potent dilator, calcitonin gene-related peptide (CGRP). In the second, work from our laboratory has demonstrated that TRPA1 is expressed in the endothelium of blood vessels exclusively in the cerebral vasculature, where its activation produces a localized Ca^{2+} signal that results in dilation of cerebral arteries. In this chapter, we provide an in-depth overview of the biophysical and pharmacological properties of TRPA1 channels and their importance in regulating vascular tone.

1. Introduction

Transient receptor potential (TRP) channels allow movement of cations across membranes to mediate various cellular functions. TRP channels were discovered during investigations of a spontaneous *Trp*-mutant line of *Drosophila melanogaster* that exhibited a transient response to light (hence the name); as a result, mutant flies could see normally in dim light, but behaved as if blind under bright light (Cosens & Manning, 1969). The genomes of most mammals encode 28 individual TRP subunits, which together constitute the TRP superfamily; in humans, only 27 TRP subunits are expressed (*TRPC2* is a non-functional pseudogene in humans). TRP subunits have been assigned into six subfamilies based on sequence homology. The ankyrin (TRPA) subfamily is the smallest, consisting of only a single member (Story et al., 2003). In this chapter, we provide a review of the current literature on TRPA1 channels that briefly introduces the structural and biophysical properties of the channel and delves deeply into the importance of TRPA1 in regulating vascular function in physiological and pathological settings. Further details regarding the structure, properties, and function of TRPA1 channels in many organ systems can be found in an outstanding recent review by Talavera et al. (2019).

The basic structure and function of TRPA1 channels are evolutionarily conserved (Kang et al., 2010; Story et al., 2003). Despite being among the last TRP channels to be discovered, TRPA1 has received a great deal of interest (Story et al., 2003). The *Trpa1* gene has been cloned from several mammalian species, including human, mouse and rat, as well as non-mammalian species, such as insects, fish, birds, and amphibians. Mammals have one *Trpa1* gene homolog, whereas several non-mammalian species have more than one homolog. For example, *D. melanogaster* has four TRPA homologs. TRPA1 was originally thought to be predominantly expressed in sensory neurons of the dorsal root ganglia, trigeminal ganglia and nodose ganglia, as well as in hair cells (Bautista et al., 2005; Nagata, Duggan, Kumar, & Garcia-Anoveros, 2005; Story et al., 2003). However, TRPA1 has since been identified in non-neuronal tissue, such as alveolar epithelial cells (Nassini et al., 2012), cardiomyocytes (Andrei, Sinharoy, Bratz, & Damron, 2016), cardiac fibroblasts (Oguri et al., 2014), and cerebral artery endothelial cells (Earley, Gonzales, & Crnich, 2009).

2. TRPA1 architecture

The functional TRPA1 ion channel is a homotetrameric protein complex (Cvetkov, Huynh, Cohen, & Moiseenkova-Bell, 2011). The high-resolution (~4 Å) structure of fully assembled TRPA1 channels, revealed by cryo-electron microscopy (cryo-EM) (Fig. 1), shows that each TRPA1 channel subunit consists of six transmembrane domains (TM1-TM6), with a pore-forming loop between TM5 and TM6, and intracellular N- and C-termini (Paulsen, Armache, Gao, Cheng, & Julius, 2015). Assembly of four TRPA1 subunits forms a Ca^{2+} permeable non-selective cation channel. The human TRPA1 monomer (hTRPA1) is a 1119-amino acid protein with a molecular mass of 127.4 kDa (Nilius, Appendino, & Owsianik, 2012; Story et al., 2003). The N-terminus forms the largest part of the protein, accounting for 64% of its total mass, whereas the C-terminus

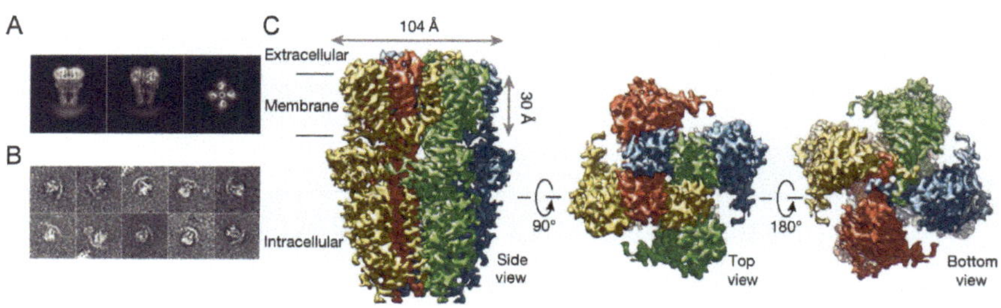

Fig. 1 High resolution cryo-EM structure of TRPA1. (A) Representative cryo-EM 2-dimensional image of TRPA1 (left and middle, side view; right, end-on view). (B) Representative image of negative stain particles. (C) 3-Dimensional density map of TRPA1 with each subunit color coded (left, side view; middle, top view; right, bottom view). *Figure derived from Paulsen, C. E., Armache, J. P., Gao, Y., Cheng, Y., & Julius, D. (2015). Structure of the TRPA1 ion channel suggests regulatory mechanisms.* Nature, *520(7548), 511–517. https://doi.org/10.1038/nature14367.*

only accounts for 14% (Cvetkov et al., 2011). The distinguishing and eponymous feature of TRPA1 is the large number of ankyrin repeat domains (ARDs) within the N-terminus. The human TRPA1 protein contains 16 ARDs—more than any other known protein (Gaudet, 2008; Mosavi, Minor, & Peng, 2002; Story et al., 2003). ARDs are 33-amino acid sequences that may be involved in elasticity, protein-protein interactions, stability, and structural integrity (Lee et al., 2006). Because ankyrin is a linear spring with a tension proportional to its extension, ARDs have a strong tendency to refold (Lee et al., 2006). Each domain forms two short anti-parallel helix-turn-helix structures, followed by a β-hairpin loop (Lee et al., 2006; Sotomayor, Corey, & Schulten, 2005). The N-terminal domain also contains a large number of cysteine residues that can interact intramolecularly or between monomers through formation of disulfide bridges (Cvetkov et al., 2011; Eberhardt et al., 2012; Gaudet, 2008; Wang, Cvetkov, Chance, & Moiseenkova-Bell, 2012). Moreover, conserved cysteine or lysine residues also provide target sites for potent electrophilic TRPA1 activators, which act through covalent modification of the receptor (Macpherson et al., 2007; Takahashi et al., 2011). Cryo-EM images have revealed that some cysteine and lysine residues are also located in the transmembrane core facing the lipid environment, where they might react with lipophilic electrophiles (Paulsen et al., 2015). Zn^{2+}, a potent TRPA1 activator, can bind to cysteine and histidine residues in the C-terminus (Andersson, Gentry, Moss, & Bevan, 2009; Hu, Bandell, Petrus, Zhu, & Patapoutian, 2009). Both N- and C-termini have been suggested to contain binding sites for Ca^{2+}, which can either activate or deactivate TRPA1 (Doerner, Gisselmann, Hatt, & Wetzel, 2007; Jordt et al., 2004; Wang, Chang, Waters, McKemy, & Liman, 2008; Zurborg, Yurgionas, Jira, Caspani, & Heppenstall, 2007).

3. Biophysical properties of the TRPA1 channel

3.1 Permeation and gating characteristics

TRPA1 is a voltage-gated non-selective cation channel that is permeable to both monovalent and divalent ions, including K^+, Na^+ and Ca^{2+}. The relative permeability for Ca^{2+} is higher than that for Na^+ or K^+ (P_{Ca}:P_{Na}:P_K=~7.9:1.0:0.98) (Bobkov, Corey, & Ache, 2011; Nilius et al., 2012; Nilius, Prenen, & Owsianik, 2011), and the Ca^{2+} fraction of the mixed cation current is approximately 20% (Karashima et al., 2010).

The mouth of the pore contains a string of negatively charged residues (E920, E924 and E930) that attract cations and repel anions (Christensen, Akyuz, & Corey, 2016; Paulsen et al., 2015). The permeation pathway has two major constriction sites. First, the upper gate is formed by the D915 residues of individual subunits, which are located at the entrance of the pore. In a tetrameric configuration, this forms a ring of negative charges that allows attraction of cations. The diameter of the pore at this region is 7 Å, which is wide enough to accommodate Ca^{2+}. The lower gate is formed by I957 and V961 residues and has a diameter of only 6 Å, which might represent the closed structure of the channel (upper gate, open; lower gate, closed) (Christensen et al., 2016; Paulsen et al., 2015).

Interestingly, agonist stimulation causes a change in the selectivity filter that results in progressive, but reversible, pore dilation, thereby allowing influx of large molecules such as *N*-methyl-D-glucamine (NMDG) and the cationic dye Yo-Pro. For example, activation of TRPA1 by allyl isothiocyanate (AITC) dilates the pore diameter up to an additional 2–3 Å over the normal open configuration (Banke, Chaplan, & Wickenden, 2010; Bobkov, Corey, & Ache, 2011; Chen et al., 2009). A separate model of the progressive increase in the permeability of large molecules over time indicates that this phenomenon may result from extensive changes in intracellular ion concentrations, rather than changes in channel permeability (Li, Toombes, Silberberg, & Swartz, 2015).

3.2 Regulation of TRPA1 by Ca^{2+}

TRPA1 channels are highly permeable to Ca^{2+} ions relative to most other TRP channels (Nilius et al., 2011). Under physiological conditions, their unitary conductance is ~70 and ~110 pS in the inward and outward directions, respectively (Nilius et al., 2012, 2011). Both intracellular and extracellular Ca^{2+} regulate TRPA1 activity. Increases in intracellular Ca^{2+} initiate and potentiate TRPA1 current responses elicited by different agonists, including AITC and cannabinoids (Jordt et al., 2004). Similarly, increase in extracellular Ca^{2+} also potentiate TRPA1 currents stimulated by electrophilic and non-electrophilic agonists (Wang et al., 2008). It has been proposed that three residues, D466, L474 and D477, within the Ca^{2+}-binding EF-hand domain located in the N-terminus are responsible for Ca^{2+}-dependent effects (Doerner et al., 2007; Hinman, Chuang, Bautista, & Julius, 2006; Nagata et al., 2005; Zurborg et al., 2007). Simulation studies have confirmed that, when bound to Ca^{2+}, the

EF-hand induces a local conformational change that stiffens the ankyrin repeats structure and therefore affects the force exerted onto the gate and thus opening of the channel (Zayats et al., 2013). A separate study demonstrated that mutation of a single amino acid residue in the pore region of hTRPA1 (D915) disrupts Ca^{2+}-induced activation and inactivation, suggesting that influx of Ca^{2+} is responsible for mediating these two effects (Wang et al., 2008).

3.3 Voltage gating of TRPA1 channels

TRPA1 channels are not voltage activated in a classical sense, but their activity can be modulated by voltage under certain conditions. The half-maximum voltage required for its activation in the absence of agonist has been reported to be +90 to +170 mV; thus, it is unlikely to be activated at physiological potentials (Karashima et al., 2007; Kremeyer et al., 2010; Meents, Fischer, & McNaughton, 2016). However, it has been demonstrated that the TRPA1 channel displays a shift in its voltage-dependent gating toward physiological membrane potentials during activation by electrophilic and non-electrophilic agonists, cold temperature, and Ca^{2+} (Fajardo, Meseguer, Belmonte, & Viana, 2008; Karashima et al., 2007). Mutations at several residues in the C-terminus (K1048, K1052, K1092 and T1099) were found to affect AITC and voltage-dependent gating (Samad et al., 2011). In general, the movement of positively charged residues in TM4 of voltage-gated ion channels is responsible for activation; however, TRPA1 does not contain any charged residues in the TM4 region, suggesting that the voltage sensor resides elsewhere. Specifically, it has been proposed that two positively charged residues, R975 and K989, within the C-terminus are responsible for voltage sensitivity. In addition, single alanine mutations of residues K969, K1009, K1071, E1077, T1078, and R1099 have been shown to abolish the voltage sensitivity of the channel (Samad et al., 2011). However, these residues are less likely to be directly affected by changes in the electric field of the cell membrane during depolarization because of their distal location, suggesting that downstream conformational changes that occur during activation of the channel could be involved in providing the connection between the distal region and the proximal putative voltage sensor. For instance, the coiled-coil domain of the C-terminus directly links the distal C-terminal to the putative voltage sensor domain, and also forms the connection with the N-terminal ankyrin repeat domain (Paulsen et al., 2015).

4. Pharmacology of TRPA1 channels

4.1 Activators

TRPA1 agonists can be categorized into two groups: electrophilic compounds (e.g., AITC, allicin and cinnamaldehyde) that covalently modulate the channel, and non-electrophilic compounds (e.g., thymol, menthol and carvacrol) that modulate the channel non-covalently. Electrophilic agonists interact with the thiol group of cysteine residues C621, C641 and C665, which are located in the N-terminal domain of TRPA1 (Bahia et al., 2016; Hinman et al., 2006). The crystal structure of TRPA1 suggests that these three cysteine residues come into close proximity, potentially forming a ligand-binding pocket (Cvetkov et al., 2011; Paulsen et al., 2015; Zayats et al., 2013). For example, JT010, a novel TRPA1 agonist ($EC_{50} \approx 65\,nM$) opens the channel by selectively and covalently binding to residue C621 (Takaya et al., 2015). A report by Cordero-Morales et al. suggests that certain ARDs in the N-terminus are involved in TRPA1 activation. AR3–8 and AR10–15 in TRPA1 from *Crotalus atrox* (western diamondback rattlesnake) are involved in heat sensation, whereas AR11–16 in hTRPA1 are sensitive to AITC (Cordero-Morales, Gracheva, & Julius, 2011).

TRPA1 responds to a wide variety of chemically diverse compounds, including environmental irritants, such as acrolein (Bautista et al., 2006) and diesel exhaust (Hazari et al., 2011), anesthetics (Leffler, Lattrell, Kronewald, Niedermirtl, & Nau, 2011), and the non-steroidal anti-inflammatory compound acetaminophen (Andersson et al., 2011), to name a few (summarized in Table 1). In addition to chemical activators, physical factors such as heat (Cordero-Morales et al., 2011; Rosenzweig et al., 2005) and cold (Kwan et al., 2006; Story et al., 2003) also activate TRPA1. However, responsiveness to cold remains controversial, with several studies failing to detect changes in TRPA1 activity in response to a cold stimulus (Bautista et al., 2006; Caspani & Heppenstall, 2009; Jordt et al., 2004).

Several studies have reported that reactive oxygen species (ROS) and their metabolites activate TRPA1 channels and are likely to be endogenous activators under physiological conditions. Macpherson et al. and Trevisani et al. were the first to report that the lipid peroxidation product, 4-hydroxy-2-nonenal (4-HNE), stimulates TRPA1 channels (Macpherson et al., 2007; Trevisani et al., 2007). Subsequently, two similar lipid peroxidation products, 4-oxononenal (4-ONE) and 4-hydroxyhexenal (4-HHE)

Table 1 List of TRPA1 channel activators.

Agonist	Source	EC_{50}; isoform, expression system, technique	References
Allyl isothiocyanate (AITC)	Mustard	33 μM; mTRPA1, CHO cells, Ca^{2+} imaging (FLIPR) 11 μM; rTRPA1, oocytes, electrophys (−60 mV) 64 μM; hTRPA1, oocytes, electrophys (+80 mV)	Bandell et al. (2004), Hinman et al. (2006), and Jordt et al. (2004)
Cinnamaldehyde	Cinnamon	100 μM; mTRPA1, CHO cells, Ca^{2+} imaging (FLIPR)	Bandell et al. (2004)
Allicin	Garlic	1.3 μM; mTRPA1, CHO cells, Ca^{2+} imaging (FLIPR) 1.9 μM; hTRPA1, CHO cells, Ca^{2+} imaging (FLIPR)	Bautista et al. (2005) and Macpherson et al. (2005)
Propofol	Anesthetic	17 μM; mTRPA1, Sf21 cells, Ca^{2+} imaging (FlexStation 3)	Woll et al. (2017)
Lidocaine	Anesthetic	5.7 mM; rTRPA1, HEK293, electrophys (−60 mV) 24 mM; hTRPA1, HEK293, electrophys (−60 mV)	Leffler et al. (2011)
4-HNE	Reactive oxygen species	13–20 μM; mTRPA1, CHO cells, Ca^{2+} imaging (FLIPR or FlexStation 3) 27 μM; rTRPA1, HEK293, Ca^{2+} imaging 5 μM; hTRPA1, HEK293, Ca^{2+} imaging	Andersson, Gentry, Moss, and Bevan (2008), Macpherson et al. (2007), Taylor-Clark et al. (2008), and Trevisani et al. (2007)

Table 1 List of TRPA1 channel activators.—cont'd

Agonist	Source	EC_{50}; isoform, expression system, technique	References
4-ONE	Reactive oxygen species	1.9 μM; mTRPA1, CHO cells, Ca^{2+} imaging (FlexStation 3) 5.8 μM; hTRPA1, HEK293, Ca^{2+} imaging	Andersson et al. (2008) and Taylor-Clark et al. (2008)
4-HHE	Reactive oxygen species	38.9 μM; mTRPA1, CHO cells, Ca^{2+} imaging (FlexStation 3) ≥4.3 μM; hTRPA1, HEK293, Ca^{2+} imaging	Andersson et al. (2008) and Taylor-Clark et al. (2008)
H_2O_2	Reactive oxygen species	230 μM; mTRPA1, CHO cells, Ca^{2+} imaging (FlexStation 3)	Andersson et al. (2008)
15-Deoxy-delta(12,14)-prostaglandin J(2) [15d-PGJ(2)]	Reactive oxygen species	5.6 μM; mTRPA1, CHO cells, Ca^{2+} imaging (FlexStation 3)	Andersson et al. (2008)
ASP7663	Synthetic	0.50 μM; mTRPA1, HEK293, Ca^{2+} imaging (FLIPR) 0.54 μM; rTRPA1, HEK293, Ca^{2+} imaging (FLIPR) 0.51 μM; hTRPA1, HEK293, Ca^{2+} imaging (FLIPR)	Kojima et al. (2014)
JT010	Synthetic	65 nM; species not specified, HEK293, Ca^{2+} imaging	Takaya et al. (2015)

were also found to activate TRPA1 (Taylor-Clark et al., 2008). Additionally, it has been reported that TRPA1 is activated by hydrogen peroxide (H_2O_2) (Andersson et al., 2008; Bessac et al., 2008), the hydroxyl radical ($^{\bullet}OH$) (Bessac et al., 2008), the cyclopentenone prostaglandin

15-deoxy-delta(12,14)-prostaglandin J(2) [15d-PGJ(2)] (Andersson et al., 2008), nitric oxide and peroxynitrite (Miyamoto, Dubin, Petrus, & Patapoutian, 2009).

4.2 Inhibitors

Several selective small-molecule TRPA1 inhibitors are available (Table 2). The first-described and most commonly used TRPA1 inhibitor is HC-030031 (McNamara et al., 2007). Chembridge-5861528, a derivative of HC-030031, has similar potency and specificity, but better solubility (Wei et al., 2009). Interestingly, AP-18, a partial agonist of TRPA1, exerts a net inhibitory effect on TRPA1 activity by desensitizing the channel (Defalco et al., 2010; Petrus et al., 2007). A-967079, Compound 10, and Compound 31 are potent available TRPA1 inhibitors with half-maximal inhibitory concentration (IC_{50}) values within the nanomolar range (Chen et al., 2011; Copeland et al., 2014; McGaraughty et al., 2010; Rooney et al., 2014). Previously, compounds such as camphor (Macpherson et al., 2006) and ruthenium red (Nagata et al., 2005) were used to block TRPA1 channels. However, these inhibitors lack specificity and thus are not suitable for experimental or clinical purposes (Story et al., 2003).

Table 2 List of TRPA1 channel antagonists.

Antagonist	**IC_{50}; isoform**	**References**
HC-030031	7.6 μM; rTRPA1 5.3–6.2 μM; hTRPA1	McNamara et al. (2007)
Chembridge-5861528	14.3–18.7 μM; hTRPA1	Wei et al. (2010) and Wei, Hamalainen, Saarnilehto, Koivisto, and Pertovaara (2009)
AP-18	4.5 μM; mTRPA1 3.1 μM; hTRPA1	Defalco et al. (2010) and Petrus et al. (2007)
A-967079	289 nM; rTRPA1 67 nM; hTRPA1	McGaraughty et al. (2010)
Compound 10	45 nM; rTRPA1 170 nM; hTRPA1	Copeland et al. (2014)
Compound 31	85 nM; rTRPA1 15 nM; hTRPA1	Rooney et al. (2014)

5. TRPA1 channels and vascular control

5.1 TRPA1 channels in sensory neurons

Bautista et al. (2005) provided the first evidence demonstrating a role for TRPA1 channels in the regulation of vascular tone through the release of sensory nerve-derived neuropeptides. They discovered that exogenous application of AITC, purified allicin, and diallyl disulfide (DADS) relaxed pre-contracted rat mesenteric arteries (Bautista et al., 2005). This response was diminished by the non-selective TRP channel antagonist ruthenium red, capsaicin-induced sensory nerve denervation, and the calcitonin gene-related peptide (CGRP) receptor antagonist $CGRP_{8-37}$, but was unaffected by the TRPV1 (TRP vanilloid 1) inhibitor capsazepine (Bautista et al., 2005). TRPA1 was found to be expressed in a subpopulation of adventitial sensory neurons surrounding rat mesenteric arteries. Furthermore, TRPA1 was detected in sensory neurons that also contain the potent vasodilator CGRP (Bautista et al., 2005; Brain, Williams, Tippins, Morris, & MacIntyre, 1985). These data suggest that TRPA1 is localized to sensory nerve terminals of the vasculature and their activation results in CGRP-dependent vasodilation, suppositions validated by Pozsgai et al. (2010). Cinnamaldehyde was found to induce a relaxation response in isolated mesenteric arteries that was unaffected by endothelium removal, but was significantly diminished in preparations from global TRPA1-knockout (KO) mice (Pozsgai et al., 2010). *In vivo*, topical application of cinnamaldehyde or AITC onto the ears of anesthetized mice increases blood flow in wild-type mice, but not TRPA1-KO mice (Pozsgai et al., 2010). A follow-up study by this group demonstrated that cinnamaldehyde-induced increases in blood flow were diminished by HC-030031 and $CGRP_{8-37}$, and were blunted in CGRP-KO mice (Aubdool et al., 2016). Intraplantar injections of 4-ONE into the hind paw of anesthetized mice was shown to increase blood flow, a response that was absent in TRPA1- and CGRP-KO mice, but was unaffected in TRPV1-KO mice (Graepel et al., 2011). Kunkler et al. investigating a role for TRPA1 channels in meningeal vasodilation, reported that AITC, cinnamaldehyde, and acrolein induced an increase in release of CGRP from cultured rat trigeminal neurons that was sensitive to TRPA1 inhibition with HC-030031 (Kunkler, Ballard, Oxford, & Hurley, 2011). Moreover, nasal administration of these TRPA1 agonists induced a transient increase in meningeal blood flow that was blocked by $CGRP_{8-37}$ and

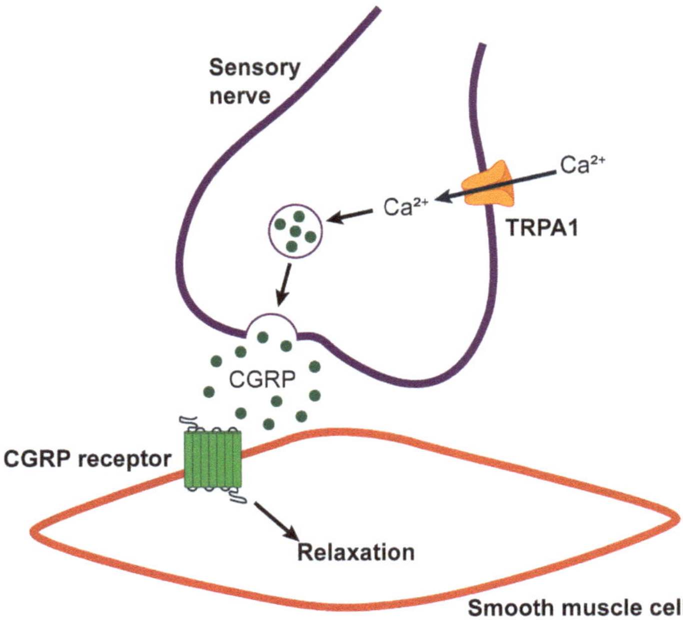

Fig. 2 Sensory nerve TRPA1-dependent vasodilation. Activation of TRPA1 channels on pre-synaptic sensory nerves leads to Ca^{2+} influx. This causes the release of CGRP, which binds CGRP receptors on vascular smooth muscle cells and mediates hyperpolarization and relaxation.

HC-030031 (Kunkler et al., 2011). Collectively, these data support the hypothesis proposed by Bautista et al. (2005) that TRPA1 channels are expressed on sensory neurons of the adventitia of blood vessels. Upon activation, Ca^{2+} influx causes CGRP to be released and bind its G protein-coupled receptor on the underlying smooth muscle cells to cause membrane hyperpolarization and relaxation (Fig. 2).

In addition to CGRP, neuronal nitric oxide (NO) and ROS are involved in TRPA1-dependent neurogenic dilation. Cinnamaldehyde-induced increases in mouse ear blood flow were attenuated by the selective neuronal nitric oxide synthase (nNOS) inhibitor, S-methyl-L-thiocitrulline (SMTC), and by ROS scavenging (Aubdool et al., 2016). The authors of this study proposed that superoxide and NO react to form peroxynitrite, which mediates dilation. Peroxynitrite is known to relax blood vessels in a cGMP-dependent manner (Casey et al., 2012; Dowell & Martin, 1997). Increased protein nitrosylation was observed following cinnamaldehyde treatment, and cinnamaldehyde-induced dilation was reduced in mice treated with the peroxynitrite scavenger FeTPPs (Aubdool et al., 2016).

However, the precise mechanism by which superoxide is produced following TRPA1 activation, which was not assessed in this study, remains unclear.

The peripheral vascular response to noxious cold has been studied for several decades (Daanen, 2003; Keatinge, 1957). Local cold exposure produces an initial rapid vasoconstriction lasting approximately 5–10 min, followed by a vasodilatory recovery phase that protects against cold-induced injury, such as chilblains (Daanen, 2003; Keatinge, 1957). Aubdool et al. (2014) proposed that TRPA1 channels are the cold sensors that initiate this vascular response. They showed that exposure of the mouse hind paw to 10 °C produced a biphasic vascular response that was absent in TRPA1-KO mice or wild-type animals treated with HC-030031 (Aubdool et al., 2014). The authors proposed that the initial vasoconstriction is dependent on TRPA1 activation-dependent production of ROS, which in turn causes release of norepinephrine from sympathetic neurons. Norepinephrine binds to α_{2C}-adrenoceptors, activating Rho-kinase to mediate vasoconstriction. However, sympathetic nerve denervation with guanethidine or blockade of α-adrenoceptors with phentolamine partially attenuated the contractile response, suggesting involvement of other contractile pathways (Aubdool et al., 2014). Moreover, neither the precise cellular location of TRPA1 responsible for mediating the contractile phase nor the mechanisms by which ROS was produced downstream of TRPA1 activation was determined. Aubdool et al. (2014) found that TRPA1 is also involved in the vasodilatory phase, showing that intravenous administration of HC-030031 immediately following the peak contraction prevented cold-induced vasodilation. Sensory nerves were found to be responsible for mediating the vasodilation phase, as evidenced by the fact that sensory nerve denervation, as well as CGRP and nNOS inhibition, attenuated the response. The role of TRPA1 channels as cold sensors remains controversial, with several reports suggesting that the channel in not activated by exposure to cold temperatures (Bautista et al., 2006; Caspani & Heppenstall, 2009; Jordt et al., 2004). Further investigation will be required to provide a definitive answer to this question.

5.2 TRPA1 channels in the endothelium

Work from our laboratory has shown that TRPA1 channels are localized to the endothelium of the cerebral vasculature of humans and rodents, but are not present in the endothelium of mesenteric, coronary, dermal, or renal arteries (Fig. 3A) (Earley et al., 2009; Sullivan et al., 2015).

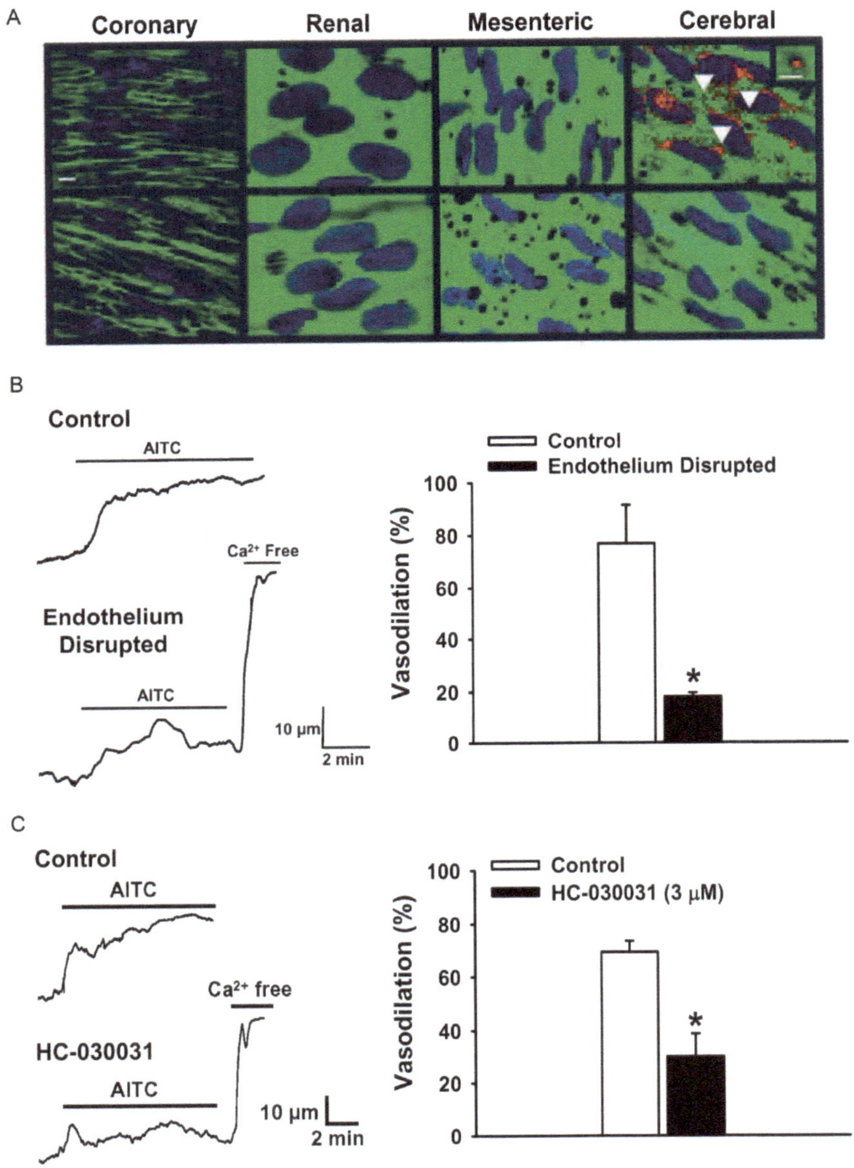

Fig. 3 TRPA1 activation causes endothelium-dependent vasodilation in cerebral arteries. (A) (Top) Immunolabeling for TRPA1 (red) in rat coronary, renal, mesenteric and cerebral arteries (scale bar, 10 μm). Green is autofluorescence produced by the internal elastic lamina. TRPA1 was only detected in the endothelium of the cerebral vasculature and not in other vascular beds. Bottom: omission of primary antibody. (B) Representative trace and summary data demonstrating that the TRPA1 agonist AITC (100 μM) induces vasodilation of rat cerebral arteries, which was significantly reduced following disruption of the endothelium ($n = 3$, $*P \leq 0.05$ vs control).

Interestingly, TRPA1 is predominantly found in endothelial cell plasma membrane projections that penetrate through the internal elastic lamina, where they are positioned proximal to vascular smooth muscle cells (Earley et al., 2009). These structures, known as myoendothelial projections (MEPs), house important signaling complexes that regulate the membrane potential and contractility of underlying smooth muscle cells (Griffith, Chaytor, Taylor, Giddings, & Edwards, 2002; Sandow & Hill, 2000). Ledoux et al. demonstrated that the endoplasmic reticulum (ER) of endothelial cells projects into MEPs and that inositol trisphosphate receptors (IP_3R) are enriched within this region (Ledoux, Bonev, & Nelson, 2008). This study also reported that spontaneous, transient subcellular Ca^{2+} events caused by Ca^{2+} release through IP_3Rs on the ER occurred at MEPs. Several studies have reported that intermediate-conductance Ca^{2+}-activated K^+ channels ($K_{Ca}3.1$) and connexin proteins 40 and 37 (Cx40 and Cx37) are also expressed at high levels within MEPs (Ledoux et al., 2008; Sandow, Neylon, Chen, & Garland, 2006). Ledoux et al. (2008) proposed that Ca^{2+} signals within MEPs hyperpolarize the plasma membrane of endothelial cells by activating nearby $K_{Ca}3.1$ channels. The resulting hyperpolarization is subsequently conducted through myoendothelial gap junctions (MEGJs) within MEPs composed of Cx40 and/or Cx37 hemichannels to underlying smooth muscle cells to evoke relaxation.

TRPA1 channels colocalize with $K_{Ca}3.1$ channels within the MEPs of cerebral arteries (Earley et al., 2009; Sullivan et al., 2015). Activation of TRPA1 in isolated rat cerebral arteries with AITC induced a relaxation response that was sensitive to inhibition of TRPA1 channels with HC-030031 and removal of the endothelium (Fig. 3B and C). Inhibition of $K_{Ca}3.1$ and small-conductance Ca^{2+}-activated K^+ channels ($K_{Ca}2.3$) with TRAM-34 and apamin, respectively, and inhibition of inwardly rectifying

Fig. 3—Cont'd (C) Representative trace and summary data showing the vasodilation to AITC is blocked by the selective TRPA1 antagonist HC-030031 (3 μM) ($n=3$, $^*P \leq 0.05$ vs control). *Panel (A) image derived from Sullivan, M. N., Gonzales, A. L., Pires, P. W., Bruhl, A., Leo, M. D., Li, W., et al. (2015). Localized TRPA1 channel Ca^{2+} signals stimulated by reactive oxygen species promote cerebral artery dilation.* Science Signaling, *8(358), ra2. doi: https://doi.org/10.1126/scisignal.2005659 (web archive link), Panels (B and C) images derived from Earley, S., Gonzales, A. L., & Crnich, R. (2009). Endothelium-dependent cerebral artery dilation mediated by TRPA1 and Ca^{2+}-activated K^+ channels.* Circulation Research, 104*(8), 987–994. https://doi.org/10.1161/CIRCRESAHA.108.189530 (web archive link).*

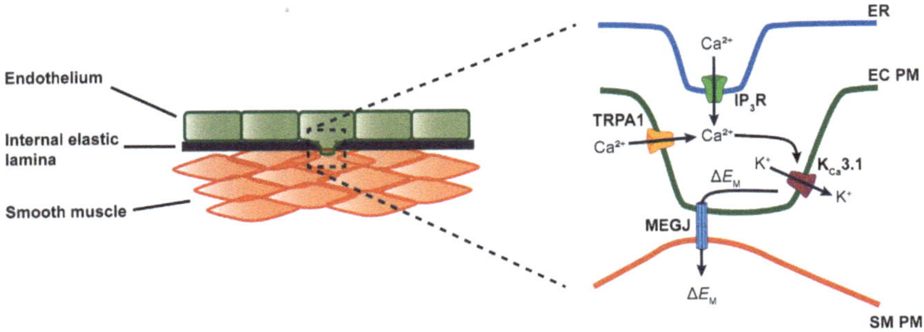

Fig. 4 Endothelial TRPA1-dependent dilation of cerebral arteries. Activation of TRPA1 channels present in myoendothelial projections (MEPs) causes Ca^{2+} influx. Ca^{2+} is released from the endothelial cell ER through IP_3Rs via a Ca^{2+}-induced Ca^{2+}-release mechanism. Intermediate conductance Ca^{2+}-activated K^+ channels ($K_{Ca}3.1$) are activated and the resulting K^+ efflux causes hyperpolarization of the endothelial cell plasma membrane (EC PM). The change in membrane potential (ΔE_M) is conducted to the underling smooth muscle through myoendothelial gap junctions (MEGJs), resulting in hyperpolarization of the smooth muscle plasma membrane (SM PM) and relaxation.

K^+ (K_{IR}) channels with Ba^{2+} also attenuated relaxation to AITC (Earley et al., 2009). These data suggest that Ca^{2+} influx through TRPA1 channels evokes endothelial cell hyperpolarization through activation of $K_{Ca}3.1$ and $K_{Ca}2.3$ (Fig. 4). This hyperpolarizing signal is propagated through MEGJs to underlying smooth muscle cells. K_{IR} channels amplify the transmitted hyperpolarizing signal, resulting in smooth muscle relaxation (Smith et al., 2008). Qian et al. further characterized the Ca^{2+}-signaling events stimulated by TRPA1 channel activity in isolated rat cerebral arteries, showing that AITC evoked a concentration-dependent increase in Ca^{2+}-signaling events that were localized to MEPs and dependent on IP_3R-mediated release of Ca^{2+} from the ER (Qian, Francis, Solodushko, Earley, & Taylor, 2013). The authors suggested that Ca^{2+} influx through TRPA1 channels present within MEPs stimulated the formation of larger Ca^{2+}-signaling events by stimulating IP_3R activity through a Ca^{2+}-induced Ca^{2+}-release mechanism. The physiological significance of this mechanism was revealed by the observation that the frequency of IP_3R-dependent Ca^{2+}-signaling events generated by stimulating TRPA1 channels was directly correlated with the vasodilatory response.

Distinctive Ca^{2+}-signaling events reflecting Ca^{2+} influx through single TRP channels can be optically recorded in isolated cells and the intact endothelium using TIRF (total internal reflection fluorescent) microscopy (Sullivan & Earley, 2013; Sullivan, Francis, Pitts, Taylor, & Earley, 2012; Sullivan et al., 2015) or high-speed high-resolution spinning-disk confocal microscopy (Sonkusare et al., 2012). These Ca^{2+}-signaling events are generically called "sparklets" and are typically associated with the name of the conducting channel (e.g., "TRPA1 sparklets"). Sullivan et al. recorded TRPA1 sparklets in native cerebral endothelial cells using TIRF microscopy (Sullivan et al., 2015), reporting that TRPA1 sparklets are large Ca^{2+} influx events, with a unitary amplitude approximately twice that of TRPV4 sparklets, probably due to the larger single-channel conductance and greater Ca^{2+} fraction conducted by TRPA1 (Story et al., 2003; Sullivan et al., 2012, 2015). The signal mass of a single TRPA1 sparklet was estimated to be 200-fold greater than that of a single L-type Ca^{2+} channel sparklet (Sullivan et al., 2015). Moreover, the most frequently occurring TRPA1 sparklets recorded had amplitudes that were consistent with the simultaneous opening of two TRPA1 channels and a doubling of the amount of Ca^{2+} influx during an event. This suggests that TRPA1 channels exist in a functionally linked binary configuration on the plasma membrane and that opening of one channel triggers cooperative gating of the adjacent channel, possibly through binding of incoming Ca^{2+} to the EF-hand domain located on the N-terminus of the adjacent TRPA1 channel (Jordt et al., 2004).

Sullivan et al. demonstrated that TRPA1 sparklet activity is low under basal conditions, and that the large localized increases in subcellular Ca^{2+} concentration induced by TRPA1 agonists occurs through recruitment of previously inactive TRPA1 channels, rather than through increases in the frequency of basally active TRPA1 channels (Sullivan et al., 2015). Notably, Sullivan et al. found that only 4–8 TRPA1 sparklets sites per cell were active following stimulation with a concentration of AITC that produced maximal dilation of isolated cerebral arteries (Sullivan et al., 2015), implying that the vast majority of TRPA1 channels are inactive even under optimal stimulation conditions. This phenomenon of "silent" channels has been observed for several types of ion channels. For example, in studies of TRPV4 sparklets in primary human endothelial cells, Sullivan et al. showed that only a few sparklet sites were detectable even during maximal stimulation with the TRPV4 agonist GSK1016790A, despite immunolabeling experiments revealing a cell-wide distribution of TRPV4 channels (Sullivan et al., 2012). An examination of L-type Ca^{2+} channels in rat cerebral artery smooth muscle cells by Navedo et al. revealed similar characteristics, further suggesting that ion channels present in the plasma membrane may exist in a "silent" state (Navedo, Amberg, Votaw, & Santana, 2005). Although the reason for this is not clear, it may reflect nanoscale association of channels with other signaling elements (Mercado et al., 2014; Navedo et al., 2005; Tajada et al., 2017).

TRPA1 is activated by the endogenously generated ROS metabolites, 4-HNE, 4-ONE, and 4-HHE (Andersson et al., 2008; Macpherson et al., 2007; Trevisani et al., 2007; Uchida, 2003). Sullivan et al. demonstrated that exogenous application of 4-HNE increased TRPA1 sparklet frequency and relaxed isolated cerebral arteries, and further showed that these responses were abolished by HC-030031 and were absent in tissues isolated from endothelial cell-specific TRPA1-KO (*Trpa1*-ecKO) mice (Sullivan et al., 2015). TRPA1 was found to colocalize with the ROS-generating enzyme, NADPH oxidase 2 (NOX2), in MEPs of isolated cerebral arteries (Sullivan et al., 2015). Exogenous administration of the NOX2 substrate NADPH increased TRPA1 sparklet frequency and dilated cerebral arteries isolated from wild-type mice, but not those from *Trpa1*-ecKO mice (Sullivan et al., 2015). Moreover, the response to NADPH was inhibited by catalase and iron chelation, indicating that extracellular H_2O_2 and hydroxyl radicals (•OH) were generated—intermediates that are involved in the production of 4-HNE. These data suggest that NOX2-derived lipid peroxidation products are endogenous activators of TRPA1 channels in the cerebral endothelium and cause endothelium-dependent dilation in this vascular segment.

5.3 TRPA1 channels in vascular smooth muscle cells

There is little evidence to support the expression of functional TRPA1 channels on vascular smooth muscle cells. In an examination of relaxation of the rat aorta to cinnamaldehyde, Yanaga et al. showed that relaxation responses were impaired by endothelium removal or NOS inhibition, but were not completely abolished (Yanaga et al., 2006). This implies that cinnamaldehyde is able to dilate arteries through direct effects on the smooth muscle. However, this study did not test the effect of TRPA1 inhibitors on this response, leaving doubt about the specific involvement of TRPA1 channels. Bodkin et al. further demonstrated that cinnamaldehyde-induced relaxation of mouse mesenteric arteries was independent of the endothelium (Bodkin et al., 2014). However, relaxation persisted in the presence of HC-030031 and in preparations from TRPA1-KO mice, indicating that cinnamaldehyde-induced relaxation occurs independently of TRPA1. TRPA1 is not found in the smooth muscle of the cerebral vasculature, and these cells do not exhibit a Ca^{2+} response to AITC (Earley et al., 2009). Thus, it remains unlikely that TRPA1 channels are expressed in vascular smooth muscle cells, and observations of direct vasorelaxant effects by TRPA1 agonists are likely attributable to off-target actions.

5.4 TRPA1 and integrative cardiovascular physiology

It is clear that activation of TRPA1 channels mediates dilation of the peripheral vasculature, but what are the integrative effects of TRPA1 channel activity on global cardiovascular hemodynamics? Pozagai et al. reported comparable mean arterial pressure (MAP) and heart rate in anesthetized wild-type and global TRPA1-KO mice (Pozsgai et al., 2010). A follow-up study from this group validated these findings in freely-moving, conscious mice using radio-telemetry devices (Bodkin et al., 2014). Interestingly, this latter study noted increased locomotor activity in global TRPA1-KO mice, although the underlying mechanism was not investigated. These studies suggest that unstressed mice are able to compensate for loss of TRPA1 activity and maintain basal cardiovascular function.

Pozagai et al. (2010) examined the effects of TRPA1 activation on the systemic circulation. The authors noted a bimodal effect of intravenous cinnamaldehyde administration in anesthetized female mice, with an initial, transient reduction in MAP and heart rate (depressor response) that was dose-independent, followed by a subsequent dose-dependent sustained rise in both parameters (pressor response) (Pozsgai et al., 2010). Further experiments in TRPA1-KO mice revealed that the depressor response was diminished in knockout mice, but the pressor response was absent following a moderate cinnamaldehyde dose (80 μM/kg body weight). However, at higher doses (320 μM/kg body weight) the pressor response, but not the depressor response, was blunted in TRPA1-KO mice. On the basis of experiments performed using TRPV1- and CGRP-KO mice and inhibitors of substance P, the authors proposed that the effects of cinnamaldehyde administration were independent of sensory nerve-derived vasodilators. Cholinergic receptor blockade diminished the pressor response to cinnamaldehyde, but had little effect on the pressor response. α-Adrenergic inhibition attenuated both pressor and depressor responses, but combined ganglionic blockade and cholinergic receptor blockade did not influence the pressor response. The authors concluded that the depressor response was due to activation of the vasovagal reflex, while the depressor response was mediated by peripheral sympathetic activation (Pozsgai et al., 2010). However, because of the transient nature of the depressor response, it is difficult to draw meaningful conclusions. Moreover, the lack of a relationship between cinnamaldehyde dose and magnitude of the reduction in MAP and heart rate, as well as the absence of a discernable difference between wild-type and TRPA1-KO mice at higher cinnamaldehyde doses further confound data interpretation. Additional experiments, in particular, investigations of both short- and long-term effects of TRPA1 agonists and antagonists on cardiovascular parameters in conscious animals, are required to better understand the significance of systemic TRPA1 activation in the regulation of cardiovascular hemodynamics.

In an examination of the role of TRPA1 channels in renal sympathetic nerve activity (RSNA), Koda et al. showed that intra-arterial injection of AITC into decerebrate rats significantly increased RSNA (Koba, Hayes, & Sinoway, 2011). This response was prevented by sectioning the sciatic nerve, indicating a muscle-based reflex response. Sustained static contraction of the hindlimb muscle, induced by electrical stimulation, resulted in increased RSNA and blood pressure that were diminished by TRPA1 blockade with HC-030031. Moreover, HC-030031 reduced RSNA in response to administration of arachidonic acid, bradykinin or diprotonated phosphate, all metabolic byproducts of muscle contraction. These findings were further supported by Xing et al. who observed similar TRPA1-dependent increases in RSNA, MAP, and heart rate (Xing, Lu, & Li, 2015). These latter authors proposed that TRPA1 channels are present on afferent nerves of muscle and may contribute to increased sympathetic nerve activity as well as associated increases in cardiovascular parameters during exercise.

6. TRPA1 in vascular pathology

6.1 Inflammation

Neuronal TRPA1 channels may play an important role in the vascular component of neurogenic inflammation by stimulating the release of substance P, CGRP, and other proinflammatory vasoactive neuropeptides. This was first shown by Trevisani et al. who reported that the TRPA1 agonist 4-HNE caused edema formation and pain-related behavior following intraplantar administration to the hind paw of rats (Trevisani et al., 2007). Further evidence was provided by Andre et al. in a model of cigarette smoke inhalation-induced inflammation, tracheal plasma extravasation was attenuated by TRPA1 inhibition with HC-030031 (Andre et al., 2008). The authors also demonstrated that capillary permeability was increased in wild-type mice following treatment with an aqueous extract of cigarette smoke, but not in global TRPA1-KO mice (Andre et al., 2008), suggesting that TRPA1 is involved in the breakdown of the endothelial cell-to-cell barrier. TRPA1 has been reported to mediate neurogenic inflammation in response to other stimuli, including *N*-acetyl-*p*-benzoquinone imine (NAPQI), a toxic by-product of acetaminophen, and the bacterial endotoxin lipopolysaccharide (Meseguer et al., 2014; Nassini et al., 2010). In contrast, Graepel et al. suggested that TRPA1 is important for the vasodilatory effect of neurogenic inflammation, but not edema formation in the mouse hind paw (Graepel et al., 2011). The reasons for these discrepancies are unclear, but these observations may suggest that TRPA1 channel activity mediates neurogenic inflammation in a tissue-specific manner. TRPA1 channels may provide a novel therapeutic target for treating pathological increases in capillary permeability and edema formation associated with inflammation.

A recent study by Zhao et al. suggested that TRPA1 channels have a protective role against the progression of atherosclerosis (Zhao, Shyue, Kou, Lu, & Lee, 2016). Atherosclerosis is a progressive, chronic inflammatory condition that is associated with an accumulation of fatty and/or fibrous material in the innermost layer of the artery that eventually impedes blood flow. It can lead to a number of other complications, such as aneurysms and stroke, and is the leading cause of vascular disease worldwide (Libby et al., 2019). Activation of TRPA1 channels with AITC has been shown to suppress atherosclerotic progression in apolipoprotein E (apoE)-KO mice, a commonly used mouse model of atherosclerosis (Plump et al., 1992; Zhang, Reddick, Piedrahita, & Maeda, 1992; Zhao et al., 2016). Moreover, apoE-KO mice treated with HC-030031 or lacking TRPA1 channels exhibit exacerbated atherosclerotic lesions, hyperlipidemia, and elevated circulating levels of inflammatory cytokines (Zhao et al., 2016). Interestingly, TRPA1 was found to be expressed in macrophages. Macrophages play a pivotal role in the pathogenesis of atherosclerosis by accumulating lipoproteins and forming foam cells, which are retained in the plaque and lead to further plaque growth (Bobryshev, Ivanova, Chistiakov, Nikiforov, & Orekhov, 2016). Moreover, infiltrated macrophages have a reduced ability to migrate, which further contributes to the proinflammatory environment, as well as plaque growth and instability. Zhao et al. proposed that TRPA1 activation prevents TNF-α (tumor necrosis factor-α)-dependent activation of macrophages, thereby reducing ingestion of lipoproteins and plaque formation (Zhao et al., 2016). These results are interesting, but further investigation is required to understand the role of TRPA1 signaling in macrophages and the effects of TRPA1 activation on macrophage recruitment and polarization.

6.2 Hypertension

Hypertension, characterized by elevated systolic and/or diastolic blood pressure, is a prevalent condition that is a major risk factor for multiple cardiovascular morbidities, such as coronary heart disease, heart failure and stroke (Taler, 2018). Studies have shown an exacerbated hypertensive response in mice lacking CGRP, a neuropeptide that is released following TRPA1 activation (Gangula et al., 2000; Smillie et al., 2014). However, Bodkin et al. demonstrated that TRPA1 may be involved in the inflammatory component, but not the blood pressure increase associated with the pathology. Blood pressure increases following subcutaneous infusion of angiotensin II over 14 days were comparable in wild-type and global TRPA1-KO mice, but the proinflammatory cytokine interleukin-6 was found to be reduced in hypertensive TRPA1-KO mice (Bodkin et al., 2014). Ma et al. (2019) further explored this concept in the context of kidney function following angiotensin II infusion over 28 days. Similar to the report of Bodkin et al. (2014), these authors found that systolic blood pressure was comparable between TRPA1-KO mice and wild-type controls (Ma, Zhang, He, Wang, & Wang, 2019). However, kidney dysfunction was exacerbated in knockout mice. Blood urea nitrogen, serum creatinine, renal fibrosis and renal inflammatory cytokine levels were enhanced in hypertensive TRPA1-KO mice (Ma et al., 2019). The authors proposed that TRPA1 inhibits macrophage-dependent production of proinflammatory mediators that provide protection against kidney dysfunction. These data imply that, although targeting TRPA1 may not reduce blood pressure in hypertensive individuals, it may provide benefit in reducing the inflammatory component of the disease.

6.3 Stroke

Stroke is a condition in which sudden interruption of regional blood flow within the brain results in tissue damage, which can lead to a number of long-lasting disabilities. The majority of strokes occur due to blockage of arteries and lead to tissue ischemia and hypoxia. Cerebral blood vessels dilate in response to hypoxia, a response that serves to enhance oxygen delivery to ischemic brain regions (Kontos et al., 1978). Takahashi et al. suggested that TRPA1 channels are maximally activated by a reduction in the partial pressure of O_2 (pO_2) from ~150 to ~80 mmHg, defined by the authors of this study as hypoxia (Takahashi et al., 2011). However, under normoxic conditions, pO_2 levels range from ~100 mmHg within the lungs to ~20–40 mmHg in the venous system (Tsai, Johnson, & Intaglietta, 2003).

According to the model proposed by Takahashi et al. (2011), this would suggest that TRPA1 channels are constitutively active, an inference inconsistent with the literature (Pires & Earley, 2017). It is more likely that, in the study by Takahashi et al. (2011), TRPA1 channels are activated by ROS generated by the transition from hyperoxic (pO_2, 150 mmHg) to normoxic (pO_2, 80 mmHg) conditions. In a previous study from our lab examining the effects of acute hypoxia on TRPA1 activity (Pires & Earley, 2018), a pO_2 of ~10–15 mmHg, which is comparable to the level in the peri-infarct area following an ischemic stroke, was used as a pathophysiologically relevant pO_2 level (Liu et al., 2004; Pires & Earley, 2018). Acute hypoxia increased TRPA1 sparklet frequency and dilated pressurized mouse cerebral arteries and penetrating parenchymal arterioles. Additionally, hypoxia increased 4-HNE levels in the cerebral endothelium (Pires & Earley, 2018). As noted above, NOX2 is responsible for producing ROS and driving TRPA1 activity under physiological conditions within the cerebral vasculature (Sullivan et al., 2015). However, hypoxia-induced dilation was unaffected by NOX inhibition with apocynin or extracellular superoxide dismutase (SOD), but was attenuated by membrane-permeant PEG-SOD and mitochondrial-targeted antioxidant mitoTEMPO (Pires & Earley, 2018). These observations suggest that hypoxia increases mitochondrial ROS generation. It has been reported that mitochondria are localized near MEPs, providing an alternative pathway of TRPA1-dependent vasodilation in the cerebral vasculature during hypoxic conditions (Maarouf, Sancho, Furstenhaupt, Tran, & Welsh, 2017). The *in vivo* importance of hypoxia-induced activation of TRPA1 channels was assessed in *Trpa1*-ecKO mice using the middle cerebral artery occlusion model of ischemic stroke (Pires & Earley, 2018). Interestingly, the infarct area produced following occlusion was significantly greater in *Trpa1*-ecKO mice, suggesting that TRPA1 protects against tissue damage associated with ischemia. Furthermore, treatment of control mice with the TRPA1 agonist cinnamaldehyde 15 min post-occlusion reduced infarct area. These data suggest that endothelial TRPA1 channels evoke vasodilation and increase blood flow in response to hypoxic conditions. TRPA1 is important for an early adaptive response to an acute reduction in pO_2 levels that serves to improve collateral blood flow to affected regions and improve tissue outcome.

7. Conclusions

This chapter has described the importance of TRPA1 channels in vascular physiology and pathophysiology. It is evident that these channels regulate vascular tone in a tissue-dependent manner. Activation of TRPA1 channels using pungent dietary compounds evokes a robust vasodilatory response either through release of vasoactive neuropeptides from sensory nerves or via endothelial cell-dependent hyperpolarization pathways. Under physiological conditions, TRPA1 is likely to be regulated endogenously by ROS and their metabolites, rather than by consumed dietary molecules. Novel therapeutic strategies may target TRPA1 channels in the treatment of certain diseases, such as neurogenic inflammation, atherosclerosis, and ischemic stroke. However, whether such treatment strategies will prevent the development of disease or provide treatment for existing conditions remains unclear. Moreover, understanding the cardiovascular effects of short- and long-term TRPA1 activation and inhibition on basal function and pathology remains unclear, and should be a focus of studies going forward.

Acknowledgments

Support for this chapter was provided by grants from the National Heart, Lung, and Blood Institute (R01HL091905, R01HL137852, and R01HL139585), the National Institute of Neurological Disorders and Stroke (RF1NS110044 and R61NS115132), and the National Institute of General Medicine (P20GM130459) to S.E.

Disclosures

The authors have no conflicts of interest, financial or otherwise, to report.

References

Andersson, D. A., Gentry, C., Alenmyr, L., Killander, D., Lewis, S. E., Andersson, A., et al. (2011). TRPA1 mediates spinal antinociception induced by acetaminophen and the cannabinoid Delta(9)-tetrahydrocannabiorcol. *Nature Communications*, *2*, 551. https://doi.org/10.1038/ncomms1559.

Andersson, D. A., Gentry, C., Moss, S., & Bevan, S. (2008). Transient receptor potential A1 is a sensory receptor for multiple products of oxidative stress. *The Journal of Neuroscience*, *28*(10), 2485–2494. https://doi.org/10.1523/JNEUROSCI.5369-07.2008.

Andersson, D. A., Gentry, C., Moss, S., & Bevan, S. (2009). Clioquinol and pyrithione activate TRPA1 by increasing intracellular Zn^{2+}. *Proceedings of the National Academy of Sciences of the United States of America*, *106*(20), 8374–8379.

Andre, E., Campi, B., Materazzi, S., Trevisani, M., Amadesi, S., Massi, D., et al. (2008). Cigarette smoke-induced neurogenic inflammation is mediated by alpha, beta-unsaturated aldehydes and the TRPA1 receptor in rodents. *The Journal of Clinical Investigation*, *118*(7), 2574–2582. https://doi.org/10.1172/JCI34886.

Andrei, S. R., Sinharoy, P., Bratz, I. N., & Damron, D. S. (2016). TRPA1 is functionally co-expressed with TRPV1 in cardiac muscle: Co-localization at z-discs, costameres and intercalated discs. *Channels (Austin, Tex.)*, *10*(5), 395–409. https://doi.org/10.1080/19336950.2016.1185579.

Aubdool, A. A., Graepel, R., Kodji, X., Alawi, K. M., Bodkin, J. V., Srivastava, S., et al. (2014). TRPA1 is essential for the vascular response to environmental cold exposure. *Nature Communications*, *5*, 5732. https://doi.org/10.1038/ncomms6732.

Aubdool, A. A., Kodji, X., Abdul-Kader, N., Heads, R., Fernandes, E. S., Bevan, S., et al. (2016). TRPA1 activation leads to neurogenic vasodilatation: Involvement of reactive oxygen nitrogen species in addition to CGRP and NO. *British Journal of Pharmacology*, *173*(15), 2419–2433. https://doi.org/10.1111/bph.13519.

Bahia, P. K., Parks, T. A., Stanford, K. R., Mitchell, D. A., Varma, S., Stevens, S. M., Jr., et al. (2016). The exceptionally high reactivity of Cys 621 is critical for electrophilic activation of the sensory nerve ion channel TRPA1. *The Journal of General Physiology*, *147*(6), 451–465.

Bandell, M., Story, G. M., Hwang, S. W., Viswanath, V., Eid, S. R., Petrus, M. J., et al. (2004). Noxious cold ion channel TRPA1 is activated by pungent compounds and bradykinin. *Neuron*, *41*(6), 849–857. https://doi.org/10.1016/s0896-6273(04)00150-3.

Banke, T. G., Chaplan, S. R., & Wickenden, A. D. (2010). Dynamic changes in the TRPA1 selectivity filter lead to progressive but reversible pore dilation. *American Journal of Physiology-Cell Physiology*, *298*(6), C1457–C1468. https://doi.org/10.1152/ajpcell.00489.2009.

Bautista, D. M., Jordt, S. E., Nikai, T., Tsuruda, P. R., Read, A. J., Poblete, J., et al. (2006). TRPA1 mediates the inflammatory actions of environmental irritants and proalgesic agents. *Cell*, *124*(6), 1269–1282. https://doi.org/10.1016/j.cell.2006.02.023.

Bautista, D. M., Movahed, P., Hinman, A., Axelsson, H. E., Sterner, O., Hogestatt, E. D., et al. (2005). Pungent products from garlic activate the sensory ion channel TRPA1. *Proceedings of the National Academy of Sciences of the United States of America*, *102*(34), 12248–12252. https://doi.org/10.1073/pnas.0505356102.

Bessac, B. F., Sivula, M., von Hehn, C. A., Escalera, J., Cohn, L., & Jordt, S. E. (2008). TRPA1 is a major oxidant sensor in murine airway sensory neurons. *The Journal of Clinical Investigation*, *118*(5), 1899–1910. https://doi.org/10.1172/JCI34192.

Bobkov, Y. V., Corey, E. A., & Ache, B. W. (2011). The pore properties of human nociceptor channel TRPA1 evaluated in single channel recordings. *Biochimica et Biophysica Acta*, *1808*(4), 1120–1128. https://doi.org/10.1016/j.bbamem.2010.12.024.

Bobryshev, Y. V., Ivanova, E. A., Chistiakov, D. A., Nikiforov, N. G., & Orekhov, A. N. (2016). Macrophages and their role in atherosclerosis: Pathophysiology and transcriptome analysis. *BioMed Research International*, *2016*, 9582430. https://doi.org/10.1155/2016/9582430.

Bodkin, J. V., Thakore, P., Aubdool, A. A., Liang, L., Fernandes, E. S., Nandi, M., et al. (2014). Investigating the potential role of TRPA1 in locomotion and cardiovascular control during hypertension. *Pharmacology Research & Perspectives*, *2*(4). e00052https://doi.org/10.1002/prp2.52.

Brain, S. D., Williams, T. J., Tippins, J. R., Morris, H. R., & MacIntyre, I. (1985). Calcitonin gene-related peptide is a potent vasodilator. *Nature*, *313*(5997), 54–56. https://doi.org/10.1038/313054a0.

Casey, D. B., Pankey, E. A., Badejo, A. M., Bueno, F. R., Bhartiya, M., Murthy, S. N., et al. (2012). Peroxynitrite has potent pulmonary vasodilator activity in the rat. *Canadian Journal of Physiology and Pharmacology*, *90*(4), 485–500. https://doi.org/10.1139/Y2012-012.

Caspani, O., & Heppenstall, P. A. (2009). TRPA1 and cold transduction: An unresolved issue? *The Journal of General Physiology*, *133*(3), 245–249. https://doi.org/10.1085/jgp.200810136.

Chen, J., Joshi, S. K., DiDomenico, S., Perner, R. J., Mikusa, J. P., Gauvin, D. M., et al. (2011). Selective blockade of TRPA1 channel attenuates pathological pain without altering noxious cold sensation or body temperature regulation. *Pain*, *152*(5), 1165–1172. https://doi.org/10.1016/j.pain.2011.01.049.

Chen, J., Kim, D., Bianchi, B. R., Cavanaugh, E. J., Faltynek, C. R., Kym, P. R., et al. (2009). Pore dilation occurs in TRPA1 but not in TRPM8 channels. *Molecular Pain, 5*, 3.

Christensen, A. P., Akyuz, N., & Corey, D. P. (2016). The outer pore and selectivity filter of TRPA1. *PLoS One, 11*(11), e0166167.

Copeland, K. W., Boezio, A. A., Cheung, E., Lee, J., Olivieri, P., Schenkel, L. B., et al. (2014). Development of novel azabenzofuran TRPA1 antagonists as in vivo tools. *Bioorganic & Medicinal Chemistry Letters, 24*(15), 3464–3468. https://doi.org/10.1016/j.bmcl.2014.05.069.

Cordero-Morales, J. F., Gracheva, E. O., & Julius, D. (2011). Cytoplasmic ankyrin repeats of transient receptor potential A1 (TRPA1) dictate sensitivity to thermal and chemical stimuli. *Proceedings of the National Academy of Sciences of the United States of America, 108*(46), E1184–E1191. https://doi.org/10.1073/pnas.1114124108.

Cosens, D. J., & Manning, A. (1969). Abnormal electroretinogram from a Drosophila mutant. *Nature, 224*(5216), 285–287. https://doi.org/10.1038/224285a0.

Cvetkov, T. L., Huynh, K. W., Cohen, M. R., & Moiseenkova-Bell, V. Y. (2011). Molecular architecture and subunit organization of TRPA1 ion channel revealed by electron microscopy. *The Journal of Biological Chemistry, 286*(44), 38168–38176.

Daanen, H. A. (2003). Finger cold-induced vasodilation: A review. *European Journal of Applied Physiology, 89*(5), 411–426. https://doi.org/10.1007/s00421-003-0818-2.

Defalco, J., Steiger, D., Gustafson, A., Emerling, D. E., Kelly, M. G., & Duncton, M. A. (2010). Oxime derivatives related to AP18: Agonists and antagonists of the TRPA1 receptor. *Bioorganic & Medicinal Chemistry Letters, 20*(1), 276–279. https://doi.org/10.1016/j.bmcl.2009.10.113.

Doerner, J. F., Gisselmann, G., Hatt, H., & Wetzel, C. H. (2007). Transient receptor potential channel A1 is directly gated by calcium ions. *The Journal of Biological Chemistry, 282*(18), 13180–13189.

Dowell, F. J., & Martin, W. (1997). The effects of peroxynitrite on rat aorta: Interaction with glucose and related substances. *European Journal of Pharmacology, 338*(1), 43–53. https://doi.org/10.1016/s0014-2999(97)01320-4.

Earley, S., Gonzales, A. L., & Crnich, R. (2009). Endothelium-dependent cerebral artery dilation mediated by TRPA1 and Ca^{2+}-activated K^{+} channels. *Circulation Research, 104*(8), 987–994. https://doi.org/10.1161/CIRCRESAHA.108.189530.

Eberhardt, M. J., Filipovic, M. R., Leffler, A., de la Roche, J., Kistner, K., Fischer, M. J., et al. (2012). Methylglyoxal activates nociceptors through transient receptor potential channel A1 (TRPA1): A possible mechanism of metabolic neuropathies. *The Journal of Biological Chemistry, 287*(34), 28291–28306.

Fajardo, O., Meseguer, V., Belmonte, C., & Viana, F. (2008). TRPA1 channels: Novel targets of 1,4-dihydropyridines. *Channels (Austin, Tex.), 2*(6), 429–438.

Gangula, P. R., Zhao, H., Supowit, S. C., Wimalawansa, S. J., Dipette, D. J., Westlund, K. N., et al. (2000). Increased blood pressure in alpha-calcitonin gene-related peptide/calcitonin gene knockout mice. *Hypertension, 35*(1 Pt. 2), 470–475. https://doi.org/10.1161/01.hyp.35.1.470.

Gaudet, R. (2008). A primer on ankyrin repeat function in TRP channels and beyond. *Molecular BioSystems, 4*(5), 372–379. https://doi.org/10.1039/b801481g.

Graepel, R., Fernandes, E. S., Aubdool, A. A., Andersson, D. A., Bevan, S., & Brain, S. D. (2011). 4-Oxo-2-nonenal (4-ONE): Evidence of transient receptor potential ankyrin 1-dependent and -independent nociceptive and vasoactive responses in vivo. *The Journal of Pharmacology and Experimental Therapeutics, 337*(1), 117–124. https://doi.org/10.1124/jpet.110.172403.

Griffith, T. M., Chaytor, A. T., Taylor, H. J., Giddings, B. D., & Edwards, D. H. (2002). cAMP facilitates EDHF-type relaxations in conduit arteries by enhancing electrotonic conduction via gap junctions. *Proceedings of the National Academy of Sciences of the United States of America, 99*(9), 6392–6397. https://doi.org/10.1073/pnas.092089799.

Hazari, M. S., Haykal-Coates, N., Winsett, D. W., Krantz, Q. T., King, C., Costa, D. L., et al. (2011). TRPA1 and sympathetic activation contribute to increased risk of triggered cardiac arrhythmias in hypertensive rats exposed to diesel exhaust. *Environmental Health Perspectives, 119*(7), 951–957. https://doi.org/10.1289/ehp.1003200.

Hinman, A., Chuang, H. H., Bautista, D. M., & Julius, D. (2006). TRP channel activation by reversible covalent modification. *Proceedings of the National Academy of Sciences of the United States of America, 103*(51), 19564–19568. https://doi.org/10.1073/pnas.0609598103.

Hu, H., Bandell, M., Petrus, M. J., Zhu, M. X., & Patapoutian, A. (2009). Zinc activates damage-sensing TRPA1 ion channels. *Nature Chemical Biology, 5*(3), 183–190.

Jordt, S. E., Bautista, D. M., Chuang, H. H., McKemy, D. D., Zygmunt, P. M., Hogestatt, E. D., et al. (2004). Mustard oils and cannabinoids excite sensory nerve fibres through the TRP channel ANKTM1. *Nature, 427*(6971), 260–265. https://doi.org/10.1038/nature02282.

Kang, K., Pulver, S. R., Panzano, V. C., Chang, E. C., Griffith, L. C., Theobald, D. L., et al. (2010). Analysis of Drosophila TRPA1 reveals an ancient origin for human chemical nociception. *Nature, 464*(7288), 597–600. https://doi.org/10.1038/nature08848.

Karashima, Y., Damann, N., Prenen, J., Talavera, K., Segal, A., Voets, T., et al. (2007). Bimodal action of menthol on the transient receptor potential channel TRPA1. *The Journal of Neuroscience, 27*(37), 9874–9884.

Karashima, Y., Prenen, J., Talavera, K., Janssens, A., Voets, T., & Nilius, B. (2010). Agonist-induced changes in Ca(2+) permeation through the nociceptor cation channel TRPA1. *Biophysical Journal, 98*(5), 773–783.

Keatinge, W. R. (1957). The effect of general chilling on the vasodilator response to cold. *The Journal of Physiology, 139*(3), 497–507. https://doi.org/10.1113/jphysiol.1957.sp005908.

Koba, S., Hayes, S. G., & Sinoway, L. I. (2011). Transient receptor potential A1 channel contributes to activation of the muscle reflex. *American Journal of Physiology. Heart and Circulatory Physiology, 300*(1), H201–H213. https://doi.org/10.1152/ajpheart.00547.2009.

Kojima, R., Nozawa, K., Doihara, H., Keto, Y., Kaku, H., Yokoyama, T., et al. (2014). Effects of novel TRPA1 receptor agonist ASP7663 in models of drug-induced constipation and visceral pain. *European Journal of Pharmacology, 723*, 288–293. https://doi.org/10.1016/j.ejphar.2013.11.020.

Kontos, H. A., Wei, E. P., Raper, A. J., Rosenblum, W. I., Navari, R. M., & Patterson, J. L., Jr. (1978). Role of tissue hypoxia in local regulation of cerebral microcirculation. *The American Journal of Physiology, 234*(5), H582–H591. https://doi.org/10.1152/ajpheart.1978.234.5.H582.

Kremeyer, B., Lopera, F., Cox, J. J., Momin, A., Rugiero, F., Marsh, S., et al. (2010). A gain-of-function mutation in TRPA1 causes familial episodic pain syndrome. *Neuron, 66*(5), 671–680.

Kunkler, P. E., Ballard, C. J., Oxford, G. S., & Hurley, J. H. (2011). TRPA1 receptors mediate environmental irritant-induced meningeal vasodilatation. *Pain, 152*(1), 38–44. https://doi.org/10.1016/j.pain.2010.08.021.

Kwan, K. Y., Allchorne, A. J., Vollrath, M. A., Christensen, A. P., Zhang, D. S., Woolf, C. J., et al. (2006). TRPA1 contributes to cold, mechanical, and chemical nociception but is not essential for hair-cell transduction. *Neuron, 50*(2), 277–289. https://doi.org/10.1016/j.neuron.2006.03.042.

Ledoux, J., Bonev, A. D., & Nelson, M. T. (2008). Ca^{2+}-activated K^{+} channels in murine endothelial cells: Block by intracellular calcium and magnesium. *The Journal of General Physiology, 131*(2), 125–135. https://doi.org/10.1085/jgp.200709875.

Lee, G., Abdi, K., Jiang, Y., Michaely, P., Bennett, V., & Marszalek, P. E. (2006). Nanospring behaviour of ankyrin repeats. *Nature, 440*(7081), 246–249. https://doi.org/10.1038/nature04437.

Leffler, A., Lattrell, A., Kronewald, S., Niedermirtl, F., & Nau, C. (2011). Activation of TRPA1 by membrane permeable local anesthetics. *Molecular Pain*, *7*, 62. https://doi.org/10.1186/1744-8069-7-62.

Li, M., Toombes, G. E., Silberberg, S. D., & Swartz, K. J. (2015). Physical basis of apparent pore dilation of ATP-activated P2X receptor channels. *Nature Neuroscience*, *18*(11), 1577–1583. https://doi.org/10.1038/nn.4120.

Libby, P., Buring, J. E., Badimon, L., Hansson, G. K., Deanfield, J., Bittencourt, M. S., et al. (2019). Atherosclerosis. *Nature Reviews Disease Primers*, *5*(1), 56. https://doi.org/10.1038/s41572-019-0106-z.

Liu, S., Shi, H., Liu, W., Furuichi, T., Timmins, G. S., & Liu, K. J. (2004). Interstitial pO_2 in ischemic penumbra and core are differentially affected following transient focal cerebral ischemia in rats. *Journal of Cerebral Blood Flow and Metabolism*, *24*(3), 343–349. https://doi.org/10.1097/01.WCB.0000110047.43905.01.

Ma, S., Zhang, Y., He, K., Wang, P., & Wang, D. H. (2019). Knockout of TRPA1 exacerbates angiotensin II-induced kidney injury. *American Journal of Physiology. Renal Physiology*, *317*(3), F623–F631. https://doi.org/10.1152/ajprenal.00069.2019.

Maarouf, N., Sancho, M., Furstenhaupt, T., Tran, C. H., & Welsh, D. G. (2017). Structural analysis of endothelial projections from mesenteric arteries. *Microcirculation*, *24*(3), e12330. https://doi.org/10.1111/micc.12330.

Macpherson, L. J., Geierstanger, B. H., Viswanath, V., Bandell, M., Eid, S. R., Hwang, S., et al. (2005). The pungency of garlic: Activation of TRPA1 and TRPV1 in response to allicin. *Current Biology*, *15*(10), 929–934. https://doi.org/10.1016/j.cub.2005.04.018.

Macpherson, L. J., Hwang, S. W., Miyamoto, T., Dubin, A. E., Patapoutian, A., & Story, G. M. (2006). More than cool: Promiscuous relationships of menthol and other sensory compounds. *Molecular and Cellular Neurosciences*, *32*(4), 335–343. https://doi.org/10.1016/j.mcn.2006.05.005.

Macpherson, L. J., Xiao, B., Kwan, K. Y., Petrus, M. J., Dubin, A. E., Hwang, S., et al. (2007). An ion channel essential for sensing chemical damage. *The Journal of Neuroscience*, *27*(42), 11412–11415.

McGaraughty, S., Chu, K. L., Perner, R. J., Didomenico, S., Kort, M. E., & Kym, P. R. (2010). TRPA1 modulation of spontaneous and mechanically evoked firing of spinal neurons in uninjured, osteoarthritic, and inflamed rats. *Molecular Pain*, *6*, 14. https://doi.org/10.1186/1744-8069-6-14.

McNamara, C. R., Mandel-Brehm, J., Bautista, D. M., Siemens, J., Deranian, K. L., Zhao, M., et al. (2007). TRPA1 mediates formalin-induced pain. *Proceedings of the National Academy of Sciences of the United States of America*, *104*(33), 13525–13530.

Meents, J. E., Fischer, M. J., & McNaughton, P. A. (2016). Agonist-induced sensitisation of the irritant receptor ion channel TRPA1. *The Journal of Physiology*, *594*(22), 6643–6660.

Mercado, J., Baylie, R., Navedo, M. F., Yuan, C., Scott, J. D., Nelson, M. T., et al. (2014). Local control of TRPV4 channels by AKAP150-targeted PKC in arterial smooth muscle. *The Journal of General Physiology*, *143*(5), 559–575. https://doi.org/10.1085/jgp.201311050.

Meseguer, V., Alpizar, Y. A., Luis, E., Tajada, S., Denlinger, B., Fajardo, O., et al. (2014). TRPA1 channels mediate acute neurogenic inflammation and pain produced by bacterial endotoxins. *Nature Communications*, *5*, 3125. https://doi.org/10.1038/ncomms4125.

Miyamoto, T., Dubin, A. E., Petrus, M. J., & Patapoutian, A. (2009). TRPV1 and TRPA1 mediate peripheral nitric oxide-induced nociception in mice. *PLoS One*, *4*(10), e7596. https://doi.org/10.1371/journal.pone.0007596.

Mosavi, L. K., Minor, D. L., Jr., & Peng, Z.-Y. (2002). Consensus-derived structural determinants of the ankyrin repeat motif. *Proceedings of the National Academy of Sciences of the United States of America*, *99*(25), 16029–16034. https://doi.org/10.1073/pnas.252537899.

Nagata, K., Duggan, A., Kumar, G., & Garcia-Anoveros, J. (2005). Nociceptor and hair cell transducer properties of TRPA1, a channel for pain and hearing. *The Journal of Neuroscience*, *25*(16), 4052–4061. https://doi.org/10.1523/JNEUROSCI.0013-05.2005.

Nassini, R., Materazzi, S., Andre, E., Sartiani, L., Aldini, G., Trevisani, M., et al. (2010). Acetaminophen, via its reactive metabolite N-acetyl-p-benzo-quinoneimine and transient receptor potential ankyrin-1 stimulation, causes neurogenic inflammation in the airways and other tissues in rodents. *The FASEB Journal*, *24*(12), 4904–4916. https://doi.org/10.1096/fj.10-162438.

Nassini, R., Pedretti, P., Moretto, N., Fusi, C., Carnini, C., Facchinetti, F., et al. (2012). Transient receptor potential ankyrin 1 channel localized to non-neuronal airway cells promotes non-neurogenic inflammation. *PLoS One*, 7(8), e42454. https://doi.org/10.1371/journal.pone.0042454.

Navedo, M. F., Amberg, G. C., Votaw, V. S., & Santana, L. F. (2005). Constitutively active L-type Ca^{2+} channels. *Proceedings of the National Academy of Sciences of the United States of America*, *102*(31), 11112–11117. https://doi.org/10.1073/pnas.0500360102.

Nilius, B., Appendino, G., & Owsianik, G. (2012). The transient receptor potential channel TRPA1: From gene to pathophysiology. *Pflügers Archiv*, *464*(5), 425–458.

Nilius, B., Prenen, J., & Owsianik, G. (2011). Irritating channels: The case of TRPA1. *The Journal of Physiology*, *589*(Pt. 7), 1543–1549.

Oguri, G., Nakajima, T., Yamamoto, Y., Takano, N., Tanaka, T., Kikuchi, H., et al. (2014). Effects of methylglyoxal on human cardiac fibroblast: Roles of transient receptor potential ankyrin 1 (TRPA1) channels. *American Journal of Physiology. Heart and Circulatory Physiology*, *307*(9), H1339–H1352. https://doi.org/10.1152/ajpheart.01021.2013.

Paulsen, C. E., Armache, J. P., Gao, Y., Cheng, Y., & Julius, D. (2015). Structure of the TRPA1 ion channel suggests regulatory mechanisms. *Nature*, *520*(7548), 511–517. https://doi.org/10.1038/nature14367.

Petrus, M., Peier, A. M., Bandell, M., Hwang, S. W., Huynh, T., Olney, N., et al. (2007). A role of TRPA1 in mechanical hyperalgesia is revealed by pharmacological inhibition. *Molecular Pain*, *3*, 40. https://doi.org/10.1186/1744-8069-3-40.

Pires, P. W., & Earley, S. (2017). Redox regulation of transient receptor potential channels in the endothelium. *Microcirculation*, *24*(3), e12329. https://doi.org/10.1111/micc.12329.

Pires, P. W., & Earley, S. (2018). Neuroprotective effects of TRPA1 channels in the cerebral endothelium following ischemic stroke. *eLife*, 7, e35316. https://doi.org/10.7554/eLife.35316.

Plump, A. S., Smith, J. D., Hayek, T., Aalto-Setala, K., Walsh, A., Verstuyft, J. G., et al. (1992). Severe hypercholesterolemia and atherosclerosis in apolipoprotein E-deficient mice created by homologous recombination in ES cells. *Cell*, *71*(2), 343–353. https://doi.org/10.1016/0092-8674(92)90362-g.

Pozsgai, G., Bodkin, J. V., Graepel, R., Bevan, S., Andersson, D. A., & Brain, S. D. (2010). Evidence for the pathophysiological relevance of TRPA1 receptors in the cardiovascular system in vivo. *Cardiovascular Research*, *87*(4), 760–768. https://doi.org/10.1093/cvr/cvq118.

Qian, X., Francis, M., Solodushko, V., Earley, S., & Taylor, M. S. (2013). Recruitment of dynamic endothelial Ca^{2+} signals by the TRPA1 channel activator AITC in rat cerebral arteries. *Microcirculation*, *20*(2), 138–148. https://doi.org/10.1111/micc.12004.

Rooney, L., Vidal, A., D'Souza, A. M., Devereux, N., Masick, B., Boissel, V., et al. (2014). Discovery, optimization, and biological evaluation of 5-(2-(trifluoromethyl)phenyl)indazoles as a novel class of transient receptor potential A1 (TRPA1) antagonists. *Journal of Medicinal Chemistry*, *57*(12), 5129–5140. https://doi.org/10.1021/jm401986p.

Rosenzweig, M., Brennan, K. M., Tayler, T. D., Phelps, P. O., Patapoutian, A., & Garrity, P. A. (2005). The Drosophila ortholog of vertebrate TRPA1 regulates thermotaxis. *Genes & Development*, *19*(4), 419–424. https://doi.org/10.1101/gad.1278205.

Samad, A., Sura, L., Benedikt, J., Ettrich, R., Minofar, B., Teisinger, J., et al. (2011). The C-terminal basic residues contribute to the chemical- and voltage-dependent activation of TRPA1. *The Biochemical Journal*, *433*(1), 197–204.

Sandow, S. L., & Hill, C. E. (2000). Incidence of myoendothelial gap junctions in the proximal and distal mesenteric arteries of the rat is suggestive of a role in endothelium-derived hyperpolarizing factor-mediated responses. *Circulation Research*, *86*(3), 341–346. https://doi.org/10.1161/01.res.86.3.341.

Sandow, S. L., Neylon, C. B., Chen, M. X., & Garland, C. J. (2006). Spatial separation of endothelial small- and intermediate-conductance calcium-activated potassium channels (K(Ca)) and connexins: Possible relationship to vasodilator function? *Journal of Anatomy*, *209*(5), 689–698. https://doi.org/10.1111/j.1469-7580.2006.00647.x.

Smillie, S. J., King, R., Kodji, X., Outzen, E., Pozsgai, G., Fernandes, E., et al. (2014). An ongoing role of alpha-calcitonin gene-related peptide as part of a protective network against hypertension, vascular hypertrophy, and oxidative stress. *Hypertension*, *63*(5), 1056–1062. https://doi.org/10.1161/HYPERTENSIONAHA.113.02517.

Smith, P. D., Brett, S. E., Luykenaar, K. D., Sandow, S. L., Marrelli, S. P., Vigmond, E. J., et al. (2008). KIR channels function as electrical amplifiers in rat vascular smooth muscle. *The Journal of Physiology*, *586*(4), 1147–1160. https://doi.org/10.1113/jphysiol.2007.145474.

Sonkusare, S. K., Bonev, A. D., Ledoux, J., Liedtke, W., Kotlikoff, M. I., Heppner, T. J., et al. (2012). Elementary Ca^{2+} signals through endothelial TRPV4 channels regulate vascular function. *Science*, *336*(6081), 597–601. https://doi.org/10.1126/science.1216283.

Sotomayor, M., Corey, D. P., & Schulten, K. (2005). In search of the hair-cell gating spring elastic properties of ankyrin and cadherin repeats. *Structure*, *13*(4), 669–682. https://doi.org/10.1016/j.str.2005.03.001.

Story, G. M., Peier, A. M., Reeve, A. J., Eid, S. R., Mosbacher, J., Hricik, T. R., et al. (2003). ANKTM1, a TRP-like channel expressed in nociceptive neurons, is activated by cold temperatures. *Cell*, *112*(6), 819–829. https://doi.org/10.1016/s0092-8674(03)00158-2.

Sullivan, M. N., & Earley, S. (2013). TRP channel Ca(2+) sparklets: Fundamental signals underlying endothelium-dependent hyperpolarization. *American Journal of Physiology. Cell Physiology*, *305*(10), C999–C1008. https://doi.org/10.1152/ajpcell.00273.2013.

Sullivan, M. N., Francis, M., Pitts, N. L., Taylor, M. S., & Earley, S. (2012). Optical recording reveals novel properties of GSK1016790A-induced vanilloid transient receptor potential channel TRPV4 activity in primary human endothelial cells. *Molecular Pharmacology*, *82*(3), 464–472. https://doi.org/10.1124/mol.112.078584.

Sullivan, M. N., Gonzales, A. L., Pires, P. W., Bruhl, A., Leo, M. D., Li, W., et al. (2015). Localized TRPA1 channel Ca^{2+} signals stimulated by reactive oxygen species promote cerebral artery dilation. *Science Signaling*, *8*(358), ra2. https://doi.org/10.1126/scisignal.2005659.

Tajada, S., Moreno, C. M., O'Dwyer, S., Woods, S., Sato, D., Navedo, M. F., et al. (2017). Distance constraints on activation of TRPV4 channels by AKAP150-bound PKCalpha in arterial myocytes. *The Journal of General Physiology*, *149*(6), 639–659. https://doi.org/10.1085/jgp.201611709.

Takahashi, N., Kuwaki, T., Kiyonaka, S., Numata, T., Kozai, D., Mizuno, Y., et al. (2011). TRPA1 underlies a sensing mechanism for O_2. *Nature Chemical Biology*, 7(10), 701–711. https://doi.org/10.1038/nchembio.640.

Takaya, J., Mio, K., Shiraishi, T., Kurokawa, T., Otsuka, S., Mori, Y., et al. (2015). A potent and site-selective agonist of TRPA1. *Journal of the American Chemical Society*, *137*(50), 15859–15864. https://doi.org/10.1021/jacs.5b10162.

Talavera, K., Startek, J. B., Alvarez-Collazo, J., Boonen, B., Alpizar, Y. A., Sanchez, A., et al. (2019). Mammalian transient receptor potential TRPA1 channels: From structure to disease. *Physiological Reviews*. https://doi.org/10.1152/physrev.00005.2019.

Taler, S. J. (2018). Initial treatment of hypertension. *The New England Journal of Medicine*, *378*(7), 636–644. https://doi.org/10.1056/NEJMcp1613481.

Taylor-Clark, T. E., McAlexander, M. A., Nassenstein, C., Sheardown, S. A., Wilson, S., Thornton, J., et al. (2008). Relative contributions of TRPA1 and TRPV1 channels in the activation of vagal bronchopulmonary C-fibres by the endogenous autacoid 4-oxononenal. *The Journal of Physiology*, *586*(14), 3447–3459. https://doi.org/10.1113/jphysiol.2008.153585.

Trevisani, M., Siemens, J., Materazzi, S., Bautista, D. M., Nassini, R., Campi, B., et al. (2007). 4-Hydroxynonenal, an endogenous aldehyde, causes pain and neurogenic inflammation through activation of the irritant receptor TRPA1. *Proceedings of the National Academy of Sciences of the United States of America*, *104*(33), 13519–13524. https://doi.org/10.1073/pnas.0705923104.

Tsai, A. G., Johnson, P. C., & Intaglietta, M. (2003). Oxygen gradients in the microcirculation. *Physiological Reviews*, *83*(3), 933–963. https://doi.org/10.1152/physrev.00034.2002.

Uchida, K. (2003). 4-Hydroxy-2-nonenal: A product and mediator of oxidative stress. *Progress in Lipid Research*, *42*(4), 318–343. doi: S0163782703000146 [pii].

Wang, Y. Y., Chang, R. B., Waters, H. N., McKemy, D. D., & Liman, E. R. (2008). The nociceptor ion channel TRPA1 is potentiated and inactivated by permeating calcium ions. *The Journal of Biological Chemistry*, *283*(47), 32691–32703.

Wang, L., Cvetkov, T. L., Chance, M. R., & Moiseenkova-Bell, V. Y. (2012). Identification of in vivo disulfide conformation of TRPA1 ion channel. *The Journal of Biological Chemistry*, *287*(9), 6169–6176.

Wei, H., Chapman, H., Saarnilehto, M., Kuokkanen, K., Koivisto, A., & Pertovaara, A. (2010). Roles of cutaneous versus spinal TRPA1 channels in mechanical hypersensitivity in the diabetic or mustard oil-treated non-diabetic rat. *Neuropharmacology*, *58*(3), 578–584. https://doi.org/10.1016/j.neuropharm.2009.12.001.

Wei, H., Hamalainen, M. M., Saarnilehto, M., Koivisto, A., & Pertovaara, A. (2009). Attenuation of mechanical hypersensitivity by an antagonist of the TRPA1 ion channel in diabetic animals. *Anesthesiology*, *111*(1), 147–154. https://doi.org/10.1097/ALN.0b013e3181a1642b.

Woll, K. A., Skinner, K. A., Gianti, E., Bhanu, N. V., Garcia, B. A., Carnevale, V., et al. (2017). Sites contributing to TRPA1 activation by the anesthetic propofol identified by photoaffinity labeling. *Biophysical Journal*, *113*(10), 2168–2172. https://doi.org/10.1016/j.bpj.2017.08.040.

Xing, J., Lu, J., & Li, J. (2015). TRPA1 mediates amplified sympathetic responsiveness to activation of metabolically sensitive muscle afferents in rats with femoral artery occlusion. *Frontiers in Physiology*, *6*, 249. https://doi.org/10.3389/fphys.2015.00249.

Yanaga, A., Goto, H., Nakagawa, T., Hikiami, H., Shibahara, N., & Shimada, Y. (2006). Cinnamaldehyde induces endothelium-dependent and -independent vasorelaxant action on isolated rat aorta. *Biological & Pharmaceutical Bulletin*, *29*(12), 2415–2418. https://doi.org/10.1248/bpb.29.2415.

Zayats, V., Samad, A., Minofar, B., Roelofs, K. E., Stockner, T., & Ettrich, R. (2013). Regulation of the transient receptor potential channel TRPA1 by its N-terminal ankyrin repeat domain. *Journal of Molecular Modeling*, *19*(11), 4689–4700. https://doi.org/10.1007/s00894-012-1505-1.

Zhang, S. H., Reddick, R. L., Piedrahita, J. A., & Maeda, N. (1992). Spontaneous hypercholesterolemia and arterial lesions in mice lacking apolipoprotein E. *Science*, *258*(5081), 468–471. https://doi.org/10.1126/science.1411543.

Zhao, J. F., Shyue, S. K., Kou, Y. R., Lu, T. M., & Lee, T. S. (2016). Transient receptor potential ankyrin 1 channel involved in atherosclerosis and macrophage-foam cell formation. *International Journal of Biological Sciences*, *12*(7), 812–823. https://doi.org/10.7150/ijbs.15229.

Zurborg, S., Yurgionas, B., Jira, J. A., Caspani, O., & Heppenstall, P. A. (2007). Direct activation of the ion channel TRPA1 by Ca^{2+}. *Nature Neuroscience*, *10*(3), 277–279.

CHAPTER SIX

Regulation of ion channels in the microcirculation by mineralocorticoid receptor activation

Laura Chambers, Anne M. Dorrance*
Department of Pharmacology and Toxicology, Michigan State University, East Lansing, MI, United States
*Corresponding author: e-mail address: dorranc3@msu.edu

Contents

Abstract

The mineralocorticoid receptor (MR) has classically been studied in the renal epithelium for its role in regulating sodium and water balance and, subsequently, blood pressure. However, the MR also plays a critical role in the microvasculature by regulating ion channel expression and function. Activation of the MR by its endogenous agonist aldosterone results in translocation of the MR into the nucleus, where it can act as a transcription factor. Although most of the actions of the aldosterone can be attributed to its genomic activity though MR activation, it can also act by nongenomic mechanisms. Activation of this ubiquitous receptor increases the expression of epithelial sodium channels (ENaC) in both the endothelium and smooth muscle cells of peripheral and cerebral vessels. MR activation also regulates activity of calcium channels,

Current Topics in Membranes, Volume 85
ISSN 1063-5823
https://doi.org/10.1016/bs.ctm.2020.02.001

calcium-activated potassium channels, and various transient receptor potential (TRP) channels. Modification of these ion channels results in a myriad of negative consequences, including impaired endothelium-dependent vasodilation, alterations in generation of myogenic tone, and increased inflammation and oxidative stress. Taken together, these studies demonstrate the importance of studying the impact of the MR on ion channel function in the vasculature. While research in this area has made advances in recent years, there are still many large gaps in knowledge that need to be filled. Crucial future directions of study include defining the molecular mechanisms involved in this interaction, as well as elucidating the potential sex differences that may exist, as these areas of understanding are currently lacking.

1. Introduction

Interest in the effects of aldosterone and activation of its cognate receptor, the type 1 mineralocorticoid receptor (MR) (Funder, Feldman, & Edelman, 1972, 1973) began a renaissance at the end of the last century with the publication of the Randomized Aldactone Evaluation Study (RALES) trial (Pitt et al., 1999). Until this publication, most studies took a nephrocentric view of the MR focusing on the effects of MR activation on the renal epithelium where MR activation increases the expression and activity of proteins involved in regulating sodium and water balance, and therefore blood pressure (Soundararajan, Pearce, Hughey, & Kleyman, 2010). The RALES trial marked a turning point in the fortunes of the MR antagonist spironolactone. This publication was one of the first to show that the MR antagonists have marked beneficial effects on the heart that lead to a reduction in morbidity and mortality in heart failure patients. This beneficial effect was confirmed using the second generation MR antagonist, eplerenone (Pitt et al., 2001). Importantly, both studies suggest the beneficial effects of the MR antagonists were not merely due to a reduction in blood pressure as the antihypertensive effects of both drugs were relatively modest. Since that time many studies have shown that aldosterone, through MR activation, impairs vascular function. These detrimental effects have been widely reviewed by others (Barrera-Chimal & Jaisser, 2019; DuPont, Hill, Bender, Jaisser, & Jaffe, 2014; Moss, Carvajal, & Jaffe, 2019) and include increased inflammation, fibrosis, and reactive oxygen species generation.

In the kidney, the MR regulates sodium and potassium transport; this should have provided the scientific community with a hint that aldosterone might also regulate ion transport in the vasculature. It is therefore surprising that the study of the effects of aldosterone on vascular ion channels is in its relative infancy (DuPont et al., 2014; Valinsky, Touyz, & Shrier, 2019). One issue associated with assessing the effects of MR activation on vascular function is that it is often difficult to dissociate the direct effects of MR activation on target genes from indirect effect of vascular injury associated with MR activation and hypertension *per se*; these indirect effects could include effects of inflammation and reactive oxygen species generation. Conversely, many studies utilize angiotensin II (AngII) hypertensive mice that exhibit marked hyperaldosteronism (Diaz-Otero et al., 2017, 2018). The majority of publications using this model attribute changes in ion channel function to the hypertensive milieu and do not pause to consider if MR activation could be a key mediator of the effects observed. This chapter will focus primarily on situations where ion channel activity or protein expression has been shown to change with MR activation and we will address the possible physiological implications of this on the vasculature. Where possible, the discussion of the effects of MR regulated ion channels on vascular function will be limited to the microcirculation: small arteries, arterioles and capillaries. Sadly, there is a limited literature regarding the effects of MR activation on ion channels in resistance arteries, so where no microvascular studies exist, we will consider studies conducted in conduit arteries and in cell culture.

2. Mineralocorticoid receptors

Before we address of the effects of MR activation on ion channels, it is important to pause and consider the MR itself and the ligands that activate it. The MR is the primary site of action for aldosterone. This nuclear receptor is encoded in humans by the *NR3C2* gene and it functions in the cell as a transcription factor. The MR is recognized to be a promiscuous receptor in terms of the ligands that can interact with it as both agonists and antagonists (Funder, 2017). Several endogenous ligands including cortisol, corticosterone, deoxycorticosterone (DOC) and progesterone bind to the MR with a similar affinity to that of aldosterone (Funder, 2017). Of this list of potential ligands, DOC appears to be entirely unimportant. Although the plasma levels of DOC are similar to those of aldosterone, DOC is largely protein bound and therefore cannot interact with the receptor (Zipser, LeBoff, Meidar, Duke, & Horton, 1980). DOC levels also do not change in response to sodium or volume depletion (Oddie, Coghlan, & Scoggins, 1972) thus it does not appear to regulate what we would consider to be normal physiological effects of the MR. Progesterone has affinity for the MR but it functions as an antagonist. Like DOC, progesterone is heavily bound to plasma protein and therefore is mostly unavailable for receptor activation. That being said, progesterone levels in the luteal phase of the menstrual cycle are high enough to produce increases in plasma aldosterone levels through the negative feedback loop (Ahmed et al., 2015). The physiological relevance of the interplay between progesterone and aldosterone is not clear at present, but with the current increased interest in sex differences in cardiovascular disease this could be a fruitful area of study.

This brings us to cortisol and corticosterone, the physiological glucocorticoids in humans and rodents, respectively. Studies of the evolution of the MR and aldosterone suggest that for thousands of years these glucocorticoids were the primary ligands responsible for MR activation (Funder, 2017). Cortisol circulates at approximately 1000-fold higher levels than aldosterone, but it is highly protein bound, which results in the levels of the hormone available for MR activation being 100-fold higher than aldosterone. Clearly, even this reduced level of available hormone still has the ability to overwhelm aldosterone such that all the MRs in the cell will be occupied by cortisol. In the vascular endothelium and smooth muscle cells, the MRs are protected from cortisol-mediated overactivation by the enzyme 11β-hydroxysteroid dehydrogenase-type 2 (11βHSD2) (Caprio et al., 2008; Christy et al., 2003; Walker et al., 1991). Classically, this enzyme was thought to shuttle cortisol to cortisone, which has no affinity for the MR. This dogma has been called into question recently. Newer studies suggest that while 11βHSD2 reduces the amount of cortisol available to bind to the MR, it only does so by a factor of 10. Thus, in most tissues the MR is occupied by cortisol where it functions as an antagonist. This antagonist function is related to high levels of nicotinamide adenine dinucleotide (NADH) which is a byproduct of the reaction catalyzed by 11βHSD2. Importantly, when 11βHSD2 fails or when the NADH levels in the cells fall, as happens in vascular disease and when there is significant tissue injury and oxidative stress, cortisol exhibits agonist effects on the MR (Funder, 2017; Mihailidou, Loan Le, Mardini, & Funder, 2009). Thus, it is possible that in many of the studies described here the MR is activated by cortisol and not by aldosterone. This change in agonist does not impact the effects of MR activation, but it is mentioned here to remind the reader that aldosterone levels do not have to be increased for there to be significant MR-mediated effects in the vasculature. Studies using the recently developed aldosterone synthesis inhibitors should provide additional information to help define which MR ligands are important for the genesis of vascular dysfunction (Bernhardt, 2016; Sakakibara et al., 2018).

Most of the effects of aldosterone occur through the classical genomic pathway whereby aldosterone binds to the MR, and this complex translocates to the nucleus where it functions as a transcription factor to regulate gene expression. There are several studies that suggest aldosterone is also capable of modulating cellular events *via* non-genomic pathways. These rapid actions of aldosterone have been proposed to occur through a membrane bound receptor and the g-protein coupled estrogen receptor (GPER or GPR30) has been suggested to be responsible for these actions (Gros et al., 2011; Wendler & Wehling, 2011).

3. Aldosterone production

Every physiology textbook describes aldosterone as a product of the adrenal zona glomerulosa. Aldosterone synthesis in this region is tightly regulated by angiotensin II, potassium, and to a certain extent adrenocorticotrophic hormone (ACTH) (MacKenzie, van Kralingen, & Davies, 2019). Several other substances modulate aldosterone production. When considering cardiovascular disease the regulation of aldosterone production by leptin (Huby et al., 2015) may be the most pertinent as is could contribute to obesity-associated vascular disease. Interestingly, this pathway exhibits significant sex differences and seems to be a more important regulator of aldosterone production in females than in males (Huby, Otvos, & Belin de Chantemèle, 2016).

In the years following the commercialization of PCR products and primers, the number of cell types reported to produce aldosterone proliferated. These studies relied on identifying tissues where aldosterone synthase, the rate limiting enzyme in the production of aldosterone, was expressed. In these studies, aldosterone was reported to be produced in the heart and in the vasculature, and as the obesity epidemic took hold aldosterone was also reported to be generated by adipocytes. Despite the excitement about potential paracrine effects of aldosterone in the vasculature, it seems entirely likely that the vasculature is not a source of aldosterone because newer studies suggest that these tissues lack the full complement of enzymes required for hormone synthesis (Ahmad, Romero, Gomez-Sanchez, & Gomez-Sanchez, 2004; Ye et al., 2005). In fact, the only two extra-adrenal sites of aldosterone production that are still widely discussed are adipose tissue and the brain (Dinh Cat, Friederich-Persson, White, & Touyz, 2016; Ye et al., 2008). The rise and ultimately the fall of the field of extra-adrenal aldosterone production has been documented by MacKenzie, Connell, and Davies (2012).

4. Epithelial sodium channels

The epithelial sodium channel (ENaC) is the quintessential MR-regulated ion channel. ENaCs were first identified in the renal epithelium; they are inwardly rectifying, amiloride-sensitive sodium channels and dysfunction of these channels is a common cause of hypertension (Ambrosius et al., 1999). Liddle syndrome, the best characterized form of hypertension associated with increased ENaC activity, is readily treatable with amiloride, an ENaC inhibitor (Tetti et al., 2018). Although discovered in the kidney, it is clear that ENaCs are expressed and functional in the vasculature (Kusche-Vihrog, Jeggle, & Oberleithner, 2014; Pérez et al., 2009) in both vascular smooth muscle (Drummond, 2007; Drummond et al., 2004, 2008; Grifoni, Gannon, Stec, & Drummond, 2006; Jernigan & Drummond, 2005) and endothelial cells (Jeggle, Hofschröer, Maase, Bertog, & Kusche-Vihrog, 2016; Kusche-Vihrog, Jeggle, & Oberleithner, 2014). The mammalian ENaC protein is closely related to a family of mechanosensitive proteins, the degenerins, that are expressed in nematodes (Folgering et al., 2008; Fronius & Clauss, 2008) and several studies suggest that vascular ENaCs function as mechanosensors. These include studies showing the ENaC opening probability is increased by increasing the laminar flow in cell culture studies using oocytes and cortical collecting ducts (Satlin, Sheng, Woda, & Kleyman, 2001). The same is true for endothelial cells where increasing sheer stress increases sodium influx (Guo et al., 2016; Wang, Meng, Mohan, Champaneri, & Gu, 2009). The proposal that ENaCs are mechanosensors is strengthened by the knowledge that in the endothelium ENaC is associated cytoskeletal proteins. Studies using cultured endothelial cells suggest that this interaction occurs primary with tubulin. This study only considered channel expression and did not assess potential changes in channel activity (Golestaneh et al., 2001). The exact role of actin binding the regulation of ENaC activity is less clear and additional studies are needed in vascular cells (Karpushev, Ilatovskaya, Pavlov, Negulyaev, & Staruschenko, 2010; Karpushev, Ilatovskaya, & Staruschenko, 2010).

ENaC is a constitutively active channel with an open probability of approximately 0.4 (Schild, 2010); therefore, the regulation of ENaC currents predominantly occurs through alterations in channel expression and membrane localization. Much of what we know about the regulation of this channel by aldosterone comes from studies conducted in the kidney where four subunits, α, β, γ and δ, have been identified. The α subunit alone is capable of forming a sodium channel, but channel activity is enhanced when either the β or γ subunit is also present and activity is maximal when all three subunits are expressed (Canessa et al., 1994). Classically the α, β, and γ subunits were thought to from a heterodimer with a 1:1:1 stoichiometry. However, a more complex structure containing up to nine subunits (Snyder, Cheng, Prince, Rogers, & Welsh, 1998) that may exist as trimers of trimers within cells has been suggested (Stewart, Haerteis, Diakov, Korbmacher, & Edwardson, 2011). In humans a δ subunit has also been reported to exist in the kidney and umbilical vein endothelial cells (HUVECs) (Downs, Johnson, Coca, & Helms, 2018; Ji et al., 2012), it seems that when present in the cell the δ subunit replaces the α subunit in the signaling complex (Ji et al., 2012). There is some evidence that the δ subunit expression is regulated by angiotensin II (AngII) via increased reactive oxygen species generation (Downs et al., 2018). It is not clear at present if aldosterone regulates the δ subunit or if this subunit is expressed in resistance arteries in either endothelial or smooth muscle cells.

The regulation of ENaC activity by aldosterone is complex and cannot simply be attributed to an increase in ENaC gene transcription and translation in response to transcription factor activation. Aldosterone increases gene transcription of ENaC through classical MR-related pathways (Alvarez de la Rosa, Li, & Canessa, 2002; May, Puoti, Gaeggeler, Horisberger, & Rossier, 1997). Interestingly the effects of aldosterone appear to be subunit and tissue specific. For example in kidney aldosterone causes increased expression of the α subunit, whereas the β and γ subunits are increased in the colon (Stokes & Sigmund, 1998). The regulation of ENaC by aldosterone involves glucocorticoid-induced leucine zipper (GILZ) protein (Robert-Nicoud et al., 2001) and connector enhancer of kinase suppressor of ras-3 (CNK3) (Ziera et al., 2009). The precise molecular mechanisms though which aldosterone regulates ENaC expression have been eloquently described by Valinsky et al. (2019). Studies of the effects of aldosterone on ENaC expression have shown that aldosterone increases ENaC expression but that the cytoskeleton in the endothelial cells must be intact for this to occur (Golestaneh et al., 2001).

The density of ENaCs in the membrane is also regulated by aldosterone-dependent signaling pathways. The c-terminal domains of all the ENaC subunits can interact with the E3 ubiquitin ligase Nedd4-2 (neural precursor cell expressed developmentally down-regulated protein) (Staub et al., 1996). Increased Nedd4-2 activity allows for ubiquitination of ENaC; this labels the subunits for proteasomal degradation. Aldosterone interrupts this degradation pathway through its ability to increase the expression of the serum and glucocorticoid regulated kinase-1 (SGK1) (Shigaev, Asher, Latter, Garty, & Reuveny, 2000). SGK1 phosphorylates Nedd4-2 which prevents it from being able to ubiquitinate ENaC, thus the channels are not marked for proteosomal degradation and they remain active in the membrane. Another early aldosterone-mediated response which occurs after 1 h of aldosterone treatment is the upregulation of ubiquitin-specific protease 2-45 (USP2-45) expression. USP2-45 deubiquitinates ENaC, this prevents ENaC from being broken down by the proteasome, leaving it in the membrane to transport sodium (Fakitsas et al., 2007).

Most of the studies regarding the regulation of ENaC by aldosterone, either directly or indirectly, have focused on classical genomic mechanisms that involve increases in gene transcription and translation of either ENaC itself or of proteins that regulate ENaC degradation. However, there is some evidence that aldosterone can regulate ENaC insertion into the membrane through nongenomic/rapid pathways. These pathways include an aldosterone-mediated phosphorylation of protein kinase D1 (McEneaney, Harvey, & Thomas, 2007), which is required for ENaC trafficking to the membrane (McEneaney, Harvey, & Thomas, 2008). These effects can be blocked by spironolactone but not by inhibitors of transcription and translation. The potential that aldosterone mediates these effects though GPER has not been considered. There is, however, evidence that estrogen regulates alveolar ENaC activity through GPER (Greenlee et al., 2013). This presents the possibility that aldosterone may regulate ENaC in a similar manner.

4.1 ENaC in the endothelium

ENaC was first identified in the vascular endothelium in the early 2000s (Fig. 1). The presence of an endothelial ENaC, sometimes referred to as EnENaC, has been documented by many groups using a combination of cell culture studies and intact and deendothelialized arteries (Golestaneh et al., 2001; Hillebrand et al., 2007; Kusche-Vihrog et al., 2008; Kusche-Vihrog, Jeggle, & Oberleithner, 2014; Oberleithner et al., 2003, 2006; Warnock, 2013). Standard electrophysiology techniques have been employed to identify amiloride-sensitive sodium currents in endothelial cells. In a study by Wang et al. using HUVECs and human microvascular endothelial cells (HMECs) this current was found to be associated only with the expression of the α subunit of ENaC. This finding was confirmed by

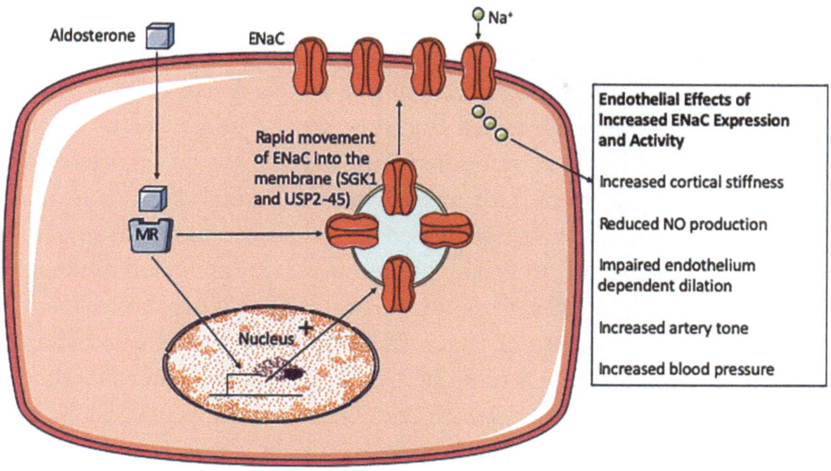

Fig. 1 Schematic representation of the proposed effects of MR activation on ENaC expression and activity in vascular endothelial cells. MR activation results in increased expression of ENaC via classical genomic mechanisms. MR activation also has rapid actions via SGK1 and USP2-45 that allow accumulation of ENaCs in the cell membrane. Activation of ENaC results in an increase in intracellular sodium, stiffening of the cell cortex, reduced eNOS-mediated NO production, impaired endothelium-dependent dilation, increased artery tone and hypertension (Jia et al., 2018; Kusche-Vihrog, Jeggle, & Oberleithner, 2014; Kusche-Vihrog, Tarjus, Fels, & Jaisser, 2014). Servier Medical Art by Servier is licensed under a Creative Commons Attribution 3.0 Unported License ("Servier Medical Art", 2004).

knockdown of the subunit using small interfering RNA techniques (Wang et al., 2009). In this study the authors were unable to detect the other subunits in the endothelial cells; other groups have since found that all three subunits are present in the endothelium (Pérez et al., 2009). Wang et al. noted that the ENaC associated current was much more difficult to identify in cell attached patches than it was in excised patches. This led the authors to propose that an intracellular compound must exert inhibitory actions on ENaC. They went on to identify this inhibitory molecule as 11,12-EET. Although not associated with aldosterone or MR activation this finding is of interest given increasing evidence that soluble epoxide hydrolase inhibitors, which increase EETs levels in the plasma, have significant protective effects on the cardiovascular system (Imig, 2012).

Much of our understanding of the effects of aldosterone on ENaC comes from studies using cultured endothelial cells. While these studies have contributed greatly to our understanding the molecular mechanisms involved in ENaC regulation it is important to remember that cell culture studies have significant limitations. Not least of these are the loss of the effects of pressure, flow, and circulating factors that would occur *in vivo*. In HUVECs aldosterone increases the number of ENaC molecules in the upper surface of the cells. These studies were conducted using quantum dot technology so that single channels could be identified, and this increase in membrane localization of ENaC was reduced by both spironolactone and by amiloride. The authors also used Western blot analysis to assess total EnaC expression and showed that this too was increased by aldosterone and prevented by amiloride. Unfortunately, these key studies were not conducted in the presence of spironolactone; therefore, it is not clear if blocking the MR would have prevented the increase in total cellular ENaC. These effects all appear to occur primarily with the α subunit; it is not clear if the other subunits are equally effected by aldosterone treatment (Kusche-Vihrog et al., 2008). One issue with this study and several other similar studies is that it was conducted using sub-confluent endothelial cells. It is not clear if the regulation of ENaC remains the same when the cells are confluent as they would be in the vasculature. It would seem particularly important that we gain additional understanding of how ENaCs are regulated in locations like the cerebral vasculature where the endothelial cells form the basis of the blood brain barrier.

Early studies of the functional effects of aldosterone and ENaC in endothelial cells focused on cell swelling. Endothelial cells were found to swell in response to aldosterone incubation. At the earliest times points studied (1 min) this effect was independent of sodium channels and MRs, but the effect observed after 5 min of aldosterone exposure could be inhibited by both spironolactone and amiloride, suggesting an MR-mediated effect on ENaCs (Oberleithner et al., 2003). Further studies focused on the concept that aldosterone-mediated increases in ENaC currents caused stiffening of the endothelial cell cortex. The cell cortex is the region just below the membrane; it is made up primarily of a mesh of actin filaments. The cortex determines cell shape and also the response to sheer stress and other mechanical stimuli (Pesen & Hoh, 2005). Pathophysiological levels of aldosterone increase endothelial cell stiffening in venous endothelial cells and human coronary artery endothelial cells. This effect appears to be MR-mediated because it could be inhibited by eplerenone (Hillebrand et al., 2007). The MR dependency of the effects of aldosterone on ENaC expression and cell stiffness has been confirmed in additional studies where spironolactone was used to block the MR (Drüppel et al., 2013). This aldosterone-mediated increase in endothelial cell stiffness can also be inhibited by inhibiting the ENaC with amiloride and is prevented in mice that lack the MR specifically in their endothelial cells (Jia et al., 2018). The movement of ENaC into the cell membrane is required for the stiffening of the endothelial cortex (Oberleithner et al., 2009). Recent studies using endothelial cell specific α-ENaC knockout mice have confirmed the role of ENaCs in endothelial cell stiffening. These studies were performed using aorta from mice after aldosterone infusion. Aldosterone increased endothelial cell stiffness in wildtype but not the knockout mice. Interestingly, the increased endothelial cell stiffness was not associated with alterations in pulse wave velocity, suggesting that the change in the vasculature was limited to the endothelium (Jia et al., 2018).

Increasing the stiffness of the vascular endothelial cells may have direct effects on vasodilation by reducing the levels of nitric oxide (NO) produced (Fels, Callies, Kusche-Vihrog, & Oberleithner, 2010; Oberleithner et al., 2007). This fateful combination of increased ENaC expression, stiffening of the endothelial cell cortex, and reduced NO production has become known as stiff endothelial cell syndrome (Lang, 2011). The effects of aldosterone on cellular stiffness and NO availability are persistent. In long-term studies using HUVECs and renal artery explants, the effects of aldosterone were found to remain throughout 24 days of treatment. In this study, aldosterone increased MR expression, ENaC expression, and cell stiffness; all of these effects were inhibited by spironolactone. Similarly, the presence of spironolactone in the aldosterone supplemented growth media also increased the release of NO in response to sheer stress (Drüppel et al., 2013). In conduit vessels, aldosterone has also been shown to reduce NO production *via* its effects on ENaC by increasing endoplasmic reticular stress. The increased ENaC expression was also associated with aldosterone-mediated increases in vascular inflammation and permeability (Jia et al., 2018). Taken together, these studies suggest that *in vivo* an increase in ENaC in the endothelium would result in impaired NO-mediated dilation. This concept fits well with the knowledge that elevated levels of aldosterone lead to endothelial injury, impaired dilation, and hypertension (Barrera-Chimal & Jaisser, 2019; DuPont et al., 2014; DuPont & Jaffe, 2017; Jaffe & Jaisser, 2014; McCurley & Jaffe, 2012).

Unfortunately, the limited number of studies in the literature that truly assess endothelial function provide a more complicated view of the situation. It is difficult to reconcile why a channel that is activated by flow (Wang et al., 2009) would in turn inhibit the production of NO when in most cases flow produces vasodilation that is mediated at least in part by NO production. Perez et al. utilized second order mesenteric arteries from adult male rats to define the functional role of ENaCs in the vasculature

(Pérez et al., 2009). Several complementary techniques were used to show that the α, β, and γ subunits are all expressed in both endothelial cells and vascular smooth muscle cells in small mesenteric arteries and pressure myography was used to show that ENaC regulates contraction in these arteries. Inhibition of ENaC with both amiloride and benzamil reduced the contraction produced by both phenylephrine and serotonin; this suggests that ENaC may produce a pro-contractile substance, or that it may inhibit the production of an anti-contractile substance, such as NO. The latter appears to be true. In an elegant series of studies, the authors showed that the effects of ENaC on contraction were largely mediated by the endothelium, where ENaC activity reduces the production of NO by endothelial nitric oxide synthase (eNOS); these findings agree with the cell culture studies described above. The effects of ENaC on eNOS activity were mediated by phosphoinositide 3 kinase through Akt (Pérez et al., 2009). Because ENaC is thought to be a mechanosensor, the authors also assessed the effects of amiloride on flow-mediated dilation. Increasing flow produced dilation in the mesenteric arteries and this was enhanced when ENaC was blocked by amiloride. These studies suggests that ENaC is not required for flow-mediated dilation to occur and, in fact, enhanced ENaC activity appears to negatively regulate this dilation (Drummond, 2009; Pérez et al., 2009). It is at present unclear if the many reports of impaired endothelium-dependent dilation induced by elevated levels of aldosterone (McCurley & Jaffe, 2012) can be attributed to this effect of ENaC on the vasculature. One clear finding from the study by Pérez et al. was that the effects of amiloride on vascular responses were not apparent in the absence of flow (Pérez et al., 2009). It is therefore entirely possible that many studies have missed this potential effect of ENaC on the endothelium because of the lack of flow in the experimental system. This raises an interesting problem for studies aimed at defining the effects of ENaC in the microcirculation; it is generally accepted that it is not possible to conduct studies with flow through small arteries and arterioles because the very small cannulas used produce a resistance to flow that is too great to overcome.

More recent studies have suggested the ENaCs actually have the opposite effect on the endothelium. In an attempt to resolve potential issues with the selectivity of the ENaC inhibitors, an endothelial cell specificic α-ENaC knockout mouse (EnENaC −/−) has been generated. These mice appear to be physiologically normal but, in keeping with the studies described above, the endothelial cortex was less stiff in the EnENaC −/− mice compaired to control. In control α-ENaC intact mice acetylcholine-mediated dilation and NO release was blocked by benzamil. This effect was not observed in the EnENaC −/− mice. Interestingly the EnENaC −/− mice had normal levels of NO because of compensation that occurs through an increase in AKT and eNOS phosphorylation. Instead of negatively regulating flow mediated dilation as the Pérez study suggests, the current study showed flow mediated dilation was markedly impaired in the absence of α-ENaC, suggesting a positive regulation of flow-mediated dilation by ENaC (Tarjus et al., 2017). It is difficult to reconcile why the compensation that occurs in the EnENaC −/− mice to maintain the acetylcholine-mediated dilation does not improve flow-mediated dilation which is at least partly NO-dependent (Looft-Wilson, Todd, Berberich, & Wolfert, 2018). The same knockout mouse strain has been used to evaluate the effects of ENaC on flow-mediated dilation *in vivo* (Sternak et al., 2018). This study made every effort to evaluate the effects of ENaC on the endothelium in a physiological manner; to that end, dilation was assessed using MRI techniques. Interestingly, the authors found that EnENaC −/− mice had marked impairments in acetylcholine-mediated dilation in the aorta and flow-mediated dilation was impaired in the femoral artery. This finding is in agreement with the the study conducted by Tarjus et al. Using the same mice, Jia et al. assessed the effects of aldosterone on endotheilial function in the presence and absence of endothelial ENaC. In these studies aldosterone impaired acetylcholine-mediated endothelium-dependent dilation in the aorta. This effect of aldosterone was not present in the endothelial cell EnENaC −/− mice, suggesting that ENaC may be responsible for the aldosterone-mediated endothelial dysfunction (Jia et al., 2018). The fact that these were conduit arteries and that dilation was assessed in the absence of flow may explain the discrepency in these results. Similar effects in aorta have been observed by the same group using mice fed a Western diet that is known to increase aldosterone levels (Sowers et al., 2019).

A recent study by Ashley, Mugloo, McDonald, and Fronius (2018) took on the challenge of dissecting the effects of ENaC on conduit and resistance arteries. Carotid and mesenteric arteries were used as models of conduit and resistance vessels. Similar to the studies described above, pressure myography was used to evaluate vascular reactivity. The authors show small but statically significant changes in artery stiffness in the small mesenteric arteries but not in carotid arteries; inhibiting ENaC with amiloride reduced artery stiffness. The magnitude of the change in the mesenteric arteries was very small, making the physiological relevance of this change difficult to assess. It is possible that this relates to the evidence that ENaC enhances endothelial cell stiffness and that this effect can be observed in small blood vessels where the ratio of endothelial cells to smooth muscle cells greater than it is in conduit arteries. Flow-mediated dilation was also assessed in both artery types. Surprisingly the authors found that while all the components of ENaC were present in both arteries the responses to ENaC inhibition were different between the vascular beds. In carotid arteries increasing the flow rate through the lumen caused a significant dilation. Inhibiting ENaC with amiloride further enhanced the dilation, suggesting that under normal conditions ENaC activation promotes vasoconstriction in carotid arteries. The opposite was true for the mesenteric arteries where increasing flow caused a significant contraction that was enhanced by amiloride, suggesting that in resistance arteries ENaC activation promotes vasodilation. In both cases this appears to be an effect of endothelial cell ENaC (Ashley et al., 2018). One major concern with this methodology is the choice of intralumenal pressure used for the vessels. In an effort to be consistent the authors chose to study both artery types at the same intralumenal pressure (60 mmHg). This pressure is on the low side for both artery types and this may be the reason why myogenic tone did not develop in the mesenteric arteries in this study in a manner similar to other reports in the literature (Greenstein et al., 2020). The way the data is presented, it is impossible to tell how much myogenic tone the vessels generate. It seems from the authors discussion that the mesenteric arteries did not generate any tone. In fact, the authors cite that as part of the reason they did not observe more dramatic effects of ENaC inhibition.

Taking all of these studies together and balancing them with the known vascular effects of aldosterone, it seems likely that enhanced ENaC activity likely impairs dilation *via* its actions on NO production (Fig. 1). There is a need for more carefully conducted studies to occur to truly compare the effects of ENaC in endothelial function in conduit and resistance arteries. It seems that it will be very important to consider flow-mediated vascular responses. At present, there are very few studies that consider the effects of ENaC in the vasculature of patients or animals with appropriate cardiovascular comorbidities. Similarly, it will be important to also assess potential effects of sex and aging on ENaC functions in the endothelium and beyond. It will also be important to conduct studies that really assess the long-term effects of ENaC blockade. It appears that the acute and chronic effects of ENaC inhibition differ significantly. In humans with type 1 diabetes, acute amiloride treatment increases NO production; this is in keeping with the studies that show increased ENaC activity impairs NO production. This is not the case in longer term studies. Unfortunately, these chronic studies were not conducted in the same group of patients; in these studies type 2 diabetic patients were used (Ydegaard et al., 2019). While the difference in comorbidities is a concern, these studies do raise an important issue that should be thoroughly addressed using preclinical models.

4.2 ENaC in vascular smooth muscle cells

While the literature regarding ENaC's effects in smooth muscle cells is small, there is consensus on how ENaC activation affects vascular function (Fig. 2). The idea that ENaC is a mechanonsensor led to the hypothesis that ENaC could be an important protein in vascular smooth muscle cells for the regulation and generation of myogenic tone. In fact, ENaC activity is essential for myogenic tone generation in renal interlobar arteries (Jernigan & Drummond, 2005, 2006) and in juxtamedullary afferent arterioles (Guan, Pollock, Cook, Hobbs, & Inscho, 2009). ENaC is also an important

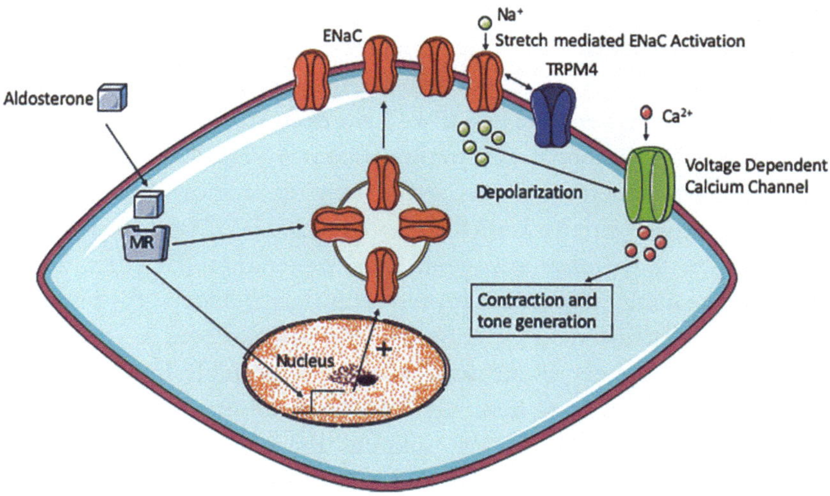

Fig. 2 Schematic representation of the proposed effects of MR activation on ENaC expression and activity in vascular smooth muscle cells. MR activation increases ENaC levels in the cell membrane. Activation of ENaC, possibly by cell stretch or increased intralumenal pressure, increases sodium transport into the cell resulting in depolarization. Cross talk between ENaC and TRPM4 appears to enhance this signaling pathway. Depolarization results in activation of the voltage-dependent calcium channels in the membrane leading to an increase in intracellular calcium that allows for phosphorylation of myosin and activation of contraction (Choi, Yeon, Kwon, Byeon, & Lee, 2017; Drummond, 2007, 2009; Drummond, Gebremedhin, & Harder, 2004; Drummond, Grifoni, & Jernigan, 2008; Kim, Ahn, Yeon, Lim, & Lee, 2012; Oyabe, Masumoto, Ueta, & Nakayama, 2000). Servier Medical Art by Servier is licensed under a Creative Commons Attribution 3.0 Unported License ("Servier Medical Art", 2004).

regulator of myogenic tone in the cerebral vasculature (Drummond, 2007; Kim, Choi, Lim, Yeon, & Lee, 2013). Initial studies using cerebral arteries could only identify the β and γ subunits of ENaC in vascular smooth muscle cells. It is possible that the α subunit is present in smooth muscle cells but at such low levels that it is not detectable by the methods utilized. Under control conditions middle cerebral arteries contract in response to increased intraluminal pressure. Amiloride and benzamil were both able to inhibit myogenic tone generation in middle cerebral arteries, suggesting that myogenic tone is regulated by ENaC activity (Drummond et al., 2004). This finding has been confirmed by others using arteries from the posterior cerebral circulation, in these studies the arteries used where deendothelialized, thereby removing any potential confounding effects of endothelial ENaC (Choi et al., 2017; Kim et al., 2012, 2013; Oyabe et al., 2000). These studies utilized a combination of pharmacological inhibition of ENaC and siRNA knockdown of the β and γ subunits of ENaC. They were also able to show that extracellular sodium was required for the ENaC-mediated increase in myogenic tone and that this led to an increase in smooth muscle cell intracellular calcium and myosin light chain phosphorylation (Kim et al., 2012). Interestingly, smooth muscle cell ENaC has been shown to associate with transient receptor potential channel melastatin 4 (TRPM4) and the two channels interact to increase myogenic tone generation at least in the posterior cerebral circulation (Kim et al., 2013). TRPM4 has been widely shown to regulate myogenic tone generation (Brayden, Earley, Nelson, & Reading, 2008). It is at present unclear if TRPM4 is important for ENaC-mediated tone generation in other vascular beds.

Increased ENaC expression appears to be responsible for the elevated myogenic tone observed in hypertensive rats. Interestingly, when comparing the expression of the ENaC subunits the authors found γ ENaC levels were preferentially increased in cerebral arteries from spontaneously hypertensive rats (SHR); the authors did not evaluate if this difference in expression was limited to vascular smooth muscle cells or if it also occurred in endothelial cells. A causative role for the increased γ EnaC expression in tone generation in SHR was identified using siRNA knockdown of γ ENaC expression. The authors also showed that phosphorylated Nedd4-2 and Sgk1, two proteins known to be involved in the aldosterone-mediated regulation of ENaC expression were increased in the hypertensive rats (Choi et al., 2017). This presents the possibility that MR activation indirectly regulates myogenic tone generation by increasing ENaC expression in vascular smooth muscle cells. This is at odds with findings from our own lab which showed that blocking the MR increased tone generation in middle cerebral arteries from hypertensive rats (Rigsby, Pollock, & Dorrance, 2007) and had no effect on tone generation in cerebral parenchymal arterioles (Pires, Jackson, & Dorrance, 2015). This difference could relate to the presence or absence on the endothelium. In the studies from Pires et al. (2015) and Rigsby et al. (Rigsby et al., 2007) the endothelium was intact, where Choi et al. (2017) utilized endothelium-denuded arteries. At present, it is not clear if MR activation directly regulates ENaC expression in either vascular smooth muscle cells or endothelial cells *in vivo*. Similarly, the interaction between endothelial and smooth muscle cell ENaCs has not been defined.

For both the endothelial and smooth muscle mediated effects of ENaC, it would seem that the literature would benefit greatly from some studies that carefully assess the effects of the channel in multiple vascular beds of different calibers. It would also seem to be important to consider the effect of this channel in males and females particularly given the evidence that it may be regulated by estrogen and the evidence that aldosterone production is differentially regulated between the sexes. The clear interplay between ENaC and NO suggests that there is also an urgent need to assess the effects of this channel on the vasculature in conditions where NO is impaired including in hypertension and diabetes.

5. Calcium activated potassium channels

5.1 Large conductance calcium-activated potassium channels (BK_{Ca})

Before commenting on the evidence that aldosterone regulates BK_{Ca} channels to alter vascular function, it is worth noting that BK_{Ca} channels are involved in the regulation of aldosterone production. BK_{Ca} channel knockout mice (BK−/−) have markedly elevated plasma aldosterone levels in both male and female mice. This hyperaldosteronism was associated with hypertension (Sausbier et al., 2005).

The literature regarding the control of BK_{Ca} channel expression and activity by aldosterone is mixed and has been recently reviewed by Zhang, Qian, and Wang (2017). Initial assessments of the effects of aldosterone on BK_{Ca} channel expression suggested that aldosterone does not regulate BK_{Ca} expression. These studies were conducted using rabbit cortical collecting ducts and aldosterone secretion was stimulated by a low sodium diet (Estilo et al., 2008). However, other studies utilizing colons from mice a fed a high potassium diet to stimulate aldosterone production suggest that MR activation increases α and β BK_{Ca} channel subunit expression. The effect of the high potassium diet could be mimicked by incubating colons with aldosterone and this effect was linked to MR activation because it could be inhibited by spironolactone (Sørensen et al., 2008).

In the vasculature, BK_{Ca} channels are primarily expressed in smooth muscle cells where they function to counter-regulate contraction. BK_{Ca} channel opening causes hyperpolarization of smooth muscle cells, which results in inactivation of the voltage-dependent calcium channels (Brayden & Nelson, 1992). Thus, BK−/− mice exhibit increased myogenic tone and impaired endothelium-dependent dilation in resistance arteries (Sausbier et al., 2005). BK_{Ca} channels have been studied in the hearts of mice that overexpress aldosterone synthase, the rate limiting enzyme in aldosterone production, in the cardiomyocytes. Endothelium-dependent dilation is impaired in coronary arteries of the aldosterone synthase overexpressing mice, and this impairment could be prevented with spironolactone. In these mice the mRNA and protein expression of the BK_{Ca} channel α and β1 subunits are reduced in the coronary arteries. Cultured coronary artery smooth muscle cells were used to confirm a direct effect of aldosterone on BK_{Ca} channel expression. The mRNA expression of both the α and β subunits was reduced by high concentrations of aldosterone, and this effect was inhibited by spironolactone (Fig. 4). Importantly, the reduction in the mRNA expression of at least the β subunit was concentration-responsive and observed at physiological aldosterone concentrations (Ambroisine et al., 2007). Clearly, these vascular studies are counter to the studies conducted in the colon where aldosterone increased BK_{Ca} channel expression (Sørensen et al., 2008) and they present the possibility that MR activation produce tissue specific regulation of BK_{Ca} channels. At present it is not clear if MR activation directly affects BK_{Ca} channel gene transcription or if a molecular intermediate is involved in the process. There are, at present, no literature reports suggesting that the BK_{Ca} channel is regulated by SGK1 or Nedd4-2 in the way ENaCs are.

5.2 Small conductance calcium-activated potassium channels (SK_{Ca})

Once again, prior to discussing how aldosterone might regulate SK_{Ca} expression it is important to consider if channel activity might regulate aldosterone production. In studies using human adrenocortical cells it has been shown that SK_{Ca} channels are a negative regulator of aldosterone production (Yang et al., 2016).

SK_{Ca} channels are calcium-activated potassium channels that are expressed in endothelial cells primarily in the myoendothelial projections. They contribute to endothelium-dependent relaxation *via* endothelium-derived hyperpolarization of smooth muscle cells. There are three subtypes of SK_{Ca}, SK1-3, and SK3 may be regulated by aldosterone. The first reports of an interaction between aldosterone and SK3 came from studies using the choroidal vasculature. These vessels lie directly under the retina and increases in permeability in these arterioles causes retinal detachment in a condition known as ocular chorioretinopathy. Aldosterone exacerbates this condition through actions on the MR. This was due to an increase in the protein expression of SK3 in response to aldosterone treatment (Zhao et al., 2012) (Fig. 3).

Studies from our own lab suggest that SK_{Ca} expression in cerebral parenchymal arterioles might be regulated by aldosterone. We showed the endothelium-dependent dilation is impaired in PAs from mice with AngII-induced hypertension. These mice also have elevated plasma aldosterone levels. Carbachol-mediated endothelium-dependent-dilation was impaired in these

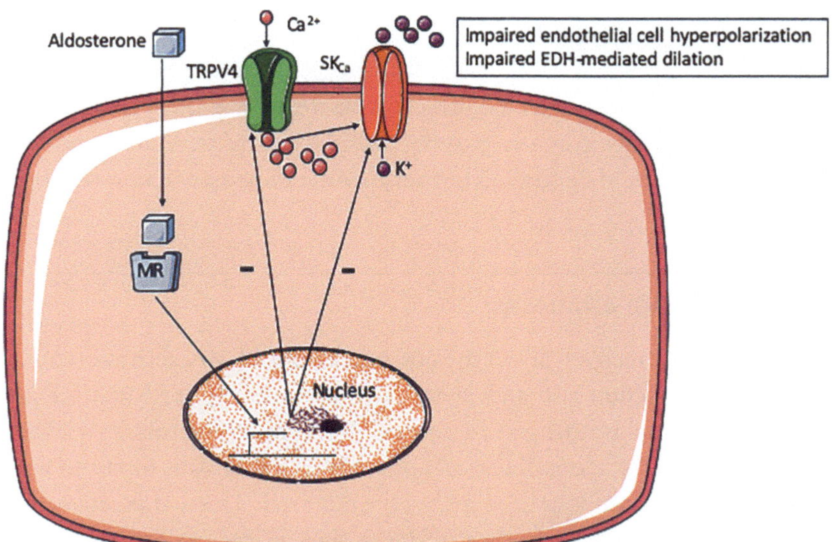

Fig. 3 Schematic representation of the proposed effects of MR activation on endothelial cell TRPV4 and SK_{Ca} channels. Activation of the MR leads to a reduced TRPV4 expression and also reduced expression of SK_{Ca}, a downstream component of the TRPV4-mediated dilation pathway. Reduced expression of these channels results in impaired endothelium dependent smooth muscle cell hyperpolarization and impaired dilation. The molecular mechanisms responsible for this effect are yet to be identified (Diaz-Otero et al., 2018). Servier Medical Art by Servier is licensed under a Creative Commons Attribution 3.0 Unported License ("Servier Medical Art", 2004).

arterioles, as was the dilation produced in response to NS309, an SK_{Ca} and intermediate calcium-activated potassium (IK_{Ca}, $K_{Ca}3.1$) channel activator. This impaired dilation was associated with a marked reduction in the expression of the mRNA for SK_{Ca}, and interestingly an increase in the mRNA expression of IK_{Ca}. SK_{Ca} and IK_{Ca} activation results in endothelium dependent hyperpolarization of the smooth muscle cells resulting in dilation (Goto, Ohtsubo, & Kitazono, 2018). The effects of AngII on dilation and mRNA expression were inhibited by eplerenone, suggesting an MR-mediated effect. Importantly, eplerenone did not reduce blood pressure in the AngII-treated mice; thus the improvement in vascular function was blood pressure-independent (Diaz-Otero et al., 2018). It is not clear if the increase in IK_{Ca} expression is the result of compensation to counteract the reduction in SK_{Ca} expression. Our studies are at odds with others carried out in DOCA-salt hypertensive rats where IK_{Ca} expression is reduced in small mesenteric resistance arteries (Kloza et al., 2017). It is unclear if this is an effect of MR activation or of hypertension. The functional significance of this reduction in expression is not clear because the EDH-mediated dilation was maintained in the resistance arteries of the DOCA-salt treated rats. These studies taken together suggest that SK_{Ca} may be more important for EDH than IK_{Ca}. However, a study by Cipolla, Smith, Kohlmeyer, and Godfrey (2009) suggested that inhibiting IK_{Ca} in parenchymal arterioles may be sufficient to reverse EDH dilation while inhibiting SK_{Ca} only partially reverses EDH. The relative importance of IK_{Ca} and SK_{Ca} should be studied further to settle these conflicting findings.

6. Calcium channels

We know much less about the potential effects of MR activation on calcium channel function although there have been several hints that calcium channel expression is changed in situations where aldosterone is elevated. The peak stimulated intracellular calcium concentration was lower in aortic endothelial cells from aldosterone salt hypertensive rats. This was associated with impaired endothelium-dependent dilation (Liu, Jones, & Sturek, 1995a). Conversely, in aortic vascular smooth muscle cells resting calcium levels were increased in aldosterone salt hypertensive rats (Liu, Jones, & Sturek, 1995b).

L-type calcium channels (Cav1.2) are expressed in vascular smooth muscle cells. These channels contain 4 subunits, α_1, β, $\alpha_2\delta$ and γ. The α_1 subunits regulate calcium permeability and are the site of action of calcium channel blockers (Singer et al., 1991). Cav1.2 channels are activated in response to SMC depolarization and their activation allows for calcium influx into the cell, which allows for contraction to occur. The function of this channel has been studied in smooth muscle cell specific MR knockout (SMMRKO) mice. The SMMRKO mice had lower blood pressure as they aged, and in mesenteric arteries the Cav1.2 channel-dependent contraction was impaired. The expression of the α_{1C} subunit was also reduced in aorta from mice lacking the MR suggesting that MR activation may positively regulate channel expression (McCurley et al., 2012). This finding was confirmed in later studies from the same group, and the regulation Cav1.2 channels was linked to the expression of a microRNA-155 (miR-155). miR-155 is down regulated with aging in control mice but not in the SMMRKO mice suggesting that miR-155 expression is regulated by the MR. The reductions in miR-155 expression resulted in an increase in Cav1.2 channel expression that could be recovered with miR-155 replacement. This interaction between MR and miR-155 was also found in humans. Patients were treated with eplerenone; a subset of these patients exhibited an increase in miR-155 levels in their plasma and this was an excellent predictor of the blood pressure lowering effects of eplerenone (DuPont et al., 2016) (Fig. 4).

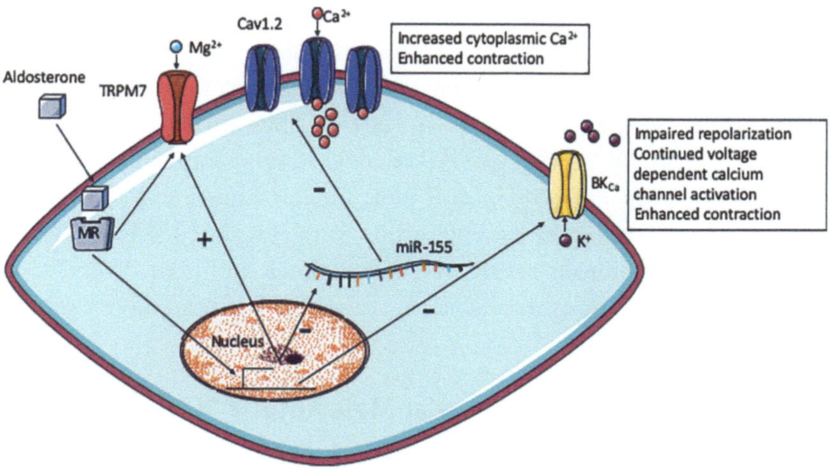

Fig. 4 Schematic representation of the proposed effects of MR activation in vascular smooth muscle cells. Aldosterone produces a nongenomic increase in TRPM7 activation (Paravicini, Yogi, Mazur, & Touyz, 2009) and increased TRPM7 (Valinsky, Jolly, Miquel, Touyz, & Shrier, 2016). At present this effect has only been shown in HK293 cells, but it is likely that a similar effect happens in smooth muscle cells. MR activation also leads to the regulation of Cav1.2; these channels are increased in response to MR activation through a mechanism involving miR-155. MR activation down regulates miR-155 and this causes increased Cav1.2 expression (DuPont et al., 2016; McCurley et al., 2012). MR activation also negatively regulates the expression of the BK_{Ca} channels. This results in a maintained contraction by preventing the inactivation of voltage dependent calcium channels (Ambroisine et al., 2007). Servier Medical Art by Servier is licensed under a Creative Commons Attribution 3.0 Unported License ("Servier Medical Art", 2004).

While these studies suggest an interaction between the MR and Cav1.2, it seems that this interaction may be vessel-dependent. Studies form our own lab using parenchymal arterioles from hypertensive rats suggest that Cav1.2 function is not affected by eplerenone treatment. In these studies, nifedipine was used to block the L-type calcium channels to produce vasodilation after myogenic tone generation. The vasodilation was similar in control and eplerenone-treated rats (Pires et al., 2015).

The interaction between the MR and calcium channels appears to be a research area that has been largely neglected. This is surprising given the clear vascular effects of aldosterone. Developing a better understanding of the potential interactions between the MR and calcium channels could provide new avenues for drug development for intransigent hypertension.

7. Transient receptor potential (TRP) channels

The MR has been shown to modulate the activity of certain members of the TRP channel family. TRP channels are a diverse superfamily of cation channels that are widely expressed on the cell membranes of both excitable and nonexcitable cell types (Moran, 2018). These channels have varying degrees of Ca^{2+} permeability and play critical roles in ion transport (Randhawa & Jaggi, 2015; Yue et al., 2015). The impacts of cation flow are physiologically far-reaching. TRP channel activity contributes to vascular function, sensations of pain and itch, inflammation, and roles in the central nervous system that have yet to be fully elucidated (Moran, 2018). Because of the ubiquitous nature of TRP channels and their diverse range of physiological functions, modulation of their activity can have extensive repercussions.

7.1 Transient receptor potential Melastatin 7 (TRPM7) channel

TRPM7 is a ubiquitously expressed cation channel permeable to both monovalent and divalent cations. TRPM7 conducts a small inward current of divalent cations and has a unique relative permeability sequence of $Zn^{2+} \approx Ni^{2+} >> Ba^{2+} > Co^{2+} > Mg^{2+} \geq Mn^{2+} \geq Sr^{2+} \geq Cd^{2+} \geq Ca^{2+}$ (Monteilh-Zoller et al., 2003). It also conducts an outward current of intracellular monovalent cations such as K^{+} and Na^{+} at positive membrane potentials. This outward current is due to a voltage-dependent block by extracellular divalent cations at more positive potentials. In the absence of divalent cations, monovalent cations are able to inwardly permeate the channel (Bae & Sun, 2011). TRPM7 is considered a "chanzyme," meaning it functions uniquely as both a cation channel and a kinase. The kinase portion of the chanzyme is structurally similar to protein kinase A in the vasculature (Ryazanova, Dorovkov, Ansari, & Ryazanov, 2004). TRPM7 is similar to its homolog TRPM6; however, TRPM7 is the most abundantly expressed TRP family channel in the majority of mouse organs. The mRNA is particularly highly expressed in the brains, kidney, lung, and intestines of mice (Kunert-Keil, Bisping, Krüger, & Brinkmeier, 2006).

TRPM7 plays several important roles in the vasculature. The chanzyme was initially shown to regulate and to be regulated by intracellular Mg^{2+} levels, with the channel being strongly activated by Mg-ATP levels below 1 mM (Nadler et al., 2001). It has since been discovered that TRPM7 plays a key role in magnesium homeostasis in several cell types, including vascular smooth muscle cells (Hamaguchi et al., 2008). TRPM7 has also been studied in mice with normal and low intracellular Mg^{2+} levels. Blood pressure was increased in the mice with low intracellular Mg^{2+} and this was associated with impaired endothelial function, as well as outward artery remodeling and vascular inflammation. TRPM7 expression was increased in the low Mg^{2+} mice but annexin-1, its downstream target, was reduced. This study revealed that TRPM7 signaling plays a critical role in vascular integrity and alterations in its expression could contribute to vascular injury (Paravicini et al., 2009). TRPM7 is also sensitive to fluid shear stress in the vasculature and has been shown to increase its presence at the plasma membrane in response to increased fluid flow rates (Oancea, Wolfe, & Clapham, 2006). A potential benefit of TRPM7 activation is that it appears to inhibit the formation of vascular calcification in aortic segments from rats (Sonou et al., 2017). However, this finding is at odds with another study that linked IL18 with TRPM7, where activation of TRPM7 enhanced vascular calcification (Zhang et al., 2017). Therefore, the role TRPM7 plays in this pathology is currently unclear.

Many of the effects of TRPM7 on the vasculature seem to be mediated directly or indirectly by MR signaling (Fig. 4). TRPM7 has been linked to the actions of angiotensin II in the process of phenotypic switching of vascular smooth muscle cells. Expression of TRPM7 is increased through an AngII type 1 receptor-dependent mechanism involving ERK1/2. This increase is associated with SMC proliferation and does not affect single channel activity (Zhang et al., 2012). Aldosterone and TRPM7 have been linked in both nongenomic and genomic mechanisms. In HEK293 cells, aldosterone stimulates Mg^{2+} influx through TRPM7 and this is dependent on activation of MR. Further, the α-kinase domain of

TRPM7 mediates proinflammatory signaling by aldosterone (Yogi et al., 2013). Aldosterone also genomically regulates TRPM7 expression in HEK293 cells. Treatment of the cells with aldosterone increased expression of TRPM7. The dose of aldosterone used was quite high (100 nM), but the effect seems to be MR-dependent as it can be completely blocked by eplerenone. Inhibition of SGK1 prevented the increased TRPM7 expression, showing SGK1 is responsible for this effect. This finding could be really important because both TRPM7 and SGK1 are ubiquitously expressed; therefore, this increase could occur in any aldosterone-sensitive cell (Valinsky et al., 2016). This could contribute to aldosterone's proinflammatory effects.

7.2 Transient receptor potential vanilloid 4 (TRPV4) channel

TRPV4 is a nonselective Ca^{2+}-permeable channel that is a critical regulator of vascular tone (Filosa, Yao, & Rath, 2013; Heathcote et al., 2019). TRPV4 is widely expressed in the endothelium and smooth muscle cells in the systemic, cerebral, and pulmonary vasculature. This channel is functionally unique due to its wide range of activators. Ca^{2+} influx through TRPV4 occurs in response to stimuli such as moderate heat, cell membrane stretching, and activation by endogenous ligands. Endogenous ligands include endocannabinoids, 4-α-phorbol esters, metabolites of arachidonic acid, and synthetic agonists like GSK1016790A (Baratchi et al., 2019). TRPV4 activity contributes greatly to endothelium-derived hyperpolarization factor (EDHF), a component of endothelium-dependent vasodilation. Small, local increases in intracellular Ca^{2+} ("sparklets") through single TRPV4 channels can activate SK_{Ca} and intermediate conductance IK_{Ca} Ca^{2+}-activated K^+ channels, thereby hyperpolarizing the endothelium (Lin et al., 2015). Hyperpolarization of the endothelium crosses to the smooth muscle cells *via* myoendothelial gap junctions and causes vasodilation (Seki et al., 2017). It is also worth noting that TRPV4 is expressed in smooth muscle cells where increases in intracellular Ca^{2+} produced by channel activation causes the stimulation of Ca^{2+} sparks through ryanodine receptors, which subsequently activate nearby large conductance Ca^{2+}-induced K^+ (BK) channels (Earley, Heppner, Nelson, & Brayden, 2005).

Expression of TRPV4 is mediated by MR activation and TRPV4 activity is impaired in various models of hypertension that are characterized by increases in plasma aldosterone. In a model of angiotensin II (AngII)-induced hypertension, male mice exhibited endothelial dysfunction in the parenchymal arterioles, which was accompanied by reduced cerebral perfusion and impaired cognition. The endothelial dysfunction was characterized by impaired dilation to carbachol and TRPV4 agonist GSK1016790A. There was also a reduction in mRNA expression of TRPV4 in the posterior cerebral arteries in AngII hypertension (Fig. 3). When AngII-hypertensive mice were treated with the mineralocorticoid receptor (MR) antagonist eplerenone, EDHF and mRNA expression of TRPV4 was restored, demonstrating the mediation of TRPV4 by MR activation (Diaz-Otero et al., 2018). Additionally, high dietary K^+, which induces the increased production of aldosterone, increases both TRPV4 mRNA expression and protein levels in the kidneys of mice. When these mice were treated with the MR antagonist spironolactone, this increase was prevented (Mamenko et al., 2017).

8. Conclusions

From the studies described above it seems clear that MR activation is involved in the regulation of ion channel function. This is most clear for ENaCs. As the field moves forward, it will be important to define the molecular mechanisms involved in the interactions between the other ion channels and the MR in the same way as has been done for ENaCs. There are many gaps in our basic knowledge at present. These include a lack of understanding of potential sex differences in how these interactions might occur, and a lack of understanding of how age might impact the effects of MR activation. Interest in the potential effects of MR-regulated miRs is also increasing and it seems possible that miR regulation of ion channels could be a fruitful research area.

References

Ahmad, N., Romero, D. G., Gomez-Sanchez, E. P., & Gomez-Sanchez, C. E. (2004). Do human vascular endothelial cells produce aldosterone? *Endocrinology*, *145*(8), 3626–3629. https://doi.org/10.1210/en.2004-0081.

Ahmed, A. H., Gordon, R. D., Ward, G., Wolley, M., Kogovsek, C., & Stowasser, M. (2015). Should aldosterone suppression tests be conducted during a particular phase of the menstrual cycle, and, if so, which phase? Results of a preliminary study. *Clinical Endocrinology*, *83*(3), 303–307. https://doi.org/10.1111/cen.12705.

Alvarez de la Rosa, D., Li, H., & Canessa, C. M. (2002). Effects of aldosterone on biosynthesis, traffic, and functional expression of epithelial sodium channels in A6 cells. *The Journal of General Physiology*, *119*(5), 427–442. https://doi.org/10.1085/jgp.20028559.

Ambroisine, M.-L., Favre, J., Oliviero, P., Rodriguez, C., Gao, J., Thuillez, C., et al. (2007). Aldosterone-induced coronary dysfunction in transgenic mice involves the calcium-activated potassium (BKCa) channels of vascular smooth muscle cells. *Circulation*, *116*(21), 2435–2443. https://doi.org/10.1161/CIRCULATIONAHA.107.722009.

Ambrosius, W. T., Bloem, L. J., Zhou, L., Rebhun, J. F., Snyder, P. M., Wagner, M. A., et al. (1999). Genetic variants in the epithelial sodium channel in relation to aldosterone and potassium excretion and risk for hypertension. *Hypertension, 34*(4 Pt. 1), 631–637. https://doi.org/10.1161/01.hyp.34.4.631.

Ashley, Z., Mugloo, S., McDonald, F. J., & Fronius, M. (2018). Epithelial Na^+ channel differentially contributes to shear stress-mediated vascular responsiveness in carotid and mesenteric arteries from mice. *American Journal of Physiology. Heart and Circulatory Physiology, 314*(5), H1022–H1032. https://doi.org/10.1152/ajpheart.00506.2017.

Bae, C. Y., & Sun, H. (2011). TRPM7 in cerebral ischemia and potential target for drug development in stroke. *Acta Pharmacologica Sinica, 32*(6), 725–733. https://doi.org/10.1038/aps.2011.60.

Baratchi, S., Keov, P., Darby, W. G., Lai, A., Khoshmanesh, K., Thurgood, P., et al. (2019). The TRPV4 agonist GSK1016790A regulates the membrane expression of TRPV4 channels. *Frontiers in Pharmacology, 10*, 6. https://doi.org/10.3389/fphar.2019.00006.

Barrera-Chimal, J., & Jaisser, F. (2019). Vascular mineralocorticoid receptor activation and disease. *Experimental Eye Research, 188*, 107796. https://doi.org/10.1016/j.exer.2019.107796.

Bernhardt, R. (2016). The potential of targeting CYP11B. *Expert Opinion on Therapeutic Targets, 20*(8), 923–934. https://doi.org/10.1517/14728222.2016.1151873.

Brayden, J. E., Earley, S., Nelson, M. T., & Reading, S. (2008). Transient receptor potential (TRP) channels, vascular tone and autoregulation of cerebral blood flow. *Clinical and Experimental Pharmacology & Physiology, 35*(9), 1116–1120. https://doi.org/10.1111/j.1440-1681.2007.04855.x.

Brayden, J. E., & Nelson, M. T. (1992). Regulation of arterial tone by activation of calcium-dependent potassium channels. *Science, 256*(5056), 532–535. https://doi.org/10.1126/science.1373909.

Canessa, C. M., Schild, L., Buell, G., Thorens, B., Gautschi, I., Horisberger, J. D., et al. (1994). Amiloride-sensitive epithelial Na^+ channel is made of three homologous subunits. *Nature, 367*(6462), 463–467. https://doi.org/10.1038/367463a0.

Caprio, M., Newfell, B. G., la Sala, A., Baur, W., Fabbri, A., Rosano, G., et al. (2008). Functional mineralocorticoid receptors in human vascular endothelial cells regulate intercellular adhesion molecule-1 expression and promote leukocyte adhesion. *Circulation Research, 102*(11), 1359–1367. https://doi.org/10.1161/CIRCRESAHA.108.174235.

Choi, S.-K., Yeon, S.-I., Kwon, Y., Byeon, S., & Lee, Y.-H. (2017). Involvement of epithelial Na^+ channel in the elevated myogenic response in posterior cerebral arteries from spontaneously hypertensive rats. *Scientific Reports*, 7, 45996. https://doi.org/10.1038/srep45996.

Christy, C., Hadoke, P. W. F., Paterson, J. M., Mullins, J. J., Seckl, J. R., & Walker, B. R. (2003). 11beta-hydroxysteroid dehydrogenase type 2 in mouse aorta: Localization and influence on response to glucocorticoids. *Hypertension, 42*(4), 580–587. https://doi.org/10.1161/01.HYP.0000088855.06598.5B.

Cipolla, M. J., Smith, J., Kohlmeyer, M. M., & Godfrey, J. A. (2009). SKCa and IKCa channels, myogenic tone, and vasodilator responses in middle cerebral arteries and parenchymal arterioles: Effect of ischemia and reperfusion. *Stroke, 40*(4), 1451–1457. https://doi.org/10.1161/STROKEAHA.108.535435.

Diaz-Otero, J. M., Fisher, C., Downs, K., Moss, M. E., Jaffe, I. Z., Jackson, W. F., et al. (2017). Endothelial mineralocorticoid receptor mediates parenchymal arteriole and posterior cerebral artery remodeling during angiotensin II-induced hypertension. *Hypertension, 70*(6), 1113–1121. https://doi.org/10.1161/HYPERTENSIONAHA.117.09598.

Diaz-Otero, J. M., Yen, T.-C., Fisher, C., Bota, D., Jackson, W. F., & Dorrance, A. M. (2018). Mineralocorticoid receptor antagonism improves parenchymal arteriole dilation via a TRPV4-dependent mechanism and prevents cognitive dysfunction in hypertension. *American Journal of Physiology. Heart and Circulatory Physiology*, *315*(5), H1304–H1315. https://doi.org/10.1152/ajpheart.00207.2018.

Dinh Cat, A. N., Friederich-Persson, M., White, A., & Touyz, R. M. (2016). Adipocytes, aldosterone and obesity-related hypertension. *Journal of Molecular Endocrinology*, *57*(1), F7–F21. https://doi.org/10.1530/JME-16-0025.

Downs, C. A., Johnson, N. M., Coca, C., & Helms, M. N. (2018). Angiotensin II regulates δ-ENaC in human umbilical vein endothelial cells. *Microvascular Research*, *116*, 26–33. https://doi.org/10.1016/j.mvr.2017.10.001.

Drummond, H. A. (2007). Enac proteins in vascular smooth muscle mechanotransduction. *Current Topics in Membranes*, *59*, 127–153. https://doi.org/10.1016/S1063-5823(06)59006-4.

Drummond, H. A. (2009). The (F)low down on the endothelial epithelial sodium channel: Epithelial sodium channel as a brake on flow-mediated vasodilation. *Hypertension*, *53*(6), 903–904. https://doi.org/10.1161/HYPERTENSIONAHA.109.128868.

Drummond, H. A., Gebremedhin, D., & Harder, D. R. (2004). Degenerin/epithelial Na^+ channel proteins: Components of a vascular mechanosensor. *Hypertension*, *44*(5), 643–648. https://doi.org/10.1161/01.HYP.0000144465.56360.ad.

Drummond, H. A., Grifoni, S. C., & Jernigan, N. L. (2008). A new trick for an old dogma: ENaC proteins as mechanotransducers in vascular smooth muscle. *Physiology*, *23*, 23–31. https://doi.org/10.1152/physiol.00034.2007.

Drüppel, V., Kusche-Vihrog, K., Grossmann, C., Gekle, M., Kasprzak, B., Brand, E., et al. (2013). Long-term application of the aldosterone antagonist spironolactone prevents stiff endothelial cell syndrome. *The FASEB Journal*, *27*(9), 3652–3659. https://doi.org/10.1096/fj.13-228312.

DuPont, J. J., Hill, M. A., Bender, S. B., Jaisser, F., & Jaffe, I. Z. (2014). Aldosterone and vascular mineralocorticoid receptors: Regulators of ion channels beyond the kidney. *Hypertension*, *63*(4), 632–637. https://doi.org/10.1161/HYPERTENSIONAHA.113.01273.

DuPont, J. J., & Jaffe, I. Z. (2017). 30 years of the mineralocorticoid receptor: The role of the mineralocorticoid receptor in the vasculature. *The Journal of Endocrinology*, *234*(1), T67–T82. https://doi.org/10.1530/JOE-17-0009.

DuPont, J. J., McCurley, A., Davel, A. P., McCarthy, J., Bender, S. B., Hong, K., et al. (2016). Vascular mineralocorticoid receptor regulates microRNA-155 to promote vasoconstriction and rising blood pressure with aging. *Journal of Clinical Investigation Insight*, *1*(14). e88942https://doi.org/10.1172/jci.insight.88942.

Earley, S., Heppner, T. J., Nelson, M. T., & Brayden, J. E. (2005). TRPV4 forms a novel Ca^{2+} signaling complex with ryanodine receptors and BKCa channels. *Circulation Research*, *97*(12), 1270–1279. https://doi.org/10.1161/01.RES.0000194321.60300.d6.

Estilo, G., Liu, W., Pastor-Soler, N., Mitchell, P., Carattino, M. D., Kleyman, T. R., et al. (2008). Effect of aldosterone on BK channel expression in mammalian cortical collecting duct. *American Journal of Physiology. Renal Physiology*, *295*(3), F780–F788. https://doi.org/10.1152/ajprenal.00002.2008.

Fakitsas, P., Adam, G., Daidié, D., van Bemmelen, M. X., Fouladkou, F., Patrignani, A., et al. (2007). Early aldosterone-induced gene product regulates the epithelial sodium channel by deubiquitylation. *Journal of the American Society of Nephrology*, *18*(4), 1084–1092. https://doi.org/10.1681/ASN.2006080902.

Fels, J., Callies, C., Kusche-Vihrog, K., & Oberleithner, H. (2010). Nitric oxide release follows endothelial nanomechanics and not vice versa. *Pflugers Archiv: European Journal of Physiology*, *460*(5), 915–923. https://doi.org/10.1007/s00424-010-0871-8.

Filosa, J. A., Yao, X., & Rath, G. (2013). TRPV4 and the regulation of vascular tone. *Journal of Cardiovascular Pharmacology*, *61*(2), 113–119. https://doi.org/10.1097/FJC.0b013e318279ba42.

Folgering, J. H. A., Sharif-Naeini, R., Dedman, A., Patel, A., Delmas, P., & Honoré, E. (2008). Molecular basis of the mammalian pressure-sensitive ion channels: Focus on vascular mechanotransduction. *Progress in Biophysics and Molecular Biology*, *97*(2–3), 180–195. https://doi.org/10.1016/j.pbiomolbio.2008.02.006.

Fronius, M., & Clauss, W. G. (2008). Mechano-sensitivity of ENaC: May the (shear) force be with you. *Pflugers Archiv: European Journal of Physiology*, *455*(5), 775–785. https://doi.org/10.1007/s00424-007-0332-1.

Funder, J. W. (2017). Aldosterone and mineralocorticoid receptors-physiology and pathophysiology. *International Journal of Molecular Sciences*, *18*(5), E1032. https://doi.org/10.3390/ijms18051032.

Funder, J. W., Feldman, D., & Edelman, I. S. (1972). Specific aldosterone binding in rat kidney and parotid. *Journal of Steroid Biochemistry*, *3*(2), 209–218. https://doi.org/10.1016/0022-4731(72)90052-0.

Funder, J. W., Feldman, D., & Edelman, I. S. (1973). The roles of plasma binding and receptor specificity in the mineralocorticoid action of aldosterone. *Endocrinology*, *92*(4), 994–1004. https://doi.org/10.1210/endo-92-4-994.

Golestaneh, N., Klein, C., Valamanesh, F., Suarez, G., Agarwal, M. K., & Mirshahi, M. (2001). Mineralocorticoid receptor-mediated signaling regulates the ion gated sodium channel in vascular endothelial cells and requires an intact cytoskeleton. *Biochemical and Biophysical Research Communications*, *280*(5), 1300–1306. https://doi.org/10.1006/bbrc.2001.4275.

Goto, K., Ohtsubo, T., & Kitazono, T. (2018). Endothelium-dependent hyperpolarization (EDH) in hypertension: The role of endothelial ion channels. *International Journal of Molecular Sciences*, *19*(1), E315. https://doi.org/10.3390/ijms19010315.

Greenlee, M. M., Mitzelfelt, J. D., Yu, L., Yue, Q., Duke, B. J., Harrell, C. S., et al. (2013). Estradiol activates epithelial sodium channels in rat alveolar cells through the G protein-coupled estrogen receptor. *American Journal of Physiology. Lung Cellular and Molecular Physiology*, *305*(11), L878–L889. https://doi.org/10.1152/ajplung.00008.2013.

Greenstein, A. S., Kadir, S. Z. A. S., Csato, V., Sugden, S. A., Baylie, R. A., Eisner, D. A., et al. (2020). Disruption of pressure-induced Ca^{2+} spark vasoregulation of resistance arteries, rather than endothelial dysfunction, underlies obesity-related hypertension. *Hypertension*, *75*, 539–548. https://doi.org/10.1161/HYPERTENSIONAHA.119.13540. HYPERTENSIONAHA11913540.

Grifoni, S. C., Gannon, K. P., Stec, D. E., & Drummond, H. A. (2006). ENaC proteins contribute to VSMC migration. *American Journal of Physiology. Heart and Circulatory Physiology*, *291*(6), H3076–H3086. https://doi.org/10.1152/ajpheart.00333.2006.

Gros, R., Ding, Q., Sklar, L. A., Prossnitz, E. E., Arterburn, J. B., Chorazyczewski, J., et al. (2011). GPR30 expression is required for the mineralocorticoid receptor-independent rapid vascular effects of aldosterone. *Hypertension*, *57*(3), 442–451. https://doi.org/10.1161/HYPERTENSIONAHA.110.161653.

Guan, Z., Pollock, J. S., Cook, A. K., Hobbs, J. L., & Inscho, E. W. (2009). Effect of epithelial sodium channel blockade on the myogenic response of rat juxtamedullary afferent arterioles. *Hypertension*, *54*(5), 1062–1069. https://doi.org/10.1161/HYPERTENSIONAHA.109.137992.

Guo, D., Liang, S., Wang, S., Tang, C., Yao, B., Wan, W., et al. (2016). Role of epithelial Na^{+} channels in endothelial function. *Journal of Cell Science*, *129*(2), 290–297. https://doi.org/10.1242/jcs.168831.

Hamaguchi, Y., Matsubara, T., Amano, T., Uetani, T., Asano, H., Iwamoto, T., et al. (2008). Na(+)-independent Mg(2+) transport sensitive to 2-aminoethoxydiphenyl borate (2-APB) in vascular smooth muscle cells: Involvement of TRPM-like channels. *Journal of Cellular and Molecular Medicine*, *12*(3), 962–974. https://doi.org/10.1111/j.1582-4934.2008.00157.x.

Heathcote, H. R., Lee, M. D., Zhang, X., Saunter, C. D., Wilson, C., & McCarron, J. G. (2019). Endothelial TRPV4 channels modulate vascular tone by Ca^{2+}-induced Ca^{2+} release at inositol 1,4,5-trisphosphate receptors. *British Journal of Pharmacology, 176*(17), 3297–3317. https://doi.org/10.1111/bph.14762.

Hillebrand, U., Schillers, H., Riethmüller, C., Stock, C., Wilhelmi, M., Oberleithner, H., et al. (2007). Dose-dependent endothelial cell growth and stiffening by aldosterone: Endothelial protection by eplerenone. *Journal of Hypertension, 25*(3), 639–647. https://doi.org/10.1097/HJH.0b013e328013f492.

Huby, A.-C., Antonova, G., Groenendyk, J., Gomez-Sanchez, C. E., Bollag, W. B., Filosa, J. A., et al. (2015). Adipocyte-derived hormone leptin is a direct regulator of aldosterone secretion, which promotes endothelial dysfunction and cardiac fibrosis. *Circulation, 132*(22), 2134–2145. https://doi.org/10.1161/CIRCULATIONAHA.115.018226.

Huby, A.-C., Otvos, L., & Belin de Chantemèle, E. J. (2016). Leptin induces hypertension and endothelial dysfunction via aldosterone-dependent mechanisms in obese female mice. *Hypertension, 67*(5), 1020–1028. https://doi.org/10.1161/HYPERTENSIONAHA.115.06642.

Imig, J. D. (2012). Epoxides and soluble epoxide hydrolase in cardiovascular physiology. *Physiological Reviews, 92*(1), 101–130. https://doi.org/10.1152/physrev.00021.2011.

Jaffe, I. Z., & Jaisser, F. (2014). Endothelial cell mineralocorticoid receptors: Turning cardiovascular risk factors into cardiovascular dysfunction. *Hypertension, 63*(5), 915–917. https://doi.org/10.1161/HYPERTENSIONAHA.114.01997.

Jeggle, P., Hofschröer, V., Maase, M., Bertog, M., & Kusche-Vihrog, K. (2016). Aldosterone synthase knockout mouse as a model for sodium-induced endothelial sodium channel up-regulation in vascular endothelium. *The FASEB Journal, 30*(1), 45–53. https://doi.org/10.1096/fj.14-259606.

Jernigan, N. L., & Drummond, H. A. (2005). Vascular ENaC proteins are required for renal myogenic constriction. *American Journal of Physiology. Renal Physiology, 289*(4), F891–F901. https://doi.org/10.1152/ajprenal.00019.2005.

Jernigan, N. L., & Drummond, H. A. (2006). Myogenic vasoconstriction in mouse renal interlobar arteries: Role of endogenous beta and gammaENaC. *American Journal of Physiology. Renal Physiology, 291*(6), F1184–F1191. https://doi.org/10.1152/ajprenal.00177.2006.

Ji, H.-L., Zhao, R.-Z., Chen, Z.-X., Shetty, S., Idell, S., & Matalon, S. (2012). δ ENaC: A novel divergent amiloride-inhibitable sodium channel. *American Journal of Physiology. Lung Cellular and Molecular Physiology, 303*(12), L1013–L1026. https://doi.org/10.1152/ajplung.00206.2012.

Jia, G., Habibi, J., Aroor, A. R., Hill, M. A., Yang, Y., Whaley-Connell, A., et al. (2018). Epithelial sodium channel in aldosterone-induced endothelium stiffness and aortic dysfunction. *Hypertension, 72*(3), 731–738. https://doi.org/10.1161/HYPERTENSIONAHA.118.11339.

Karpushev, A. V., Ilatovskaya, D. V., Pavlov, T. S., Negulyaev, Y. A., & Staruschenko, A. (2010). Intact cytoskeleton is required for small G protein dependent activation of the epithelial Na^{+} channel. *PLoS One, 5*(1), e8827. https://doi.org/10.1371/journal.pone.0008827.

Karpushev, A. V., Ilatovskaya, D. V., & Staruschenko, A. (2010). The actin cytoskeleton and small G protein RhoA are not involved in flow-dependent activation of ENaC. *BMC Research Notes, 3*, 210. https://doi.org/10.1186/1756-0500-3-210.

Kim, E.-C., Ahn, D.-S., Yeon, S.-I., Lim, M., & Lee, Y.-H. (2012). Epithelial Na^{+} channel proteins are mechanotransducers of myogenic constriction in rat posterior cerebral arteries. *Experimental Physiology, 97*(4), 544–555. https://doi.org/10.1113/expphysiol.2011.062232.

Kim, E.-C., Choi, S.-K., Lim, M., Yeon, S.-I., & Lee, Y.-H. (2013). Role of endogenous ENaC and TRP channels in the myogenic response of rat posterior cerebral arteries. *PLoS One*, *8*(12), e84194. https://doi.org/10.1371/journal.pone.0084194.

Kloza, M., Baranowska-Kuczko, M., Malinowska, B., Karpińska, O., Harasim-Symbor, E., Kasacka, I., et al. (2017). The influence of DOCA-salt hypertension and chronic administration of the FAAH inhibitor URB597 on $K_{Ca}2.3/K_{Ca}3.1$-EDH-type relaxation in rat small mesenteric arteries. *Vascular Pharmacology*, *99*, 65–73. https://doi.org/10.1016/j.vph.2017.10.001.

Kunert-Keil, C., Bisping, F., Krüger, J., & Brinkmeier, H. (2006). Tissue-specific expression of TRP channel genes in the mouse and its variation in three different mouse strains. *BMC Genomics*, 7, 159.

Kusche-Vihrog, K., Jeggle, P., & Oberleithner, H. (2014). The role of ENaC in vascular endothelium. *Pflugers Archiv: European Journal of Physiology*, *466*(5), 851–859. https://doi.org/10.1007/s00424-013-1356-3.

Kusche-Vihrog, K., Sobczak, K., Bangel, N., Wilhelmi, M., Nechyporuk-Zloy, V., Schwab, A., et al. (2008). Aldosterone and amiloride alter ENaC abundance in vascular endothelium. *Pflugers Archiv: European Journal of Physiology*, *455*(5), 849–857. https://doi.org/10.1007/s00424-007-0341-0.

Kusche-Vihrog, K., Tarjus, A., Fels, J., & Jaisser, F. (2014). The epithelial Na^+ channel: A new player in the vasculature. *Current Opinion in Nephrology and Hypertension*, *23*(2), 143–148. https://doi.org/10.1097/01.mnh.0000441054.88962.2c.

Lang, F. (2011). Stiff endothelial cell syndrome in vascular inflammation and mineralocorticoid excess. *Hypertension*, *57*(2), 146–147. https://doi.org/10.1161/HYPERTENSIONAHA.110.164558.

Lin, M. T., Jian, M.-Y., Taylor, M. S., Cioffi, D. L., Yap, F. C., Liedtke, W., et al. (2015). Functional coupling of TRPV4, IK, and SK channels contributes to Ca(2+)-dependent endothelial injury in rodent lung. *Pulmonary Circulation*, *5*(2), 279–290. https://doi.org/10.1086/680166.

Liu, Y., Jones, A. W., & Sturek, M. (1995a). Attenuated Ca^{2+} response to acetylcholine in endothelial cells from aorta of aldosterone-salt hypertensive rats. *American Journal of Hypertension*, *8*(4 Pt. 1), 404–408. https://doi.org/10.1016/0895-7061(94)00205-p.

Liu, Y., Jones, A. W., & Sturek, M. (1995b). Ca(2^+)-dependent K^+ current in arterial smooth muscle cells from aldosterone-salt hypertensive rats. *The American Journal of Physiology*, *269*(4 Pt. 2), H1246–H1257. https://doi.org/10.1152/ajpheart.1995.269.4.H1246.

Looft-Wilson, R. C., Todd, S. E., Berberich, K. M., & Wolfert, M. R. (2018). Flow does not alter eNOS phosphoryation at Ser1179 or Thr495 in preconstricted mouse mesenteric arteries. *Physiological Reports*, *6*(17). e13864https://doi.org/10.14814/phy2.13864.

MacKenzie, S. M., Connell, J. M. C., & Davies, E. (2012). Non-adrenal synthesis of aldosterone: A reality check. *Molecular and Cellular Endocrinology*, *350*(2), 163–167. https://doi.org/10.1016/j.mce.2011.06.026.

MacKenzie, S. M., van Kralingen, J. C., & Davies, E. (2019). Regulation of aldosterone secretion. *Vitamins and Hormones*, *109*, 241–263. https://doi.org/10.1016/bs.vh.2018.07.001.

Mamenko, M. V., Boukelmoune, N., Tomilin, V. N., Zaika, O. L., Jensen, V. B., O'Neil, R. G., et al. (2017). The renal TRPV4 channel is essential for adaptation to increased dietary potassium. *Kidney International*, *91*(6), 1398–1409. https://doi.org/10.1016/j.kint.2016.12.010.

May, A., Puoti, A., Gaeggeler, H. P., Horisberger, J. D., & Rossier, B. C. (1997). Early effect of aldosterone on the rate of synthesis of the epithelial sodium channel alpha subunit in A6 renal cells. *Journal of the American Society of Nephrology*, *8*(12), 1813–1822.

McCurley, A., & Jaffe, I. Z. (2012). Mineralocorticoid receptors in vascular function and disease. *Molecular and Cellular Endocrinology, 350*(2), 256–265. https://doi.org/10.1016/j.mce.2011.06.014.

McCurley, A., Pires, P. W., Bender, S. B., Aronovitz, M., Zhao, M. J., Metzger, D., et al. (2012). Direct regulation of blood pressure by smooth muscle cell mineralocorticoid receptors. *Nature Medicine, 18*(9), 1429–1433. https://doi.org/10.1038/nm.2891.

McEneaney, V., Harvey, B. J., & Thomas, W. (2007). Aldosterone rapidly activates protein kinase D via a mineralocorticoid receptor/EGFR trans-activation pathway in the M1 kidney CCD cell line. *The Journal of Steroid Biochemistry and Molecular Biology, 107*(3–5), 180–190. https://doi.org/10.1016/j.jsbmb.2007.03.043.

McEneaney, V., Harvey, B. J., & Thomas, W. (2008). Aldosterone regulates rapid trafficking of epithelial sodium channel subunits in renal cortical collecting duct cells via protein kinase D activation. *Molecular Endocrinology, 22*(4), 881–892. https://doi.org/10.1210/me.2007-0225.

Mihailidou, A. S., Loan Le, T. Y., Mardini, M., & Funder, J. W. (2009). Glucocorticoids activate cardiac mineralocorticoid receptors during experimental myocardial infarction. *Hypertension, 54*(6), 1306–1312. https://doi.org/10.1161/HYPERTENSIONAHA.109.136242.

Monteilh-Zoller, M. K., Hermosura, M. C., Nadler, M. J. S., Scharenberg, A. M., Penner, R., & Fleig, A. (2003). TRPM7 provides an ion channel mechanism for cellular entry of trace metal ions. *The Journal of General Physiology, 121*(1), 49–60. https://doi.org/10.1085/jgp.20028740.

Moran, M. M. (2018). TRP channels as potential drug targets. *Annual Review of Pharmacology and Toxicology, 58*, 309–330. https://doi.org/10.1146/annurev-pharmtox-010617-052832.

Moss, M. E., Carvajal, B., & Jaffe, I. Z. (2019). The endothelial mineralocorticoid receptor: Contributions to sex differences in cardiovascular disease. *Pharmacology & Therapeutics, 203*, 107387. https://doi.org/10.1016/j.pharmthera.2019.06.009.

Nadler, M. J., Hermosura, M. C., Inabe, K., Perraud, A. L., Zhu, Q., Stokes, A. J., et al. (2001). LTRPC7 is a Mg.ATP-regulated divalent cation channel required for cell viability. *Nature, 411*(6837), 590–595. https://doi.org/10.1038/35079092.

Oancea, E., Wolfe, J. T., & Clapham, D. E. (2006). Functional TRPM7 channels accumulate at the plasma membrane in response to fluid flow. *Circulation Research, 98*(2), 245–253. https://doi.org/10.1161/01.RES.0000200179.29375.cc.

Oberleithner, H., Callies, C., Kusche-Vihrog, K., Schillers, H., Shahin, V., Riethmüller, C., et al. (2009). Potassium softens vascular endothelium and increases nitric oxide release. *Proceedings of the National Academy of Sciences of the United States of America, 106*(8), 2829–2834. https://doi.org/10.1073/pnas.0813069106.

Oberleithner, H., Riethmüller, C., Ludwig, T., Shahin, V., Stock, C., Schwab, A., et al. (2006). Differential action of steroid hormones on human endothelium. *Journal of Cell Science, 119*(Pt. 9), 1926–1932. https://doi.org/10.1242/jcs.02886.

Oberleithner, H., Riethmüller, C., Schillers, H., MacGregor, G. A., de Wardener, H. E., & Hausberg, M. (2007). Plasma sodium stiffens vascular endothelium and reduces nitric oxide release. *Proceedings of the National Academy of Sciences of the United States of America, 104*(41), 16281–16286. https://doi.org/10.1073/pnas.0707791104.

Oberleithner, H., Schneider, S. W., Albermann, L., Hillebrand, U., Ludwig, T., Riethmüller, C., et al. (2003). Endothelial cell swelling by aldosterone. *The Journal of Membrane Biology, 196*(3), 163–172. https://doi.org/10.1007/s00239-003-0635-6.

Oddie, C. J., Coghlan, J. P., & Scoggins, B. A. (1972). Plasma desoxycorticosterone levels in man with simultaneous measurement of aldosterone, corticosterone, cortisol and 11-deoxycortisol. *The Journal of Clinical Endocrinology and Metabolism, 34*(6), 1039–1054. https://doi.org/10.1210/jcem-34-6-1039.

Oyabe, A., Masumoto, N., Ueta, K., & Nakayama, K. (2000). Amiloride-sensitive pressure-induced myogenic contraction in rat cerebral artery. *Fundamental & Clinical Pharmacology, 14*(4), 369–377. https://doi.org/10.1111/j.1472-8206.2000.tb00418.x.

Paravicini, T. M., Yogi, A., Mazur, A., & Touyz, R. M. (2009). Dysregulation of vascular TRPM7 and annexin-1 is associated with endothelial dysfunction in inherited hypomagnesemia. *Hypertension, 53*(2), 423–429. https://doi.org/10.1161/HYPERTENSIONAHA.108.124651.

Pérez, F. R., Venegas, F., González, M., Andrés, S., Vallejos, C., Riquelme, G., et al. (2009). Endothelial epithelial sodium channel inhibition activates endothelial nitric oxide synthase via phosphoinositide 3-kinase/Akt in small-diameter mesenteric arteries. *Hypertension, 53*(6), 1000–1007. https://doi.org/10.1161/HYPERTENSIONAHA.108.128520.

Pesen, D., & Hoh, J. H. (2005). Micromechanical architecture of the endothelial cell cortex. *Biophysical Journal, 88*(1), 670–679. https://doi.org/10.1529/biophysj.104.049965.

Pires, P. W., Jackson, W. F., & Dorrance, A. M. (2015). Regulation of myogenic tone and structure of parenchymal arterioles by hypertension and the mineralocorticoid receptor. *American Journal of Physiology. Heart and Circulatory Physiology, 309*(1), H127–H136. https://doi.org/10.1152/ajpheart.00168.2015.

Pitt, B., Williams, G., Remme, W., Martinez, F., Lopez-Sendon, J., Zannad, F., et al. (2001). The EPHESUS trial: Eplerenone in patients with heart failure due to systolic dysfunction complicating acute myocardial infarction. Eplerenone post-AMI heart failure efficacy and survival study. *Cardiovascular Drugs and Therapy, 15*(1), 79–87. https://doi.org/10.1023/a:1011119003788.

Pitt, B., Zannad, F., Remme, W. J., Cody, R., Castaigne, A., Perez, A., et al. (1999). The effect of spironolactone on morbidity and mortality in patients with severe heart failure. Randomized aldactone evaluation study investigators. *The New England Journal of Medicine, 341*(10), 709–717. https://doi.org/10.1056/NEJM199909023411001.

Randhawa, P. K., & Jaggi, A. S. (2015). TRPV4 channels: Physiological and pathological role in cardiovascular system. *Basic Research in Cardiology, 110*(6), 54. https://doi.org/10.1007/s00395-015-0512-7.

Rigsby, C. S., Pollock, D. M., & Dorrance, A. M. (2007). Spironolactone improves structure and increases tone in the cerebral vasculature of male spontaneously hypertensive stroke-prone rats. *Microvascular Research, 73*(3), 198–205. https://doi.org/10.1016/j.mvr.2006.12.001.

Robert-Nicoud, M., Flahaut, M., Elalouf, J. M., Nicod, M., Salinas, M., Bens, M., et al. (2001). Transcriptome of a mouse kidney cortical collecting duct cell line: Effects of aldosterone and vasopressin. *Proceedings of the National Academy of Sciences of the United States of America, 98*(5), 2712–2716. https://doi.org/10.1073/pnas.051603198.

Ryazanova, L. V., Dorovkov, M. V., Ansari, A., & Ryazanov, A. G. (2004). Characterization of the protein kinase activity of TRPM7/ChaK1, a protein kinase fused to the transient receptor potential ion channel. *Journal of Biological Chemistry, 279*(5), 3708–3716.

Sakakibara, R., Sasaki, W., Onda, Y., Yamaguchi, M., Ushirogochi, H., Hiraga, Y., et al. (2018). Discovery of novel pyrazole-based selective aldosterone synthase (CYP11B2) inhibitors: A new template to coordinate the heme-iron motif of CYP11B2. *Journal of Medicinal Chemistry, 61*(13), 5594–5608. https://doi.org/10.1021/acs.jmedchem.8b00328.

Satlin, L. M., Sheng, S., Woda, C. B., & Kleyman, T. R. (2001). Epithelial Na(+) channels are regulated by flow. *American Journal of Physiology. Renal Physiology, 280*(6), F1010–F1018. https://doi.org/10.1152/ajprenal.2001.280.6.F1010.

Sausbier, M., Arntz, C., Bucurenciu, I., Zhao, H., Zhou, X.-B., Sausbier, U., et al. (2005). Elevated blood pressure linked to primary hyperaldosteronism and impaired vasodilation in BK channel-deficient mice. *Circulation, 112*(1), 60–68. https://doi.org/10.1161/01.CIR.0000156448.74296.FE.

Schild, L. (2010). The epithelial sodium channel and the control of sodium balance. *Biochimica et Biophysica Acta, 1802*(12), 1159–1165. https://doi.org/10.1016/j.bbadis.2010.06.014.

Seki, T., Goto, K., Kiyohara, K., Kansui, Y., Murakami, N., Haga, Y., et al. (2017). Downregulation of endothelial transient receptor potential Vanilloid type 4 channel and small-conductance of Ca^{2+}-activated K^+ channels underpins impaired endothelium-dependent hyperpolarization in hypertension. *Hypertension, 69*(1), 143–153. https://doi.org/10.1161/HYPERTENSIONAHA.116.07110.

Servier Medical Art, (2004). Retrieved January 14, 2020, from https://smart.servier.com/.

Shigaev, A., Asher, C., Latter, H., Garty, H., & Reuveny, E. (2000). Regulation of sgk by aldosterone and its effects on the epithelial Na(+) channel. *American Journal of Physiology. Renal Physiology, 278*(4), F613–F619. https://doi.org/10.1152/ajprenal.2000.278.4.F613.

Singer, D., Biel, M., Lotan, I., Flockerzi, V., Hofmann, F., & Dascal, N. (1991). The roles of the subunits in the function of the calcium channel. *Science, 253*(5027), 1553–1557. https://doi.org/10.1126/science.1716787.

Snyder, P. M., Cheng, C., Prince, L. S., Rogers, J. C., & Welsh, M. J. (1998). Electrophysiological and biochemical evidence that DEG/ENaC cation channels are composed of nine subunits. *The Journal of Biological Chemistry, 273*(2), 681–684. https://doi.org/10.1074/jbc.273.2.681.

Sonou, T., Ohya, M., Yashiro, M., Masumoto, A., Nakashima, Y., Ito, T., et al. (2017). Magnesium prevents phosphate-induced vascular calcification via TRPM7 and Pit-1 in an aortic tissue culture model. *Hypertension Research, 40*(6), 562–567. https://doi.org/10.1038/hr.2016.188.

Sørensen, M. V., Matos, J. E., Sausbier, M., Sausbier, U., Ruth, P., Praetorius, H. A., et al. (2008). Aldosterone increases KCa1.1 (BK) channel-mediated colonic K+ secretion. *The Journal of Physiology, 586*(17), 4251–4264. https://doi.org/10.1113/jphysiol.2008.156968.

Soundararajan, R., Pearce, D., Hughey, R. P., & Kleyman, T. R. (2010). Role of epithelial sodium channels and their regulators in hypertension. *The Journal of Biological Chemistry, 285*(40), 30363–30369. https://doi.org/10.1074/jbc.R110.155341.

Sowers, J. R., Habibi, J., Aroor, A. R., Yang, Y., Lastra, G., Hill, M. A., et al. (2019). Epithelial sodium channels in endothelial cells mediate diet-induced endothelium stiffness and impaired vascular relaxation in obese female mice. *Metabolism: Clinical and Experimental, 99*, 57–66. https://doi.org/10.1016/j.metabol.2019.153946.

Staub, O., Dho, S., Henry, P., Correa, J., Ishikawa, T., McGlade, J., et al. (1996). WW domains of Nedd4 bind to the proline-rich PY motifs in the epithelial Na+ channel deleted in Liddle's syndrome. *The EMBO Journal, 15*(10), 2371–2380.

Sternak, M., Bar, A., Adamski, M. G., Mohaissen, T., Marczyk, B., Kieronska, A., et al. (2018). The deletion of endothelial sodium channel α (αENaC) impairs endothelium-dependent vasodilation and endothelial barrier integrity in indotoxemia in vivo. *Frontiers in Pharmacology, 9*, 178. https://doi.org/10.3389/fphar.2018.00178.

Stewart, A. P., Haerteis, S., Diakov, A., Korbmacher, C., & Edwardson, J. M. (2011). Atomic force microscopy reveals the architecture of the epithelial sodium channel (ENaC). *The Journal of Biological Chemistry, 286*(37), 31944–31952. https://doi.org/10.1074/jbc.M111.275289.

Stokes, J. B., & Sigmund, R. D. (1998). Regulation of rENaC mRNA by dietary NaCl and steroids: Organ, tissue, and steroid heterogeneity. *The American Journal of Physiology, 274*(6), C1699–C1707. https://doi.org/10.1152/ajpcell.1998.274.6.C1699.

Tarjus, A., Maase, M., Jeggle, P., Martinez-Martinez, E., Fassot, C., Loufrani, L., et al. (2017). The endothelial αENaC contributes to vascular endothelial function in vivo. *PLoS One*, *12*(9), e0185319. https://doi.org/10.1371/journal.pone.0185319.

Tetti, M., Monticone, S., Burrello, J., Matarazzo, P., Veglio, F., Pasini, B., et al. (2018). Liddle syndrome: Review of the literature and description of a new case. *International Journal of Molecular Sciences*, *19*(3), E812. https://doi.org/10.3390/ijms19030812.

Valinsky, W. C., Jolly, A., Miquel, P., Touyz, R. M., & Shrier, A. (2016). Aldosterone upregulates transient receptor potential melastatin 7 (TRPM7). *The Journal of Biological Chemistry*, *291*(38), 20163–20172. https://doi.org/10.1074/jbc.M116.735175.

Valinsky, W. C., Touyz, R. M., & Shrier, A. (2019). Aldosterone and ion channels. *Vitamins and Hormones*, *109*, 105–131. https://doi.org/10.1016/bs.vh.2018.10.004.

Walker, B. R., Yau, J. L., Brett, L. P., Seckl, J. R., Monder, C., Williams, B. C., et al. (1991). 11 beta-hydroxysteroid dehydrogenase in vascular smooth muscle and heart: Implications for cardiovascular responses to glucocorticoids. *Endocrinology*, *129*(6), 3305–3312. https://doi.org/10.1210/endo-129-6-3305.

Wang, S., Meng, F., Mohan, S., Champaneri, B., & Gu, Y. (2009). Functional ENaC channels expressed in endothelial cells: A new candidate for mediating shear force. *Microcirculation*, *16*(3), 276–287. https://doi.org/10.1080/10739680802653150.

Warnock, D. G. (2013). The amiloride-sensitive endothelial sodium channel and vascular tone. *Hypertension*, *61*(5), 952–954. https://doi.org/10.1161/HYPERTENSIONAHA.113.00768.

Wendler, A., & Wehling, M. (2011). Is GPR30 the membrane aldosterone receptor postulated 20 years ago? *Hypertension*, *57*(5). e16. https://doi.org/10.1161/HYPERTENSIONAHA.111.170977. author reply e17.

Yang, T., Zhang, H.-L., Liang, Q., Shi, Y., Mei, Y.-A., Barrett, P. Q., et al. (2016). Small-conductance Ca^{2+}-activated potassium channels negatively regulate aldosterone secretion in human adrenocortical cells. *Hypertension*, *68*(3), 785–795. https://doi.org/10.1161/HYPERTENSIONAHA.116.07094.

Ydegaard, R., Andersen, H., Oxlund, C. S., Jacobsen, I. A., Hansen, P. B. L., Jürgensen, J. F., et al. (2019). The acute blood pressure-lowering effect of amiloride is independent of endothelial ENaC and eNOS in humans and mice. *Acta Physiologica*, *225*(1), e13189. https://doi.org/10.1111/apha.13189.

Ye, P., Kenyon, C. J., MacKenzie, S. M., Jong, A. S., Miller, C., Gray, G. A., et al. (2005). The aldosterone synthase (CYP11B2) and 11beta-hydroxylase (CYP11B1) genes are not expressed in the rat heart. *Endocrinology*, *146*(12), 5287–5293. https://doi.org/10.1210/en.2005-0370.

Ye, P., Kenyon, C. J., Mackenzie, S. M., Nichol, K., Seckl, J. R., Fraser, R., et al. (2008). Effects of ACTH, dexamethasone, and adrenalectomy on 11beta-hydroxylase (CYP11B1) and aldosterone synthase (CYP11B2) gene expression in the rat central nervous system. *The Journal of Endocrinology*, *196*(2), 305–311. https://doi.org/10.1677/JOE-07-0439.

Yogi, A., Callera, G. E., O'Connor, S., Antunes, T. T., Valinsky, W., Miquel, P., et al. (2013). Aldosterone signaling through transient receptor potential melastatin 7 cation channel (TRPM7) and its α-kinase domain. *Cellular Signalling*, *25*(11), 2163–2175. https://doi.org/10.1016/j.cellsig.2013.07.002.

Yue, Z., Xie, J., Yu, A. S., Stock, J., Du, J., & Yue, L. (2015). Role of TRP channels in the cardiovascular system. *American Journal of Physiology. Heart and Circulatory Physiology*, *308*(3), H157–H182. https://doi.org/10.1152/ajpheart.00457.2014.

Zhang, Z.-Y., Qian, L.-L., & Wang, R.-X. (2017). Molecular mechanisms underlying renin-angiotensin-aldosterone system mediated regulation of BK channels. *Frontiers in Physiology*, *8*, 698. https://doi.org/10.3389/fphys.2017.00698.

Zhang, Z., Wang, M., Fan, X.-H., Chen, J.-H., Guan, Y.-Y., & Tang, Y.-B. (2012). Upregulation of TRPM7 channels by angiotensin II triggers phenotypic switching of vascular smooth muscle cells of ascending aorta. *Circulation Research, 111*(9), 1137–1146. https://doi.org/10.1161/CIRCRESAHA.112.273755.

Zhang, K., Zhang, Y., Feng, W., Chen, R., Chen, J., Touyz, R. M., et al. (2017). Interleukin-18 enhances vascular calcification and osteogenic differentiation of vascular smooth muscle cells through TRPM7 activation. *Arteriosclerosis, Thrombosis, and Vascular Biology, 37*(10), 1933–1943. https://doi.org/10.1161/ATVBAHA.117.309161.

Zhao, M., Célérier, I., Bousquet, E., Jeanny, J.-C., Jonet, L., Savoldelli, M., et al. (2012). Mineralocorticoid receptor is involved in rat and human ocular chorioretinopathy. *The Journal of Clinical Investigation, 122*(7), 2672–2679. https://doi.org/10.1172/JCI61427.

Ziera, T., Irlbacher, H., Fromm, A., Latouche, C., Krug, S. M., Fromm, M., et al. (2009). Cnksr3 is a direct mineralocorticoid receptor target gene and plays a key role in the regulation of the epithelial sodium channel. *The FASEB Journal, 23*(11), 3936–3946. https://doi.org/10.1096/fj.09-134759.

Zipser, R. D., LeBoff, M., Meidar, V., Duke, R., & Horton, R. (1980). Deoxycorticosterone and aldosterone clearance and binding in normal and hypertensive man. *The Journal of Clinical Endocrinology and Metabolism, 51*(5), 1085–1088. https://doi.org/10.1210/jcem-51-5-1085.

CHAPTER SEVEN

Ion channels and myogenic activity in retinal arterioles

Peter Barabas, Josy Augustine, José A. Fernández, J. Graham McGeown, Mary K. McGahon, Tim M. Curtis*
Wellcome-Wolfson Institute for Experimental Medicine, Queen's University of Belfast, Belfast, United Kingdom
*Corresponding author: e-mail address: t.curtis@qub.ac.uk

Contents

Abstract

Retinal pressure autoregulation is an important mechanism that protects the retina by stabilizing retinal blood flow during changes in arterial or intraocular pressure. Similar to other vascular beds, retinal pressure autoregulation is thought to be mediated largely through the myogenic response of small arteries and arterioles which constrict when transmural pressure increases or dilate when it decreases. Over recent years, we and others have investigated the signaling pathways underlying the myogenic response in retinal arterioles, with particular emphasis on the involvement of different ion channels expressed in the smooth muscle layer of these vessels. Here, we review and extend previous work on the expression and spatial distribution of the plasma membrane and sarcoplasmic reticulum ion channels present in retinal vascular smooth muscle cells (VSMCs) and discuss their contribution to pressure-induced myogenic tone in retinal arterioles. This includes new data demonstrating that several key players and modulators of the myogenic response show distinctively heterogeneous expression along the length of the retinal arteriolar network, suggesting differences in myogenic signaling

Current Topics in Membranes, Volume 85
ISSN 1063-5823
https://doi.org/10.1016/bs.ctm.2020.01.008

between larger and smaller pre-capillary arterioles. Our immunohistochemical investigations have also highlighted the presence of actin-containing microstructures called myobridges that connect the retinal VSMCs to one another. Although further work is still needed, studies to date investigating myogenic mechanisms in the retina have contributed to a better understanding of how blood flow is regulated in this tissue. They also provide a basis to direct future research into retinal diseases where blood flow changes contribute to the pathology.

1. Introduction

The retina is one of the most metabolically demanding tissues of the body, requiring a constant supply of oxygen and nutrients to maintain normal function (Niven & Laughlin, 2008). The metabolic needs of the retina are met through a dual blood supply from the retinal and choroidal circulations. The retinal circulation supplies the bulk of the inner retina, while the choroidal circulation nourishes the photoreceptors and the retinal pigment epithelium (RPE). Over recent decades, considerable efforts have been made to elucidate the complex molecular and cellular mechanisms that govern blood flow and oxygen delivery to the retina. These studies have been motivated by the fact that changes in retinal haemodynamics are thought to play a critical role in the pathophysiology of several of the leading causes of vision loss worldwide, including diabetic retinopathy, glaucoma and retinal vascular occlusive disease (Kur, Newman, & Chan-Ling, 2012; Pournaras, Rungger-Brandle, Riva, Hardarson, & Stefansson, 2008). Much of the work to date has focused on retinal arterioles, which are the main resistance vessels of the retina and therefore determine retinal capillary perfusion. Since the retinal circulatory system lacks an autonomic nerve supply, retinal blood flow is mainly regulated through local signaling mechanisms. One of the most important of these is pressure autoregulation, where the arterioles adjust their diameter to keep blood flow constant during changes in arterial blood pressure. As in other vascular beds, pressure autoregulation in the retina occurs via a myogenic mechanism resulting from the intrinsic ability of the vascular smooth muscle cells (VSMCs) to contract, and hence the vessels to constrict, when arterial or transmural pressure rises (Pournaras et al., 2008; Scholfield, McGeown, & Curtis, 2007). In recent years, we have focused our attention on understanding the signaling mechanisms that underlie the myogenic response in retinal arterioles, and this work forms the basis of the current review. We begin by describing the topography of the retinal vascular supply and some of the key anatomical features of retinal arterioles. These discussions are followed by an overview of pressure autoregulation in the retina as well as the methods and software that we have developed to study myogenic signaling mechanisms in ex vivo isolated retinal arteriole segments. Our data, and those of others, supporting the involvement of a multitude of ion channels in the development and modulation of the retinal arteriolar myogenic response are then discussed. Finally, we highlight regional heterogeneity in ion channel expression among different order arterioles of the retinal circulatory system suggesting that myogenic mechanisms may vary along the length of the arteriolar network.

2. Retinal vascular supply

Human retinal blood vessels are formed by two distinct cellular processes of vasculogenesis and angiogenesis (Hughes, Yang, & Chan-Ling, 2000). Vasculogenesis is the formation of de novo vessels from vascular mesenchymal precursor cells and angioblasts, while angiogenesis is the extension of the vascular network by proliferating endothelial cells from pre-existing blood vessels. The retinal vasculature originates from the central retinal artery, which enters the optic disc through a gap in the lamina cribrosa of the sclera (Fig. 1A), where it divides into four major intra-retinal arteries, each supplying one quadrant of the retina. It has an end artery arrangement, with all of the blood from the retinal arteries and arterioles passing through the capillaries to the retinal venules and finally the veins (Kuwabara & Cogan, 1960). The vessels are arranged as a superficial vascular plexus located in the ganglion cell and nerve fiber layers (GCL and NFL), an intermediate vascular plexus which extends down to the inner nuclear layer (INL) and a deep vascular plexus which is located at the junction of the INL and outer plexiform layers (OPL) (Fig. 1B). The superficial plexus is ensheathed by macroglia and contains arteries, arterioles, capillaries, post-capillary venules and veins (Kur et al., 2012). The intermediate and deep vascular beds are ensheathed solely by Müller glia (the principal glial cells of the retina) and consist predominantly of capillary-sized vessels (Kur et al., 2012). Vascular density is greatest in the macular region of the retina which has the highest density of photoreceptors and is involved in central vision (Curtis, Gardiner, & Stitt, 2009).

The retinal arteries are primarily situated in the GCL layer, directly beneath the inner limiting membrane (ILM) (Scholfield et al., 2007).

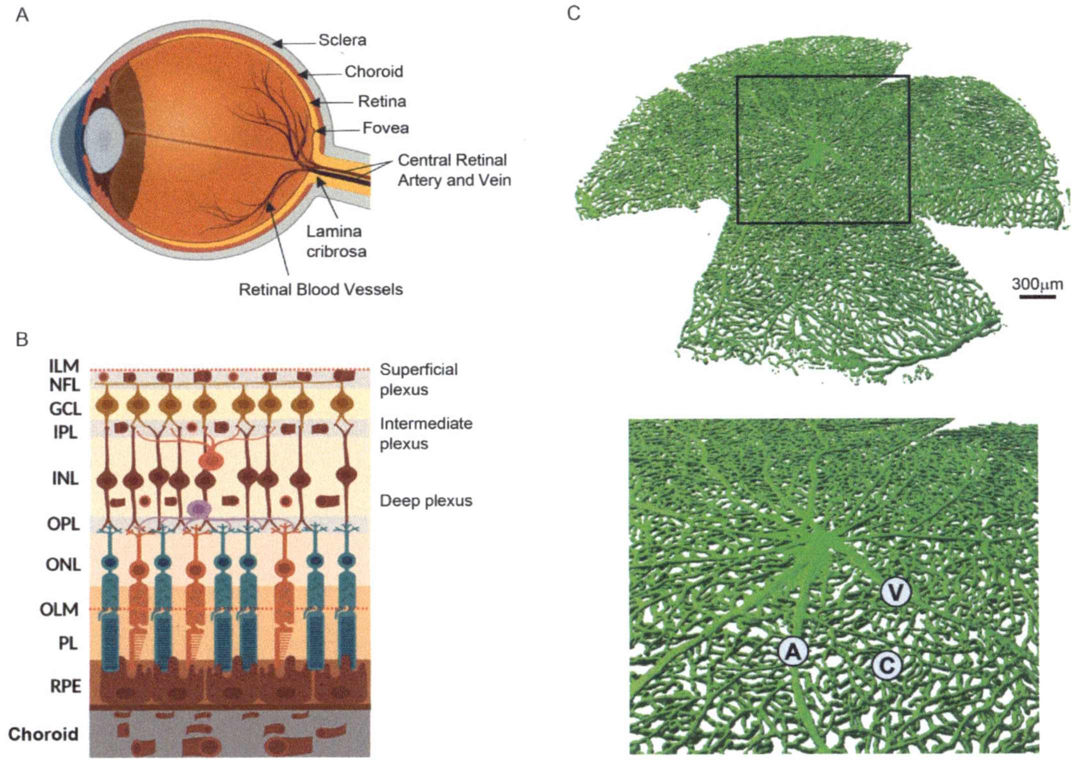

Fig. 1 (A) Diagrammatic cross-section of the human eye showing how the blood vessels enter at the optic disk and spread across the retinal surface. (B) Schematic showing the trilaminar vascular network of the retina, comprising of the superficial, intermediate and deep capillary plexi. ILM, inner limiting membrane; NFL, nerve fiber layer; GCL, ganglion cell layer; IPL, inner plexiform layer; INL, inner nuclear layer; OPL, outer plexiform layer; ONL, outer nuclear layer; OLM, outer limiting membrane; PL, photoreceptor layer; RPE, retinal pigment epithelium. Panels (A) and (B) were created with BioRender (https://app.biorender.com). (C) Three-dimensional reconstruction of the retinal vasculature from a FITC-albumin perfused mouse retina. The retina was cut into a maltese cross and wholemounted with the GCL layer facing upwards. Confocal z-stacks were captured using a Leica SP8 confocal microscope and volume rendering was performed using Imaris v9.5 (Bitplane AG, Zurich, Switzerland). The upper panel shows the vasculature of the entire mouse retina tilted backwards by approximately 60°. The lower panel is a higher magnification view of the boxed region in the upper panel. A, arteriole; C, capillaries; V, venule.

The major arteries are composed of a single layer of endothelial cells lining the vessel and approximately 5–7 layers of VSMCs called the tunica media (Lutty et al., 2012). The VSMCs are orientated both circularly and longitudinally in the retinal arteries, with each being surrounded by a basal lamina comprised of an increasing amount of collagen and elastin fibers toward the acellular tunica externa. The tunica intima is composed of a single sheet of endothelial cells that line the lumen of the vessel and are oriented along the longitudinal axis of the vessel (Pournaras et al., 2008). The arteries divide into smaller arterioles and then terminal pre-capillary arterioles, which feed into the capillary beds of the retina (Scholfield et al., 2007).

Capillaries are found in all three vascular plexuses of the retina. They are 5–6 μm in diameter and visible as a meshwork of small caliber vessels (Curtis et al., 2018; Scholfield et al., 2007). They are composed of a continuous single layer of endothelium and are characterized by the presence of intramural pericytes, which extend circumferentially around the endothelium of the capillary lumen. Pericytes form a discontinuous layer and there is a high density of pericytes in the retinal circulatory system, with an almost one-to-one ratio with endothelial cells (Curtis et al., 2009). Retinal endothelial cells and pericytes are in direct communication via gap junctional complexes and share a common basement membrane (Gardiner, Archer, Curtis, & Stitt, 2007; Puro, 2007). Fluid exchange across the retinal capillary beds is controlled by the inner blood-retinal-barrier (iBRB), which is formed by tight and adheren junctions between adjacent endothelial cells of the retinal blood vessels (Klaassen, Van Noorden, & Schlingemann, 2013). As is the case with the blood–brain barrier, the iBRB is characterized by a high transendothelial electrical resistance and greatly restricts the paracellular flow of water, proteins, lipids, pathogens and immune cells into the retina. In general, retinal capillary density is greater in the central retina, decreasing toward the periphery (Hill, 1977). The extreme retinal periphery, fovea and outer retinal layers are avascular and receive their nutrition from the choroid (Engerman, 1976; Provis, 2001).

The venous system of the retina has a similar arrangement to that of the arteries. The retinal veins interdigitate with the retinal arteries and blood is drained via the central retinal vein which leaves the eye at the optic disc (Lutty et al., 2012). Retinal veins are composed of endothelial cells and only 1–2 layers of VSMCs. The endothelial cells are again orientated longitudinally along the axis of the veins. They share their basement membrane with the adjacent VSMCs and collagen fibrils are seen in the outer layers of the basement membrane. The post-capillary venules are thin walled and have relatively sparse smooth muscle cell coverage (Curtis et al., 2018). The terminal parts of the branching systems in the retina such as pre-capillary arterioles and post-capillary venules, are connected through the anastomotic capillaries (Scholfield et al., 2007).

The retinal vascular supply differs markedly among different species. The retina of birds, for example, is completely avascular and dependent on the choroidal circulation for its blood supply (Chase, 1982; Wolburg, Liebner, Reichenbach, & Gerhardt, 1999). In the rabbit, horse and guinea pig, the retina is only partially vascularized, with only a portion of the inner retina containing blood vessels (Lutty et al., 2012). Similar to humans, other mammals such as primates, pigs, cows, dogs, rats and mice have retinas that are richly vascularized (Provis, 2001) (for rodent retina, see Fig. 1C). Even though the rodent and human retinal circulatory systems are organized in a similar pattern, the rodent retinal vasculature has more side branches at the level of the arterial bed and the larger arterioles branch directly off the central retinal artery (Pannarale, Onori, Ripani, & Gaudio, 1996). Also, in contrast to humans, the total vascular volume in the rodent retina can be modified by the opening and closing of sphincters that are located at the junctions of the arterioles (Pannarale et al., 1996). Finally, because rodent retinas do not contain a macula, they lack a foveal avascular zone.

3. Retinal arterioles

Retinal arterioles are the small arterial branches of the retina and have a structure similar to the arteries with reduced diameter size. Retinal arterioles play a key role in the regulation of blood flow in retina by dilating or constricting their luminal diameter, as determined by relaxation or contraction of the VSMCs that encircle these vessels. In humans, they comprise of 1–2 layers of VSMCs and an inner lining of longitudinally arranged endothelial cells. The structure of retinal arterioles in the mouse and rat retina is similar, but these vessels have just a single layer of VSMCs that wrap around the endothelium (Curtis et al., 2018). Rat retinal arterioles can be divided into first order, second order and pre-capillary arterioles based on their diameter (Curtis et al., 2018). In situ measurements of non-pressurized vessels within rat retinal wholemount preparations have shown that the first order arterioles are 30–45 μm in diameter, while the second order and pre-capillary arterioles have diameter ranges of 20–30 μm and 8–20 μm, respectively. Following their isolation from the retinal tissue, primary, secondary and pre-capillary arterioles can be identified under bright-field microscopy based

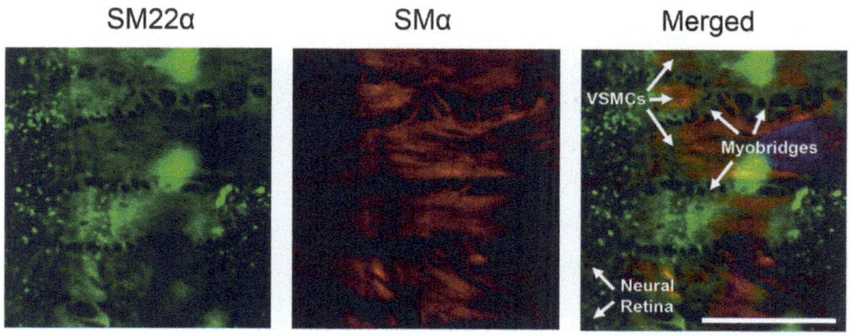

Fig. 2 Confocal images of a short segment of a rat retinal arteriole within a wholemount preparation labeled with smooth muscle protein 22α (SM22α; green) and α-smooth muscle actin (SMα; red) antibodies. Nuclei are stained with TO-PRO3 (pseudo-colored blue) in the merged image. Myobridges have been labeled that connect adjacent VSMCs together. Scale bar = 10 μm.

on their size, caliber and the arrangement of the VSMCs (Curtis et al., 2018; Pannarale et al., 1996). Studies from our own laboratory have found that adjacent VSMCs of retinal arterioles are connected by thin fibers called myobridges, which contain actin myofilaments and presumably act to help stabilize and maintain the structural integrity of the vessels (Fig. 2). These myobridges may also represent important sites for intercellular communication between neighboring VSMCs, but this possibility has yet to be evaluated experimentally.

4. Retinal pressure autoregulation

Pressure autoregulation has been extensively studied in many organs (Johnson, 1986). As stated earlier, pressure autoregulation is the innate ability of an organ to maintain blood flow at a constant level despite changes in arterial or transmural pressure. This autoregulation of blood flow protects the capillary network from excessive pressure and ensures that oxygen delivery and waste removal is maintained at a steady level. The myogenic response of small arteries and arterioles is thought to be the principal mechanism behind pressure autoregulation in most vascular beds (Faraci, Baumbach, & Heistad, 1989; Muller, Davis, & Chilian, 1996). This refers to the inherent ability of the small arteries and arterioles to constrict when perfusion pressure increases or dilate when it decreases (Meininger & Davis, 1992). Studying pressure autoregulation in vivo requires isolation of the tissue from neural and hormonal influences. As such, pressure autoregulation may be more easily investigated in the retina than in many other vascular beds because the retinal arterial vessels lack autonomic innervation (Ye, Laties, & Stone, 1990) and the endothelial cells are tightly connected via tight and adheren junctions, which limits exposure of the VSMCs to circulating hormones (Klaassen et al., 2013). Due to their visibility through the optics of the eye, retinal vessels also provide the opportunity for investigators to directly and non-invasively image autoregulatory responses in vivo.

Pressure autoregulation in the retina can be studied by modifying the ocular perfusion pressure (OPP), which is defined as the difference between the pressure in the retinal arterial vessels and the intraocular pressure (IOP) (Harris, Ciulla, Chung, & Martin, 1998). Physiological alterations in OPP occur regularly due to nocturnal-, exercise-, stress- and posture-induced changes in arterial blood pressure (Dumskyj, Eriksen, Dore, & Kohner, 1996; Hague & Hill, 1988; Scholfield et al., 2007) and diurnal variations in IOP (Kitazawa & Horie, 1975; Liu et al., 1998). Experimentally, several techniques have been used to modify OPP to investigate pressure autoregulatory responses in the retina. These include the use of scleral suction cups which decrease OPP by increasing IOP (Grunwald, Sinclair, & Riva, 1982; Riva, Grunwald, & Petrig, 1986; Riva, Sinclair, & Grunwald, 1981) and the use of isometric and dynamic exercises, posture changes and the cold pressor test (CPT) to increase OPP by raising arterial blood pressure (Dumskyj et al., 1996; Iester, Torre, Bricola, Bagnis, & Calabria, 2007; Lorenzi et al., 2010; Nagaoka, Mori, & Yoshida, 2002; Robinson, Riva, Grunwald, Petrig, & Sinclair, 1986). These studies have allowed estimates to be made of the boundaries of pressure autoregulation in the human retinal circulation. Using the scleral suction cup technique, it has been demonstrated that human retinal blood flow remains constant until the average OPP is decreased by ~36–38% below baseline (Riva et al., 1986, 1981). The upper limit for human retinal blood flow autoregulation appears to lie ~34–60% above the baseline OPP as determined by isometric exercise testing (Pournaras et al., 2008). Similar to humans, pressure autoregulatory responses to increases and decreases in OPP have been reported in the retinas of several animal species, including rats, cats, pigs and nonhuman primates (Davies, Sullivan, Morris, Caldwell, & Kohner, 1991; Liang, Fortune, Cull, Cioffi, & Wang, 2010; Tani, Nagaoka, Nakabayashi, Yoshioka, & Yoshida, 2014; Wong et al., 2015). The upper limit of retinal pressure autoregulation in miniature pigs lies ~31–74% above the baseline OPP (Davies et al., 1991); however, in general the boundaries of the retinal autoregulatory range in different animal species has yet to be studied in detail. In clinical studies, during isometric exercise the pressure-induced constriction of retinal arterial vessels increases with decreasing diameter of the vessels (Jeppesen, Sanye-Hajari, & Bek, 2007). These results suggest that the distal retinal arterioles are the most reactive vascular segments within the retinal circulation and therefore act as key effectors of pressure autoregulation in the retina.

As pressure autoregulation is the only homeostatic mechanism protecting the retinal capillary beds from overt increases in input pressure, a significant reduction or loss of autoregulation could be expected to lead to capillary damage and pathology. Indeed, diminished pressure autoregulation is an early feature of diabetic retinopathy that precedes the development of clinically detectable vascular disease (Rassam, Patel, & Kohner, 1995; Sinclair et al., 1982). The Lorenzi group recently showed that 4-years after testing their autoregulatory control, diabetic patients with a defective autoregulatory response developed diabetic retinopathy at an accelerated rate compared to diabetic individuals who had normal autoregulation, and that this effect was independent of their glycaemic status (Tecilazich, Feke, Mazzantini, Sobrin, & Lorenzi, 2016). These findings provide some evidence that loss of retinal pressure autoregulation is a contributing factor in the development of diabetic retinopathy, although exactly why this process breaks down in diabetes remains to be determined. To address this question, we first need to identify and understand the fundamental molecular and cellular mechanisms that drive pressure autoregulation in the retina, and this comprises the main focus of the remainder of this chapter.

5. Measurement of the retinal arteriolar myogenic response

To properly understand the process of pressure autoregulation in the retina, it is necessary to determine the signal transduction mechanisms that underlie the myogenic contraction of retinal arterioles. This has required the development of techniques for isolating and cannulating retinal arterioles for use in pressure myography experiments. Most studies to date have used porcine retinal arterioles that can be surgically excised from the freshly isolated retina and are sufficiently large (>50 μm maximal diameter) to mount on commercially available pressure myograph systems (Hein et al., 2009; Hessellund, Jeppesen, Aalkjaer, & Bek, 2003; Jeppesen, Aalkjaer, & Bek, 2003; Nagaoka, Kuo, Ren, Yoshida, & Hein, 2008; Omae, Nagaoka, Tanano, & Yoshida, 2013; Otani et al., 2016; Potts, Bradley, Xu, Kuo, & Hein, 2013). Using this methodology, porcine retinal arterioles display a robust myogenic response at intravascular pressures ranging between ~25 and 70 mmHg (Rosa, Hein, & Kuo, 2004). Consistent with studies in other vascular beds, the myogenic response in porcine retinal arterioles is not dependent on the presence of an intact endothelium (Jeppesen et al., 2003; Nagaoka et al., 2008; Omae et al., 2013; Potts et al., 2013). Pressure myography has also been successfully applied to human retinal arterioles (Hein, Rosa, Yuan, Roberts, & Kuo, 2010). This work has shown that arterioles isolated from the human retina exhibit myogenic tone and dilate to endothelial cell agonists and increased shear stress in a manner similar to that observed in porcine retinal vessels (Hein et al., 2010).

A more limited number of studies have attempted to use isolated retinal arterioles from the rat and mouse retina for pressure myography experiments, as their smaller size (≤45 μm maximal diameter) makes their use with traditional pressure myograph systems more challenging (Curtis et al., 2018). There are several benefits, however, of using retinal arterioles from experimental rodents, including their ease and cost of maintenance and the wide availability of gene knockout and ocular disease models. Over recent years, we have developed bespoke methods that enable the routine application of pressure myography to intact arteriole segments isolated from the rodent retina (Fig. 3). Although a detailed description of these methods is beyond the scope of this chapter, step-by-step written and video protocols from our laboratory have recently been published online (Curtis et al., 2018). Pressure myography requires the accurate tracking of changes in vessel diameter throughout the course of the experiment. For larger arteries and arterioles this is usually accomplished using semi-automated or automated thresholding algorithms that are capable of localizing the inner and outer boundaries of the VSMC wall (Lawton et al., 2019). However, because arterioles isolated from the rodent retina contain just a single layer of VSMCs, these algorithms often fail because there is insufficient contrast between the vessel edges and other areas of the image to enable the vessel walls to be detected. To circumvent this problem, we have recently developed new software called MyoTracker for the automatic tracking and measurement of the diameter of small isolated arterioles from the rodent retina. The software is freely available as a Fiji plug-in and uses a combination of edge enhancement and cost analysis to reliably and precisely detect the vessel walls (Fernandez, Bankhead, Zhou, McGeown, & Curtis, 2014). It is also amenable for use with retinal artery and arteriole preparations isolated from larger animals, including pigs and cows (Fernandez et al., 2014).

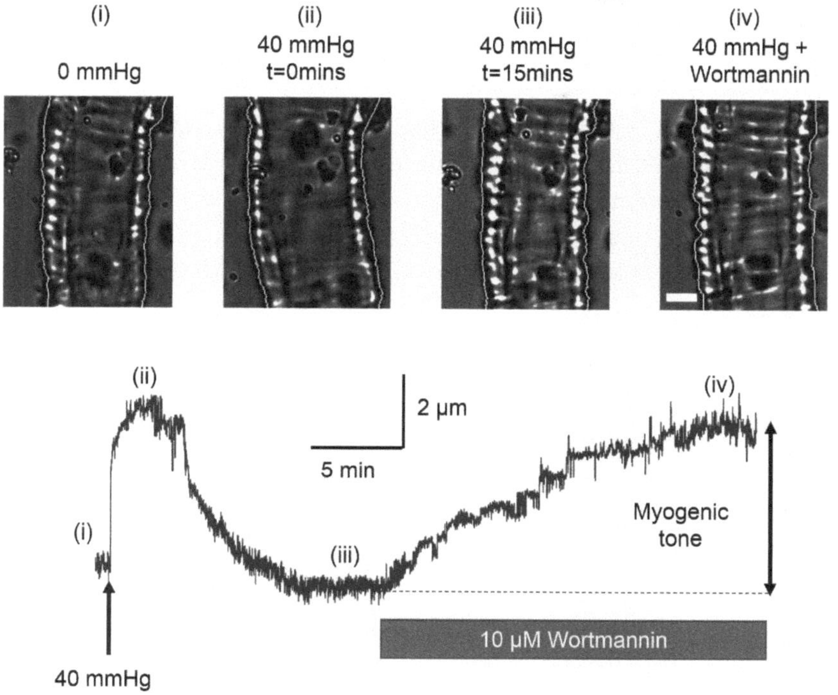

Fig. 3 Arteriolar pressure myography. Top panels show images of an isolated, cannulated, rat retinal arteriole prior to and following pressurization to 40 mmHg. At the end of the experiment the arteriole was superfused with zero Ca^{2+} solution containing the MLCK inhibitor, wortmannin (10 μM), to obtain the passive diameter of the vessel. Scale bar = 10 μm. The lower panel shows a time course record of the vessel diameter for the experiment shown in the top panels (sample rate, 2.5 Hz). Following pressurization, myogenic tone developed and stabilized over a period of ~15 min. *This figure is adapted from Curtis, T. M., McLaughlin, D., O'Hare, M., Kur, J., Barabas, P., Revolta, G., et al. (2018). Isolation of retinal arterioles for ex vivo cell physiology studies.* Journal of Visualized Experiments, *(137). doi:10.3791/57944.*

As with other smooth muscle cells, the contraction of the VSMCs on retinal arterioles depends on an increase in the intracellular Ca^{2+} concentration ($[Ca^{2+}]_i$) leading to myosin light chain kinase (MLCK) activation and actin–myosin cross-bridge formation. To study how changes in intravascular pressure alter intracellular Ca^{2+} signaling in retinal VSMCs, retinal arteriolar pressure myography can be combined with conventional or confocal microscopy-based Ca^{2+} imaging techniques (Curtis et al., 2018; Kur, Bankhead, Scholfield, Curtis, & McGeown, 2013). Recordings are restricted to the retinal VSMCs, as the endothelial cells of these vessels fail to load with the Ca^{2+} indicator dyes under normal conditions

(Scholfield & Curtis, 2000; Scholfield et al., 2007). It is also possible to make patch-clamp electrophysiology recordings from retinal VSMCs still embedded within their parent arterioles, allowing the different ionic currents present in these cells to be characterized. These experiments, however, require the isolated arterioles to be enzymatically digested to remove the vascular basement membranes and electrically uncouple the VSMCs from one another and the underlying endothelium (Curtis et al., 2018; McGahon, Dash, et al., 2007; McGahon, Dawicki, et al., 2007; McGahon, Dawicki, Scholfield, McGeown, & Curtis, 2005). Thus, to date, it has not been possible to combine patch-clamp with pressure myography studies due to the loss of vessel integrity during the enzymatic dissociation process.

6. Mechanosensitive ion channels

Understanding of the molecular underpinnings of the myogenic response and pressure autoregulation in the retina requires an appreciation of the mechanisms through which the retinal arteriolar VSMCs detect and respond to changes in arterial or transmural pressure. The prevailing view in many vascular beds is that the myogenic response results from the activation of stretch-activated cation channels on the plasma membrane of the VSMCs, which trigger membrane depolarization and an increase in voltage-dependent Ca^{2+} influx, thereby leading to cell contraction (Baek & Kim, 2011; Davis & Hill, 1999). While the above has been proven experimentally, only recently has progress been made in identifying the molecular make-up of the channels that mediate stretch-activated cation currents in VSMCs. Below, we briefly review the contribution of different stretch-activated cation channels in mediating the vascular myogenic response and discuss our recent work identifying TRPV2 as the main channel underlying stretch-activated signaling in retinal arteriolar VSMCs.

Over the past two decades, several different mechanosensitive cation channels have been identified at a molecular level. These include Piezo1/2, OSCA/TMEM63, TMC1/2 (Douguet & Honore, 2019), ENaC (Drummond, Gebremedhin, & Harder, 2004) and TRP channels (Liu & Montell, 2015). While most of these channels may be directly mechanically gated, the response of some TRP channels to membrane stretch has been shown to be indirect, resulting from the upstream activation of mechanosensitive G-protein coupled receptors (GCPRs) (Gonzales et al., 2014). Piezo1 has been found in VSMCs (Retailleau et al., 2015), as well as endothelial cells and has been identified as an important component in vascular development, remodeling and as a shear stress sensor driving endothelium-dependent vasodilation (Wang et al., 2016). While Piezo1 does not seem to participate in myogenic signaling in small-diameter arteries, it does contribute to their structural remodeling during hypertension (Retailleau et al., 2015). Piezo2 is the mechanotransducer necessary for touch (Poole, Herget, Lapatsina, Ngo, & Lewin, 2014; Ranade et al., 2014; Woo et al., 2014), proprioception (Florez-Paz, Bali, Kuner, & Gomis, 2016; Woo et al., 2014) and pain sensation (Murthy et al., 2018; Szczot et al., 2018). This channel may have some involvement in developmental angiogenesis in the retina (Yang et al., 2016), but its main role in the cardiovascular system seems to be limited to driving the baroreceptor reflex together with Piezo1 (Zeng et al., 2018). TMEM63 is a highly conserved mechanotransducer recently found in plants, insects and mice (Murthy et al., 2018; Yue, Sheng, Kang, & Xiao, 2019), while TMC1/2 proteins form the mechanosensory transduction channels in auditory and vestibular hair cells (Kawashima et al., 2011; Kurima et al., 2002). The expression of these channels in VSMCs and their possible involvement in the myogenic constriction of small arteries and arterioles has yet to be studied in detail. Currently only ENaC (Choi, Yeon, Kwon, Byeon, & Lee, 2017; Drummond et al., 2004) and several transient receptor potential (TRP) channel family members (Earley, Waldron, & Brayden, 2004; McGahon et al., 2016; Narayanan et al., 2013; Soni, Peixoto-Neves, Matthews, & Adebiyi, 2017; Welsh, Morielli, Nelson, & Brayden, 2002) have been clearly linked to stretch-activated signaling and myogenic activity in vascular tissues. Strong expression of ENaC has been detected in blood vessels of the choroid and iris of the eye, but not in those of the retina (Mirshahi et al., 1999), suggesting that these channels are unlikely to play a significant role in the myogenic signaling pathway of retinal arterioles.

Of the different classes of mechanosensitive cation channels, TRP channels stand out due to their almost ubiquitous expression across many tissues (Kunert-Keil, Bisping, Kruger, & Brinkmeier, 2006; Nilius & Owsianik, 2011). TRP channels act as multimodal ion channels, not only responding to mechanical forces, but also to a wide range of chemical and biophysical intracellular and extracellular stimuli (Muller, Morales, & Reggio, 2018; Nilius & Owsianik, 2011; Pedersen, Owsianik, & Nilius, 2005; Putney, 2004). Of the 28 known mammalian TRP channels, a subset of 14 have been shown to respond to direct mechanical force or osmotic driven membrane stretch (Moore & Liedtke, 2017); these include TRPA1, TRPC1/3/5/6, TRPM3/4/7, TRPML3, TRPP1/2 and TRPV1/2/4. With the exception of TRPA1, all of these channels have been reported to be expressed in

VSMCs (Albert & Large, 2003; Alonso-Carbajo et al., 2017; Earley, Heppner, Nelson, & Brayden, 2005; Earley et al., 2004; Griffin, Torres, Grande, & Kumar, 1997; Inoue, Jian, & Kawarabayashi, 2009; McGahon et al., 2016, 2012; Naylor et al., 2010; Noorani, Noel, & Marrelli, 2011; Oancea, Wolfe, & Clapham, 2006; Soni et al., 2017; Torres et al., 2001; Yip et al., 2004). There is significant variation in TRP channel expression among different types of VSMCs, suggesting vascular bed-specific regulation of TRP channel gene expression (Inoue et al., 2009). Notably, retinal blood vessels might reasonably be expected to be very similar to those of the brain, given that their anatomical location means that both are subjected to relatively similar changes of input pressure and both provide a tight barrier function for a metabolically demanding neuroglial tissue (Ames, Li, Heher, & Kimble, 1992; Feke & Pasquale, 2008; Howard, Blakeslee, & Laughlin, 1987; Lam, Haunton, Robinson, & Panerai, 2019; Toda, Kawazu, Oyabu, Miyazaki, & Kiuchi, 2011; Wang et al., 2010). Yet, the myogenic response of small cerebral arteries has been shown to be dependent on TRPC6 and TRPM4, which are not present in retinal VSMCs (Earley et al., 2004; McGahon et al., 2016; Welsh et al., 2002). Of the mechanosensitive TRP channels characterized to date, VSMCs of retinal arterioles have been found to contain: TRPC1, TRPV1/2/4, TRPM7 and TRPP1 (McGahon et al., 2016, 2012). Although retinal VSMCs express a range of mechanosensitive TRP channels, our recent evidence suggests that TRPV2 is the principal channel that underlies stretch-activated cation currents and myogenic contraction in retinal arterioles. In particular, we found that blockade of TRPC1, TRPV1/4 and TRPM7 had no effect on hypotonic stretch-induced Ca^{2+} influx in retinal VSMCs and that the permeability properties of the stretch-activated cation currents in these cells were inconsistent with TRPP1 (McGahon et al., 2016). Moreover, direct plasma membrane stretch elicited cation currents that could be blocked by TRPV2 pharmacological inhibitors and specific TRPV2 pore-blocking antibodies and pre-incubation of the arterioles with these agents prevented the development and maintenance of myogenic tone (Fig. 4) (McGahon et al., 2016). It is perhaps surprising that retinal arteriolar VSMCs contain a range of TRP channels implicated in mechanotransduction, yet only TRPV2 contributes to the myogenic response of these vessels. This could be in part explained by the fact that several of the other mechanosensitive channels including TRPM7, TRPP1, TRPV1 and TRPV4 were found to be mainly localized to cytoplasmic or nuclear domains of the cells, as opposed to the plasma membrane (McGahon et al., 2016). Our results for

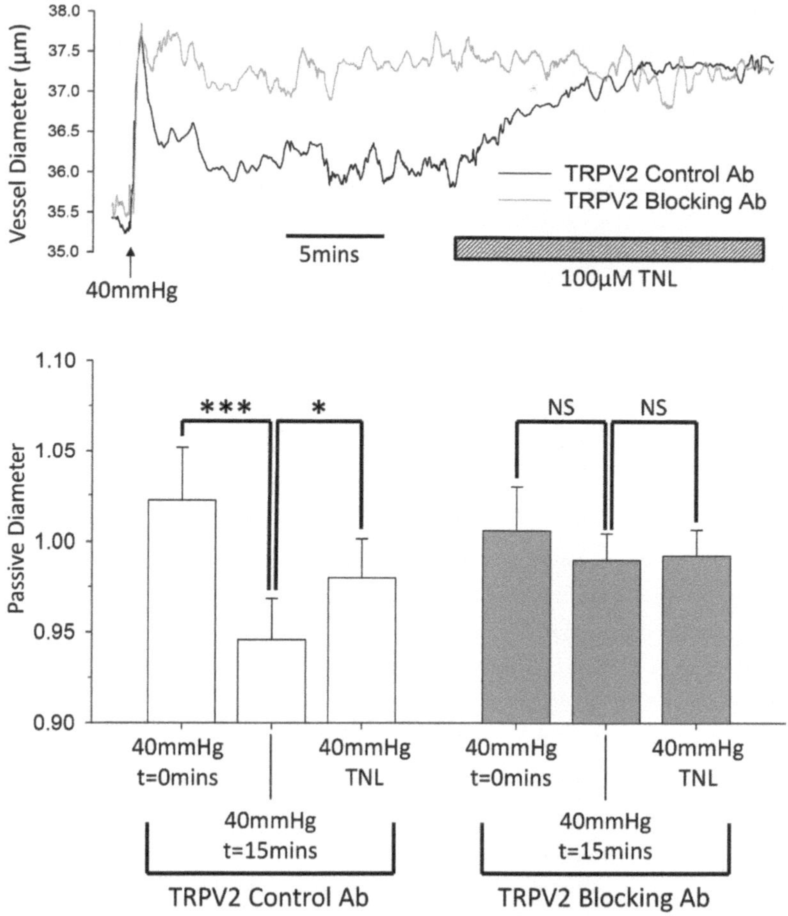

Fig. 4 Effects of TRPV2 channel inhibition on the myogenic constriction of isolated retinal arterioles. Top panel, representative traces showing the effects of pore-blocking and negative control anti-TRPV2 antibodies on myogenic tone development in isolated retinal arterioles. The TRPV2 channel inhibitor, tranilast (TNL), was applied 15 min after pressurization. Bottom panel, mean diameter data (normalized to the maximal passive diameter) showing that anti-TRPV2 pore blocking antibodies abolished myogenic tone development and that tranilast caused a significant dilation of arterioles exposed to the negative control antibodies under conditions of steady-state myogenic tone ($n = 11$ arterioles per group). $^*P < 0.05$ and $^{***}P < 0.001$. *This figure is adapted from McGahon, M. K., Fernandez, J. A., Dash, D. P., McKee, J., Simpson, D. A., Zholos, A. V., et al. (2016). TRPV2 channels contribute to stretch-activated cation currents and myogenic constriction in retinal arterioles.* Investigative Ophthalmology & Visual Science, *57(13), 5637–5647. doi:10.1167/iovs.16-20279.*

TRPC1 concur with those reported using cerebral vessels from TRPC1 knockout mice, which suggested that these channels are not an obligatory component of stretch-activated cation channel complexes in VSMCs (Dietrich et al., 2007).

7. Voltage-gated Ca^{2+} channels (VGCCs)

VGCCs are considered to be a major route of extracellular Ca^{2+} entry during the myogenic response (Jackson & Boerman, 2018). As with TRP channels, there seems to be some heterogeneity in VGCC expression patterns among VSMCs from different organs as well as between larger and smaller diameter vessels, although myogenic tone in most vascular beds appears to be mainly driven by Ca^{2+} entry through L-type $Ca_v1.2$ channels with a smaller contribution from other types of VGCCs (Moosmang et al., 2003; Nikitina et al., 2007). The importance of L-type $Ca_v1.2$ channels is perhaps most clearly demonstrated in smooth muscle cell-specific $Ca_v1.2$ knockout mice, which entirely lacked a myogenic response in their excised tibialis arteries, and had a significant drop in their mean arterial pressure (Moosmang et al., 2003), most likely due to a reduced constriction of resistance vessels and a drop in total peripheral resistance. Retinal VSMCs express L-type $Ca_v1.2$ VGCCs (Fernandez, McGahon, McGeown, & Curtis, 2015; McGahon, Zhang, Scholfield, Curtis, & McGeown, 2007) (Fig. 5) and we and others have previously shown that these channels play a key role in myogenic signaling in rat and porcine retinal arterioles (Jeppesen et al., 2003; Kur et al., 2013). The myogenic response provides a level of basal tone upon which vasoactive molecules can act and the L-type $Ca_v1.2$ channels of retinal VSMCs appear to be the main target underlying the vasodilatory actions of arachidonic acid in the retina (Kur et al., 2014).

In addition to $Ca_v1.2$, it has been reported that other L-type Ca^{2+} channel isoforms ($Ca_v1.3$), P-/Q-type ($Ca_v2.1$), N-type ($Ca_v2.2$) and T-type ($Ca_v3.1$–3.3) VGCCs contribute significantly to myogenic vasoconstriction in different arterial beds (Andreasen et al., 2006; Cazade, Bidaud, Hansen, Lory, & Chemin, 2014; Jensen, Salomonsson, Jensen, & Holstein-Rathlou, 2004; Nikitina et al., 2007). R-type ($Ca_v2.3$) VGCCs have also been implicated in the enhanced myogenic constriction and reduced blood flow seen in cerebral arteries following subarachnoid hemorrhage (Ishiguro et al., 2005). We have recently shown that T-type $Ca_v3.1$ channels are expressed at a molecular and functional level in retinal VSMCs and are involved in the myogenic reactivity of retinal arterioles (Fernandez et al., 2015) (Fig. 5).

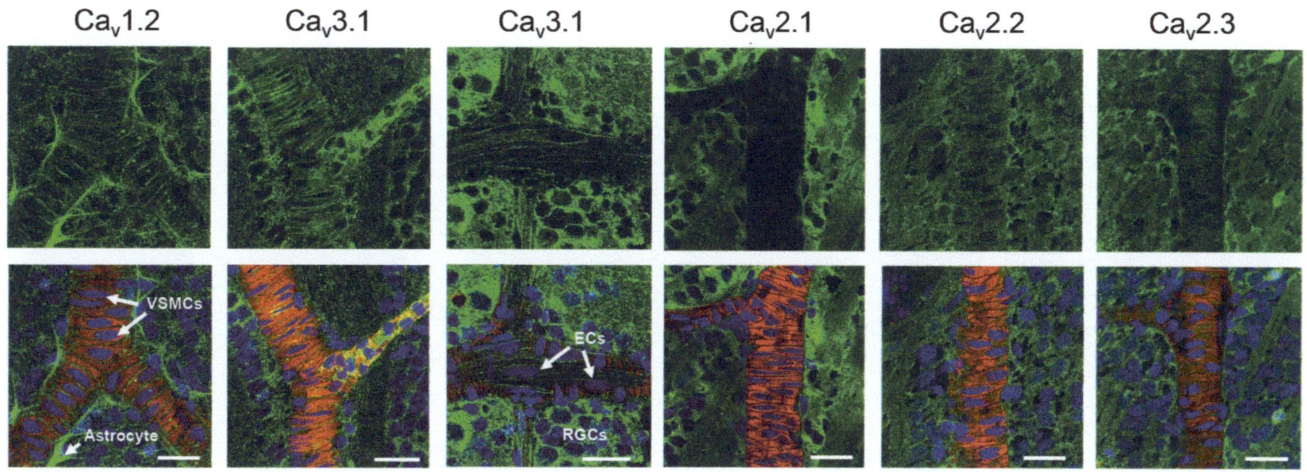

Fig. 5 Confocal images of rat retinal arterioles within retinal wholemount preparations immunolabeled for L-type ($Ca_v1.2$), T-type ($Ca_v3.1$), P-/Q-type ($Ca_v2.1$), N-type ($Ca_v2.2$) and R-type ($Ca_v2.3$) VGCCs (green). Lower panels show merged images with SMα (red) and TO-PRO3 nuclear dye (pseudo-colored blue). In addition to the retinal VSMCs, strong $Ca_v1.2$-associated immunofluorescence was seen in retinal astrocytes. $Ca_v3.1$ is present in both the VSMC and endothelial cell (EC) layers of retinal arterioles, as well as retinal ganglion cells (RGCs). Scale bars = 30 μm.

Consistent with reports in other vascular beds, $Ca_v3.1$ channels are also strongly expressed in the vascular endothelial cells of retinal arterioles and capillaries (Fig. 5). In more recent work, we have undertaken confocal immunolabeling studies to investigate the possible expression and localization of P-/Q-type, N-type and R-type VGCCs in retinal arteriolar VSMCs. Although P-/Q-type channels are found in the retinal neuropile, these channels were absent in the VSMCs of retinal arterioles (Fig. 5). We did, however, find that N-type $Ca_v2.2$ and R-type $Ca_v2.3$ channels are expressed in retinal VSMCs, distributed to plasma membrane, cytosolic and perinuclear regions (Fig. 5). The possible role of these channels in the myogenic activity of retinal arterioles warrants future investigation.

8. Intracellular Ca^{2+} release channels

Although Ca^{2+} influx via VGCCs is essential for the myogenic response, several studies have shown that Ca^{2+} mobilization from sarcoplasmic reticulum (SR) Ca^{2+} stores is also required (Inscho, Cook, Mui, & Imig, 1998; Kotecha & Hill, 2005; Westcott, Goodwin, Segal, & Jackson, 2012). Compared to the cytoplasm, the SR contains about a thousand times higher concentration of free Ca^{2+} ions, and the opening of SR Ca^{2+} release channels causes the efflux of Ca^{2+} from the SR into the cytoplasm of the cell (Hill, Zou, Potocnik, Meininger, & Davis, 2001). The main SR Ca^{2+} release channels in VSMCs are ryanodine receptors (RyRs) and inositol 1,4,5-trisphosphate (IP_3) receptors (IP_3Rs). There are three isoforms of each receptor type and different combinations are co-expressed in different VSMCs (Narayanan, Adebiyi, & Jaggar, 2012; Zheng et al., 2008). RyR2, IP_3R1 and IP_3R2 appear to be the main SR Ca^{2+} release channels expressed by retinal arteriolar VSMCs (Curtis et al., 2008), although RyR3 is also present, albeit at much reduced levels (Fig. 6). RyR1 channels are not expressed in retinal VSMCs, and there have been no studies to date examining the expression of IP_3R3 channels in these cells. In addition to Ca^{2+} release channels, the SR membrane of retinal VSMCs also contains sarcoplasmic/endoplasmic reticulum Ca^{2+} ATPase pumps (SERCA) that actively transport intracellular Ca^{2+} back into the SR (Curtis, Major, Trimble, & Scholfield, 2003).

Ca^{2+} sparks are highly localized Ca^{2+} release events that arise due to the opening of clusters of RyRs on the SR (Yin, Blayney, & Lai, 2005). We have previously shown that retinal VSMCs generate spontaneous Ca^{2+} sparks with spatiotemporal properties resembling those described in other

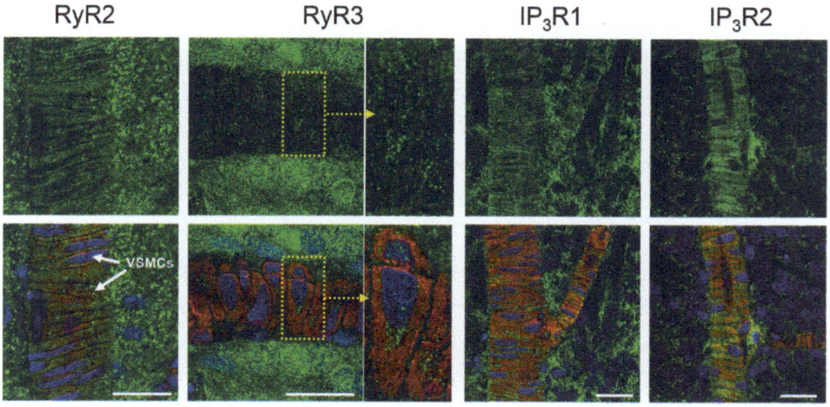

Fig. 6 Immunohistochemical staining of RyR2, RyR3, IP_3R1 and IP_3R2 (green) in rat retinal arterioles within retinal wholemount preparations. Lower panels show merged images with SMα (red) and TO-PRO3 nuclear dye (pseudo-colored blue). Scale bars = 30 μm.

types of VSMCs (Curtis, Tumelty, Dawicki, Scholfield, & McGeown, 2004). In cardiac muscle, Ca^{2+} sparks summate to produce global Ca^{2+} transients that elicit cell contraction (Cheng, Lederer, Lederer, & Cannell, 1996; Hoang-Trong, Ullah, & Jafri, 2015). We have made similar observations in retinal arterioles, where Ca^{2+} sparks seem to act as building blocks for more prolonged global Ca^{2+} signals that can initiate retinal VSMC contraction (Curtis et al., 2004; Tumelty, Scholfield, Stewart, Curtis, & McGeown, 2007). In cardiac muscle, Ca^{2+} entry through L-type VGCCs directly triggers the generation of Ca^{2+} sparks by opening RyRs through a process termed Ca^{2+}-induced Ca^{2+} release (CICR) (Hymel, Inui, Fleischer, & Schindler, 1988). In retinal VSMCs, however, Ca^{2+} spark production is only loosely coupled to Ca^{2+} influx via VGCCs (Tumelty et al., 2007). While inhibition of the L-type $Ca_v1.2$ VGCCs reduces both the frequency and amplitude of the Ca^{2+} sparks and global Ca^{2+} signals in these cells, this seems to arise due to depletion of the SR Ca^{2+} content, rather than reflecting inhibition of CICR (Tumelty et al., 2007). We have recently examined the role of Ca^{2+} sparks in the myogenic activity of retinal arterioles. Following pressurization, the myogenic constriction of retinal arterioles was associated with an increase in the frequency of Ca^{2+} sparks and global Ca^{2+} oscillations (Kur et al., 2013). Inhibition of Ca^{2+} sparks with RyR, SERCA pump and L-type VGCC inhibitors produced a reversible dilatation of pressurized retinal arterioles, suggesting that Ca^{2+} sparks play an important role in mediating the myogenic response in these vessels (Kur et al., 2013).

Based on our work to date, it seems likely that the increased Ca^{2+} spark and Ca^{2+} oscillation activity in myogenically active retinal arterioles results from an increase in the SR Ca^{2+} content due to the activation of Ca^{2+} influx through VGCCs, although direct CICR cannot be ruled out. It is worth noting that our results demonstrating a constrictor role for Ca^{2+} sparks in retinal arterioles contrasts with findings in many other blood vessels where these events have been shown to exclusively play a vasodilatory role. Ca^{2+} sparks can activate large conductance Ca^{2+}-activated K^+ channels (BK_{Ca}) in the VSMC membrane (see Section 9), and thereby cause hyperpolarization, deactivation of VGCCs and vasodilatation (Jaggar, Porter, Lederer, & Nelson, 2000; Nelson et al., 1995). An excitatory, constrictor function for Ca^{2+} sparks, however, has also been reported in VSMCs of the portal vein and afferent renal arterioles (Balasubramanian, Ahmed, Lo, Sham, & Yip, 2007; Bolton & Gordienko, 1998; Gordienko, Greenwood, & Bolton, 2001).

A number of studies have highlighted the importance of IP_3Rs in the myogenic response. IP_3R inhibitors, for example, attenuate Ca^{2+} waves, global Ca^{2+} and myogenic tone in arteries and arterioles supplying the cremaster muscle (Jackson & Boerman, 2017; Westcott et al., 2012). IP_3R-induced Ca^{2+} release is also essential for pressure-induced TRPM4 channel activity, VSMC depolarization, and the generation of myogenic tone in cerebral arteries (Gonzales et al., 2014). We have previously shown that endothelin-1(ET-1) induced stimulation of Ca^{2+} spark and Ca^{2+} oscillation activity in retinal arterioles depends upon activation and cross-talk between IP_3Rs and RyRs (Tumelty et al., 2011), but the contribution of IP_3Rs to myogenic responses in these vessels remains to be determined.

9. K^+ channels

The opening of K^+ channels in the VSMC membrane leads to K^+ efflux and membrane hyperpolarization, leading to deactivation and closure of VGCCs. Thus K^+ channels are important negative regulators of myogenic tone since their activation causes a decrease in Ca^{2+} influx leading to vasodilation (Armstead, 1997; Jaggar et al., 1998). Several types of K^+ channels are expressed in arterial VSMCs including: voltage-activated K^+ (K_v) channels, large-conductance Ca^{2+}-activated K^+ channels (BK_{Ca}), ATP-sensitive K^+ channels (K_{ATP}) and inward rectifier K^+ (K_{IR}) channels (Brayden, 1996; Tykocki, Boerman, & Jackson, 2017). Of these, K_v and BK_{Ca} channels are expressed by retinal VSMCs and have been documented to regulate myogenic activity (Curtis et al., 2018; McGahon, Dash, et al., 2007; McGahon, Dawicki, et al., 2007; McGahon et al., 2005). K_{ATP} channel expression in retinal VSMCs appears to be species-specific, with these channels being strongly expressed in porcine, but not rat, retinal arterioles (Hein, Yuan, Rosa, & Kuo, 2005; Ishizaki, Fukumoto, & Puro, 2009; Jeppesen, Aalkjaer, & Bek, 2002). K_{ATP} channels in porcine retinal arterioles have been reported to play a prominent role in mediating vasodilator responses to adenosine, lactate, NO donors and phosphodiesterase inhibitors (Hein, Xu, & Kuo, 2006; Hein et al., 2005; Yuan, Hein, Rosa, & Kuo, 2008), but do not appear to be involved in the myogenic constriction of these vessels (Jeppesen et al., 2002). K_{ir} channels are mainly localized to retinal capillary pericytes with minimal expression in the arteriolar VSMCs (Matsushita & Puro, 2006; Scholfield et al., 2007).

Mammalian K_v channels are encoded by 40 genes, and are divided into 12 families (K_v1–K_v12) according to their sequence similarity (Gutman et al., 2005). K_v channels are broadly classified as either slow delayed rectifier K^+ channels or transient A-type K^+ channels (Gutman et al., 2005; Yellen, 2002). The delayed rectifier K^+ currents exhibit a delayed onset of activation followed by little or slow inactivation, while the transient A-type K^+ currents are generally characterized by their rapid rate of inactivation (Yellen, 2002). Previously, we identified an A-type K^+ current in retinal VSMC cells with biophysical and pharmacological properties that in general were similar those described in other types of VSMCs (McGahon et al., 2005). Using molecular, pharmacological and blocking antibody approaches we identified $K_v1.5$ as being a principal component underlying the A-type K^+ current in retinal VSMCs (McGahon, Dawicki, et al., 2007). Based on our findings, it seems likely that $K_v1.5$ forms a heterotetrameric complex with other TEA-sensitive $K_v1\alpha$ subunits and that the transient nature of the current arises through association of the channels with $K_v\beta1$ subunits (McGahon, Dawicki, et al., 2007). Rapidly inactivating A-type K^+ currents have also been identified and characterized in arterial VSMCs from the aorta (Halliday, Aaronson, Evans, & Gurney, 1995), mesentery (Smirnov & Aaronson, 1992) and kidney (Gordienko, Clausen, & Goligorsky, 1994), although the molecular make-up of the channels that underlie these currents is unknown. We have recently shown that the K_v1 channel inhibitor, correolide, causes constriction of pressurized retinal arterioles (Curtis et al., 2018), suggesting that these channels serve as an important negative feedback mechanism limiting myogenic activity in these vessels.

BK_{Ca} channels are Ca^{2+}- and voltage-dependent K^+ channels whose activation reduces cellular excitability (Dopico, Bukiya, & Jaggar, 2018; Jaggar et al., 2000). BK_{Ca} channels are ubiquitously expressed in VSMCs and the importance of these channels in reducing myogenic tone has been demonstrated in many vascular beds using BK_{Ca} channel inhibitors like iberiotoxin and penitrem A (Bae, Park, Lee, Ho, & Earm, 1999; Brayden & Nelson, 1992; Curtis et al., 2018; McGahon, Dash, et al., 2007; Nelson, Patlak, Worley, & Standen, 1990). In VSMCs, Ca^{2+} spark activation of BK_{Ca} channels in the spark microdomain causes generation of spontaneous transient outward currents (STOCs) that hyperpolarize the cell membrane potential and thereby reduce Ca^{2+} currents via VGCCs (Jaggar et al., 2000). BK_{Ca} channels have also been shown to participate in the regulation of VSMCs by intercellular signaling. The effects of endothelium-derived vasorelaxant agents (CO, NO, EETs, PGE2) are thought to be mediated by activation of BK_{Ca} channels following simulation of the protein kinase A (PKA) and protein kinase G (PKG) signaling pathways (Schubert & Nelson, 2001). BK_{Ca} channels are also modulated by a number of small molecule steroid hormones (Dick & Sanders, 2001; Hu et al., 2011), inflammatory cytokines (Levin, Konakov, & Godukhin, 2016; Martin et al., 2008) and reactive oxygen species (ROS) (Hermann, Sitdikova, & Weiger, 2015), providing a common target molecule for the regulation of myogenic tone by endothelial, immune-, and neuroglial signals in the retina. Although Ca^{2+} sparks have a predominantly excitatory, pro-contractile role in retinal VSMCs (see Section 8), BK_{Ca} channel activation acts as a built-in braking mechanism for spark-driven contraction (Kur et al., 2013; McGahon, Dash, et al., 2007) and these channels, together with K_v1 channels, play an important role in promoting basal blood flow in the retina (Needham et al., 2014).

10. Cl^- channels

Chloride (Cl^-) is the most abundant permeant anion in cells and plays an important role in a variety of cellular functions (Huang, Wong, & Jan, 2012). The movement of Cl^- through the cell membrane is tightly regulated and mainly mediated by Cl^- channels and transporters (Jentsch, Stein, Weinreich, & Zdebik, 2002). Cl^- channels can be broadly divided into five categories based on their regulation: ligand-gated Cl^- channels (glycine-/GABA-activated), cAMP-dependent Cl^- channels (CFTR), Ca^{2+}-activated Cl^- channels (CaCCs), voltage-gated Cl^- channels (ClCs) and volume-regulated Cl^- channels (Verkman & Galietta, 2009).

With the exception of the ligand-gated Cl^- channels, all known Cl^- channel families have been identified in VSMCs (Matchkov, Secher Dam, Bodtkjer, & Aalkjaer, 2013). Because the Cl^- equilibrium potential is positive to the resting membrane potential, activation of Cl^- channels within the plasma membrane of VSMCs results in Cl^- efflux and membrane depolarization (Large & Wang, 1996). Cl^- channels therefore play an excitatory role in VSMCs, enhancing vessel contractility through the activation of VGCCs (Large & Wang, 1996).

Studies from our laboratory and others have shown that CaCCs are expressed in retinal arteriolar VSMCs (Heinze et al., 2014; McGahon, Needham, Scholfield, McGeown, & Curtis, 2009; Needham et al., 2014). These channels have biophysical and permeability properties similar to those reported in large vessel VSMCs (McGahon et al., 2009). The proteins that have been proposed as the main components of CaCCs include CLC-3, bestrophins, CLCA and the recently identified transmembrane proteins, TMEM16A and TMEM16B (Caputo et al., 2008; Gandhi et al., 1998; Gruber, Schreur, Ji, Fuller, & Pauli, 1999; Hartzell et al., 2005; Huang et al., 2001; Pifferi, Dibattista, & Menini, 2009; Sun, Tsunenari, Yau, & Nathans, 2002). TMEM16A appears to be the main component underlying CaCCs in retinal arteriolar VSMCs (Heinze et al., 2014) (Fig. 7). In the rat retinal vasculature, CaCC currents are present in the VSMCs throughout the retinal vascular tree (McGahon et al., 2009), while in mice their expression appears to be restricted mainly to the second and third order arterioles (Heinze et al., 2014). Retinal arteriolar VSMCs also express the large conductance "maxi" Cl^- channel, Tweety 3 (Fig. 7). Although this channel was originally proposed to be a CaCC (Suzuki, 2006), more recent work has revealed that it forms the swell-dependent volume-regulated anion channel, $VRAC_{swell}$ (Han et al., 2019). Both TMEM16A and Tweety 3 are sensitive to blockade by the Cl^- channel inhibitor, disodium 4,4′diisothioclanatostilbene-2, 2′-disulfonate (DIDS). Previously, we have shown that high concentrations of DIDS (1 mM) have no effect on the diameter of pressurized retinal arterioles, suggesting that Cl^- channels do not appear to contribute to the myogenic activity of these vessels, at least under ex vivo conditions (McGahon et al., 2009). These results contrast with those reported in cerebral arteries and renal arterioles, where the activation of Cl^- channels constitutes a major mechanism for pressure-induced myogenic tone (Nelson, Conway, Knot, & Brayden, 1997; Yip et al., 2018). In rat renal arterioles, the Cl^- efflux that contributes to the membrane depolarization and pressure-dependent vasoconstriction has been shown to be mediated through Ca^{2+} spark-induced activation of TMEM16A channels

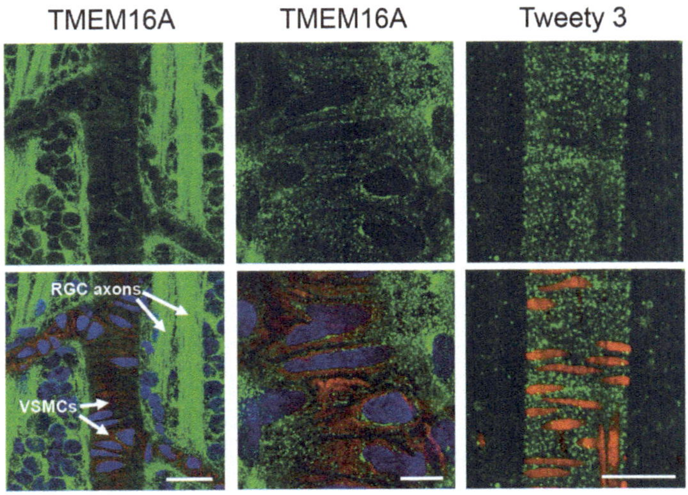

Fig. 7 Confocal images of rat retinal arterioles within wholemount preparations immunolabeled for the Cl^- channel proteins, TMEM16A and Tweety 3 (green). The low and high magnification images for TMEM16A are from different arterioles. The merged images for TMEM16A (lower panels) include SMα (red) and TO-PRO3 (pseudo-colored blue), while for Tweety 3, the wholemount was counterstained with propidium iodide nuclear stain (red). In addition to the retinal VSMCs, TMEM16A is strongly localized to retinal ganglion cell (RGC) axons. Scale bars for the low magnification TMEM16A and Tweety 3 images = 30 μm. Scale bar for the high magnification TMEM16A image = 10 μm.

(Yip et al., 2018). Although Cl^- channels do not appear to be involved in myogenic signaling in retinal arterioles, these channels do play an important role in agonist-induced responses. ET-1, for example, stimulates the activation of repetitive spontaneous transient inward currents (STICs) in retinal arteriolar VSMCs mediated through the activation of CaCCs (Scholfield et al., 2007). These currents contribute to the vasoconstrictor effects of ET-1 by promoting the generation of phasic Ca^{2+} waves and oscillations through a mechanism involving ET_A receptor activation, stimulation of intracellular Ca^{2+} release via IP3Rs and RyRs and Ca^{2+} influx through voltage-dependent and -independent mechanisms (McGahon et al., 2009; McGeown, 2010; Scholfield et al., 2007; Stewart et al., 2012).

11. Heterogeneity of myogenic signaling components in retinal arterioles

Although not widely recognized in the literature, the phenotype of retinal VSMCs can vary according to their location within the arteriolar network and even between adjacent cells within the same arteriole segment. This is perhaps most clearly illustrated by immunolabeling retinal arterioles within retinal wholemount preparations with antibodies against a range of VSMC markers. α-Smooth muscle actin (SMα) is one of six actin isoforms expressed in mammalian cells and is the predominant actin isoform found in VSMCs (Herman, 1993). In first order retinal arterioles, SMα is arranged into fiber bundles running parallel to the VSMCs' long axis, but these become considerably more disorganized in non-primary arterioles (Fig. 8). Calponin is an actin-binding protein that plays a role in VSMC contraction by inhibiting actomyosin ATPase activity (Liu & Jin, 2016). Calponin is strongly expressed in VSMCs of the first order retinal arterioles, but its expression is variable in the VSMCs of second order arterioles, giving a checkerboard-like staining pattern, and virtually absent in the pre-capillary arterioles (Fig. 8). SM22α is an actin-binding protein that is involved in Ca^{2+}-independent VSMC contraction (Je & Sohn, 2007).

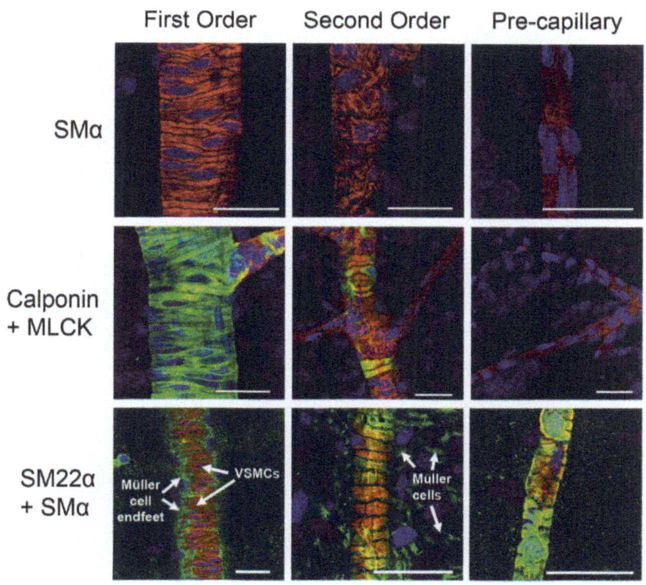

Fig. 8 Phenotypic heterogeneity among VSMCs of the retinal arteriolar network. Confocal images of first order, second order and pre-capillary rat retinal arterioles embedded within wholemount preparations and immunolabeled for SMα (red; top and bottom panels), calponin (green; middle panels), MLCK (red; middle panels) and SM22α (green; bottom panels). SMα organization and calponin and SM22α expression varies considerably in different sized arterioles of the retinal microcirculation. Although SM22α is commonly regarded as a smooth muscle cell-specific protein, it is also strongly expressed in Müller (macroglial) cells of the retina. Scale bars = 30 μm.

SM22α immunofluorescence is strongest in the VSMCs of the pre-capillary retinal arterioles with much weaker expression in the VSMCs of the larger arterioles (Fig. 8). Despite these obvious differences, MLCK seems to be homogeneously distributed in retinal VSMCs along the arteriolar network (Fig. 8).

Regional heterogeneity in ion channel expression is also evident in retinal arteriolar VSMCs, with important implications for our understanding of myogenic mechanisms, particularly in the small pre-capillary retinal arterioles. Our earlier work showed that while first and second order arterioles of the rat retina display large global $[Ca^{2+}]_i$ transients following activation of L-type VGCCs and RyRs with high K^+ and caffeine, respectively, these responses were absent in the VSMCs of pre-capillary retinal arterioles (Scholfield & Curtis, 2000). These functional data strongly suggest that L-type $Ca_v1.2$ and RyR2 channels are not expressed in the VSMC layer of pre-capillary retinal arterioles, and we have recently confirmed these findings using immunohistochemistry (Fig. 9). Consistent with these results, we have also been unable to detect Ca^{2+} sparks in the VSMCs of the small pre-capillary retinal arterioles. In contrast, other myogenic signaling components, including TRPV2, $Ca_v3.1$, $K_v1.5$ and BK_{Ca} channels seem to be uniformly distributed across the different arteriolar vessels of the retina. Most of our pressure myography work to date has used first order retinal arterioles, although we have been able to record myogenic responses in arterioles from all branching orders of the retina, including the pre-capillary vessels (Fig. 10). Taken as a whole, our results imply that myogenic mechanisms

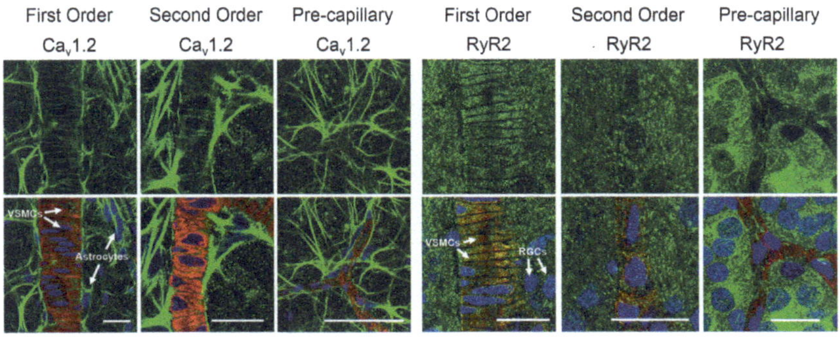

Fig. 9 Heterogeneity in $Ca_v1.2$ and RyR2 expression in arterioles from different branching orders of the retina. Confocal images of first order, second order and pre-capillary rat retinal arterioles embedded within wholemount preparations and immunolabeled for $Ca_v1.2$ and RyR2 (green). Lower panels show merged images with SMα (red) and TO-PRO3 nuclear dye (pseudo-colored blue). A gradient of expression for $Ca_v1.2$ and RyR2 is evident, with both proteins highly expressed in the VSMCs of first order arterioles and eventually disappearing in the pre-capillary vessels. Scale bars = 30 μm.

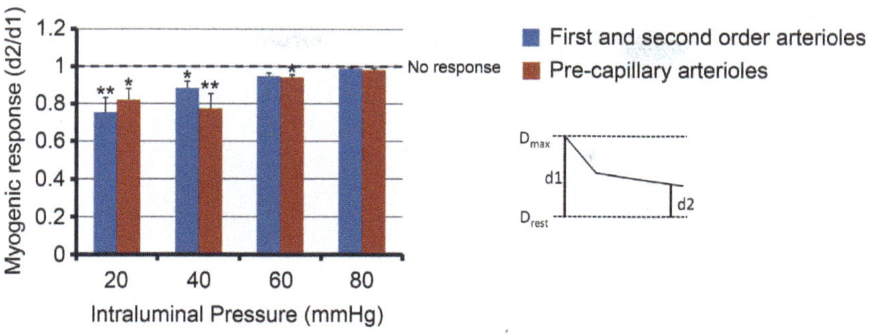

Fig. 10 Myogenic activity in different sized arterioles of the rat retinal microcirculation. Myogenic responses were measured in first and second order retinal arterioles and compared with those in pre-capillary arterioles across a range of intraluminal pressures (20–80 mmHg). The steady-state myogenic response is expressed as a ratio of the passive diameter measured immediately following pressurization (see explanatory inset). Myogenic responses in small pre-capillary arterioles were similar to those of first and second order retinal arterioles for all intraluminal pressures tested. The pressure range for myogenic constriction in rat retinal arterioles (~20–60 mmHg) is comparable to that previously measured in porcine retinal arterioles (see Section 5). Data were analysed by ANOVA and shown as mean ± SEM. $N = 9$ arterioles per group. Statistical significance relative to no response (i.e., d2 = d1) is indicated as follows: NS, $P > 0.05$; *$P < 0.05$; **$P < 0.01$.

in pre-capillary retinal arterioles are likely to differ substantially from those in the larger arteriolar vessels. Whether the myogenic response in these vessels is driven mainly by Ca^{2+} influx through non-$Ca_v1.2$ VGCCs or depends more on Ca^{2+}-independent contractile pathways is worthy of future investigation.

12. Concluding remarks

Here, we have reviewed evidence gathered over the past two decades regarding the involvement of different VSMC ion channels in mediating the myogenic response in retinal arterioles. Of the vascular systems studied so far, the myogenic response in retinal arterioles appears to be uniquely dependent upon TRPV2 stretch sensors triggering increased Ca^{2+} spark and global Ca^{2+} oscillation activity through the opening of VGCCs. Similar to other vessels, however, K_v and BK_{Ca} channels act as a braking mechanism that limits the degree of myogenic vasoconstriction. A diagram summarizing the ionic mechanisms responsible for myogenic signaling in retinal arterioles is shown in Fig. 11.

One of the limitations of the work undertaken to date is that it has relied mainly upon the use of pharmacological approaches to determine the involvement of specific ion channels in the myogenic activity of retinal

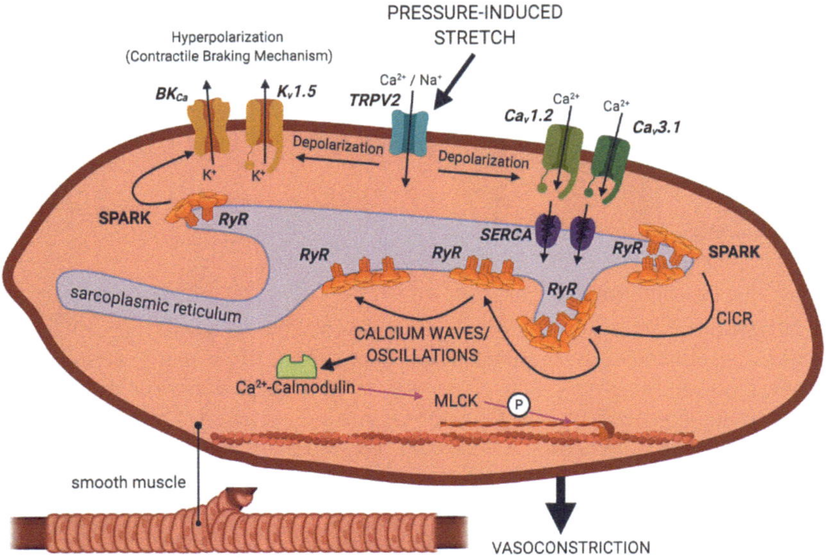

Fig. 11 Summary diagram showing the ionic mechanisms responsible for myogenic signaling in first and second order retinal arterioles. The myogenic response in large retinal arterioles is initiated by TRPV2 channel activation in the retinal VSMCs resulting from pressure-induced mechanical stretch on the vessel wall. TRPV2 channel activation triggers cell membrane potential depolarization, which increases the open probability of $Ca_V1.2$ and $Ca_V3.1$ VGCCs, providing Ca^{2+} influx to fill the SR. High SR Ca^{2+} levels stimulate RyR-mediated Ca^{2+} spark activity that spreads to generate propagating global Ca^{2+} waves and Ca^{2+} oscillations that initiate cell contraction. At the same time, negative feedback loops engage, operating mainly through Kv1.5 and BK_{Ca} channels, limiting the extent of myogenic tone development. This figure was created with BioRender.

arterioles. An important aim of future studies should be to validate the above findings using arterioles isolated from VSMC-specific gene knockout rodent models. Such models would also enable assessment of the contribution of individual myogenic signaling components in controlling retinal pressure autoregulatory responses in vivo. Armed with a clearer knowledge of the mechanisms underlying myogenic signaling in retinal arterioles, we are now well-positioned to better understand how endothelial, glial and neuronal signals modulate vascular tone and blood flow in the retina.

References

Albert, A. P., & Large, W. A. (2003). Store-operated Ca^{2+}-permeable non-selective cation channels in smooth muscle cells. *Cell Calcium*, *33*(5–6), 345–356. https://doi.org/10.1016/s0143-4160(03)00048-4.

Alonso-Carbajo, L., Kecskes, M., Jacobs, G., Pironet, A., Syam, N., Talavera, K., et al. (2017). Muscling in on TRP channels in vascular smooth muscle cells and cardiomyocytes. *Cell Calcium*, *66*, 48–61. https://doi.org/10.1016/j.ceca.2017.06.004.

Ames, A., 3rd, Li, Y. Y., Heher, E. C., & Kimble, C. R. (1992). Energy metabolism of rabbit retina as related to function: High cost of Na+ transport. *The Journal of Neuroscience*, *12*(3), 840–853.

Andreasen, D., Friis, U. G., Uhrenholt, T. R., Jensen, B. L., Skott, O., & Hansen, P. B. (2006). Coexpression of voltage-dependent calcium channels $Ca_v1.2$, 2.1a, and 2.1b in vascular myocytes. *Hypertension*, *47*(4), 735–741. https://doi.org/10.1161/01.HYP.0000203160.80972.47.

Armstead, W. M. (1997). Role of activation of calcium-sensitive K+ channels in NO- and hypoxia-induced pial artery vasodilation. *The American Journal of Physiology*, *272*(4 Pt. 2), H1785–H11790. https://doi.org/10.1152/ajpheart.1997.272.4.H1785.

Bae, Y. M., Park, M. K., Lee, S. H., Ho, W. K., & Earm, Y. E. (1999). Contribution of Ca^{2+}-activated K^+ channels and non-selective cation channels to membrane potential of pulmonary arterial smooth muscle cells of the rabbit. *The Journal of Physiology*, *514*(Pt. 3), 747–758. https://doi.org/10.1111/j.1469-7793.1999.747ad.x.

Baek, E. B., & Kim, S. J. (2011). Mechanisms of myogenic response: Ca(2+)-dependent and -independent signaling. *Journal of Smooth Muscle Research*, *47*(2), 55–65. https://doi.org/10.1540/jsmr.47.55.

Balasubramanian, L., Ahmed, A., Lo, C. M., Sham, J. S., & Yip, K. P. (2007). Integrin-mediated mechanotransduction in renal vascular smooth muscle cells: Activation of calcium sparks. *American Journal of Physiology. Regulatory, Integrative and Comparative Physiology*, *293*(4), R1586–R1594. https://doi.org/10.1152/ajpregu.00025.2007.

Bolton, T. B., & Gordienko, D. V. (1998). Confocal imaging of calcium release events in single smooth muscle cells. *Acta Physiologica Scandinavica*, *164*(4), 567–575. https://doi.org/10.1046/j.1365-201X.1998.00464.x.

Brayden, J. E. (1996). Potassium channels in vascular smooth muscle. *Clinical and Experimental Pharmacology & Physiology*, *23*(12), 1069–1076. https://doi.org/10.1111/j.1440-1681.1996.tb01172.x.

Brayden, J. E., & Nelson, M. T. (1992). Regulation of arterial tone by activation of calcium-dependent potassium channels. *Science*, *256*(5056), 532–535. https://doi.org/10.1126/science.1373909.

Caputo, A., Caci, E., Ferrera, L., Pedemonte, N., Barsanti, C., Sondo, E., et al. (2008). TMEM16A, a membrane protein associated with calcium-dependent chloride channel activity. *Science*, *322*(5901), 590–594. https://doi.org/10.1126/science.1163518.

Cazade, M., Bidaud, I., Hansen, P. B., Lory, P., & Chemin, J. (2014). 5,6-EET potently inhibits T-type calcium channels: Implication in the regulation of the vascular tone. *Pflügers Archiv*, *466*(9), 1759–1768. https://doi.org/10.1007/s00424-013-1411-0.

Chase, J. (1982). The evolution of retinal vascularization in mammals. A comparison of vascular and avascular retinae. *Ophthalmology*, *89*(12), 1518–1525. https://doi.org/10.1016/s0161-6420(82)34608-4.

Cheng, H., Lederer, M. R., Lederer, W. J., & Cannell, M. B. (1996). Calcium sparks and [Ca2+]i waves in cardiac myocytes. *The American Journal of Physiology*, *270*(1 Pt. 1), C148–C159. https://doi.org/10.1152/ajpcell.1996.270.1.C148.

Choi, S. K., Yeon, S. I., Kwon, Y., Byeon, S., & Lee, Y. H. (2017). Involvement of epithelial Na(+) channel in the elevated myogenic response in posterior cerebral arteries from spontaneously hypertensive rats. *Scientific Reports*, 7, 45996. https://doi.org/10.1038/srep45996.

Curtis, T. M., Gardiner, T. A., & Stitt, A. W. (2009). Microvascular lesions of diabetic retinopathy: Clues towards understanding pathogenesis? *Eye (London, England)*, *23*(7), 1496–1508. https://doi.org/10.1038/eye.2009.108.

Curtis, T. M., Major, E. H., Trimble, E. R., & Scholfield, C. N. (2003). Diabetes-induced activation of protein kinase C inhibits store-operated Ca2+ uptake in rat retinal microvascular smooth muscle. *Diabetologia*, *46*(9), 1252–1259. https://doi.org/10.1007/s00125-003-1178-5.

Curtis, T. M., McLaughlin, D., O'Hare, M., Kur, J., Barabas, P., Revolta, G., et al. (2018). Isolation of retinal arterioles for ex vivo cell physiology studies. *Journal of Visualized Experiments, 137*. https://doi.org/10.3791/57944.

Curtis, T. M., Tumelty, J., Dawicki, J., Scholfield, C. N., & McGeown, J. G. (2004). Identification and spatiotemporal characterization of spontaneous Ca2+ sparks and global Ca2+ oscillations in retinal arteriolar smooth muscle cells. *Investigative Ophthalmology & Visual Science, 45*(12), 4409–4414. https://doi.org/10.1167/iovs.04-0719.

Curtis, T. M., Tumelty, J., Stewart, M. T., Arora, A. R., Lai, F. A., McGahon, M. K., et al. (2008). Modification of smooth muscle Ca2+-sparks by tetracaine: Evidence for sequential RyR activation. *Cell Calcium, 43*(2), 142–154. https://doi.org/10.1016/j.ceca.2007.04.016.

Davies, G., Sullivan, P. M., Morris, A. H., Caldwell, G., & Kohner, E. M. (1991). Retinal blood flow following a tyramine-induced increase in blood pressure. *Journal of Hypertension, 9*(5), 425–430. https://doi.org/10.1097/00004872-199105000-00006.

Davis, M. J., & Hill, M. A. (1999). Signaling mechanisms underlying the vascular myogenic response. *Physiological Reviews, 79*(2), 387–423.

Dick, G. M., & Sanders, K. M. (2001). (Xeno)estrogen sensitivity of smooth muscle BK channels conferred by the regulatory beta1 subunit: A study of beta1 knockout mice. *The Journal of Biological Chemistry, 276*(48), 44835–44840. https://doi.org/10.1074/jbc.M106851200.

Dietrich, A., Kalwa, H., Storch, U., Mederos y Schnitzler, M., Salanova, B., Pinkenburg, O., et al. (2007). Pressure-induced and store-operated cation influx in vascular smooth muscle cells is independent of TRPC1. *Pflügers Archiv, 455*(3), 465–477. https://doi.org/10.1007/s00424-007-0314-3.

Dopico, A. M., Bukiya, A. N., & Jaggar, J. H. (2018). Calcium- and voltage-gated BK channels in vascular smooth muscle. *Pflügers Archiv, 470*(9), 1271–1289. https://doi.org/10.1007/s00424-018-2151-y.

Douguet, D., & Honore, E. (2019). Mammalian mechanoelectrical transduction: Structure and function of force-gated ion channels. *Cell, 179*(2), 340–354. https://doi.org/10.1016/j.cell.2019.08.049.

Drummond, H. A., Gebremedhin, D., & Harder, D. R. (2004). Degenerin/epithelial Na+ channel proteins: Components of a vascular mechanosensor. *Hypertension, 44*(5), 643–648. https://doi.org/10.1161/01.HYP.0000144465.56360.ad.

Dumskyj, M. J., Eriksen, J. E., Dore, C. J., & Kohner, E. M. (1996). Autoregulation in the human retinal circulation: Assessment using isometric exercise, laser Doppler velocimetry, and computer-assisted image analysis. *Microvascular Research, 51*(3), 378–392. https://doi.org/10.1006/mvre.1996.0034.

Earley, S., Heppner, T. J., Nelson, M. T., & Brayden, J. E. (2005). TRPV4 forms a novel Ca2+ signaling complex with ryanodine receptors and BKCa channels. *Circulation Research, 97*(12), 1270–1279. https://doi.org/10.1161/01.RES.0000194321.60300.d6.

Earley, S., Waldron, B. J., & Brayden, J. E. (2004). Critical role for transient receptor potential channel TRPM4 in myogenic constriction of cerebral arteries. *Circulation Research, 95*, 922–929.

Engerman, R. L. (1976). Development of the macular circulation. *Investigative Ophthalmology, 15*(10), 835–840.

Faraci, F. M., Baumbach, G. L., & Heistad, D. D. (1989). Myogenic mechanisms in the cerebral circulation. *Journal of Hypertension. Supplement*, 7(4), S61–S64. discussion S65.

Feke, G. T., & Pasquale, L. R. (2008). Retinal blood flow response to posture change in glaucoma patients compared with healthy subjects. *Ophthalmology, 115*(2), 246–252. https://doi.org/10.1016/j.ophtha.2007.04.055.

Fernandez, J. A., Bankhead, P., Zhou, H., McGeown, J. G., & Curtis, T. M. (2014). Automated detection and measurement of isolated retinal arterioles by a combination of edge enhancement and cost analysis. *PLoS One*, *9*(3), e91791. https://doi.org/10.1371/journal.pone.0091791.

Fernandez, J. A., McGahon, M. K., McGeown, J. G., & Curtis, T. M. (2015). $Ca_v3.1$ T-type Ca2+ channels contribute to myogenic signaling in rat retinal arterioles. *Investigative Ophthalmology & Visual Science*, *56*(9), 5125–5132. https://doi.org/10.1167/iovs.15-17299.

Florez-Paz, D., Bali, K. K., Kuner, R., & Gomis, A. (2016). A critical role for Piezo2 channels in the mechanotransduction of mouse proprioceptive neurons. *Scientific Reports*, *6*, 25923. https://doi.org/10.1038/srep25923.

Gandhi, R., Elble, R. C., Gruber, A. D., Schreur, K. D., Ji, H. L., Fuller, C. M., et al. (1998). Molecular and functional characterization of a calcium-sensitive chloride channel from mouse lung. *The Journal of Biological Chemistry*, *273*(48), 32096–32101. https://doi.org/10.1074/jbc.273.48.32096.

Gardiner, T. A., Archer, D. B., Curtis, T. M., & Stitt, A. W. (2007). Arteriolar involvement in the microvascular lesions of diabetic retinopathy: Implications for pathogenesis. *Microcirculation*, *14*(1), 25–38. https://doi.org/10.1080/10739680601072123.

Gonzales, A. L., Yang, Y., Sullivan, M. N., Sanders, L., Dabertrand, F., Hill-Eubanks, D. C., et al. (2014). A PLCgamma1-dependent, force-sensitive signaling network in the myogenic constriction of cerebral arteries. *Science Signaling*, 7(327), ra49. https://doi.org/10.1126/scisignal.2004732.

Gordienko, D. V., Clausen, C., & Goligorsky, M. S. (1994). Ionic currents and endothelin signaling in smooth muscle cells from rat renal resistance arteries. *The American Journal of Physiology*, *266*(2 Pt. 2), F325–F341. https://doi.org/10.1152/ajprenal.1994.266.2.F325.

Gordienko, D. V., Greenwood, I. A., & Bolton, T. B. (2001). Direct visualization of sarcoplasmic reticulum regions discharging Ca(2+)sparks in vascular myocytes. *Cell Calcium*, *29*(1), 13–28. https://doi.org/10.1054/ceca.2000.0180.

Griffin, M. D., Torres, V. E., Grande, J. P., & Kumar, R. (1997). Vascular expression of polycystin. *Journal of the American Society of Nephrology*, *8*(4), 616–626.

Gruber, A. D., Schreur, K. D., Ji, H. L., Fuller, C. M., & Pauli, B. U. (1999). Molecular cloning and transmembrane structure of hCLCA2 from human lung, trachea, and mammary gland. *The American Journal of Physiology*, *276*(6), C1261–C1270. https://doi.org/10.1152/ajpcell.1999.276.6.C1261.

Grunwald, J. E., Sinclair, S. H., & Riva, C. E. (1982). Autoregulation of the retinal circulation in response to decrease of intraocular pressure below normal. *Investigative Ophthalmology & Visual Science*, *23*(1), 124–127.

Gutman, G. A., Chandy, K. G., Grissmer, S., Lazdunski, M., McKinnon, D., Pardo, L. A., et al. (2005). International Union of Pharmacology. LIII. Nomenclature and molecular relationships of voltage-gated potassium channels. *Pharmacological Reviews*, *57*(4), 473–508. https://doi.org/10.1124/pr.57.4.10.

Hague, S., & Hill, D. W. (1988). Postural changes in perfusion pressure and retinal arteriolar calibre. *The British Journal of Ophthalmology*, *72*(4), 253–257. https://doi.org/10.1136/bjo.72.4.253.

Halliday, F. C., Aaronson, P. I., Evans, A. M., & Gurney, A. M. (1995). The pharmacological properties of K+ currents from rabbit isolated aortic smooth muscle cells. *British Journal of Pharmacology*, *116*(8), 3139–3148. https://doi.org/10.1111/j.1476-5381.1995.tb15116.x.

Han, Y. E., Kwon, J., Won, J., An, H., Jang, M. W., Woo, J., et al. (2019). Tweety-homolog (Ttyh) family encodes the pore-forming subunits of the swelling-dependent volume-regulated anion channel (VRACswell) in the brain. *Experimental Neurobiology*, *28*(2), 183–215. https://doi.org/10.5607/en.2019.28.2.183.

Harris, A., Ciulla, T. A., Chung, H. S., & Martin, B. (1998). Regulation of retinal and optic nerve blood flow. *Archives of Ophthalmology, 116*(11), 1491–1495. https://doi.org/10.1001/archopht.116.11.1491.

Hartzell, C., Qu, Z., Putzier, I., Artinian, L., Chien, L. T., & Cui, Y. (2005). Looking chloride channels straight in the eye: Bestrophins, lipofuscinosis, and retinal degeneration. *Physiology (Bethesda), 20,* 292–302. https://doi.org/10.1152/physiol.00021.2005.

Hein, T. W., Ren, Y., Yuan, Z., Xu, W., Somvanshi, S., Nagaoka, T., et al. (2009). Functional and molecular characterization of the endothelin system in retinal arterioles. *Investigative Ophthalmology & Visual Science, 50*(7), 3329–3336. https://doi.org/10.1167/iovs.08-3129.

Hein, T. W., Rosa, R. H., Jr., Yuan, Z., Roberts, E., & Kuo, L. (2010). Divergent roles of nitric oxide and rho kinase in vasomotor regulation of human retinal arterioles. *Investigative Ophthalmology & Visual Science, 51*(3), 1583–1590. https://doi.org/10.1167/iovs.09-4391.

Hein, T. W., Xu, W., & Kuo, L. (2006). Dilation of retinal arterioles in response to lactate: Role of nitric oxide, guanylyl cyclase, and ATP-sensitive potassium channels. *Investigative Ophthalmology & Visual Science, 47*(2), 693–699. https://doi.org/10.1167/iovs.05-1224.

Hein, T. W., Yuan, Z., Rosa, R. H., Jr., & Kuo, L. (2005). Requisite roles of A2A receptors, nitric oxide, and KATP channels in retinal arteriolar dilation in response to adenosine. *Investigative Ophthalmology & Visual Science, 46*(6), 2113–2119. https://doi.org/10.1167/iovs.04-1438.

Heinze, C., Seniuk, A., Sokolov, M. V., Huebner, A. K., Klementowicz, A. E., Szijarto, I. A., et al. (2014). Disruption of vascular Ca2+-activated chloride currents lowers blood pressure. *The Journal of Clinical Investigation, 124*(2), 675–686. https://doi.org/10.1172/JCI70025.

Herman, I. M. (1993). Actin isoforms. *Current Opinion in Cell Biology, 5*(1), 48–55. https://doi.org/10.1016/s0955-0674(05)80007-9.

Hermann, A., Sitdikova, G. F., & Weiger, T. M. (2015). Oxidative stress and maxi calcium-activated potassium (BK) channels. *Biomolecules, 5*(3), 1870–1911. https://doi.org/10.3390/biom5031870.

Hessellund, A., Jeppesen, P., Aalkjaer, C., & Bek, T. (2003). Characterization of vasomotion in porcine retinal arterioles. *Acta Ophthalmologica Scandinavica, 81*(3), 278–282. https://doi.org/10.1034/j.1600-0420.2003.00063.x.

Hill, D. W. (1977). The regional distribution of retinal circulation. *Annals of the Royal College of Surgeons of England, 59*(6), 470–475.

Hill, M. A., Zou, H., Potocnik, S. J., Meininger, G. A., & Davis, M. J. (2001). Invited review: Arteriolar smooth muscle mechanotransduction: Ca(2+) signaling pathways underlying myogenic reactivity. *Journal of Applied Physiology (Bethesda, MD: 1985), 91*(2), 973–983. https://doi.org/10.1152/jappl.2001.91.2.973.

Hoang-Trong, T. M., Ullah, A., & Jafri, M. S. (2015). Calcium sparks in the heart: Dynamics and regulation. *Research and Reports in Biology, 6,* 203–214. https://doi.org/10.2147/RRB.S61495.

Howard, J., Blakeslee, B., & Laughlin, S. B. (1987). The intracellular pupil mechanism and photoreceptor signal: Noise ratios in the fly Lucilia cuprina. *Proceedings of the Royal Society of London—Series B: Biological Sciences, 231*(1265), 415–435. https://doi.org/10.1098/rspb.1987.0053.

Hu, X. Q., Xiao, D., Zhu, R., Huang, X., Yang, S., Wilson, S., et al. (2011). Pregnancy upregulates large-conductance Ca(2+)-activated K(+) channel activity and attenuates myogenic tone in uterine arteries. *Hypertension, 58*(6), 1132–1139. https://doi.org/10.1161/HYPERTENSIONAHA.111.179952.

Huang, P., Liu, J., Di, A., Robinson, N. C., Musch, M. W., Kaetzel, M. A., et al. (2001). Regulation of human CLC-3 channels by multifunctional Ca2+/calmodulin-dependent protein kinase. *The Journal of Biological Chemistry*, *276*(23), 20093–20100. https://doi.org/10.1074/jbc.M009376200.

Huang, F., Wong, X., & Jan, L. Y. (2012). International union of basic and clinical pharmacology. LXXXV: Calcium-activated chloride channels. *Pharmacological Reviews*, *64*(1), 1–15. https://doi.org/10.1124/pr.111.005009.

Hughes, S., Yang, H., & Chan-Ling, T. (2000). Vascularization of the human fetal retina: Roles of vasculogenesis and angiogenesis. *Investigative Ophthalmology & Visual Science*, *41*(5), 1217–1228.

Hymel, L., Inui, M., Fleischer, S., & Schindler, H. (1988). Purified ryanodine receptor of skeletal muscle sarcoplasmic reticulum forms Ca^{2+}-activated oligomeric Ca^{2+} channels in planar bilayers. *Proceedings of the National Academy of Sciences of the United States of America*, *85*(2), 441–445. https://doi.org/10.1073/pnas.85.2.441.

Iester, M., Torre, P. G., Bricola, G., Bagnis, A., & Calabria, G. (2007). Retinal blood flow autoregulation after dynamic exercise in healthy young subjects. *Ophthalmologica*, *221*(3), 180–185. https://doi.org/10.1159/000099298.

Inoue, R., Jian, Z., & Kawarabayashi, Y. (2009). Mechanosensitive TRP channels in cardiovascular pathophysiology. *Pharmacology & Therapeutics*, *123*, 371–385.

Inscho, E. W., Cook, A. K., Mui, V., & Imig, J. D. (1998). Calcium mobilization contributes to pressure-mediated afferent arteriolar vasoconstriction. *Hypertension*, *31*(1 Pt. 2), 421–428. https://doi.org/10.1161/01.hyp.31.1.421.

Ishiguro, M., Wellman, T. L., Honda, A., Russell, S. R., Tranmer, B. I., & Wellman, G. C. (2005). Emergence of a R-type Ca^{2+} channel (Ca_v 2.3) contributes to cerebral artery constriction after subarachnoid hemorrhage. *Circulation Research*, *96*(4), 419–426.

Ishizaki, E., Fukumoto, M., & Puro, D. G. (2009). Functional K(ATP) channels in the rat retinal microvasculature: Topographical distribution, redox regulation, spermine modulation and diabetic alteration. *The Journal of Physiology*, *587*(Pt. 10), 2233–2253. https://doi.org/10.1113/jphysiol.2009.169003.

Jackson, W. F., & Boerman, E. M. (2017). Regional heterogeneity in the mechanisms of myogenic tone in hamster arterioles. *American Journal of Physiology. Heart and Circulatory Physiology*, *313*(3), H667–H675. https://doi.org/10.1152/ajpheart.00183.2017.

Jackson, W. F., & Boerman, E. M. (2018). Voltage-gated Ca(2+) channel activity modulates smooth muscle cell calcium waves in hamster cremaster arterioles. *American Journal of Physiology. Heart and Circulatory Physiology*, *315*(4), H871–H878. https://doi.org/10.1152/ajpheart.00292.2018.

Jaggar, J. H., Porter, V. A., Lederer, W. J., & Nelson, M. T. (2000). Calcium sparks in smooth muscle. *American Journal of Physiology. Cell Physiology*, *278*(2), C235–C256. https://doi.org/10.1152/ajpcell.2000.278.2.C235.

Jaggar, J. H., Wellman, G. C., Heppner, T. J., Porter, V. A., Perez, G. J., Gollasch, M., et al. (1998). Ca^{2+} channels, ryanodine receptors and Ca(2+)-activated K+ channels: A functional unit for regulating arterial tone. *Acta Physiologica Scandinavica*, *164*(4), 577–587. https://doi.org/10.1046/j.1365-201X.1998.00462.x.

Je, H. D., & Sohn, U. D. (2007). SM22alpha is required for agonist-induced regulation of contractility: Evidence from SM22alpha knockout mice. *Molecules and Cells*, *23*(2), 175–181.

Jensen, L. J., Salomonsson, M., Jensen, B. L., & Holstein-Rathlou, N. H. (2004). Depolarization-induced calcium influx in rat mesenteric small arterioles is mediated exclusively via mibefradil-sensitive calcium channels. *British Journal of Pharmacology*, *142*(4), 709–718. https://doi.org/10.1038/sj.bjp.0705841.

Jentsch, T. J., Stein, V., Weinreich, F., & Zdebik, A. A. (2002). Molecular structure and physiological function of chloride channels. *Physiological Reviews*, *82*(2), 503–568. https://doi.org/10.1152/physrev.00029.2001.

Jeppesen, P., Aalkjaer, C., & Bek, T. (2002). Adenosine relaxation in small retinal arterioles requires functional Na-K pumps and K(ATP) channels. *Current Eye Research*, *25*(1), 23–28. https://doi.org/10.1076/ceyr.25.1.23.9966.

Jeppesen, P., Aalkjaer, C., & Bek, T. (2003). Myogenic response in isolated porcine retinal arterioles. *Current Eye Research*, *27*(4), 217–222. https://doi.org/10.1076/ceyr.27.4.217.16597.

Jeppesen, P., Sanye-Hajari, J., & Bek, T. (2007). Increased blood pressure induces a diameter response of retinal arterioles that increases with decreasing arteriolar diameter. *Investigative Ophthalmology & Visual Science*, *48*(1), 328–331. https://doi.org/10.1167/iovs.06-0360.

Johnson, P. C. (1986). Autoregulation of blood flow. *Circulation Research*, *59*(5), 483–495. https://doi.org/10.1161/01.res.59.5.483.

Kawashima, Y., Géléoc, G. S., Kurima, K., Labay, V., Lelli, A., Asai, Y., et al. (2011). Mechanotransduction in mouse inner ear hair cells requires transmembrane channel-like genes. *The Journal of Clinical Investigation*, *121*(12), 4796–4809.

Kitazawa, Y., & Horie, T. (1975). Diurnal variation of intraocular pressure in primary open-angle glaucoma. *American Journal of Ophthalmology*, *79*(4), 557–566. https://doi.org/10.1016/0002-9394(75)90792-8.

Klaassen, I., Van Noorden, C. J., & Schlingemann, R. O. (2013). Molecular basis of the inner blood-retinal barrier and its breakdown in diabetic macular edema and other pathological conditions. *Progress in Retinal and Eye Research*, *34*, 19–48. https://doi.org/10.1016/j.preteyeres.2013.02.001.

Kotecha, N., & Hill, M. A. (2005). Myogenic contraction in rat skeletal muscle arterioles: Smooth muscle membrane potential and Ca(2+) signaling. *American Journal of Physiology. Heart and Circulatory Physiology*, *289*(4), H1326–H1334. https://doi.org/10.1152/ajpheart.00323.2005.

Kunert-Keil, C., Bisping, F., Kruger, J., & Brinkmeier, H. (2006). Tissue-specific expression of TRP channel genes in the mouse and its variation in three different mouse strains. *BMC Genomics*, *7*, 159. https://doi.org/10.1186/1471-2164-7-159.

Kur, J., Bankhead, P., Scholfield, C. N., Curtis, T. M., & McGeown, J. G. (2013). Ca(2+) sparks promote myogenic tone in retinal arterioles. *British Journal of Pharmacology*, *168*(7), 1675–1686. https://doi.org/10.1111/bph.12044.

Kur, J., McGahon, M. K., Fernandez, J. A., Scholfield, C. N., McGeown, J. G., & Curtis, T. M. (2014). Role of ion channels and subcellular Ca2+ signaling in arachidonic acid-induced dilation of pressurized retinal arterioles. *Investigative Ophthalmology & Visual Science*, *55*(5), 2893–2902. https://doi.org/10.1167/iovs.13-13511.

Kur, J., Newman, E. A., & Chan-Ling, T. (2012). Cellular and physiological mechanisms underlying blood flow regulation in the retina and choroid in health and disease. *Progress in Retinal and Eye Research*, *31*(5), 377–406. https://doi.org/10.1016/j.preteyeres.2012.04.004.

Kurima, K., Peters, L. M., Yang, Y., Riazuddin, S., Ahmed, Z. M., Naz, S., et al. (2002). Dominant and recessive deafness caused by mutations of a novel gene, TMC1, required for cochlear hair-cell function. *Nature Genetics*, *30*(3), 277–284.

Kuwabara, T., & Cogan, D. G. (1960). Studies of retinal vascular patterns. I. Normal architecture. *Archives of Ophthalmology*, *64*, 904–911. https://doi.org/10.1001/archopht.1960.01840010906012.

Lam, M. Y., Haunton, V. J., Robinson, T. G., & Panerai, R. B. (2019). Dynamic cerebral autoregulation measurement using rapid changes in head positioning: experiences in acute ischemic stroke and healthy control populations. *American Journal of Physiology. Heart and Circulatory Physiology*, *316*(3), H673–H683. https://doi.org/10.1152/ajpheart.00550.2018.

Large, W. A., & Wang, Q. (1996). Characteristics and physiological role of the Ca(2+) -activated Cl- conductance in smooth muscle. *The American Journal of Physiology, 271*(2 Pt. 1), C435–C454. https://doi.org/10.1152/ajpcell.1996.271.2.C435.

Lawton, P. F., Lee, M. D., Saunter, C. D., Girkin, J. M., McCarron, J. G., & Wilson, C. (2019). VasoTracker, a low-cost and open source pressure myograph system for vascular physiology. *Frontiers in Physiology, 10,* 99. https://doi.org/10.3389/fphys.2019.00099.

Levin, S. G., Konakov, M. V., & Godukhin, O. V. (2016). Role of BK(Ca) potassium channels in the mechanisms of modulatory effects of IL-10 on hypoxia-induced changes in activity of hippocampal neurons. *Bulletin of Experimental Biology and Medicine, 160*(5), 643–645. https://doi.org/10.1007/s10517-016-3238-0.

Liang, Y., Fortune, B., Cull, G., Cioffi, G. A., & Wang, L. (2010). Quantification of dynamic blood flow autoregulation in optic nerve head of rhesus monkeys. *Experimental Eye Research, 90*(2), 203–209. https://doi.org/10.1016/j.exer.2009.10.009.

Liu, R., & Jin, J. P. (2016). Calponin isoforms CNN1, CNN2 and CNN3: Regulators for actin cytoskeleton functions in smooth muscle and non-muscle cells. *Gene, 585*(1), 143–153. https://doi.org/10.1016/j.gene.2016.02.040.

Liu, J. H., Kripke, D. F., Hoffman, R. E., Twa, M. D., Loving, R. T., Rex, K. M., et al. (1998). Nocturnal elevation of intraocular pressure in young adults. *Investigative Ophthalmology & Visual Science, 39*(13), 2707–2712.

Liu, C., & Montell, C. (2015). Forcing open TRP channels: Mechanical gating as a unifying activation mechanism. *Biochemical and Biophysical Research Communications, 460*(1), 22–25.

Lorenzi, M., Feke, G. T., Pitler, L., Berisha, F., Kolodjaschna, J., & McMeel, J. W. (2010). Defective myogenic response to posture change in retinal vessels of well-controlled type 1 diabetic patients with no retinopathy. *Investigative Ophthalmology & Visual Science, 51*(12), 6770–6775. https://doi.org/10.1167/iovs.10-5785.

Lutty, G. A., Bhutto, I., & McLeod, D. S. (2012). Anatomy of the ocular vasculatures. In L. Schmetterer & J. Kiel (Eds.), *Ocular blood flow* (pp. 3–21). Springer.

Martin, G., O'Connell, R. J., Pietrzykowski, A. Z., Treistman, S. N., Ethier, M. F., & Madison, J. M. (2008). Interleukin-4 activates large-conductance, calcium-activated potassium (BKCa) channels in human airway smooth muscle cells. *Experimental Physiology, 93*(7), 908–918. https://doi.org/10.1113/expphysiol.2008.042432.

Matchkov, V. V., Secher Dam, V., Bodtkjer, D. M., & Aalkjaer, C. (2013). Transport and function of chloride in vascular smooth muscles. *Journal of Vascular Research, 50*(1), 69–87. https://doi.org/10.1159/000345242.

Matsushita, K., & Puro, D. G. (2006). Topographical heterogeneity of K(IR) currents in pericyte-containing microvessels of the rat retina: Effect of diabetes. *The Journal of Physiology, 573*(Pt. 2), 483–495. https://doi.org/10.1113/jphysiol.2006.107102.

McGahon, M. K., Dash, D. P., Arora, A., Wall, N., Dawicki, J., Simpson, D. A., et al. (2007). Diabetes downregulates large-conductance Ca2+-activated potassium beta 1 channel subunit in retinal arteriolar smooth muscle. *Circulation Research, 100*(5), 703–711. https://doi.org/10.1161/01.RES.0000260182.36481.c9.

McGahon, M. K., Dawicki, J. M., Arora, A., Simpson, D. A., Gardiner, T. A., Stitt, A. W., et al. (2007). Kv1.5 is a major component underlying the A-type potassium current in retinal arteriolar smooth muscle. *American Journal of Physiology. Heart and Circulatory Physiology, 292*(2), H1001–H1008. https://doi.org/10.1152/ajpheart.01003.2006.

McGahon, M. K., Dawicki, J. M., Scholfield, C. N., McGeown, J. G., & Curtis, T. M. (2005). A-type potassium current in retinal arteriolar smooth muscle cells. *Investigative Ophthalmology & Visual Science, 46*(9), 3281–3287. https://doi.org/10.1167/iovs.04-1465.

McGahon, M. K., Fernandez, J. A., Dash, D. P., McKee, J., Simpson, D. A., Zholos, A. V., et al. (2016). TRPV2 channels contribute to stretch-activated cation currents and myogenic constriction in retinal arterioles. *Investigative Ophthalmology & Visual Science, 57*(13), 5637–5647. https://doi.org/10.1167/iovs.16-20279.

McGahon, M. K., McKee, J., Dash, D. P., Brown, E., Simpson, D. A., Curtis, T. M., et al. (2012). Pharmacological profiling of store-operated Ca^{2+} entry in retinal arteriolar smooth muscle. *Microcirculation, 19*(7), 586–597. https://doi.org/10.1111/j.1549-8719.2012.00192.x.
McGahon, M. K., Needham, M. A., Scholfield, C. N., McGeown, J. G., & Curtis, T. M. (2009). Ca2+-activated Cl- current in retinal arteriolar smooth muscle. *Investigative Ophthalmology & Visual Science, 50*(1), 364–371. https://doi.org/10.1167/iovs.08-2524.
McGahon, M. K., Zhang, X., Scholfield, C. N., Curtis, T. M., & McGeown, J. G. (2007). Selective downregulation of the BKbeta1 subunit in diabetic arteriolar myocytes. *Channels (Austin, Tex), 1*(3), 141–143. https://doi.org/10.4161/chan.4596.
McGeown, J. G. (2010). Seeing is believing! Imaging Ca2+-signalling events in living cells. *Experimental Physiology, 95*(11), 1049–1060. https://doi.org/10.1113/expphysiol.2010.052456.
Meininger, G. A., & Davis, M. J. (1992). Cellular mechanisms involved in the vascular myogenic response. *The American Journal of Physiology, 263*(3 Pt. 2), H647–H659. https://doi.org/10.1152/ajpheart.1992.263.3.H647.
Mirshahi, M., Nicolas, C., Mirshahi, S., Golestaneh, N., d'Hermies, F., & Agarwal, M. K. (1999). Immunochemical analysis of the sodium channel in rodent and human eye. *Experimental Eye Research, 69*(1), 21–32. https://doi.org/10.1006/exer.1999.0675.
Moore, C., & Liedtke, W. B. (2017). Osmomechanical-sensitive TRPV channels in mammals. In T. L. R. Emir (Ed.), *Neurobiology of TRP channels* (pp. 85–94). Boca Raton (FL): CRC Press/Taylor & Francis.
Moosmang, S., Schulla, V., Welling, A., Feil, R., Feil, S., Wegener, J. W., et al. (2003). Dominant role of smooth muscle L-type calcium channel $Ca_v1.2$ for blood pressure regulation. *The EMBO Journal, 22*(22), 6027–6034.
Muller, J. M., Davis, M. J., & Chilian, W. M. (1996). Integrated regulation of pressure and flow in the coronary microcirculation. *Cardiovascular Research, 32*(4), 668–678.
Muller, C., Morales, P., & Reggio, P. H. (2018). Cannabinoid ligands targeting TRP channels. *Frontiers in Molecular Neuroscience, 11*, 487. https://doi.org/10.3389/fnmol.2018.00487.
Murthy, S. E., Loud, M. C., Daou, I., Marshall, K. L., Schwaller, F., Kuhnemund, J., et al. (2018). The mechanosensitive ion channel Piezo2 mediates sensitivity to mechanical pain in mice. *Science Translational Medicine, 10*(462), eaat9897. https://doi.org/10.1126/scitranslmed.aat9897.
Nagaoka, T., Kuo, L., Ren, Y., Yoshida, A., & Hein, T. W. (2008). C-reactive protein inhibits endothelium-dependent nitric oxide-mediated dilation of retinal arterioles via enhanced superoxide production. *Investigative Ophthalmology & Visual Science, 49*(5), 2053–2060. https://doi.org/10.1167/iovs.07-1387.
Nagaoka, T., Mori, F., & Yoshida, A. (2002). Retinal artery response to acute systemic blood pressure increase during cold pressor test in humans. *Investigative Ophthalmology & Visual Science, 43*(6), 1941–1945.
Narayanan, D., Adebiyi, A., & Jaggar, J. H. (2012). Inositol trisphosphate receptors in smooth muscle cells. *American Journal of Physiology. Heart and Circulatory Physiology, 302*(11), H2190–H2210. https://doi.org/10.1152/ajpheart.01146.2011.
Narayanan, D., Bulley, S., Leo, M. D., Burris, S. K., Gabrick, K. S., Boop, F. A., et al. (2013). Smooth muscle cell transient receptor potential polycystin-2 (TRPP2) channels contribute to the myogenic response in cerebral arteries. *The Journal of Physiology, 591*(20), 5031–5046.
Naylor, J., Li, J., Milligan, C. J., Zeng, F., Sukumar, P., Hou, B., et al. (2010). Pregnenolone sulphate- and cholesterol-regulated TRPM3 channels coupled to vascular smooth muscle secretion and contraction. *Circulation Research, 106*(9), 1507–1515. https://doi.org/10.1161/CIRCRESAHA.110.219329.

Needham, M., McGahon, M. K., Bankhead, P., Gardiner, T. A., Scholfield, C. N., Curtis, T. M., et al. (2014). The role of K+ and Cl- channels in the regulation of retinal arteriolar tone and blood flow. *Investigative Ophthalmology & Visual Science, 55*(4), 2157–2165. https://doi.org/10.1167/iovs.13-12948.

Nelson, M. T., Cheng, H., Rubart, M., Santana, L. F., Bonev, A. D., Knot, H. J., et al. (1995). Relaxation of arterial smooth muscle by calcium sparks. *Science, 270*(5236), 633–637. https://doi.org/10.1126/science.270.5236.633.

Nelson, M. T., Conway, M. A., Knot, H. J., & Brayden, J. E. (1997). Chloride channel blockers inhibit myogenic tone in rat cerebral arteries. *The Journal of Physiology, 502*(Pt. 2), 259–264. https://doi.org/10.1111/j.1469-7793.1997.259bk.x.

Nelson, M. T., Patlak, J. B., Worley, J. F., & Standen, N. B. (1990). Calcium channels, potassium channels, and voltage dependence of arterial smooth muscle tone. *American Journal of Physiology. Cell Physiology, 259*, C3–C18.

Nikitina, E., Zhang, Z. D., Kawashima, A., Jahromi, B. S., Bouryi, V. A., Takahashi, M., et al. (2007). Voltage-dependent calcium channels of dog basilar artery. *The Journal of Physiology, 580*(Pt. 2), 523–541. https://doi.org/10.1113/jphysiol.2006.126128.

Nilius, B., & Owsianik, G. (2011). The transient receptor potential family of ion channels. *Genome Biology, 12*(3), 218.

Niven, J. E., & Laughlin, S. B. (2008). Energy limitation as a selective pressure on the evolution of sensory systems. *The Journal of Experimental Biology, 211*(Pt. 11), 1792–1804. https://doi.org/10.1242/jeb.017574.

Noorani, M. M., Noel, R. C., & Marrelli, S. P. (2011). Upregulated TRPC3 and downregulated TRPC1 channel expression during hypertension is associated with increased vascular contractility in rat. *Frontiers in Physiology, 2*, 42.

Oancea, E., Wolfe, J. T., & Clapham, D. E. (2006). Functional TRPM7 channels accumulate at the plasma membrane in response to fluid flow. *Circulation Research, 98*(2), 245–253.

Omae, T., Nagaoka, T., Tanano, I., & Yoshida, A. (2013). Adiponectin-induced dilation of isolated porcine retinal arterioles via production of nitric oxide from endothelial cells. *Investigative Ophthalmology & Visual Science, 54*(7), 4586–4594. https://doi.org/10.1167/iovs.13-11756.

Otani, S., Nagaoka, T., Omae, T., Tanano, I., Kamiya, T., Ono, S., et al. (2016). Histamine-induced dilation of isolated porcine retinal arterioles: Role of endothelium-derived hyperpolarizing factor. *Investigative Ophthalmology & Visual Science, 57*(11), 4791–4798. https://doi.org/10.1167/iovs.15-19038.

Pannarale, L., Onori, P., Ripani, M., & Gaudio, E. (1996). Precapillary patterns and perivascular cells in the retinal microvasculature. A scanning electron microscope study. *Journal of Anatomy, 188*(Pt. 3), 693–703.

Pedersen, S. F., Owsianik, G., & Nilius, B. (2005). TRP channels: An overview. *Cell Calcium, 38*(3-4), 233–252.

Pifferi, S., Dibattista, M., & Menini, A. (2009). TMEM16B induces chloride currents activated by calcium in mammalian cells. *Pflügers Archiv, 458*(6), 1023–1038. https://doi.org/10.1007/s00424-009-0684-9.

Poole, K., Herget, R., Lapatsina, L., Ngo, H. D., & Lewin, G. R. (2014). Tuning Piezo ion channels to detect molecular-scale movements relevant for fine touch. *Nature Communications, 5*, 3520. https://doi.org/10.1038/ncomms4520.

Potts, L. B., Bradley, P. D., Xu, W., Kuo, L., & Hein, T. W. (2013). Role of endothelium in vasomotor responses to endothelin system and protein kinase C activation in porcine retinal arterioles. *Investigative Ophthalmology & Visual Science, 54*(12), 7587–7594. https://doi.org/10.1167/iovs13-13178.

Pournaras, C. J., Rungger-Brandle, E., Riva, C. E., Hardarson, S. H., & Stefansson, E. (2008). Regulation of retinal blood flow in health and disease. *Progress in Retinal and Eye Research, 27*(3), 284–330. https://doi.org/10.1016/j.preteyeres.2008.02.002.

Provis, J. M. (2001). Development of the primate retinal vasculature. *Progress in Retinal and Eye Research*, *20*(6), 799–821. https://doi.org/10.1016/s1350-9462(01)00012-x.
Puro, D. G. (2007). Physiology and pathobiology of the pericyte-containing retinal microvasculature: New developments. *Microcirculation*, *14*(1), 1–10. https://doi.org/10.1080/10739680601072099.
Putney, J. W., Jr. (2004). The enigmatic TRPCs: Multifunctional cation channels. *Trends in Cell Biology*, *14*(6), 282–286. https://doi.org/10.1016/j.tcb.2004.04.002.
Ranade, S. S., Woo, S. H., Dubin, A. E., Moshourab, R. A., Wetzel, C., Petrus, M., et al. (2014). Piezo2 is the major transducer of mechanical forces for touch sensation in mice. *Nature*, *516*(7529), 121–125. https://doi.org/10.1038/nature13980.
Rassam, S. M., Patel, V., & Kohner, E. M. (1995). The effect of experimental hypertension on retinal vascular autoregulation in humans: A mechanism for the progression of diabetic retinopathy. *Experimental Physiology*, *80*(1), 53–68. https://doi.org/10.1113/expphysiol.1995.sp003834.
Retailleau, K., Duprat, F., Arhatte, M., Ranade, S. S., Peyronnet, R., Martins, J. R., et al. (2015). Piezo1 in smooth muscle cells is involved in hypertension-dependent arterial remodeling. *Cell Reports*, *13*(6), 1161–1171. https://doi.org/10.1016/j.celrep.2015.09.072.
Riva, C. E., Grunwald, J. E., & Petrig, B. L. (1986). Autoregulation of human retinal blood flow. An investigation with laser Doppler velocimetry. *Investigative Ophthalmology & Visual Science*, *27*(12), 1706–1712.
Riva, C. E., Sinclair, S. H., & Grunwald, J. E. (1981). Autoregulation of retinal circulation in response to decrease of perfusion pressure. *Investigative Ophthalmology & Visual Science*, *21*(1 Pt. 1), 34–38.
Robinson, F., Riva, C. E., Grunwald, J. E., Petrig, B. L., & Sinclair, S. H. (1986). Retinal blood flow autoregulation in response to an acute increase in blood pressure. *Investigative Ophthalmology & Visual Science*, *27*(5), 722–726.
Rosa, R. H., Hein, T. W., & Kuo, L. (2004). Retinal autoregulation: The myogenic response in retinal arterioles. *Investigative Ophthalmology & Visual Science*, *45*, 2335.
Scholfield, C. N., & Curtis, T. M. (2000). Heterogeneity in cytosolic calcium regulation among different microvascular smooth muscle cells of the rat retina. *Microvascular Research*, *59*(2), 233–242. https://doi.org/10.1006/mvre.1999.2227.
Scholfield, C. N., McGeown, J. G., & Curtis, T. M. (2007). Cellular physiology of retinal and choroidal arteriolar smooth muscle cells. *Microcirculation*, *14*(1), 11–24. https://doi.org/10.1080/10739680601072115.
Schubert, R., & Nelson, M. T. (2001). Protein kinases: Tuners of the BKCa channel in smooth muscle. *Trends in Pharmacological Sciences*, *22*(10), 505–512. https://doi.org/10.1016/s0165-6147(00)01775-2.
Sinclair, S. H., Grunwald, J. E., Riva, C. E., Braunstein, S. N., Nichols, C. W., & Schwartz, S. S. (1982). Retinal vascular autoregulation in diabetes mellitus. *Ophthalmology*, *89*(7), 748–750. https://doi.org/10.1016/s0161-6420(82)34720-x.
Smirnov, S. V., & Aaronson, P. I. (1992). Ca(2+)-activated and voltage-gated K+ currents in smooth muscle cells isolated from human mesenteric arteries. *The Journal of Physiology*, *457*, 431–454. https://doi.org/10.1113/jphysiol.1992.sp019386.
Soni, H., Peixoto-Neves, D., Matthews, A. T., & Adebiyi, A. (2017). TRPV4 channels contribute to renal myogenic autoregulation in neonatal pigs. *American Journal of Physiology. Renal Physiology*, *313*(5), F1136–F1148. https://doi.org/10.1152/ajprenal.00300.2017.
Stewart, M., Needham, M., Bankhead, P., Gardiner, T. A., Scholfield, C. N., Curtis, T. M., et al. (2012). Feedback via Ca(2)(+)-activated ion channels modulates endothelin 1 signaling in retinal arteriolar smooth muscle. *Investigative Ophthalmology & Visual Science*, *53*(6), 3059–3066. https://doi.org/10.1167/iovs.11-9192.

Sun, H., Tsunenari, T., Yau, K. W., & Nathans, J. (2002). The vitelliform macular dystrophy protein defines a new family of chloride channels. *Proceedings of the National Academy of Sciences of the United States of America*, *99*(6), 4008–4013. https://doi.org/10.1073/pnas.052692999.

Suzuki, M. (2006). The Drosophila tweety family: Molecular candidates for large-conductance Ca2+-activated Cl- channels. *Experimental Physiology*, *91*(1), 141–147. https://doi.org/10.1113/expphysiol.2005.031773.

Szczot, M., Liljencrantz, J., Ghitani, N., Barik, A., Lam, R., Thompson, J. H., et al. (2018). PIEZO2 mediates injury-induced tactile pain in mice and humans. *Science Translational Medicine*, *10*(462), aat9892. https://doi.org/10.1126/scitranslmed.aat9892.

Tani, T., Nagaoka, T., Nakabayashi, S., Yoshioka, T., & Yoshida, A. (2014). Autoregulation of retinal blood flow in response to decreased ocular perfusion pressure in cats: Comparison of the effects of increased intraocular pressure and systemic hypotension. *Investigative Ophthalmology & Visual Science*, *55*(1), 360–367. https://doi.org/10.1167/iovs.13-12591.

Tecilazich, F., Feke, G. T., Mazzantini, S., Sobrin, L., & Lorenzi, M. (2016). Defective myogenic response of retinal vessels is associated with accelerated onset of retinopathy in type 1 diabetic individuals. *Investigative Ophthalmology & Visual Science*, *57*(4), 1523–1529. https://doi.org/10.1167/iovs.15-18356.

Toda, R., Kawazu, K., Oyabu, M., Miyazaki, T., & Kiuchi, Y. (2011). Comparison of drug permeabilities across the blood-retinal barrier, blood-aqueous humor barrier, and blood-brain barrier. *Journal of Pharmaceutical Sciences*, *100*(9), 3904–3911. https://doi.org/10.1002/jps.22610.

Torres, V. E., Cai, Y., Chen, X., Wu, G. Q., Geng, L., Cleghorn, K. A., et al. (2001). Vascular expression of polycystin-2. *Journal of the American Society of Nephrology*, *12*(1), 1–9.

Tumelty, J., Hinds, K., Bankhead, P., McGeown, N. J., Scholfield, C. N., Curtis, T. M., et al. (2011). Endothelin 1 stimulates Ca2+-sparks and oscillations in retinal arteriolar myocytes via IP3R and RyR-dependent Ca2+ release. *Investigative Ophthalmology & Visual Science*, *52*(6), 3874–3879. https://doi.org/10.1167/iovs.10-6029.

Tumelty, J., Scholfield, N., Stewart, M., Curtis, T., & McGeown, G. (2007). Ca2+-sparks constitute elementary building blocks for global Ca2+-signals in myocytes of retinal arterioles. *Cell Calcium*, *41*(5), 451–466. https://doi.org/10.1016/j.ceca.2006.08.005.

Tykocki, N. R., Boerman, E. M., & Jackson, W. F. (2017). Smooth muscle ion channels and regulation of vascular tone in resistance arteries and arterioles. *Comprehensive Physiology*, *7*(2), 485–581. https://doi.org/10.1002/cphy.c160011.

Verkman, A. S., & Galietta, L. J. (2009). Chloride channels as drug targets. *Nature Reviews. Drug Discovery*, *8*(2), 153–171. https://doi.org/10.1038/nrd2780.

Wang, S., Chennupati, R., Kaur, H., Iring, A., Wettschureck, N., & Offermanns, S. (2016). Endothelial cation channel PIEZO1 controls blood pressure by mediating flow-induced ATP release. *The Journal of Clinical Investigation*, *126*(12), 4527–4536. https://doi.org/10.1172/JCI87343.

Wang, Z., Ying, Z., Bosy-Westphal, A., Zhang, J., Schautz, B., Later, W., et al. (2010). Specific metabolic rates of major organs and tissues across adulthood: Evaluation by mechanistic model of resting energy expenditure. *The American Journal of Clinical Nutrition*, *92*(6), 1369–1377. https://doi.org/10.3945/ajcn.2010.29885.

Welsh, D. G., Morielli, A. D., Nelson, M. T., & Brayden, J. E. (2002). Transient receptor potential channels regulate myogenic tone of resistance arteries. *Circulation Research*, *90*, 248–250.

Westcott, E. B., Goodwin, E. L., Segal, S. S., & Jackson, W. F. (2012). Function and expression of ryanodine receptors and inositol 1,4,5-trisphosphate receptors in smooth muscle cells of murine feed arteries and arterioles. *The Journal of Physiology*, *590*(8), 1849–1869. https://doi.org/10.1113/jphysiol.2011.222083.

Wolburg, H., Liebner, S., Reichenbach, A., & Gerhardt, H. (1999). The pecten oculi of the chicken: A model system for vascular differentiation and barrier maturation. *International Review of Cytology*, *187*, 111–159. https://doi.org/10.1016/s0074-7696(08)62417-1.

Wong, V. H., Armitage, J. A., He, Z., Hui, F., Vingrys, A. J., & Bui, B. V. (2015). Chronic intraocular pressure elevation impairs autoregulatory capacity in streptozotocin-induced diabetic rat retina. *Ophthalmic & Physiological Optics*, *35*(2), 125–134. https://doi.org/10.1111/opo.12174.

Woo, S. H., Ranade, S., Weyer, A. D., Dubin, A. E., Baba, Y., Qiu, Z., et al. (2014). Piezo2 is required for Merkel-cell mechanotransduction. *Nature*, *509*(7502), 622–626. https://doi.org/10.1038/nature13251.

Yang, H., Liu, C., Zhou, R. M., Yao, J., Li, X. M., Shen, Y., et al. (2016). Piezo2 protein: A novel regulator of tumor angiogenesis and hyperpermeability. *Oncotarget*, *7*(28), 44630–44643. https://doi.org/10.18632/oncotarget.10134.

Ye, X. D., Laties, A. M., & Stone, R. A. (1990). Peptidergic innervation of the retinal vasculature and optic nerve head. *Investigative Ophthalmology & Visual Science*, *31*(9), 1731–1737.

Yellen, G. (2002). The voltage-gated potassium channels and their relatives. *Nature*, *419*(6902), 35–42. https://doi.org/10.1038/nature00978.

Yin, C. C., Blayney, L. M., & Lai, F. A. (2005). Physical coupling between ryanodine receptor-calcium release channels. *Journal of Molecular Biology*, *349*(3), 538–546. https://doi.org/10.1016/j.jmb.2005.04.002.

Yip, K. P., Balasubramanian, L., Kan, C., Wang, L., Liu, R., Ribeiro-Silva, L., et al. (2018). Intraluminal pressure triggers myogenic response via activation of calcium spark and calcium-activated chloride channel in rat renal afferent arteriole. *American Journal of Physiology. Renal Physiology*, *315*(6), F1592–F1600. https://doi.org/10.1152/ajprenal.00239.2018.

Yip, H., Chan, W. Y., Leung, P. C., Kwan, H. Y., Liu, C., Huang, Y., et al. (2004). Expression of TRPC homologs in endothelial cells and smooth muscle layers of human arteries. *Histochemistry and Cell Biology*, *122*(6), 553–561.

Yuan, Z., Hein, T. W., Rosa, R. H., Jr., & Kuo, L. (2008). Sildenafil (Viagra) evokes retinal arteriolar dilation: Dual pathways via NOS activation and phosphodiesterase inhibition. *Investigative Ophthalmology & Visual Science*, *49*(2), 720–725. https://doi.org/10.1167/iovs.07-1208.

Yue, X., Sheng, Y., Kang, L., & Xiao, R. (2019). Distinct functions of TMC channels: A comparative overview. *Cellular and Molecular Life Sciences*, *76*(21), 4221–4232. https://doi.org/10.1007/s00018-019-03214-1.

Zeng, W. Z., Marshall, K. L., Min, S., Daou, I., Chapleau, M. W., Abboud, F. M., et al. (2018). PIEZOs mediate neuronal sensing of blood pressure and the baroreceptor reflex. *Science*, *362*(6413), 464–467. https://doi.org/10.1126/science.aau6324.

Zheng, Y. M., Wang, Q. S., Liu, Q. H., Rathore, R., Yadav, V., & Wang, Y. X. (2008). Heterogeneous gene expression and functional activity of ryanodine receptors in resistance and conduit pulmonary as well as mesenteric artery smooth muscle cells. *Journal of Vascular Research*, *45*(6), 469–479. https://doi.org/10.1159/000127438.

CHAPTER EIGHT

K_{IR} channels in the microvasculature: Regulatory properties and the lipid-hemodynamic environment

Maria Sancho, Donald G. Welsh*

Robarts Research Institute and the Department of Physiology & Pharmacology, University of Western Ontario, London, ON, Canada

*Corresponding author: e-mail address: dwelsh@robarts.ca

Contents

Abstract

Basal tone and perfusion control is set in cerebral arteries by the sensing of pressure and flow, key hemodynamic stimuli. These forces establish a contractile foundation within arterial networks upon which local neurovascular stimuli operate. This fundamental process is intimately tied to arterial V_M and the rise in cytosolic [Ca^{2+}] by the graded opening of voltage-operated Ca^{2+} channels. Arterial V_M is in turn controlled by a dynamic interaction among several resident ion channels, K_{IR} being one of particular significance. As the name suggests, K_{IR} displays strong inward rectification, retains a small outward component, potentiated by extracellular K^+ and blocked by micromolar Ba^{2+}. Cerebrovascular K_{IR} is unique from other K^+ currents as it is present in both smooth muscle and endothelium yet lacking in classical regulatory modulation. Such observations have fostered the view that K_{IR} is nothing more than a background conductance,

Current Topics in Membranes, Volume 85
ISSN 1063-5823
https://doi.org/10.1016/bs.ctm.2020.01.006

activated by extracellular K^+ and which passively facilitates dilation. Recent work in cell model systems has; however, identified two membrane lipids, phosphatidylinositol 4,5-bisphosphate (PIP_2) and cholesterol, that interact with $K_{IR}2.x$, to stabilize the channel in the preferred open or silent state, respectively. Translating this unique form of regulation, recent studies have demonstrated that specific lipid-protein interactions enable unique K_{IR} populations to sense distinct hemodynamic stimuli and set basal tone. This review summarizes the current knowledge of vascular K_{IR} channels and how the lipid and hemodynamic impact their activity.

1. Introduction

Cerebral resistance arteries form a complex vascular network to tune blood flow delivery in support to neural activity and brain function (Segal, 2000; Segal & Duling, 1986). Under dynamic conditions, tone within this arterial network is set in a coordinated manner in response to multiple vasoactive stimuli derived from tissue metabolism (Filosa et al., 2006), neural activity (Shih et al., 2012; Si & Lee, 2002), blood flow (Drouin & Thorin, 2009; Garcia-Roldan & Bevan, 1990), and intravascular pressure (Harder, Alkayed, Lange, Gebremedhin, & Roman, 1998; Knot & Nelson, 1998). Irrespective of the origin, these vasoactive stimuli initiate vasomotor responses by activating receptors and linked transduction pathways that alter smooth muscle $[Ca^{2+}]$ and the sensitivity of the contractile apparatus to this divalent cation (Somlyo & Somlyo, 1994, 2003; Sward et al., 2003). The concentration of cytosolic Ca^{2+} is, in turn, tightly coupled to membrane potential (V_M) and the graded opening of voltage-operated Ca^{2+} channels (Knot & Nelson, 1998; Welsh, Morielli, Nelson, & Brayden, 2002; Welsh, Nelson, Eckman, & Brayden, 2000).

In resistance vasculature, steady state V_M is controlled by the sharing of charge and the dynamic interaction among depolarizing inward, and hyperpolarizing outward currents. In general, inward conductance is defined by a range of currents including Ca^{2+}-activated Cl^- (Ledoux, Greenwood, Villeneuve, & Leblanc, 2003); voltage-operated Ca^{2+} (Gollasch & Nelson, 1997), and TRP-like non-selective cation-currents (Welsh et al., 2000, 2002). In contrast, K^+ channels are mainly involved in setting outward K^+ conductance leading to hyperpolarization, which causes voltage-operated Ca^{2+} channels to close, intracellular $[Ca^{2+}]$ to decrease, smooth muscle to relax, and lastly vasodilation. These K^+ channels mainly includes voltage-gated (K_v), Ca^{2+}-activated (BK_{Ca}, IK_{Ca} and SK_{Ca}), ATP-sensitive (K_{ATP}), and inward rectifying (K_{IR}) K^+ channels (Nelson & Quayle, 1995; Quayle, Nelson, & Standen, 1997). In particular, K_{IR} channels, aptly named for their distinctive ability to pass current more readily in the inward direction, constitute a key subset of potassium selective channels that possess crucial properties and functions. Cerebral arterial K_{IR} channels are traditionally viewed as a resting conductance with limited-to-no-regulatory control (Edwards, Dora, Gardener, Garland, & Weston, 1998; Quayle et al., 1997; Sonkusare, Dalsgaard, Bonev, & Nelson, 2016). However, its distinct biophysical signature attributes to the channel an important role in arterial myogenic tone development and vascular pathobiology (Goto, Rummery, Grayson, & Hill, 2004; Marrelli, Johnson, Khorovets, Childres, & Bryan, 1998; Mayhan, Mayhan, Sun, & Patel, 2004; Park, Ko, et al., 2007). Although an unlikely target of classic receptor signaling, recent studies in cell model systems have reported that surrounding membrane lipids directly interact with K_{IR}, placing the channel in an active or silent state (Enkvetchakul, Jeliazkova, & Nichols, 2005; Rohacs, Chen, Prestwich, & Logothetis, 1999; Rosenhouse-Dantsker, Noskov, Durdagi, Logothetis, & Levitan, 2013; Rosenhouse-Dantsker, Noskov, Logothetis, & Levitan, 2013). It has been also documented that K_{IR} channels respond to specific hemodynamic stimuli, such as pressure and flow, which set basal tone and brain perfusion control (Sancho et al., 2019; Wu et al., 2007). This review will provide readers an overview of the basic properties, functional role and lipid/hemodynamic regulation of vascular K_{IR} channels, with a particular focus on the cerebral circulation.

2. Inward rectifying potassium channels: Structure, classification and functional roles

On a molecular level, K_{IR} channels are tetramers of α-subunits (Raab-Graham & Vandenberg, 1998; Yang, Jan, & Jan, 1995), each comprised of two membrane-spanning domains (M1 and M2), an intracellular N- and C-terminal domain, and a highly conserved extracellular pore loop (P) that operates as a K^+ selectivity filter (Bichet, Haass, & Jan, 2003; Lu, 2004), critical for channel modulation and setting inward rectification (Fig. 1). This defining characteristic of K_{IR} channels emerges from the interaction with divalent cations (Mg^{2+}) and intracellular polyamines (e.g., spermidine and spermine), which block the outward-flowing K^+ current at

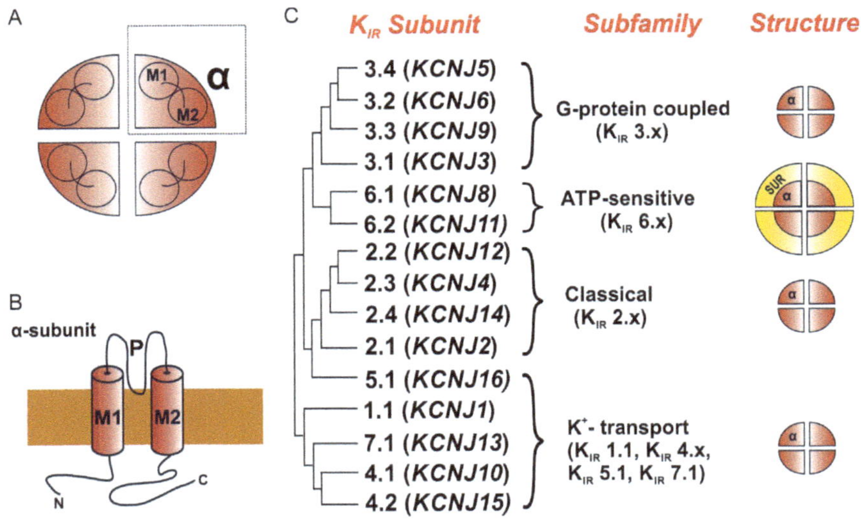

Fig. 1 Molecular structure and classification of K_{IR} channels. (A) K_{IR} channels constitute tetramers of alpha-subunits. (B) Each α-subunit is comprised by two membrane-spanning domains (M1 and M2), an N- and C-terminal domain, and an ion selectivity pore (P) region. (C) Classification of K_{IR} channels according to their phylogenic tree and functional role. All seven subfamilies consist on four α-subunits and K_{IR}6.x subfamily require an additional β-subunit, the sulfonylurea (SUR), to form a functional channel.

depolarized membrane potentials (Kubo & Murata, 2001; Matsuda, Oishi, & Omori, 2003; Robertson, Bonev, & Nelson, 1996).

To date, cloning techniques have identified 15 K_{IR} subunit genes in mammalian tissues (see Table 1 for a detailed tissue distribution of each subunit), grouped into seven structurally distinct subfamilies (K_{IR} 1.x to K_{IR} 7.x) and represented in a phylogenetic tree based by their amino acid sequence identity (~40% homology among subfamilies and ~60% between subfamily members). Based on their functional role, K_{IR} channels are classified into four principal groups: (1) classical strong inward rectifiers (K_{IR} 2.x subfamily), governing the excitability of neuronal and non-neuronal cells; (2) G-protein coupled inward rectifiers (K_{IR} 3.x subfamily), which reduce membrane excitability via G-protein coupled receptors (GPCR); (3) ATP-sensitive K^+ channels, composed by K_{IR} 6.x subunits and an accessory sulfonylurea receptor (SUR) and; (4) pH-sensitive K^+-transport channels (K_{IR} 1.x, K_{IR} 4.x, K_{IR} 5.x, and K_{IR} 7.x subfamilies), typically distributed in kidney cells and astrocytes, where they are involved in K^+ re-uptake following by membrane repolarization (Bichet et al., 2003).

Table 1 Tissue distribution of the different K_{IR} subunits.

K_{IR} subunit	Human gene	Tissue distribution	References
K_{IR} 1.1	*KCNJ1*	Brain, kidney, skeletal muscle, pancreas, heart, liver	Lee and Hebert (1995), Shuck et al. (1994)
K_{IR} 2.1	*KCNJ2*	Brain, blood vessels, skeletal muscle, heart, macrophage cells	Fang et al. (2005), Kubo, Baldwin, Jan, and Jan (1993), Sancho et al. (2019), Smith et al. (2008)
K_{IR} 2.2	*KCNJ12*	Brain, blood vessels, skeletal muscle	Fang et al. (2005), Kubo, Baldwin, et al. (1993), Sancho et al. (2019), Smith et al. (2008)
K_{IR} 2.3	*KCNJ4*	Brain, heart, kidney, astrocytes	Olsen, Liu, Wade, Merot, and Welling (2002), Périer, Radeke, and Vandenberg (1994), Perillan et al. (2000)
K_{IR} 2.4	*KCNJ14*	Lung, retina, blood vessels	Fang et al. (2005), Hughes et al. (2000), Tennant, Cui, Tinker, and Clapp (2006)
K_{IR} 3.1	*KCNJ3*	Forebrain, cerebellum, atrium, Purkinje neurons	Aguado et al. (2008), Kubo, Reuveny, Slesinger, Jan, and Jan (1993)
K_{IR} 3.2	*KCNJ6*	Brain, testis, pancreas	Inanobe, Horio, et al. (1999), Inanobe, Yoshimoto, et al. (1999), Stoffel et al. (1995)
K_{IR} 3.3	*KCNJ9*	Brain, testis, small intestine, kidney, skeletal muscle	Lesage et al. (1994), Vaughn, Wolford, Prochazka, and Permana (2000)
K_{IR} 3.4	*KCNJ5*	Brain, heart, pancreas	Wickman, Karschin, Karschin, Picciotto, and Clapham (2000), Yoshimoto et al. (1999)
K_{IR} 4.1	*KCNJ10*	Brain, stomach, retina, kidney	Fujita et al. (2002), Ishii et al. (2003), Tucker, Imbrici, Salvatore, D'Adamo, and Pessia (2000)
K_{IR} 4.2	*KCNJ15*	Kidney, pancreas, lung, prostate	Ohira et al. (1997), Shuck et al. (1997)
K_{IR} 5.1	*KCNJ16*	Kidney, pancreas, retina	Ishii et al. (2003), Liu et al. (2000)

Continued

Table 1 Tissue distribution of the different K_{IR} subunits.—cont'd

K_{IR} subunit	Human gene	Tissue distribution	References
K_{IR} 6.1	*KCNJ8*	Brain, heart, skeletal muscle, lung, colon	Inagaki, Gonoi, et al. (1995), Kane et al. (2006)
K_{IR} 6.2	*KCNJ11*	Brain, heart, pancreas, skeletal muscle	Inagaki, Tsuura, et al. (1995), Miki et al. (1998)
K_{IR} 7.1	*KCNJ13*	Brain, kidney, small intestine, stomach, retina	Krapivinsky et al. (1998), Nakamura et al. (1999), Partiseti, Collura, Agnel, Culouscou, and Graham (1998), Yang, Zhang, and Hughes (2008)

3. Classical strong K_{IR} channels

Classical strong K_{IR} channels (K_{IR} 2.x subfamily) are highly expressed in excitable and non-excitable cells including but not limited to heart, skeletal muscle, nervous system, and vascular tissue (Bradley et al., 1999; Dhamoon et al., 2004; Jantzi et al., 2006; Liu et al., 2001; Wu et al., 2007). There are four distinct members of this subfamily (K_{IR} 2.1–2.4) and they share ~50–70% amino acid identity (Liu et al., 2001; Schram, Pourrier, Wang, White, & Nattel, 2003). Fig. 2 summarizes the key properties of K_{IR}2.x channels in regard to their subunit composition. In brief, homomultimers of K_{IR}2.1 or K_{IR}2.2 differ considerably from K_{IR}2.3 or K_{IR}2.4 with respect to: (1) negative slope conductance; (2) unitary conductance; (3) $BaCl_2$ sensitivity; and (4) K^+-induced potentiation of outward current. Heteromultimeric channels often display properties distinct from their homomultimeric counterparts (Fig. 2). For example, the Ba^{2+} sensitivity of K_{IR}2.1/2.2 channels rests between the two homomultimers. This distinctiveness has been routinely exploited as a means of defining the molecular basis of tissue specific K_{IR} currents (Liu et al., 2001; Schram et al., 2003).

3.1 K_{IR}2.x channel distribution in resistance vasculature

There is strong molecular evidence that K_{IR}2.x channels are robustly expressed in vascular tissue including the cerebral (Longden et al., 2017; Sancho et al., 2017; Wu et al., 2007), coronary (Knot, Zimmermann, & Nelson, 1996; Smith et al., 2008), mesenteric (Ahn et al., 2017;

K_{IR} Subunit	Negative Slope Conductance	Unitary Conductance (pS)	Ba^{2+} Sensitivity (IC_{50})	K+-induced Potentiation of Outward Current	Whole-Cell Current Shape
K_{IR} 2.1	Strong	23- 30.6	~ 16 µM (-120 mV)	Yes	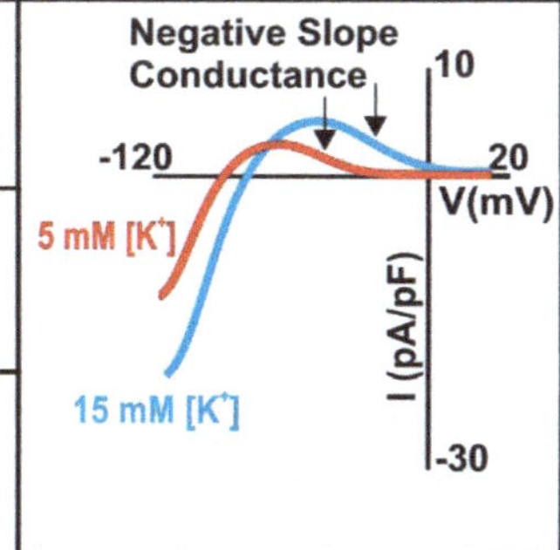
K_{IR}2.2	Strong	34.2- 40	~ 2.3 µM (-120 mV)	Yes	
Heteromer K_{IR}2.1-2.2	Strong	30	~ 0.7-5 µM (-100 to -120 mV)	Yes	
K_{IR} 2.3	Weak	13-14.2	~ 18 µM (-120 mV)	No	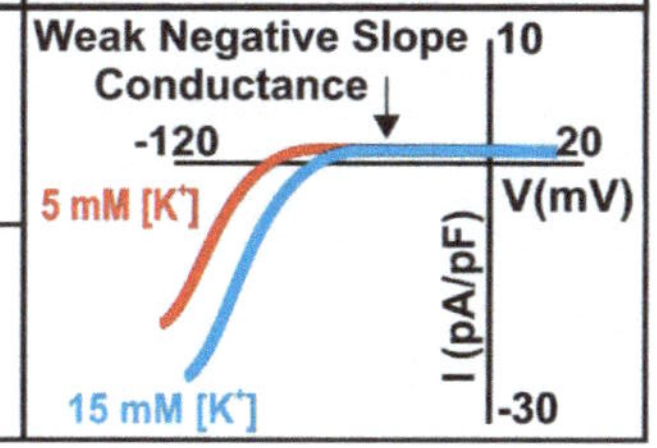
K_{IR} 2.4	Weak	15	~ 72 µM (-120 mV)	No	

Fig. 2 Key biophysical properties of K_{IR} homo- and heteromultimeric channels. *Data were selected from Bradley, K. K., Jaggar, J. H., Bonev, A. D., Heppner, T. J., Flynn, E. R., Nelson, M. T., et al. (1999). Kir2.1 encodes the inward rectifier potassium channel in rat arterial smooth muscle cells.* The Journal of Physiology, 515, *639–651; Dhamoon, A. S., Pandit, S. V., Sarmast, F., Parisian, K. R., Guha, P., Li, Y., et al. (2004). Unique Kir2.1 properties determine regional and species in the cardiac inward rectifier K^+ current.* Circulation Research, 94, *1332–1339; Hughes, B. A., Kumar, G., Yuan, Y., Swaminathan, A., Yan, D., Sharma, A., et al. (2000). Cloning and functional expression of human retinal kir2.4, a pH-sensitive inwardly rectifying K (+) channel.* Americam Journal Physiology-Cell Physiology, 279, *C771–C784; Liu, G. X., Derst, C., Schlichthorl, G., Heinen, S., Seebohm, G., Bruggemann, A., et al. (2001). Comparison of cloned Kir2 channels with native inward rectifier K^+ channels from guinea-pig cardiomyocytes.* The Journal of Physiology, 532, *115–126; Schram, G., Melnyk, P., Pourrier, M., Wang, Z., & Nattel, S. (2002). Kir2.4 and Kir2.1 K(+) channel subunits co-assemble: A potential new contributor to inward rectifier current heterogeneity.* The Journal of Physiology, 544, *337–349; Schram, G., Pourrier, M., Wang, Z., White, M., & Nattel, S. (2003). Barium block of Kir2 and human cardiac inward rectifier currents: Evidence for subunit-heteromeric contribution to native currents.* Cardiovascular Research, 59, *328–338; Shieh, R. C., Chang, J. C., & Arreola, J. (1998). Interaction of Ba2+ with the pores of the cloned inward rectifier K^+ channels Kir2.1 expressed in Xenopus oocytes.* Biophysical Journal, 75, *2313–2322.*

Sonkusare et al., 2016), and renal (Chilton et al., 2008) circulation. Generally speaking, the expression and influence of K_{IR} channels on membrane potential is inversely proportional to the vessel diameter, thus predominating in small resistance vessels and capillaries (Cao, Goo, Lee-Kwon, & Pallone, 2006; Chilton & Loutzenhiser, 2001; Edwards, Hirst, & Silverberg, 1988; Prior, Webster, Quinn, Beech, & Yates, 1998; Quayle, Dart, & Standen, 1996). Recently, Longden et al. (2017) have reported a central role for capillary endothelial K_{IR}2.1 channels in propagating electrical signals upstream to parenchymal arterioles and therefore constituting an effective mechanism for cerebral blood flow regulation. There is also evidence for differential expression of K_{IR}2.x among vascular beds; while they are strongly expressed in both smooth muscle and endothelium from cerebral (Sancho et al., 2019) or cremaster arterioles (Burns, Cohen, & Jackson, 2004), K_{IR}2.x channels are absent from mesenteric vascular smooth muscle (Goto et al., 2004; Smith et al., 2008; Sonkusare et al., 2016). This diversity may explain the existence of unique cell-specific pools of K_{IR} channels that sense different stimuli to drive vasomotor responses and set basal perfusion.

3.2 Crystal structure

Over the last decade, numerous studies have described the high-resolution X-ray crystal structures of bacterial and mammalian K_{IR}2.1 and K_{IR}2.2 subunits, providing the foundation for understanding the functionality of these channels. K_{IR}2.x channels are made up of a transmembrane domain; essential for ion selectivity and gating; and a large cytoplasmic domain which acts as a gating regulator (Fig. 3). The transmembrane domain is comprised of two segments (M1 and M2) linked by an extracellular turret region, a short pore helix and the selectivity filter (Hansen, Tao, & MacKinnon, 2011; Tao, Avalos, Chen, & MacKinnon, 2009) that contains the signature GYG motif, with similar sequence in all inward rectifiers. The presence of negatively charged amino acid residues in M2 (D172 in K_{IR} 2.1; D173 in K_{IR}2.2) is essential for the high-affinity binding of cytoplasmic ions such as polyamines and Mg^{2+} (Lopatin, Makhina, & Nichols, 1994).

The pore-forming domain consists of a selectivity filter, a water central cavity and a bundle crossing region where M2 helixes associate with each other. The existence of a conserved glycine residue located in M2 helix (G168 in K_{IR} 2.1) is crucial to provide flexibility to the structure and act as movable hinge. As it will be discussed below, the dynamic phospholipid

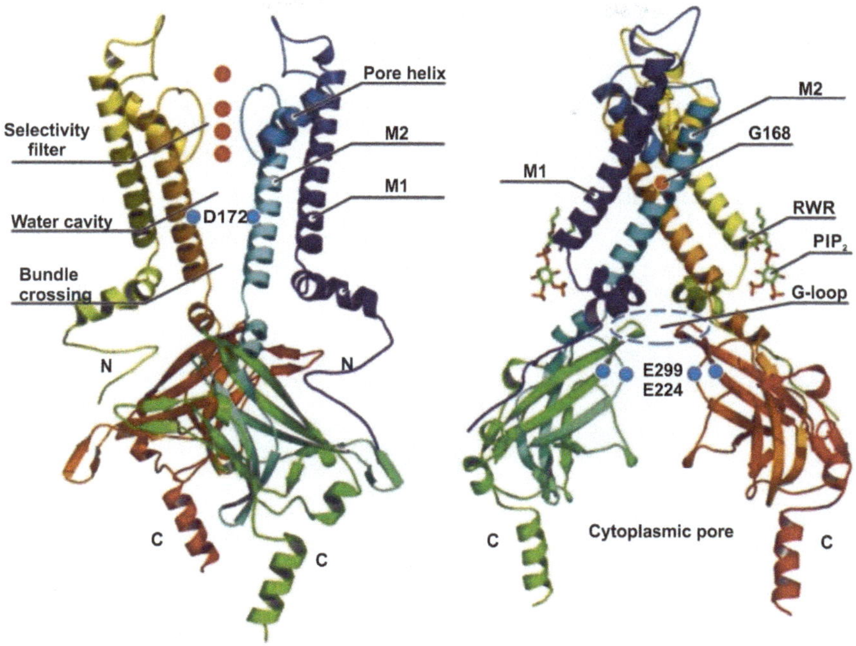

Fig. 3 Crystal architecture of $K_{IR}2.x$ channels, and the location of the most important regions and residues essential to polyamine/Mg^{2+} block. *Modified from Lopatin, A. N., & Anumonwo, J. M. B. (2014). Structural and molecular bases of cardiac inward rectifier potassium channel function. In* Cardiac electrophysiology: From cell to bedside *(6th ed.). (pp. 33–41).*

component of cell membranes, phosphatidylinositol 4,5-bisphosphate (PIP_2) is one the primary activators of strong inward rectifiers by binding at the interface between the transmembrane and the cytoplasmic domains (Hansen et al., 2011). In this sense, high-resolution structural studies have located that PIP_2-binding site in a conserved non-specific RWR motif in the M1 helix as well as in a specific region in the cytoplasmic domain. Finally, the cytoplasmic domain retains a conserved motif that acts as a second physical gate (G-loop) and several residues (E299, E224) constitute a low-affinity polyamine/Mg^{2+} block site (Tao et al., 2009; Whorton & MacKinnon, 2013).

3.3 Basic properties of vascular $K_{IR}2.x$

3.3.1 Inward rectification

As their name suggests, inward rectifiers possess the inherent ability to pass more current inward (into the cell) than outward. K_{IR} channels generate a large K^+ conductance at V_M negative to the potassium equilibrium potential (E_K) while they are able to pass small outward currents at voltages positive to it (Fig. 4A) (Smith et al., 2008; Wu et al., 2007). The property of inward rectification, originally known as "anomalous rectification" was first defined in skeletal muscle (Katz, 1949), and contributes to the maintenance of a resting membrane potential. Electrophysiology established that inward rectification arises as a consequence of strong voltage-dependent pore

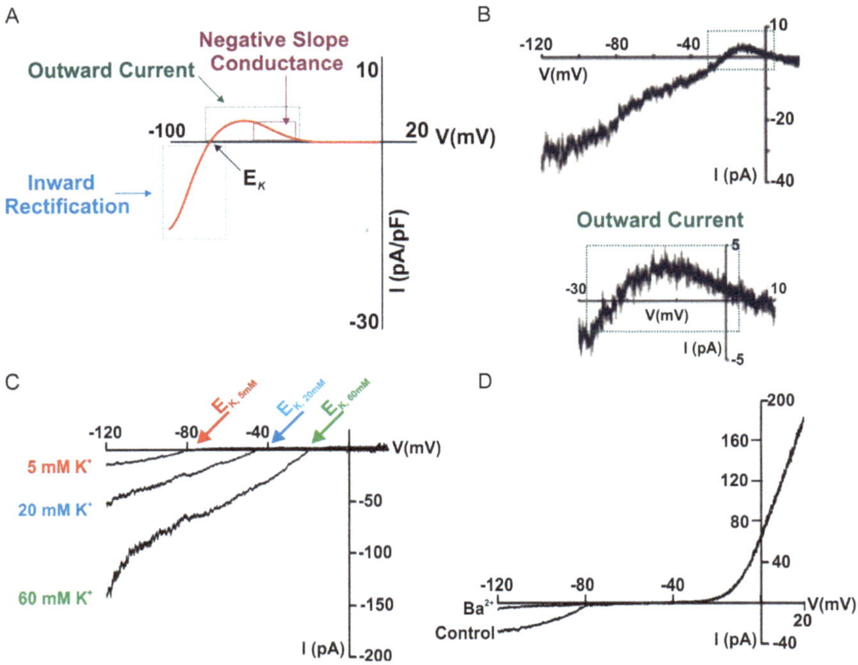

Fig. 4 Properties of vascular K_{IR} conductance. (A) Illustrative I–V relationship of a K_{IR} current comprised of $K_{IR}2.1$ and/or $K_{IR}2.2$ subunits. Note the existence of an outward current with the characteristic negative slope conductance that spans over the physiological voltage range and the inward rectification at hyperpolarized V_M. (B) Whole-cell currents recorded in rat cerebral smooth muscle cells bathed in 60 mM $[K^+]$ solution (top). Note that a small (<5 pA) outward current with a negative slope conductance was successfully detected (bottom; magnified from top). (C) Effect of extracellular $[K^+]$ on K_{IR} current. (D) Effect of $BaCl_2$ on cerebral arterial smooth muscle whole-cell current. *Panel (B–D): Modified from Smith, P. D., Brett, S. E., Luykenaar, K. D., Sandow, S. L., Marrelli, S. P., Vigmond, E. J., et al. (2008). KIR channels function as electrical amplifiers in rat vascular smooth muscle.* The Journal of Physiology, 586, *1147–1160; Wu, B. N., Luykenaar, K. D., Brayden, J. E., Giles, W. R., Corteling, R. L., Wiehler, W. B., et al. (2007). Hyposmotic challenge inhibits inward rectifying K^+ channels in cerebral arterial smooth muscle cells.* American Journal of Physiology—Heart and Circulatory Physiology, 292, *H1085–H1094, respectively.*

blockade by intracellular cations such as polyamines and Mg^{2+}. At hyperpolarizing membrane voltages, the blocking cations are absent from the pore so that K^+ flows, whereas at depolarizing V_M polyamines and Mg^{2+} are driven into the pore from the cytosol so that K^+ movement across the membrane is blocked (Lopatin & Nichols, 1996; Schram et al., 2003).

3.3.2 Outward current and negative slope conductance

While inward rectification is the electrophysiological fingerprint of K_{IR} channels, it is the small outward component which is crucial for arterial V_M/tone regulation and consequently blood flow control. These outward currents are known as the region of negative slope conductance, a key biophysical property that spans over the physiological voltage range (Dhamoon et al., 2004; Matsuda et al., 2003; Schram et al., 2003), and enables K_{IR} channels to amplify responses initiated by other K^+ conductances (Smith et al., 2008). Negative slope conductance is typically observed in channels comprised by $K_{IR}2.1$ and $K_{IR}2.2$ subunits, constituents often expressed in vascular smooth muscle and endothelium (Sancho et al., 2017, 2019; Wu et al., 2007). Under physiological conditions, the characterization of the outward current has proven challenging as its amplitude is a few picoamperes, close to the resolution limits of a patch-clamp amplifier (Quayle, McCarron, Brayden, & Nelson, 1993; Wu et al., 2007). However, some studies have successfully documented this property in vascular tissue by enhancing K_{IR} activity (Smith et al., 2008) (Fig. 4B).

3.3.3 Extracellular K^+ potentiation

One more key feature of K_{IR} channels is their dependence on extracellular $[K^+]$ (Figs. 2 and 4C). Specifically, an elevation of $[K^+]_o$ augments K_{IR} inward current and elicits a shift of the activation curve to more positive potentials accompanied by the consequent rightward shift in the E_K (Filosa et al., 2006; Wu et al., 2007). Physiologically, smooth muscle and endothelial V_M are normally positive to the E_K, thus a $[K^+]$-induced rightward shift in K_{IR} activation curve would also enhance the outward component of K_{IR} conductance. The resulting extracellular $[K^+]$—K_{IR} activation enables these channels to contribute to functional hyperemia in excitable tissues such as brain or skeletal muscle (Amstrong, Dua, & Murrant, 2007; Filosa et al., 2006).

3.3.4 Ba^{2+}-blockade

The characterization of vascular smooth muscle and endothelial $K_{IR}2.x$ currents has been facilitated by using extracellular $BaCl_2$ (micromolar range);

a potent and selective voltage-dependent blocker of K_{IR} channels (Knot et al., 1996). As shown in Fig. 2, heteromeric constructs of $K_{IR}2.1$–$K_{IR}2.2$ subunits (highly expressed on vascular cells) exhibits different Ba^{2+} blocking kinetics (IC_{50} = 0.7–5 μM, at −100 to −120 mV) from that of homomeric $K_{IR}2.1$ or $K_{IR}2.2$ channels (IC_{50} = 16 μM for $K_{IR}2.1$ and 2.3 μM for $K_{IR}2.2$, at −120 mV), suggesting that heteromeric assembles display unique electrophysiological properties that might be involved in the functionality of these channels. Note that although $BaCl_2$ also blocks K_{ATP} current, it's efficacy is much lower (IC_{50} = 100 μM) and full blockade is only achieved in the millimolar range (Bonev & Nelson, 1993).

In vitro studies using intact arteries of various vascular beds have demonstrated a robust depolarization and vasoconstriction at rest after Ba^{2+} superfusion (Park, Kang, et al., 2007; Park, Ko, et al., 2007), an effect particularly pronounced at low intravascular pressure when myogenic tone was limited and V_M negative (Smith et al., 2008; Wu et al., 2007). Importantly, Ba^{2+} selectivity of K_{IR} channels is demonstrated by two key evidences: (1) Ba^{2+} only blocks the inward K_{IR} component at voltages negative to the apparent E_K, and (2) Ba^{2+}-sensitive $[K^+]_o$-induced vasodilation is only apparent in tissues expressing K_{IR} channels (Smith et al., 2008; Wu et al., 2007). In alignment with these findings, in vivo approaches have also showed a strong constriction of rodent cremaster (Loeb, Godeny, & Longnecker, 2000) and pial (Mayhan et al., 2004) arterioles in response to micromolar concentrations of Ba^{2+}. More remarkably, studies in humans have reliably revealed that Ba^{2+} infusion into forearm increased Ba^{2+} plasma levels (~50 μM) and caused a strong vasoconstriction accompanied by reduced blood flow without eliciting systemic effects (Crecelius, Richards, Luckasen, Larson, & Dinenno, 2013; Dawes et al., 2002).

3.3.5 Electrical amplification

Studies in rodent vascular tissue have defined K_{IR} channels as amplifiers of electrical/vasomotor responses originating from other smooth muscle and endothelial K^+ conductances. In this context, the key property of negative slope conductance enhances outward K conductance to amplify the initial hyperpolarizing/vasodilation response (Jantzi et al., 2006; Smith et al., 2008; Sonkusare et al., 2016). These remarkable results highlight how subtle channel properties can play a critical role in the regulation of arterial blood flow. Interestingly, these findings have translated into human skeletal muscle microvasculature, with hand grip exercise augmented acetylcholine-mediated vasodilation (hyperpolarizing stimulus) while sodium nitroprusside (an agent that relaxes smooth muscle independent on membrane potential) having no measurable effect on vascular hyperemia (Hearon et al., 2019). Subsequent $BaCl_2$ infusion effectively abolished acetylcholine-induced amplification of hyperemic responses, findings indicative of K_{IR} channels acting as electrical amplifiers.

4. Regulation of vascular $K_{IR}2.x$ channels

4.1 Membrane lipids

K_{IR} channels are often considered poor regulatory targets as stimuli that mobilize classic G-coupled receptors routinely fail to alter intrinsic activity (Quayle et al., 1997; Sonkusare et al., 2016). However, we and others consistently demonstrated that the channel is strongly regulated by two signaling lipid regulators that constitute an integral part of the plasma membrane (Harraz, Longden, Dabertrand, Hill-Eubanks, & Nelson, 2018; Hilgemann, 2007; Huang, Feng, & Hilgemann, 1998; Rohacs et al., 2003; Romanenko et al., 2004; Romanenko, Rothblat, & Levitan, 2002; Zhang, He, Yan, Mirshahi, & Logothetis, 1999). These include phosphatidylinositol 4,5-bisphosphate (PIP_2) and cholesterol, a minor and major lipid component of the plasma membrane that facilitates and suppresses the activity of $K_{IR}2.x$ channels, respectively (McLaughlin, Wang, Gambhir, & Murray, 2002; Yeagle, 1985).

4.1.1 Phosphatidylinositol 4,5-bisphosphate

PIP_2 is the major polyphosphoinositide in mammalian cells and constitutes up to 1% of the phospholipids in the plasma membrane inner, cytoplasmic leaflet (McLaughlin et al., 2002). Accurate measurements of PIP_2 concentration in a cell has proven difficult; however recent studies using constructs of PH domains linked to green fluorescent protein (GFP) estimated that cellular PIP_2 levels ranges from 2 to 10 μM. Despite the relative low abundance, PIP_2 retains the ability to impact a wide range of downstream signaling pathways, and represents the most important phospholipid regulator of a variety of enzymes, transporters and ion channels. Structurally, it comprises an inositol head group, a phosphoglycerol backbone and two acyl chains (Fig. 5A). While the molecule contains four negative charges, its net charge depends on several factors such as the local pH, lipid environment or specific interactions with proteins (McLaughlin et al., 2002). PIP_2 is synthetized by its precursor phosphatidylinositol (PI) in two consecutive phosphorylation steps by phosphatidylinositol 4 (PI4K)

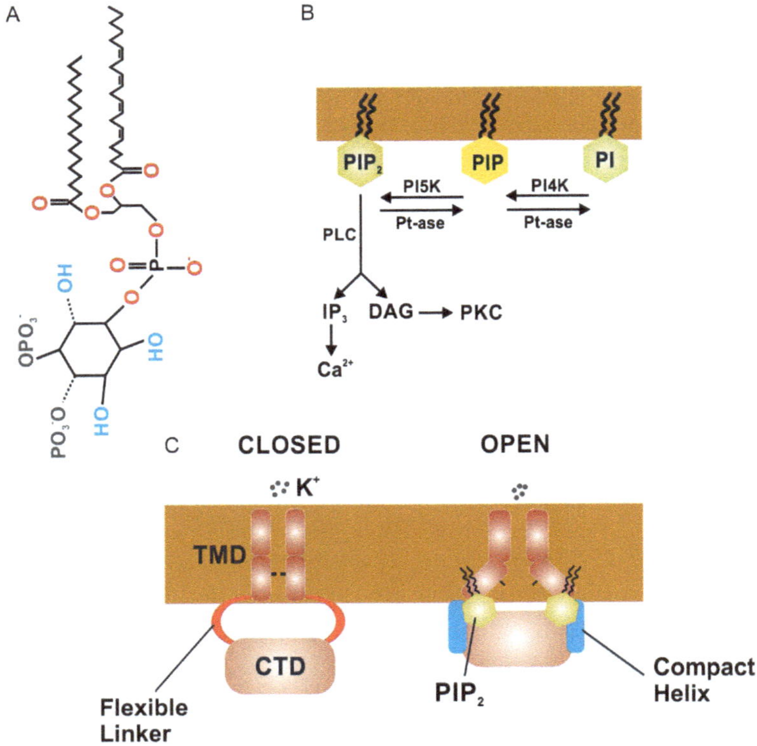

Fig. 5 Molecular structure, metabolism and mechanism of $K_{IR}2.x$ activation by PIP_2. (A) A PIP_2 molecule is constituted by an inositol head group, a phosphoglycerol backbone and two acyl chains. (B) PIP_2 synthesis and degradation pathways. PIP_2 is synthetized from its precursor phosphatidylinositol (PI) in two consecutive phosphorylation steps by phosphatidylinositol 4 and 5 kinases (PI4K and PI5K, respectively). This process is reversible by the action of lipid phosphatases (Pt-ase). Hydrolysis of PIP_2 by PLC yields DAG and IP_3. Both are essential second messengers; DAG activates protein kinase C (PKC) and IP_3 stimulates the release of Ca^{2+} from intracellular stores. (C) Upon PIP_2 binding, a flexible linker contracts to a compact helical structure, the cytoplasmic domain translates toward and become tethered to the transmembrane domain, and the channel results mechanically open.

and 5 kinases (PI5K). The level of PIP_2 also depends on metabolism as it can be hydrolyzed by different isoforms of phospholipase C enzymes, which generates diacylglycerol (DAG) and inositol (1,4,5) triphosphate (IP_3), two second intracellular messengers involved in essential cellular processes (Fig. 5B).

Accumulating evidence supports the hypothesis that PIP_2 is absolutely required in the membrane to stabilize $K_{IR}2.x$ channel in a preferred open state (Enkvetchakul et al., 2005; Rohacs et al., 1999). In the past decade, crystallographic and structure studies elegantly defined the precise binding sites by which negative phosphate groups of PIP_2 electrostatically interact with positively charged amino acid residues of K_{IR}2.x. In particular, the majority of these residues are located in proximity to the interface between the transmembrane and the cytoplasmic domains (Fig. 3). While the inositol head group seems to bind to a specific PI binding region in the cytoplasmic domain, the acyl chains and phosphoglycerol backbone make interactions with a conserved non-specific phospholipid binding region (RWR) in the transmembrane domain. Upon PIP_2 binding, a flexible linker contracts to a compact helical structure, the cytoplasmic domain translates toward and becomes tethered to the transmembrane domain, thus providing a tangential force that mechanically opens the channel and initiates the ion conduction pathway (Hansen et al., 2011) (Fig. 5C). Previous studies using K_{IR}2.1 channel constructs expressed in *Xenopus* oocytes demonstrated that the interaction of a single subunit with PIP_2 is enough to retain the channel in a fully open state (Xie, John, Ribalet, & Weiss, 2008), and interestingly, the apparent affinity of channel-PIP_2 interactions is crucial in defining the modulation of K_{IR} channels. In this context, K_{IR}2.1 channels present high affinity for PIP_2 so their activity is not affected by the lipid depletion; however K_{IR}2.2 and 2.3 subunits are strongly regulated by PIP_2 hydrolysis as they possess relatively low PIP_2 affinity (Du et al., 2004). Furthermore, the strength of channel-PIP_2 interactions also determines the sensitivity of K_{IR} channels to regulation by other modulators including pH, Mg^{2+} or phosphorylation (Du et al., 2004).

Moving to the cerebral vasculature, and consistent with PIP_2 facilitating K_{IR}2.x activity, we have recently demonstrated that endothelial K_{IR}2.1–2.2 conductance was impacted in a time-dependent fashion by dialyzing the cells with different drugs that alter cellular membrane PIP_2 levels. While PIP_2 depletion by intracellular [ATP] modifications, PIK blockade or direct sequestration (using neomycin as a PIP_2 scavenger) decreased inward K_{IR} conductance, cell dialysis with the water soluble PIP_2 analog diC8-PIP_2 enhanced channel activity (Fig. 6) (Sancho et al., 2019). In contrast, smooth muscle K_{IR} channel population seems to be insensitive to [PIP_2] manipulations, an indicative observation of each cellular K_{IR} population being distinctly regulated (Sancho et al., 2019), with PIP_2 being a dominant regulator of endothelial K_{IR}2.x by placing the channel in a preferred active state.

At the capillary level, Harraz, Longden, Dabertrand, et al. (2018) consistently identified PIP_2 as an intrinsic modulator of endothelial K_{IR}2.1 channels,

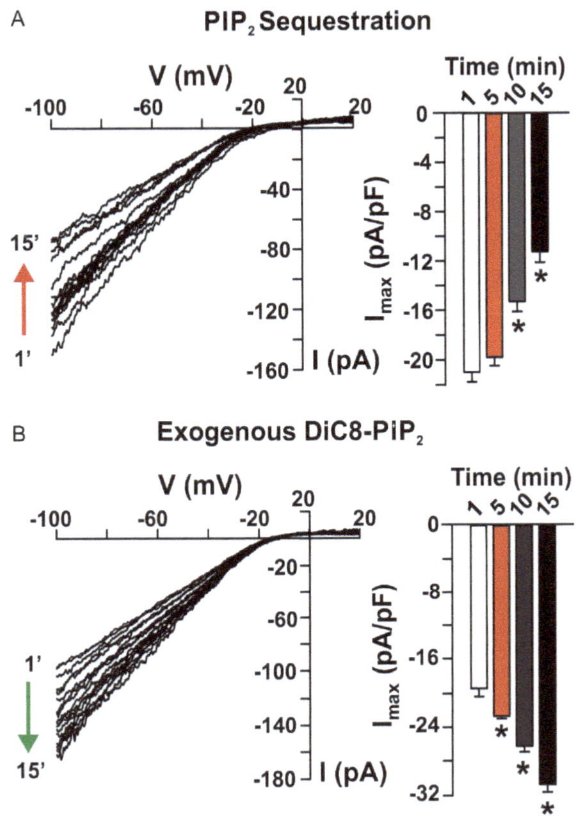

Fig. 6 Phosphatidylinositol 4,5-bisphosphate (PIP_2) drives cerebral vascular endothelial $K_{IR}2.1$–2.2 channel activity. Representative traces and summary data of peak inward current (−100 mV) in native endothelial cells during the dialysis (15 min) of (A) neomycin (50 μM, $n = 9$, repeated measurements [RM]-ANOVA, Tukey post hoc analysis) or (B) the water soluble PIP_2 analog diC8-PIP_2 (50 μM, $n = 8$, repeated measurements [RM]-ANOVA, Tukey post hoc analysis). Data are means ± SE and * significant difference at $P \leq 0.05$. *Modified from Sancho, M., Fabris, S., Hald, B. O., Brett, S. E., Sandow, S. L., Poepping, T. L., et al. (2019). Membrane lipid-KIR2.x channel interactions enable hemodynamic sensing in cerebral arteries.* Arteriosclerosis, Thrombosis, and Vascular Biology, 39, *1072–1087.*

being thus essential for the capillary-to arteriole electrical signaling transmission and cerebral blood flow control. Interestingly, this signaling lipid causes opposite effects on capillary endothelial TRPV4 activity, therefore it has been suggested to act as a bidirectional regulatory pathway to sustain and suppress K_{IR} (hyperpolarizing) and TRPV4 (depolarizing) activity, respectively (Harraz, Longden, Hill-Eukbanks, & Nelson, 2018).

4.1.2 Cholesterol

Cholesterol is one of the main lipid components in mammalian cells, constituting 35–45% of total lipid molecules (Yeagle, 1985), being more than the 90% of cellular cholesterol non-uniformly located at the plasma membrane (Lange, Swaisgood, Ramos, & Steck, 1989). As an essential component of plasma membranes, cholesterol maintains the integrity and functionality of lipid rafts, as well as regulates multiple signaling pathways including receptors, enzymes and ion channels (Gimpi, Burger, & Fahrenholz, 1997). As a consequence, an imbalance of the cellular cholesterol level may lead to tissue or cell dysfunction, and may contribute to the pathology of hypercholesterolemia.

Cholesterol molecules consist of four fused hydrocarbon rings, which are strongly hydrophobic, and a polar 3β-hydroxyl (OH) group that confers the molecule an amphipathic character (Fig. 7A), and plays an essential role in cholesterol orientation with the hydrophobic regions reaching the hydrophobic core of the bilayer (Ohvo-Rekila, Ramstedt, Leppimaki, & Slotte, 2002). Furthermore, the OH group can also mediate the interaction of cholesterol with water or other membrane phospholipids influencing their behavior. The level of membrane cholesterol is determined by de novo biosynthesis from acetyl coenzyme A in the endoplasmic reticulum and subsequent transport to the membrane by caveolins or via uptake from plasma lipoproteins, both pathways being tightly controlled by feedback mechanisms (Fielding & Fielding, 1995; Reinhart, Billheimer, Faust, & Gaylor, 1987).

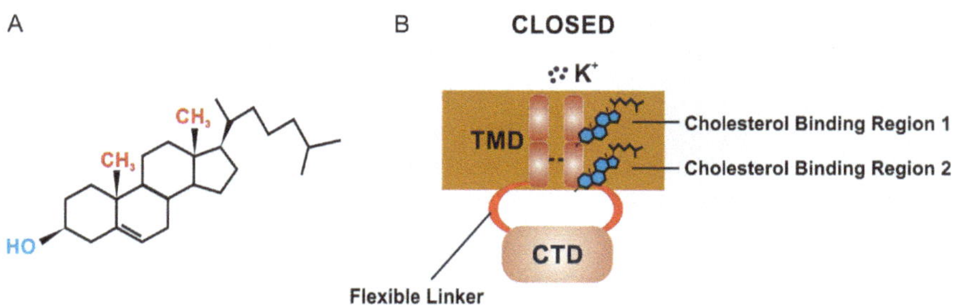

Fig. 7 Molecular architecture and location of presumed cholesterol binding regions in $K_{IR}2.1$ channels. (A) The cholesterol molecule comprises four fused non-polar hydrocarbon rings and a polar 3β-hydroxyl (OH) group. (B) Two non-annular hydrophobic cholesterol binding regions are identified in $K_{IR}2.1$ channels, one in the center of the transmembrane domain (1) and the second at the interface between transmembrane and cytoplasmic domains of the channel (2). Upon direct cholesterol binding, the hinging motion of the transmembrane pore helix changes, promoting thus channel stabilization in a preferred silent state.

In contrast to PIP_2, membrane cholesterol suppresses the activity of K_{IR} by stabilizing the channel in a preferred silent state. This idea came from studies documenting a decrease in whole-cell $K_{IR}2.x$ current after increasing membrane cholesterol levels, while open probability, unitary conductance or channel protein expression remained slightly altered (Romanenko et al., 2004, 2002). With regard to cholesterol-K_{IR} regulatory cores, recent studies from Levitan's group identified several cytosolic residues located in the G-loop, N-terminus and the connecting segment between the C-terminus and the inner transmembrane helix of the $K_{IR}2.1$ channel. All these residues form a cytosolic belt around the cytoplasmic putative pore that modulates the cholesterol sensitivity of the channel. Although the cholesterol sensitivity belt does not constitute a cholesterol binding site, it may enable cholesterol to bind to another part of the channel controlling the gating apparatus (Rosenhouse-Dantsker, Logothetis, & Levitan, 2011). Interestingly, some residues of the cytosolic belt are also important for PIP_2 sensitivity, supporting the idea that both membrane lipids could share a common regulatory switch to control the channel state. Studies of endothelial $K_{IR}2.x$ channel activity substituting native cholesterol by its optical isomer epicholesterol revealed opposing effects (suppression and activation, respectively), demonstrating that cholesterol regulates K_{IR} channels by specific binding rather than by modifying plasma membrane structure (Singh, Rosenhouse-Dantsker, Nichols, Enkvetchakul, & Levitan, 2009). In particular, two non-annular hydrophobic binding pockets were identified in $K_{IR}2.1$ channels, one is located in the center of the transmembrane domain and the second at the interface between transmembrane and cytoplasmic membrane of the channel (Fig. 7B). The idea is that cholesterol binds directly to the channel, modifies the hinging motion of the transmembrane pore helix, and therefore promotes its stabilization in a preferred closed state (Rosenhouse-Dantsker, Noskov, Durdagi, et al., 2013).

Numerous studies using endothelial expression systems revealed an inhibition or activation of $K_{IR}2.x$ channels after increasing or depleting membrane cholesterol, respectively (Epshtein et al., 2009; Fang et al., 2006; Romanenko et al., 2002; Rosenhouse-Dantsker, Noskov, Durdagi, et al., 2013; Rosenhouse-Dantsker, Noskov, Logothetis, et al., 2013), being $K_{IR}2.1$ and 2.2 subunits predominantly sensitive to cholesterol level manipulations (Romanenko et al., 2004). In an effort to translate these significant findings into native cerebral vascular cells, we recently depleted membrane cholesterol by dialyzing smooth muscle and endothelial cells with methyl-β-cyclodextrin (MβCD). Intriguingly, cholesterol depletion impacted K_{IR}

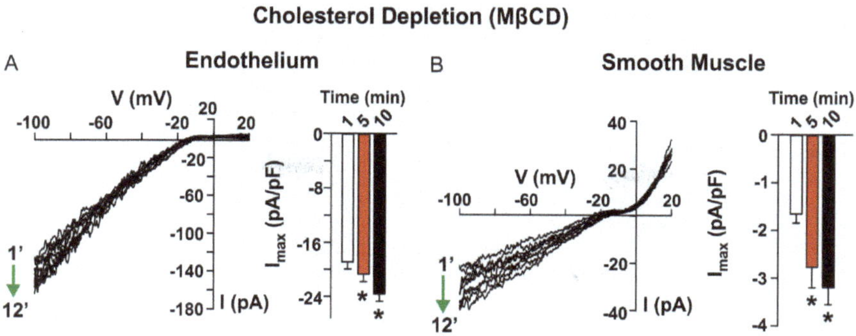

Fig. 8 Cholesterol is a dominant regulator of cerebral arterial smooth muscle but not endothelial K_{IR}2.1–2.2 channels. Representative traces and summary data of peak inward current (−100 mV) in native endothelial (A) and smooth muscle cells (B) over the 10 min dialysis period of methyl-β-cyclodextrin (5 mM, n = 7–8, repeated measurements [RM]-ANOVA, Tukey post hoc analysis). Data are means ± SE and * significant difference at $P \leq 0.05$. *Modified from Sancho, M., Fabris, S., Hald, B. O., Brett, S. E., Sandow, S. L., Poepping, T. L., et al. (2019). Membrane lipid-KIR2.x channel interactions enable hemodynamic sensing in cerebral arteries.* Arteriosclerosis, Thrombosis, and Vascular Biology, 39, *1072–1087.*

activity in a time-dependent manner, an effect that was abrogated when exposing the cells to cholesterol-saturated MβCD. The more remarkable discovery here is that these effects were particularly prominent in smooth muscle rather than endothelium (Fig. 8), suggesting that cholesterol is a dominant regulator of smooth muscle K_{IR}2.x by promoting channel stabilization in a preferred silent state (Sancho et al., 2019).

A variety of ion channels including K_{IR}2.x have been also commonly associated to cholesterol-rich membrane domains or caveolae, which are flask-shaped invaginations of the plasma membrane found in a plethora of cell types (Maguy, Hebert, & Nattel, 2006). Specifically, one of the major structural proteins of caveolae, Caveolin-1 (Cav-1) physically interacts with K_{IR}2.1 and decreases current density without any variations on the single channel properties or protein expression on the plasma membrane. It has been demonstrated that Cav-1 stabilizes the channel in a silent state by a shared mechanism with cholesterol (Han et al., 2014), although this protein is not necessary to confer cholesterol sensitivity to the channel. Two Cav-1 binding motifs were described with the aid of crystal structures, one at the interface between the outer transmembrane helix and the N-terminus of the channel, and the second in the outer transmembrane helix close to the extracellular domain of the channel (Han et al., 2014). In the cerebral circulation, we consistently showed a negative impact of Cav-1 in smooth muscle K_{IR}

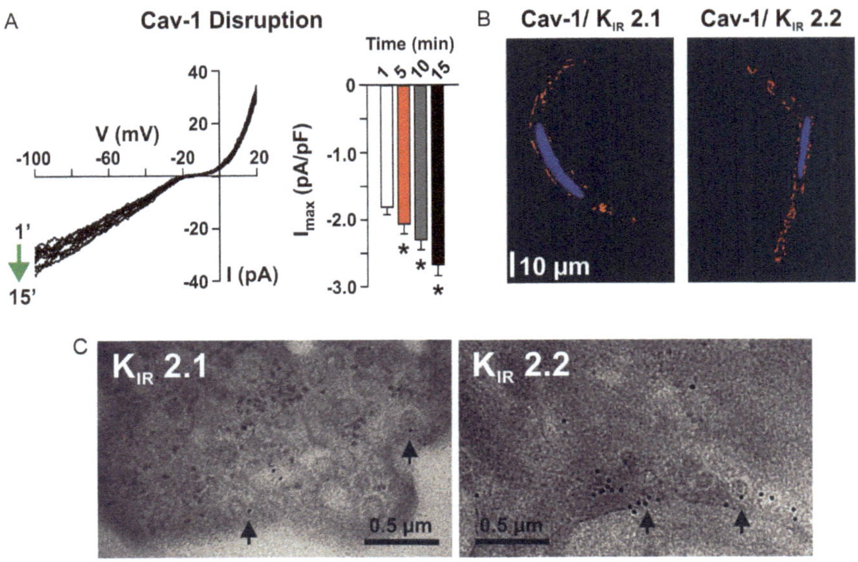

Fig. 9 Caveolin-1 is a negative regulator of cerebral arterial smooth muscle $K_{IR}2.1–2.2$ channels. (A) Representative traces and summary data of peak inward current (−100 mV) in native smooth muscle cells (B) over the 10 min dialysis period of Cav-1 scaffolding domain peptide (10 µM, $n = 6$, repeated measurements [RM]-ANOVA, Tukey post hoc analysis). Data are means ± SE and * significant difference at $P \leq 0.05$. (B) Protein ligation assay shows the presence and close apposition (⊠40 nm) of $K_{IR}2.x$ subunits and Cav-1 in isolated smooth muscle cells. (C) $K_{IR}2.x$ subunits are confined to the smooth muscle membrane in association with caveolae. *Modified from Sancho, M., Fabris, S., Hald, B. O., Brett, S. E., Sandow, S. L., Poepping, T. L., et al. (2019). Membrane lipid-KIR2.x channel interactions enable hemodynamic sensing in cerebral arteries.* Arteriosclerosis, Thrombosis, and Vascular Biology, 39, *1072–1087.*

activity as Cav-1 disruption (dialyzing a blocking peptide against Cav-1) resulted in a significant increase in $K_{IR}2.x$ activity. Further microscopy approaches confirmed that K_{IR} channel subunits reside in close association with smooth muscle caveolae (Sancho et al., 2019) (Fig. 9).

4.2 Hemodynamic forces and lipid-K_{IR} interactions

Smooth muscle and endothelial cells are the principal constituents of the vascular wall and they sense two critical hemodynamic stimuli (pressure and/or shear stress) to set basal tone and brain perfusion control (Fujiii, Heistad, & Faraci, 1991; Meininger & Davis, 1992). While flow induces dilation of small arteries or arterioles, intraluminal pressure causes vasoconstriction (Davis & Hill, 1999), and therefore an increase or decrease of K^+ conductance, respectively (Lu & Kassab, 2011). In this framework, several studies suggested $K_{IR}2.x$ channels act as cellular mechanosensitive elements, being involved in the early response of both mechanical-mediated responses (Hoger, Ilyin, Forsyth, & Hoger, 2002; Sancho et al., 2019; Wu et al., 2007).

4.2.1 Pressure

The study by Wu et al. (2007) was the first to indicate that cerebral arterial smooth muscle K_{IR} channels were mechanically sensitive, as channel inhibition was evident after hyposmotic challenge. Swelling stimuli stretch the cell membrane and have been reported to activate mechanically sensitive inward currents involved in myogenic tone development response (Ellershaw, Greenwood, & Large, 2002; Welsh et al., 2000; Yamazaki et al., 1998). Sancho et al. (2019) replicated these observations and further showed that reductions in cell stretch, by negative pipette pressurization via pneumatic transducer (−15 to −45 mmHg), induced a marked rise in K_{IR} conductance. These findings functionally imply that smooth muscle K_{IR} works in concert with pressure-sensitive transient receptor potential channels (Earley, Waldron, & Brayden, 2004; Welsh et al., 2002) to facilitate myogenic development in cerebral arteries, hypothesis that was subsequently verified as Ba^{2+}-induced constriction (used as an index of K_{IR} activity) was greater at low rather than high intravascular pressures. Linking the lipid microenvironment regulation of K_{IR} channels to hemodynamic sensing, we further demonstrated that cholesterol depletion (but not PIP_2) prevents pressure stimuli from inhibiting smooth muscle K_{IR} as well as facilitates Ba^{2+}-induced constriction at high intravascular pressures. These findings are consistent with cholesterol conferring pressure sensitivity to smooth muscle K_{IR} channels (Sancho et al., 2019) (Fig. 10).

4.2.2 Shear stress

An increase in flow (i.e., shear stress) provokes hyperpolarization of the endothelial cell membrane by increasing the open probability of K^+ channels, and subsequent vasodilation. In this sense, a rich observational base of studies using cell model systems identifies vascular endothelial $K_{IR}2.1$ channel as the major driver of the endothelial shear stress-induced K^+ response as it resulted blocked by extracellular Ba^{2+} (Ahn et al., 2017; Hoger et al., 2002; Lieu, Pappone, & Barakat, 2004; Olensen, Clapham, & Davies, 1988). A transcriptional remodeling of coronary endothelial $K_{IR}2.x$ channel subunits was also documented as a mechanism of cellular flow-adaptation (Kefaloyianni & Coetzee, 2011). With regard to native tissue, our group

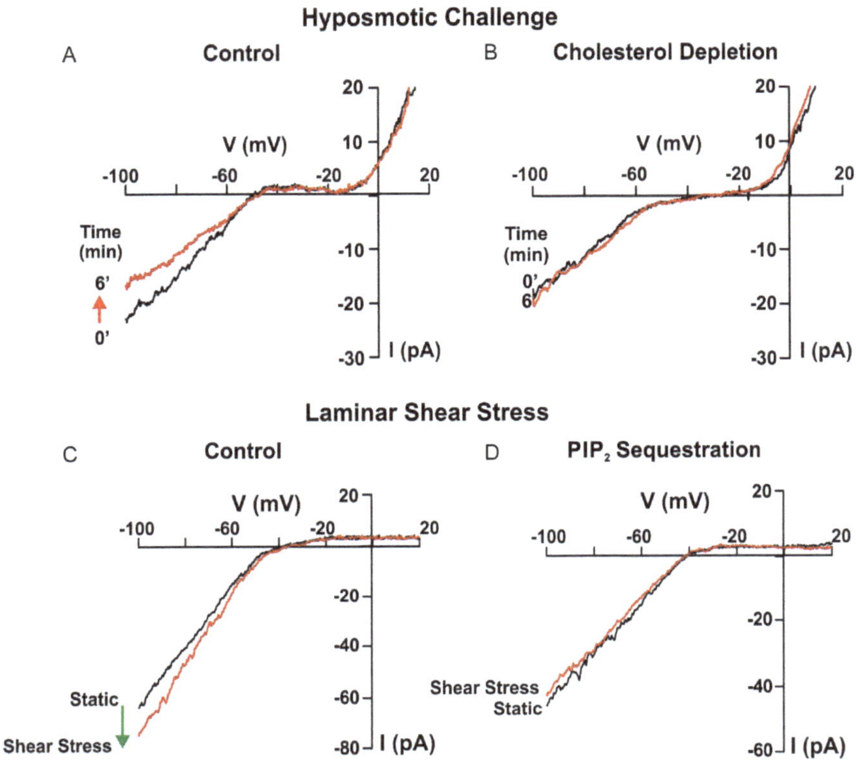

Fig. 10 Cholesterol confers pressure sensitivity to smooth muscle K_{IR} and PIP_2 enables endothelial K_{IR} to sense laminar shear stress. (A) Representative traces of smooth muscle K_{IR} currents (20 mM K^+ solution) before and after pressure simulation by hyposmotic challenge. (B) Cholesterol depletion by MβCD dialysis prevents pressure from suppressing smooth muscle K_{IR} channel activity. (C) Representative traces of endothelial K_{IR} currents (20 mM K^+ solution) prior to and following the induction of shear stress. (D) PIP_2 sequestration by dialyzing the cells with neomycin impairs shear stress-induced potentiation of endothelial K_{IR}. *Modified from Sancho, M., Fabris, S., Hald, B. O., Brett, S. E., Sandow, S. L., Poepping, T. L., et al. (2019). Membrane lipid-KIR2.x channel interactions enable hemodynamic sensing in cerebral arteries.* Arteriosclerosis, Thrombosis, and Vascular Biology, 39, *1072–1087.*

(Sancho et al., 2019) was the first to demonstrate that fluid shear stress potentiates cerebral vascular endothelial $K_{IR}2.x$ currents, and this potentiation facilitates shear stress-mediated vasodilation as intraluminal Ba^{2+} perfusion impaired this hemodynamic response. Furthermore, linking cellular mechanosensitivity to membrane lipid-protein interactions, we found that PIP_2 sequestration markedly decreases shear stress-induced potentiation of endothelial K_{IR} activity, and intraluminal perfusion of neomycin

(PIP_2 scavenger) efficiently abolished shear stress-mediated dilation of endothelium-intact cerebral arteries. This provocative finding suggests that PIP_2-endothelial K_{IR} interactions are essential for hemodynamic sensing of shear stress in the cerebral vasculature (Sancho et al., 2019) (Fig. 10). Alternatively, as PIP_2 binding is essential for sustaining K_{IR} activity, one could also consider cellular PIP_2 depletion as a brake on K_{IR} activation. Intriguingly, recent work by Levitan and colleagues has reported that the endothelial K_{IR} response to shear stress is markedly suppressed when membrane cholesterol is elevated (Fancher et al., 2018).

5. K_{IR} channels and disease

K_{IR} channels are impacted and involved in a number of pathophysiological disorders including hypertension, ischemia, obesity, diabetes, stress, hypertrophy and traumatic brain injury. It has been documented a decrease of K_{IR} current density and K_{IR}-mediated responses in cerebral arteries of hypertensive rats (McCarron & Halpern, 1990; Seo et al., 2014) as well as a downregulation of $K_{IR}2.1$ expression in mice mesenteric smooth muscle vascular cells (Tajada, Cidad, Moreno-Dominguez, Perez-Garcia, & Lopez-Lopez, 2012). Marrelli et al. (1998) reported vascular smooth muscle K_{IR} altered function after ischemia/reperfusion as isolated arteries no longer dilated in response to extracellular K^+ concentrations that activate K_{IR} channels. While little is known about the impact of obesity on K_{IR} channel function, endothelium-derived hyperpolarization was diminished in mesenteric arteries from diet-induced obese rats as a result of alterations in the underlying potassium channel signaling mechanisms (Haddock et al., 2011). Reduced K_{IR} channel activity was also reported in pial arterioles from diabetic rats as K^+-induced vasodilation were significantly decreased under this pathological condition (Mayhan et al., 2004). Longden, Dabertrand, Hill-Eubanks, Hammack, and Nelson (2014) observed loss of $K_{IR}2.1$ channel function in amygdala parenchymal arteriole myocytes after chronic stress and Park, Ko, et al. (2007) documented reduced K_{IR} channel function during hypertrophy. Significant new insights have been also obtained in elucidating whether endothelial $K_{IR}2.x$ channels are targeted in diet-induced hypercholesterolemia. In this regard, $K_{IR}2.x$ current suppression has been reported in aortic endothelial cells harvested from hypercholesterolemic pigs, an effect that could be reversed by depleting membrane cholesterol (Fang et al., 2006). Similar findings were reported in porcine bone marrow endothelial progenitor cells (Mohler et al., 2007). More interestingly, hypercholesterolemia provoked endothelial membrane depolarization, and markedly impaired shear stress-induced vasodilation (Fang et al., 2006). These findings are significant in that they suggest that K_{IR} current suppression is a major contributor in hypercholesterolemia-induced endothelial dysfunction and potentially arteriosclerotic vascular disease.

6. Conclusions and future directions

In cerebral arteries, $K_{IR}2.x$ resides in both smooth muscle and endothelium and is regulated by key lipid signaling molecules that sense the surrounding hemodynamic environment (Fig. 11). In detail, $K_{IR}2.x$ channels are uniquely modulated by PIP_2 and cholesterol depending on their cellular location; and these specific lipid-protein interactions confer shear stress- and

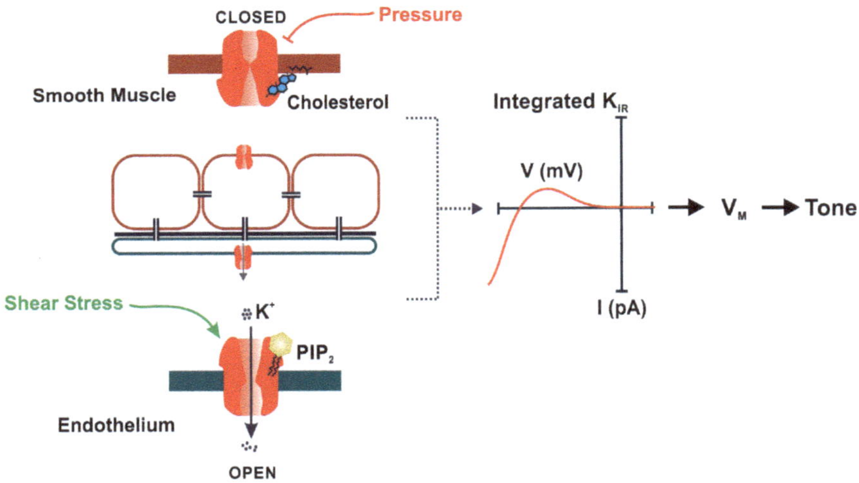

Fig. 11 Cerebrovascular K_{IR} channels and the facilitation of hemodynamic control. Schematic diagram illustrating the unique regulation of each pool of K_{IR} channels in cerebral arterial smooth muscle and endothelium. Membrane lipids such as PIP_2 and cholesterol are dominant regulators of K_{IR}, with the former prominently activating endothelial K_{IR} while the latter preferentially inhibits smooth muscle K_{IR}. These specific lipid-K_{IR} interactions help confer pressure and flow sensing, hemodynamic forces that impair and activate smooth muscle and endothelial K_{IR} channels, respectively. Both channel pools together drive an integrated K_{IR} current through which hemodynamic stimuli cooperate to control V_M and arterial tone. *Modified from Sancho, M., Fabris, S., Hald, B. O., Brett, S. E., Sandow, S. L., Poepping, T. L., et al. (2019). Membrane lipid-KIR2.x channel interactions enable hemodynamic sensing in cerebral arteries.* Arteriosclerosis, Thrombosis, and Vascular Biology, 39, *1072–1087.*

pressure-sensitivity to the endothelial and smooth muscle K_{IR} channels, respectively. These two channel subsets have been suggested to interact each other and drive an integrated vascular $K_{IR}2.x$ conductance that ultimately control V_M and basal tone in the cerebral circulation.

It is intriguing to consider translational implications in the context of dyslipidemia where plasma LDL-cholesterol is elevated. Under this condition, endothelial cells would be particularly exposed to higher LDL-cholesterol levels, a response that could alter lipid-protein interactions and promote K_{IR} channel stabilization to a preferred silent state. We hypothesize that with dyslipidemia a regulatory switch (PIP_2 to cholesterol) would reduce endothelial K_{IR} activity, impair shear stress-induced activation, and ultimately impact brain blood flow. This future approach could initiate the process of connecting K_{IR} to endothelial dysfunction prior to atherosclerosis development. This linkage may extend to humans and explain the diminished brachial artery flow-induced dilation in patients who suffer early cardiovascular disease (Anderson et al., 2011).

Acknowledgments

This work was supported by operating grants to Dr. Welsh from the Natural Science and Engineering Research Council of Canada and the Canadian Institute of Health Research. Dr. Welsh is the Rorabeck Chair in Neuroscience and Vascular Biology at the University of Western Ontario.

References

Aguado, C., Colon, J., Ciruela, F., Schlaudraff, F., Cabañero, M. J., Perry, C., et al. (2008). Cell type-specific subunit composition of G protein-gated potassium channels in the cerebellum. *Journal of Neurochemistry*, *105*, 497–511.

Ahn, S. J., Fancher, I. S., Bian, J. T., Zhang, C. X., Schwab, S., Gaffin, R., et al. (2017). Inwardly rectifying K^+ channels are major contributors to flow-induced vasodilation in resistance arteries. *The Journal of Physiology*, *595*, 2339–2354.

Amstrong, M. L., Dua, A. K., & Murrant, C. L. (2007). Potassium initiates vasodilatation induced by a single skeletal muscle contraction in hamster cremaster muscle. *The Journal of Physiology*, *581*, 841–852.

Anderson, T. J., Charbonneau, F., Title, L. M., Buithieu, J., Rose, M. S., Conradson, H., et al. (2011). Microvascular function predicts cardiovascular events in primary prevention: Long-term results from the firefighters and their endothelium (FATE) study. *Circulation*, *123*, 163–169.

Bichet, D., Haass, F. A., & Jan, L. Y. (2003). Merging functional studies with structures of inward-rectifier K^+ channels. *Nature Reviews Neuroscience*, *4*, 957–967.

Bonev, A. D., & Nelson, M. T. (1993). ATP-sensitive potassium channels in smooth muscle cells from guinea pig urinary bladder. *American Journal of Physiology*, *264*, C1190–C1200.

Bradley, K. K., Jaggar, J. H., Bonev, A. D., Heppner, T. J., Flynn, E. R., Nelson, M. T., et al. (1999). Kir2.1 encodes the inward rectifier potassium channel in rat arterial smooth muscle cells. *The Journal of Physiology*, *515*, 639–651.

Burns, W. R., Cohen, K. D., & Jackson, W. F. (2004). K^+-induced dilation of hamster cremasteric arterioles involves both the Na+/K+-ATPase and inward-rectifier K^+ channels. *Microcirculation*, *11*, 279–293.

Cao, C., Goo, J. H., Lee-Kwon, W., & Pallone, T. L. (2006). Vasa recta pericytes express a strong inward rectifier K^+ conductance. *American Journal of Physiology—Regulatory, Integrative and Comparative Physiology*, *290*, 1601–1607.

Chilton, L., & Loutzenhiser, R. (2001). Functional evidence of inward rectifier potassium channel current in rat afferent arterioles. *Circulation Research*, *88*, 152–158.

Chilton, L., Loutzenhiser, K., Morales, E., Breaks, J., Kargacin, G. J., & Loutzenhiser, R. (2008). Inward rectifier K(+) currents and Kir2.1 expression in renal afferent and efferent arterioles. *Journal of the American Society of Nephrology*, *19*, 69–76.

Crecelius, A. R., Richards, J. C., Luckasen, G. J., Larson, D. G., & Dinenno, F. A. (2013). Reactive hyperemia occurs via activation of inwardly rectifying potassium channels and Na+/K+-ATPase in humans. *Circulation Research*, *113*, 1023–1032.

Davis, M. J., & Hill, M. A. (1999). Signaling mechanisms underlying the vascular myogenic response. *Physiological Reviews*, *79*, 387–423.

Dawes, M., Sieniawska, C., Delves, T., Dwivedi, R., Chowienczyk, P. J., & Ritter, J. M. (2002). Barium reduces resting blood flow and inhibits potassium-induced vasodilation in the human forearm. *Circulation*, *105*, 1323–1328.

Dhamoon, A. S., Pandit, S. V., Sarmast, F., Parisian, K. R., Guha, P., Li, Y., et al. (2004). Unique Kir2.1 properties determine regional and species in the cardiac inward rectifier K^+ current. *Circulation Research*, *94*, 1332–1339.

Drouin, A., & Thorin, E. (2009). Flow-induced dilation is mediated by Akt-dependent activation of endothelial nitric oxide synthase-derived hydrogen peroxide in mouse cerebral arteries. *Stroke*, *40*, 1827–1833.

Du, X., Zhang, H., Lopes, C., Mirshahi, T., Rohacs, T., & Logothetis, D. E. (2004). Characteristic interactions with phosphatidylinositol 4,5-bisphosphate determine regulation of Kir channels by diverse modulator. *The Journal of Biological Chemistry*, *279*, 37271–37281.

Earley, S., Waldron, B. J., & Brayden, J. E. (2004). Critical role for transient receptor potential channel TRPM4 in myogenic constriction of cerebral arteries. *Circulation Research*, *95*, 922–929.

Edwards, G., Dora, K. A., Gardener, M. J., Garland, C. J., & Weston, A. H. (1998). K^+ is an endothelium-derived hyperpolarizing factor in rat arteries. *Nature*, *396*, 269–272.

Edwards, F. R., Hirst, G. D., & Silverberg, G. D. (1988). Inward rectification in rat cerebral arterioles; involvement of potassium ions in autoregulation. *The Journal of Physiology*, *404*, 455–466.

Ellershaw, D. C., Greenwood, I. A., & Large, W. A. (2002). Modulation of volume-sensitive chloride current by noradrenaline in rabbit portal vein myocytes. *The Journal of Physiology*, *542*, 537–547.

Enkvetchakul, D., Jeliazkova, I., & Nichols, C. G. (2005). Direct modulation of Kir channel gating by membrane phosphatidylinositol 4,5-bisphosphate. *The Journal of Biological Chemistry*, *280*, 35785–35788.

Epshtein, Y., Chopra, A. P., Rosenhouse-Dantsker, A., Kowalsky, G. B., Logothetis, D. E., & Levitan, I. (2009). Identification of a C-terminus domain critical for sensitivity of Kir2.1 to cholesterol. *Proceedings of the National Academy of Sciences of the United States of America*, *106*, 8055–8060.

Fancher, I. S., Ahn, S. J., Adamos, C., Osborn, C., Oh, M.-J., Fang, Y., et al. (2018). Hypercholesterolemia-induced loss of flow-induced vasodilation and lesion formation in apolipoprotein E-deficient mice critically depend on inwardly rectifying K^+ channels. *Journal of the American Heart Association, 7,* e007430.

Fang, Y., Mohler, E. R., 3rd., Hsieh, E., Osman, H., Hashemi, S. M., Davies, P. F., et al. (2006). Hypercholesterolemia suppresses inwardly rectifying K^+ channels in aortic endothelium in vitro and in vivo. *Circulation Research, 98,* 1064–1071.

Fang, Y., Schram, G., Romanenko, V. G., Shi, C., Conti, L., Vandenberg, C. A., et al. (2005). Functional expression of Kir2.x in human aortic endothelial cells: The dominant role of Kir2.2. *American Journal of Physiology-Cell Physiology, 289,* C1134–C1144.

Fielding, C. J., & Fielding, P. E. (1995). Molecular physiology of reverse cholesterol transport. *The Journal of Lipid Research, 36,* 211–228.

Filosa, J. A., Bonev, A. D., Straub, S. V., Meredith, A. L., Wilkerson, M. K., Aldrich, R. W., et al. (2006). Local potassium signaling couples neuronal activity to vasodilation in the brain. *Nature Neuroscience, 9,* 1397–1403.

Fujiii, K., Heistad, D. D., & Faraci, F. M. (1991). Flow-mediated dilatation of the basilar artery in vivo. *Circulation Research, 69,* 697–705.

Fujita, A., Horio, Y., Higashi, K., Mouri, T., Hata, F., Takeguchi, N., et al. (2002). Specific localization of an inwardly rectifying K(+) channel, Kir4.1, at the apical membrane of rat gastric parietal cells; its possible involvement in K(+) recycling for the H(+)-K(+)-pump. *The Journal of Physiology, 540,* 85–92.

Garcia-Roldan, J. L., & Bevan, J. A. (1990). Flow-induced constriction and dilation of cerebral resistance arteries. *Circulation Research, 66,* 1445–1448.

Gimpi, G., Burger, K., & Fahrenholz, F. (1997). Cholesterol as modulator of receptor function. *Biochemistry, 36,* 10959–10974.

Gollasch, M., & Nelson, M. T. (1997). Voltage-dependent Ca2+ channels in arterial smooth muscle cells. *Kidney and Blood Pressure Research, 20,* 355–371.

Goto, K., Rummery, N. M., Grayson, T. H., & Hill, C. E. (2004). Attenuation of conducted vasodilation in rat mesenteric arteries during hypertension: Role of inwardly rectifying potassium channels. *The Journal of Physiology, 561,* 215–231.

Haddock, R. E., Grayson, T. H., Morris, M. J., Howitt, L., Chadha, P. S., & Sandow, S. L. (2011). Diet-induced obesity impairs endothelium-derived hyperpolarization via altered potassium channel signaling mechanisms. *PLoS One, 6,* e16423.

Han, H., Rosenhouse-Dantsker, A., Gnanasambandam, R., Epshtein, Y., Chen, Z., Sachs, F., et al. (2014). Silencing of Kir2 channels by caveolin-1: Cross-talk with cholesterol. *The Journal of Physiology, 592,* 4025–4038.

Hansen, S. B., Tao, X., & MacKinnon, R. (2011). Structural basis of PIP_2 activation of the classical inward rectifier K^+ channel Kir2.2. *Nature, 477,* 495–498.

Harder, D. R., Alkayed, N. J., Lange, A. R., Gebremedhin, D., & Roman, R. J. (1998). Functional hyperemia in the brain: Hypothesis for astrocyte-derived vasodilator metabolites. *Stroke, 29,* 229–234.

Harraz, O. F., Longden, T. A., Dabertrand, F., Hill-Eubanks, D., & Nelson, M. T. (2018). Endothelial GqPCR activity controls capillary electrical signaling and brain blood flow through PIP_2 depletion. *Proceedings of the National Academy of Sciences of the United States of America, 115,* 3569–3577.

Harraz, O. H., Longden, T. A., Hill-Eukbanks, D., & Nelson, M. T. (2018). PIP_2 depletion promotes TRPV4 channel activity in mouse brain capillary endothelial cells. *eLife, 7,* e38689.

Hearon, C. M., Richards, J. C., Racine, M. L., Luckasen, G. J., Larson, D. G., & Dinenno, F. A. (2019). Amplification of endothelium-dependent vasodilation in contracting human skeletal muscle: Role of KIR channels. *The Journal of Physiology, 597,* 1321–1335.

Hilgemann, D. W. (2007). Local PIP(2) signals: When, where, and how? *Pflugers Archiv: European Journal of Physiology, 455,* 55–67.

Hoger, J. H., Ilyin, V. I., Forsyth, S., & Hoger, A. (2002). Shear stress regulates the endothelial Kir2.1 ion channel. *Proceedings of the National Academy of Sciences of the United States of America*, *99*, 7780–7785.
Huang, C. L., Feng, S., & Hilgemann, D. W. (1998). Direct activation of inward rectifier potassium channels by PIP2 and its stabilization by Gbetagamma. *Nature*, *391*, 803–806.
Hughes, B. A., Kumar, G., Yuan, Y., Swaminathan, A., Yan, D., Sharma, A., et al. (2000). Cloning and functional expression of human retinal Kir2.4, a pH-sensitive inwardly rectifying K(+) channel. *American Journal of Physiology. Cell Physiology*, *279*, C771–C784.
Inagaki, N., Gonoi, T., Clement, J. P., Namba, N., Inazawa, J., Gonzalez, G., et al. (1995). Reconstitution of IKATP: An inward rectifier subunit plus the sulfonylurea receptor. *Science*, *270*(5239), 1166–1170.
Inagaki, N., Tsuura, Y., Namba, N., Masuda, K., Gonoi, T., Horie, M., et al. (1995). Cloning and functional characterization of a novel ATP-sensitive potassium channel ubiquitously expressed in rat tissues, including pancreatic islets, pituitary, skeletal muscle, and heart. *The Journal of Biological Chemistry*, *270*, 5691–5694.
Inanobe, A., Horio, Y., Fujita, A., Tanemoto, M., Hibino, H., Inageda, K., et al. (1999). Molecular cloning and characterization of a novel splicing variant of the Kir3.2 subunit predominantly expressed in mouse testis. *The Journal of Physiology*, *521*, 19–30.
Inanobe, A., Yoshimoto, Y., Horio, Y., Morishige, K. I., Hibino, H., Matsumoto, S., et al. (1999). Characterization of G-protein-gated K^+ channels composed of Kir3.2 subunits in dopaminergic neurons of the substantia nigra. *Journal of Neuroscience*, *19*, 1006–1017.
Ishii, M., Fujita, A., Iwai, K., Kusaka, S., Higashi, K., Inanobe, A., et al. (2003). Differential expression and distribution of Kir5.1 and Kir4.1 inwardly rectifying K^+ channels in retina. *American Journal of Physiology-Cell Physiology*, *285*, C260–C267.
Jantzi, M. C., Brett, S. E., Jackson, W. F., Corteling, R. L., Vigmond, E. J., & Welsh, D. G. (2006). Inward rectifying potassium channels facilitate cell-to-cell communication in hámster retractor muscle feed arteries. *American Journal of Physiology—Heart and Circulatory Physiology*, *291*, H1319–H1328.
Kane, G. C., Lam, C. F., O'Cochlain, F., Hodgson, D. M., Reyes, S., Liu, X. K., et al. (2006). Gene knockout of the KCNJ8-encoded Kir6.1 K(ATP) channel imparts fatal susceptibility to endotoxemia. *The FASEB Journal*, *20*, 2271–2280.
Katz, B. (1949). Les Constantes electriques de la membrane du muscle. *Archives des Science Physiologiques*, *3*, 285–299.
Kefaloyianni, E., & Coetzee, W. A. (2011). Transcriptional remodeling of ion channel subunits by flow adaptation in human coronary artery endothelial cells. *Journal of Vascular Research*, *48*, 357–367.
Knot, H. J., & Nelson, M. T. (1998). Regulation of arterial diameter and wall [Ca2+] in cerebral arteries of rat by membrane potential and intravascular pressure. *The Journal of Physiology*, *508*, 199–209.
Knot, H. J., Zimmermann, P. A., & Nelson, M. T. (1996). Extracellular K(+)-induced hyperpolarizations and dilations of rat coronary and cerebral arteries involve inward rectifier K(+) channels. *The Journal of Physiology*, *492*, 419–430.
Krapivinsky, G., Medina, I., Eng, L., Krapivinsky, L., Yang, Y., & Clapham, D. E. (1998). A novel inward rectifier K^+ channel with unique pore properties. *Neuron*, *20*, 995–1005.
Kubo, Y., Baldwin, T. J., Jan, Y. N., & Jan, L. Y. (1993). Primary structure and functional expression of a mouse inward rectifier potassium channel. *Nature*, *362*, 127–133.
Kubo, Y., & Murata, Y. (2001). Control of rectification and permeation by two distinct sites after the second transmembrane region in Kir2.1 K^+ channel. *The Journal of Physiology*, *531*, 645–660.
Kubo, Y., Reuveny, E., Slesinger, P. A., Jan, Y. N., & Jan, L. Y. (1993). Primary structure and functional expression of a rat G-protein-coupled muscarinic potassium channel. *Nature*, *364*(6440), 802–806.

Lange, Y., Swaisgood, M. H., Ramos, B. V., & Steck, T. L. (1989). Plasma membranes contain half the phospholipid and 90% of the cholesterol and sphingomyelin in cultured human fibroblasts. *The Journal of Biological Chemistry*, *264*, 3786–3793.

Ledoux, J., Greenwood, I., Villeneuve, L. R., & Leblanc, N. (2003). Modulation of Ca^{2+}-dependent Cl^- channels by calcineurin in rabbit coronary arterial myocytes. *The Journal of Physiology*, *552*, 701–714.

Lee, W. S., & Hebert, S. C. (1995). ROMK inwardly rectifying ATP-sensitive K^+ channel. I. Expression in rat distal nephron segments. *American Journal of Physiology*, *268*, 1124–1131.

Lesage, F., Duprat, F., Fink, M., Guillemare, E., Coppola, T., Lazdunski, M., et al. (1994). Cloning provides evidence for a family of inward rectifier and G-protein coupled K^+ channels in the brain. *FEBS Letters*, *353*, 37–42.

Lieu, D. K., Pappone, P. A., & Barakat, A. I. (2004). Differential membrane potential and ion current responses to different types of shear stress in vascular endothelial cells. *American Journal of Physiology—Cell Physiology*, *286*, 1367–1375.

Liu, G. X., Derst, C., Schlichthorl, G., Heinen, S., Seebohm, G., Bruggemann, A., et al. (2001). Comparison of cloned Kir2 channels with native inward rectifier K^+ channels from guinea-pig cardiomyocytes. *The Journal of Physiology*, *532*, 115–126.

Liu, Y., McKenna, E., Figueroa, D. J., Blevins, R., Austin, C. P., Bennett, P. B., et al. (2000). The human inward rectifier K(+) channel subunit kir5.1 (KCNJ16) maps to chromosome 17q25 and is expressed in kidney and pancreas. *Cytogenetics and Cell Genetics*, *90*, 60–63.

Loeb, A. L., Godeny, I., & Longnecker, D. E. (2000). Functional evidence for inward-rectifier potassium channels in rat cremaster muscle arterioles. *Microvascular Research*, *59*, 1–6.

Longden, T. A., Dabertrand, F., Hill-Eubanks, D. C., Hammack, S. E., & Nelson, M. T. (2014). Stress-induced glucocorticoid signaling remodels neurovascular coupling through impairment of cerebrovascular inwardly rectifying K^+ channel function. *Proceedings of the National Academy of Sciences of the United States of America*, *111*, 7462–7477.

Longden, T. A., Dabertrand, F., Koide, M., Gonzalez, A. L., Tykocki, N. R., Brayden, J. E., et al. (2017). Capillary K^+-sensing initiates retrograde hyperpolarization to increase local cerebral blood flow. *Nature Neuroscience*, *20*, 717–726.

Lopatin, A. N., Makhina, E. N., & Nichols, C. G. (1994). Potassium channel block by cytoplasmic polyamines as the mechanism of intrinsic rectification. *Nature*, *372*, 366–369.

Lopatin, A. N., & Nichols, C. G. (1996). [K] dependence of polyamine-induced rectification in inward rectifier potassium channels (IRK1, Kir2.1). *Journal of General Physiology*, *108*, 105–113.

Lu, Z. (2004). Mechanism of rectification in inward-rectifier K^+ channels. *Annual Review of Physiology*, *66*, 103–129.

Lu, D., & Kassab, G. S. (2011). Role of shear stress and stretch in vascular mechanobiology. *Journal of the Royal Society Interface*, *8*, 1379–1385.

Maguy, A., Hebert, T. E., & Nattel, S. (2006). Involvement of lipid rafts and caveolae in cardiac ion channel function. *Cardiovascular Research*, *69*, 798–807.

Marrelli, S. P., Johnson, T. D., Khorovets, A., Childres, W. F., & Bryan, R. M., Jr. (1998). Altered function of inward rectifier potassium channels in cerebrovascular smooth muscle after ischemia reperfusion. *Stroke*, *29*, 1469–1474.

Matsuda, H., Oishi, K., & Omori, K. (2003). Voltage-dependent gating and block by internal spermine of the murine inwardly rectifying K^+ channel, Kir2.1. *The Journal of Physiology*, *548*, 361–371.

Mayhan, W. G., Mayhan, J. F., Sun, H., & Patel, K. P. (2004). In vivo properties of potassium channels in cerebral blood vessels during diabetes mellitus. *Microcirculation*, *11*, 605–613.

McCarron, J. G., & Halpern, W. (1990). Impaired potassium-induced dilation in hypertensive rat cerebral arteries does not reflect altered Na+, K(+)-ATPase dilation. *Circulation Research*, *67*, 1035–1039.

McLaughlin, S., Wang, J., Gambhir, A., & Murray, D. (2002). PIP(2) and proteins: Interactions, organization and information flow. *Annual Review of Biophysics and Biomolecular Structure*, *31*, 151–175.

Meininger, G. A., & Davis, M. J. (1992). Cellular mechanisms involved in the vascular myogenic response. *American Journal of Physiology*, *263*, H647–H659.

Miki, T., Nagashima, K., Tashiro, F., Kotake, K., Yoshitomi, H., Tamamoto, A., et al. (1998). Defective insulin secretion and enhanced insulin action in KATP channel-deficient mice. *Proceedings of the National Academy of Sciences of the United States of America*, *95*, 10402–10406.

Mohler, E. R., Fang, Y., Shaffer, R. G., Moore, J., Wilensky, R. L., Parmacek, M., et al. (2007). Hypercholesterolemia suppresses Kir channels in porcine bone marrow progenitor cells in vivo. *Biochemical and Biophysical Research Communications*, *358*, 317–324.

Nakamura, N., Suzuki, Y., Sakuta, H., Ookata, K., Kawahara, K., & Hirose, S. (1999). Inwardly rectifying K^+ channel Kir7.1 is highly expressed in thyroid follicular cells, intestinal epithelial cells and choroid plexus epithelial cells: Implication for a functional coupling with Na+,K+-ATPase. *Biochemical Journal*, *342*, 329–336.

Nelson, M. T., & Quayle, J. M. (1995). Physiological roles and properties of potassium channels in arterial smooth muscle. *American Journal of Physiology*, *268*, C799–C822.

Ohira, M., Seki, N., Nagase, T., Suzuki, E., Nomura, N., Ohara, O., et al. (1997). Gene identification in 1.6-Mb region of the Down syndrome region on chromosome 21. *Genome Research*, 7, 47–58.

Ohvo-Rekila, H., Ramstedt, B., Leppimaki, P., & Slotte, J. P. (2002). Cholesterol interactions with phospholipids in membranes. *Progress in Lipid Research*, *41*, 66–97.

Olensen, S. P., Clapham, D. E., & Davies, P. F. (1988). Haemodynamic shear stress activates a K^+ current in vascular endothelial cells. *Nature*, *331*, 168–170.

Olsen, O., Liu, H., Wade, J. B., Merot, J., & Welling, P. A. (2002). Basolateral membrane expression of the Kir 2.3 channel is coordinated by PDZ interaction with Lin-7/CASK complex. *American Journal of Physiology-Cell Physiology*, *282*, C183–C195.

Park, W. S., Kang, S. H., Son, Y. K., Kim, N. R., Ko, J. H., Kim, H. K., et al. (2007). The mitochondrial Ca2+-activated K^+ channel activator, NS 1619 inhibits L-type Ca2+ channels in rat ventricular myocytes. *Biochemical and Biophysical Research Communications*, *362*, 31–36.

Park, W. S., Ko, J. H., Son, Y. K., Kang, S. H., Warda, M., Jung, I. D., et al. (2007). Increased inhibition of inward rectifier K^+ channels by angiotensin II in small-diameter coronary artery of isoproterenol-induced hypertrophied model. *Arteriosclerosis, Thrombosis and Vascular Biology*, *27*, 1768–1775.

Partiseti, M., Collura, V., Agnel, M., Culouscou, J. M., & Graham, D. (1998). Cloning and characterization of a novel human inwardly rectifying potassium channel predominantly expressed in small intestine. *FEBS Letters*, *434*, 171–176.

Périer, F., Radeke, C. M., & Vandenberg, C. A. (1994). Primary structure and characterization of a small-conductance inwardly rectifying potassium channel from human hippocampus. *Proceedings of the National Academy of Sciences of the United States of America*, *91*, 6240–6244.

Perillan, P. R., Li, X., Potts, E. A., Chen, M., Bredt, D. S., & Simard, J. M. (2000). Inward rectifier K(+) channel Kir2.3 (IRK3) in reactive astrocytes from adult rat brain. *Glia*, *31*, 181–192.

Prior, H. M., Webster, N., Quinn, K., Beech, D. J., & Yates, M. S. (1998). K(+)-induced dilation of a small renal artery: No role for inward rectifier K^+ channels. *Cardiovascular Research*, *37*, 780–790.

Quayle, J. M., Dart, C., & Standen, N. B. (1996). The properties and distribution of inward rectifier potassium currents in pig coronary arterial smooth muscle. *The Journal of Physiology*, *494*, 715–726.

Quayle, J. M., McCarron, J. G., Brayden, J. E., & Nelson, M. T. (1993). Inward rectifier K^+ currents in smooth muscle cells from resistance-sized cerebral arteries. *American Journal of Physiology-Cell Physiology*, *265*, C1363–C1370.

Quayle, J. M., Nelson, M. T., & Standen, N. B. (1997). ATP-sensitive and inwardly rectifying potassium channels in smooth muscle. *Physiological Reviews*, 77, 1165–1232.

Raab-Graham, K. F., & Vandenberg, C. A. (1998). Tetrameric subunit structure of the native brain inwardly rectifying potassium channel Kir2.2. *Journal of Biological Chemistry*, *273*, 19699–19707.

Reinhart, M. P., Billheimer, J. T., Faust, J. R., & Gaylor, J. L. (1987). Subcellular localization of the enzymes of cholesterol biosynthesis and metabolism in rat liver. *The Journal of Biological Chemistry*, *262*, 9649–9655.

Robertson, B. E., Bonev, A. D., & Nelson, M. T. (1996). Inward rectifier K^+ currents in smooth muscle cells from rat coronary arteries: Block by Mg^{2+}, Ca^{2+}, and Ba^{2+}. *American Journal of Physiology*, *271*, H696–H705.

Rohacs, T., Chen, J., Prestwich, G. D., & Logothetis, D. E. (1999). Distinct specificities of inwardly rectifying K(+) channels for phosphoinositides. *The Journal of Biological Chemistry*, *274*, 36065–36072.

Rohacs, T., Lopes, C. M., Jin, T., Ramdya, P. P., Molnár, Z., & Logothetis, D. E. (2003). Specificity of activation by phosphoinositides determines lipid regulation of Kir channels. *Proceedings of the National Academy of Sciences of the United States of America*, *100*, 745–750.

Romanenko, V. G., Fang, Y., Byfield, F., Travis, A. J., Vandenberg, C. A., Rothblat, G. H., et al. (2004). Cholesterol sensitivity and lipid raft targeting of Kir2.1 channels. *Biophysical Journal*, *87*, 3850–3861.

Romanenko, V. G., Rothblat, G. H., & Levitan, I. (2002). Modulation of endothelial inward-rectifier K^+ current by optical isomers of cholesterol. *Biophysical Journal*, *83*, 3211–3222.

Rosenhouse-Dantsker, A., Logothetis, D. E., & Levitan, I. (2011). Cholesterol sensitivity of KIT2.1 is controlled by a belt of residues around the cytosolic pore. *Biophysical Journal*, *100*, 381–389.

Rosenhouse-Dantsker, A., Noskov, S., Durdagi, S., Logothetis, D. E., & Levitan, I. (2013). Identification of novel cholesterol-binding regions in Kir2 channels. *The Journal of Biological Chemistry*, *288*, 31154–31164.

Rosenhouse-Dantsker, A., Noskov, S., Logothetis, D. E., & Levitan, I. (2013). Cholesterol sensitivity of KIR2.1 depends on functional inter-links between the N and C termini. *Channels (Austin, Texas)*, 7, 303–312.

Sancho, M., Fabris, S., Hald, B. O., Brett, S. E., Sandow, S. L., Poepping, T. L., et al. (2019). Membrane lipid-K_{IR}2.x channel interactions enable hemodynamic sensing in cerebral arteries. *Arteriosclerosis, Thrombosis, and Vascular Biology*, *39*, 1072–1087.

Sancho, M., Samson, N. C., Hald, B. O., Hashad, A. M., Marrelli, S. P., Brett, S. E., et al. (2017). K_{IR} channels tune electrical communication in cerebral arteries. *Journal of Cerebral Blood Flow & Metabolism*, *37*, 2171–2184.

Schram, G., Pourrier, M., Wang, Z., White, M., & Nattel, S. (2003). Barium block of Kir2 and human cardiac inward rectifier currents: Evidence for subunit-heteromeric contribution to native currents. *Cardiovascular Research*, *59*, 328–338.

Segal, S. S. (2000). Integration of blood flow control to skeletal muscle: Key role of feed arteries. *Acta Physiologica Scandinavica*, *168*, 511–518.

Segal, S. S., & Duling, B. R. (1986). Flow control among microvessels coordinated by intercellular conduction. *Science*, *234*, 868–870.

Seo, E. Y., Kim, H. J., Zhao, Z. H., Jang, J. H., Jin, C. Z., Yoo, H. Y., et al. (2014). Low K^+ current in arterial myocytes with impaired K^+-vasodilation and its recovery by exercise in hypertensive rats. *Pflugers Archiv—European Journal of Physiology*, *466*, 2101–2111.

Shih, A. Y., Driscoll, J. D., Drew, P. J., Nishimura, N., Schaffer, C. B., & Kleinfeld, D. (2012). Two-photon microscopy as a tool to study blood flow and neurovascular coupling in the rodent brain. *Journal of Cerebral Blood Flow & Metabolism*, *32*, 1277–1309.

Shuck, M. E., Bock, J. H., Benjamin, C. W., Tsai, T. D., Lee, K. S., Slightom, J. L., et al. (1994). Cloning and characterization of multiple forms of the human kidney ROM-K potassium channel. *The Journal of Biological Chemistry*, *269*, 24261–24270.

Shuck, M. E., Piser, T. M., Bock, J. H., Slightom, J. L., Lee, K. S., & Bienkowski, M. J. (1997). Cloning and characterization of two K^+ inward rectifier (Kir) 1.1 potassium channel homologs from human kidney (Kir1.2 and Kir1.3). *The Journal of Biological Chemistry*, *272*, 586–593.

Si, M. L., & Lee, T. J. (2002). Alpha7-nicotinic acetylcholine receptors on cerebral perivascular sympathetic nerves mediate choline-induced nitrergic neurogenic vasodilation. *Circulation Research*, *91*, 62–69.

Singh, D. K., Rosenhouse-Dantsker, A., Nichols, C. G., Enkvetchakul, D., & Levitan, I. (2009). Direct regulation of prokaryotic Kir channel by cholesterol. *The Journal of Biological Chemistry*, *284*, 30727–30736.

Smith, P. D., Brett, S. E., Luykenaar, K. D., Sandow, S. L., Marrelli, S. P., Vigmond, E. J., et al. (2008). KIR channels function as electrical amplifiers in rat vascular smooth muscle. *The Journal of Physiology*, *586*, 1147–1160.

Somlyo, A. P., & Somlyo, A. V. (1994). Signal transduction and regulation in smooth muscle. *Nature*, *372*, 231–236.

Somlyo, A. P., & Somlyo, A. V. (2003). Ca^{2+} sensitivity of smooth muscle and nonmuscle myosin II: Modulated by G proteins, kinases, and myosin phosphatase. *Physiological Reviews*, *83*, 1325–1358.

Sonkusare, S. K., Dalsgaard, T., Bonev, A. D., & Nelson, M. T. (2016). Inward rectifier potassium (Kir2.1) channels as end-stage boosters of endothelium-dependent vasodilators. *The Journal of Physiology*, *594*, 3271–3285.

Stoffel, M., Tokuyama, Y., Trabb, J. B., German, M. S., Tsaar, M. L., Jan, L. Y., et al. (1995). Cloning of rat KATP-2 channel and decreased expression in pancreatic islets of male Zucker diabetic fatty rats. *Biochemical and Biophysical Research Communications*, *212*, 894–899.

Sward, K., Mita, M., Wilson, D. P., Deng, J. T., Susnjar, M., & Walsh, M. P. (2003). The role of RhoA and Rho-associated kinase in vascular smooth muscle contraction. *Current Hypertension Reports*, *5*, 66–72.

Tajada, S., Cidad, P., Moreno-Dominguez, A., Perez-Garcia, M. T., & Lopez-Lopez, J. R. (2012). High blood pressure associates with the remodeling of inward rectifier K^+ channels in mice mesenteric vascular smooth muscle cells. *The Journal of Physiology*, *590*, 6075–6091.

Tao, X., Avalos, J. L., Chen, J., & MacKinnon, R. (2009). Crystal structure of the eukaryotic strong inward-rectifier K^+ channel Kir2.2 at 3.1 A resolution. *Science*, *326*, 1668–1674.

Tennant, B. P., Cui, Y., Tinker, A., & Clapp, L. H. (2006). Functional expression of inward rectifier potassium channels in cultured human pulmonary smooth muscle cells: Evidence for a major role of Kir2.4 subunits. *The Journal of Membrane Biology*, *213*, 19–29.

Tucker, S. J., Imbrici, P., Salvatore, L., D'Adamo, M. C., & Pessia, M. (2000). pH dependence of the inwardly rectifying potassium channel, Kir5.1, and localization in renal tubular epithelia. *The Journal of Biological Chemistry*, *275*, 16404–16407.

Vaughn, J., Wolford, J. K., Prochazka, M., & Permana, P. A. (2000). Genomic structure and expression of human KCNJ9 (Kir3.3/GIRK3). *Biochemical and Biophysical Research Communications*, *274*, 302–309.

Welsh, D. G., Morielli, A. D., Nelson, M. T., & Brayden, J. E. (2002). Transient receptor potential changes regulate myogenic tone of resistance arteries. *Circulation Research*, *90*, 248–250.

Welsh, D. G., Nelson, M. T., Eckman, D. M., & Brayden, J. E. (2000). Swelling-activated cation channels mediate depolarization of rat cerebrovascular smooth muscle by hyposmolarity and intravascular pressure. *The Journal of Physiology*, *527*, 139–148.

Whorton, M. R., & MacKinnon, R. (2013). X-ray structure of the mammalian GIRK2-betagamma G-protein complex. *Nature*, *498*, 190–197.

Wickman, K., Karschin, C., Karschin, A., Picciotto, M. R., & Clapham, D. E. (2000). Brain localization and behavioral impact of the G-protein-gated K^+ channel subunit GIRK4. *Journal of Neuroscience*, *20*, 5608–5615.

Wu, B. N., Luykenaar, K. D., Brayden, J. E., Giles, W. R., Corteling, R. L., Wiehler, W. B., et al. (2007). Hyposmotic challenge inhibits inward rectifying K^+ channels in cerebral arterial smooth muscle cells. *American Journal of Physiology—Heart and Circulatory Physiology*, *292*, H1085–H1094.

Xie, L.-H., John, S. A., Ribalet, B., & Weiss, J. N. (2008). Phosphatidylinositol-4,5-bisphosphate (PIP_2) regulation of strong inward rectifier Kir2.1 channels: Multilevel positive cooperativity. *The Journal of Physiology*, *586*, 1833–1848.

Yamazaki, J., Duan, D., Janiak, R., Kuenzli, K., Horowitz, B., & Hume, J. R. (1998). Functional and molecular expression of volume-regulated chloride channels in canine vascular smooth muscle cells. *The Journal of Physiology*, *507*, 729–736.

Yang, J., Jan, Y. N., & Jan, L. Y. (1995). Determination of the subunit stoichiometry of an inwardly rectifying potassium channel. *Neuron*, *15*, 1441–1447.

Yang, D., Zhang, X., & Hughes, B. A. (2008). Expression of inwardly rectifying potassium channel subunits in native human retinal pigment epithelium. *Experimental Eye Research*, *87*, 176–183.

Yeagle, P. L. (1985). Cholesterol and the cell membrane. *Biochimica et Biophysica Acta*, *822*, 267–287.

Yoshimoto, Y., Fukuyama, Y., Horio, Y., Inanobe, A., Gotoh, M., & Kurachi, Y. (1999). Somatostatin induces hyperpolarization in pancreatic islet alpha cells by activating a G protein-gated K^+ channel. *FEBS Letters*, *444*, 265–269.

Zhang, H., He, C., Yan, X., Mirshahi, T., & Logothetis, D. E. (1999). Activation of inwardly rectifying K^+ channels by distinct PtdIns(4,5)P_2 interactions. *Nature Cell Biology*, *1*, 183–188.

CHAPTER NINE

Ion channels in capillary endothelium

Daniela C.G. Garcia, Thomas A. Longden*
Department of Physiology, School of Medicine, University of Maryland, Baltimore, MD, United States
*Corresponding author: e-mail address: thomas.longden@som.umaryland.edu

Contents

Current Topics in Membranes, Volume 85
ISSN 1063-5823
https://doi.org/10.1016/bs.ctm.2020.01.005

Abstract

Vascular beds are anatomically and functionally compartmentalized into arteries, capillaries, and veins. The bulk of the vasculature consists of the dense, anastomosing capillary network, composed of capillary endothelial cells (cECs) that are intimately associated with the parenchyma. Despite their abundance, the ion channel expression and function and Ca^{2+} signaling behaviors of capillaries have only recently begun to be explored in detail. Here, we discuss the established and emerging roles of ion channels and Ca^{2+} signaling in cECs. By mining a publicly available RNA-seq dataset, we outline the wide variety of ion channel genes that are expressed in these cells, which potentially imbue capillaries with a broad range of sensing and signal transduction capabilities. We also underscore subtle but critical differences between cEC and arteriolar EC ion channel expression that likely underlie key functional differences in ECs at these different levels of the vascular tree. We focus our discussion on the cerebral vasculature, but the findings and principles being elucidated in this area likely generalize to other vascular beds.

1. Introduction

The human body is estimated to contain almost 5000 miles of capillaries, composed of 160 billion endothelial cells (ECs), which make up 2% of the total cell count in the body—second only to erythrocytes and platelets, which account for a staggering 89% of all cells (Sender, Fuchs, & Milo, 2016). Capillaries are the finest vessels of the vascular system, consisting of end-to-end ECs organized into tubular structures that wrap around to form the vessel lumen, making self-contact along their lateral edges. The lumen is wide enough to allow red blood cells to pass in single-file, and the ECs exceptional thinness of ~100 nm allows efficient exchange of gases, nutrients, and waste with local tissues.

A great body of work has established the barrier and transport functions of cECs (see Mann, Yudilevich, & Sobrevia, 2003; Rodrigues & Granger, 2015, for reviews), but comparatively little is known of their ion channel complement or the contributions of functional channels to their physiological roles. Recent work, however, has begun to build on foundations laid by early studies and to shed light on the question of which ion channels are functionally expressed by cECs.

In this chapter, we emphasize the use of recent developments in expression profiling as a springboard to accelerate progress in the field, which can be deployed to identify viable candidates for functional ion channel investigations. To this end, we mine a publicly available RNA-Seq dataset

(He et al., 2018; Vanlandewijck et al., 2018) and use these data to guide our discussion of the ion channel repertoire of cECs. Our aim is twofold: (i) to highlight the roles of ion channels that are known to be functional in cECs, and; (ii) to leverage these expression data to provide a roadmap for future experiments aiming to elucidate further roles for ion channels in these cells. Functional data obtained by us and others corroborate this approach, but two important caveats should be emphasized before going further. First, it is widely known that gene expression per se does not always translate into detectable functional channels. Second, it is possible that very low turnover of channel proteins underlain by gene expression below the threshold for detection can still result in formation of functional channels. Thus, this approach can be used to help focus investigations, but cannot substitute for direct functional experiments (i.e., patch clamp electrophysiology as the gold-standard to detect channel currents in native cells).

As we are just beginning to scratch the surface in terms of our understanding of cEC ion channels, in the majority of cases function awaits confirmation by measurement in native cells. Our discussion is primarily focused on the brain circulation, but it is likely that findings here generalize to other vascular beds. Space constraints preclude a full discussion of every ion channel gene expressed by capillary ECs, and thus we focus on ion channels for which a vascular role is well understood, those whose genes are highly expressed, or cases in which expression suggests a possible unique role in the context of cECs (see Table 1 for a detailed outline).

2. Anatomy of a typical microcirculatory network

The microcirculation is divided into three major segments, with distinct cell types populating each level. Arteries are large-diameter muscularized vessels that carry oxygenated blood into the tissue. These consist of several layers of contractile smooth muscle cells (SMCs), which regulate vessel diameter, and an inner lining of arteriolar ECs (aECs) organized in a cobblestone-like pattern and oriented in the direction of blood flow (Haas & Duling, 1997). SMCs and aECs are separated by a fenestrated internal elastic lamina (IEL), composed mainly of elastin and collagen secreted by SMCs (Jones, Scott-Burden, & Gevers, 1979; Schwartz, Adamany, & Blumenfeld, 1981), through which aECs extend small processes to make direct contact with the adjacent SM (Aydin, Rosenblum, & Povlishock, 1991). These myoendothelial junctions (MEJs; Straub, Zeigler, & Isakson, 2014) are signaling hubs through which critical intercellular crosstalk takes

Table 1 Ion channels expressed by capillary endothelial cells.

Channel name	Gene	Selectivity	mRNA FPKM	Functional?	Key properties of functional channels	Key references
IP_3R1	*Itpr1*	Ca^{2+}	34.53	Yes	Endoplasmic reticulum ligand-gated Ca^{2+} channels that are sensitized to Ca^{2+} by binding of IP_3. Key roles in Ca^{2+}-dependent processes in all cells	Foskett, White, Cheung, and Mak (2007)
IP_3R2	*Itpr2*	Ca^{2+}	132.92	Yes		
IP_3R3	*Itpr3*	Ca^{2+}	23.7	Yes		
Orai 1	*Orai1*	Ca^{2+}	29.99	n.d[a]	SOCE channels that replenish ER Ca^{2+} levels upon depletion	Prakriya and Lewis (2015)
Orai 3	*Orai3*	Ca^{2+}	36.81	n.d		
TPC1	*Tpcn1*	Na^+ Ca^{2+}	15.91	n.d	Role in protein trafficking in the endolysosomal system	Ruas et al. (2014)
TPC2	*Tpcn2*	Na^+ Ca^{2+}	12.1	n.d	Mediates NAADP-dependent Ca^{2+}-release from lysosomal stores. Role in endolysosomal trafficking	Zong et al. (2009)
TRPML1	*Mcoln1*	Cations	77.61	n.d	Role in endolysosomal system. Ubiquitous Preferentially conducts Na^+ and K^+ over Ca^{2+}	Venkatachalam, Wong, and Zhu (2015)
TRPP2	*Pkd2*	Cations	25.87	n.d	Mechanosensitive in complex with TRPP1	Du, Fu, Xia and Shen (2016)
TRPM3	*Trpm3*	Cations	1.82	n.d	Activated by the neurosteroid pregnenolone sulfate	Wagner et al. (2008)
TRPM4	*Trpm4*	Monovalent cations	9.51	n.d	Activated by Ca^{2+} release from ER. Conducts Na^+ and K^+ and depolarizes Vm	Gonzales et al. (2014)
TRPM6	*Trpm6*	Cations	14.45	n.d	Both channels are Ca^{2+}, Zn^{2+} and Mg^{2+} permeable. Ubiquitous	Schlingmann, Waldegger, Konrad, Chubanov, and Gudermann (2007)
TRPM7	*Trpm7*	Cations	59.57	n.d		
TRPC3	*Trpc3*	Cations	2.47	n.d	Preferentially conducts Ca^{2+} over Na^+. DAG-activated	Trebak, Bird, McKay, Birnbaumer, and Putney (2003)

TRPV4	*Trpv4*	Cations	1.47	Yes	Allows significant Ca^{2+} entry. Tonically inhibited by PIP_2	Harraz, Longden, Dabertrand, Hill-Eubanks, and Nelson (2018) and Harraz, Longden, Hill-Eubanks, and Nelson (2018)
Kir2.1	*Kcnj2*	K^+	9.12	Yes	Strong inward rectifier. K^+ activated. Hyperpolarization-activated. Role in capillary-to-arteriole electrical signaling	Longden et al. (2017)
Kir4.2	*Kcnj15*	K^+	5.75	n.d	Intermediate inward rectifier. K^+ activated	Edvinsson, Shah, and Palmer (2011)
Kir6.1	*Kcnj8*	K^+	9.1	Yes	K_{ATP} channel subunit. Associates with SUR β-subunits to form functional channels. Functional channels observed in heart cECs	Mederos y Schnitzler, Derst, Daut, and Preisig-Müller (2000)
Kir7.1	*Kcnj13*	K^+	2.32	n.d	Weak inward rectifier	Krapivinsky et al. (1998)
$K_V1.5$	*Kcna5*	K^+	1.52	Yes?	Voltage-activated K^+ currents not observed in brain cECs to date, but K_V currents have been recorded in heart cECs	Dittrich and Daut (1999)
$K_V7.1$	*Kcnq1*	K^+	22.04	Yes?		
ClC-2	*Clcn2*	Cl^-	1.24	n.d	Voltage-dependent Cl^- channel. Hyperpolarization-activated	Nilius and Droogmans (2003)
Piezo1	*Piezo1*	Cations	21.02	n.d	Large mechanosensitive ion channel with key roles in vascular development and EC orientation to shear stress. Also regulates vascular diameter	Coste et al. (2010)
P2X4	*P2rx4*	Cations	17.9	Yes	Mediates Ca^{2+} influx and is sensitive to nM ATP. Currents recorded in cECs of the cochlea	Wu, Dai, Shi, Jiang, and Nuttall (2011)
P2X7	*P2rx7*	Cations	7.72	n.d	Mediates Ca^{2+} influx. Sensitive to ATP	

Continued

Table 1 Ion channels expressed by capillary endothelial cells.—cont'd

Channel name	Gene	Selectivity	mRNA FPKM	Functional?	Key properties of functional channels	Key references
HCN2	*Hcn2*	Cations	12.29	n.d	Hyperpolarization-activated and cyclic nucleotide-gated cation channel	Wahl-Schott and Biel (2008)
Nav1.2	*Scn2a*	Na^+	1.98	n.d	Voltage-dependent Voltage-activated Na^+ currents not observed in cECs to date	Yu and Catterall (2003)
αENaC	*Scnn1a*	Na^+	1.41	n.d	Tetrameric channel, typically with an α2βγ stoichiometry. Functional α4 homomers reported	Wang, Meng, Mohan, Champaneri, and Gu (2009)
GABA ρ2	*Gabrr2*	Cl^-	1.12	n.d	Forms homopentameric ligand-gated $GABA_C$ receptors	Naffaa, Hung, Chebib, Johnston, and Hanrahan (2017)

[a]n.d., no data. Measurement of a current with the expected characteristics in native cECs is taken as the gold-standard for functional channel confirmation.
β/accessory subunits that are expressed above the threshold of 1 FPKM in the absence of detectable expression of α-subunits were not considered further (for example, the nicotinic acetylcholine receptor β1 subunit gene is highly expressed by cECs, but expression of other nicotinic receptor subunits is minimal). *Snc2a* encodes rapidly-activating, voltage-dependent Na^+ channels. Due to a lack of classic Na_V currents in cEC recordings to date, these channels are not considered further in the main text.
Abbreviations: ATP, adenosine triphosphate; DAG, diacylglycerol; SOCE, store-operated Ca^{2+} entry; SUR, sulphonylurea receptor; Vm, membrane potential.

place. Arteries give way to smaller arterioles, with similar features but fewer layers of SMCs. The finest "pre-capillary" arterioles consist of just a single layer of SMCs, before giving way to the capillary bed. It is worthy of note that the transition from arteriole to capillary is not clear-cut. In the cerebral circulation in particular, difficulty in delineating this change has hampered understanding of blood flow regulation in this zone (Box 1), and further work is needed to clearly identify the functional and morphological criteria that define the transition from arteriole to capillary. The capillary bed itself is composed of cECs (see below) and overlying pericytes that extend long thin-strand processes that cover the majority of the capillary network (Hartmann et al., 2015). Capillaries bifurcate many times over to form a

BOX 1 Where do arterioles end, and capillaries begin?

This seemingly straightforward question has been surprisingly difficult to answer in studies of the brain circulation. The confusion that has arisen can be attributed primarily to a population of mural cells which reside on the first ~4 branches of the vessels arising from the parenchymal arteriole. Here, cells with ovoid cell bodies and a bump-on-a-log morphology and multiple processes that enwrap the vessel have been observed (Attwell, Mishra, Hall, O'Farrell, & Dalkara, 2016; Grant et al., 2017). These mural cells lack morphological features consistent with traditional smooth muscle (which are banded, relatively flat cells that concentrically encircle the vessel without processes; Grant et al., 2017; Hill et al., 2015), but nonetheless possess contractile α-actin which allows them to regulate the diameter of the underlying vessel. As these cells thus possess characteristics that would allow them to be classified as either pericytes or smooth muscle cells, this has led to debate over their identity (Attwell et al., 2016; Hill et al., 2015). Recent RNA-Seq screens (Vanlandewijck et al., 2018) have classed these cells as SMCs due to their robust expression of the *Acta2* gene, a commonly used SMC marker. However, the possibility of subtle differences between arteriolar SMCs and these mural cells cannot yet be ruled out, and thus further precise work is needed to fully clarify their identity. Techniques to isolate only these initial branches, or to extract or visualize genetic material directly from these cells where they reside (such as seqFISH+; Eng et al., 2019), would allow for direct and unambiguous comparison with the expression profiles of upstream SMCs and downstream thin-strand pericytes. The results of such a screen would facilitate the delineation of whether these cells are indeed genetically indistinguishable from SMCs, in which case the vessels they cover are fine pre-capillary arterioles. If instead their expression profile is closer to those of pericytes, or indicates a hybrid SMC-pericyte genotype, then the vessels they cover would be identified as capillaries or transitional vessels between the arterioles and true capillaries. This issue likely extends to other vascular beds too.

complex, anastomosing, and arborescent network that is in close apposition with cells of the local tissue. These drain into venules, formed of venous ECs with a similar cobblestone morphology to aECs (O'Nejll, 1947), covered in sparse and disorganized venous SMCs (Hill et al., 2015). Venules lack an IEL and in many vascular beds have specialized luminal valves to prevent the backflow of blood. Venules, in turn, give way to larger veins as blood is returned toward the heart.

2.1 The vascular anatomy of the brain

The cerebral circulation begins with the internal carotid and vertebral arteries. Blood arriving through the internal carotid arteries supplies the anterior cerebrum, while the vertebral arteries converge to form the basilar artery which provides blood to the remainder of the cerebrum and the cerebellum and brain stem (Cipolla, 2009). The carotid and basilar arteries meet at the base of the brain, forming an arterial ring known as the circle of Willis, and from this the major cerebral arteries branch out over the brain's surface, giving rise to the posterior, middle, and anterior cerebral arteries that supply regions of the cerebral cortex (Purves et al., 2001).

Pial arteries on the cortex's surface are highly anastomosing and form a collateral network, ensuring that loss of blood flow through an occluded vessel can be readily buffered by compensatory changes in flow through patent connected arteries (Schaffer et al., 2006). Notably, pial collateralization has been shown to exhibit wide variation due to natural genetic differences, with the extent of collateralization being inversely correlated with infarct volume after vessel occlusion (Zhang, Prabhakar, Sealock, & Faber, 2010). These arteries are extrinsically innervated by the peripheral nervous system (Cipolla, Li, & Vitullo, 2004) and typically follow the brains sulci, where present. Though the size of these arteries varies, which influences the number of SMC layers, a general structure can be defined: an inner layer of cobblestone aECs with an overlying IEL, a middle 2–3 layers of SMCs embedded in elastin and collagen fibers, and an outer layer with an extracellular matrix primarily composed of collagen, fibroblasts, and perivascular nerves (Lee, 1995).

From the pial arteries, parenchymal arterioles (PAs) dive orthogonally into the brain. In the murine cortex, these vessels branch at regular intervals as they dive toward the white matter, in turn giving way to the capillary bed (Blinder et al., 2013). The PAs act as bottlenecks to blood flowing into the brain (Nishimura, Schaffer, Friedman, Lyden, & Kleinfeld, 2007), with occlusion of one PA causing reduction in blood flow and damage to

$\sim$0.22 mm^3 of surrounding tissue, thus defining the minimal area that these vessels are responsible for nourishing (Shih et al., 2013). PAs have higher myogenic tone (contraction to intravascular pressure (Bayliss, 1902)) than pial arteries (Longden, Hill-Eubanks, & Nelson, 2015) and in contrast lack extensive extrinsic innervation (Cipolla et al., 2004).

After a transitional zone (see Box 1), arteries give rise to the capillary network. The total length of capillaries in the human brain is estimated to be hundreds of miles, accounting for more than 90% of the total intracerebral vascular volume (i.e., not including surface vessels; Gould, Tsai, Kleinfeld, & Linninger, 2016) and providing a vast surface area to enable exchange between brain and blood (Zlokovic, 2008) (Fig. 1).

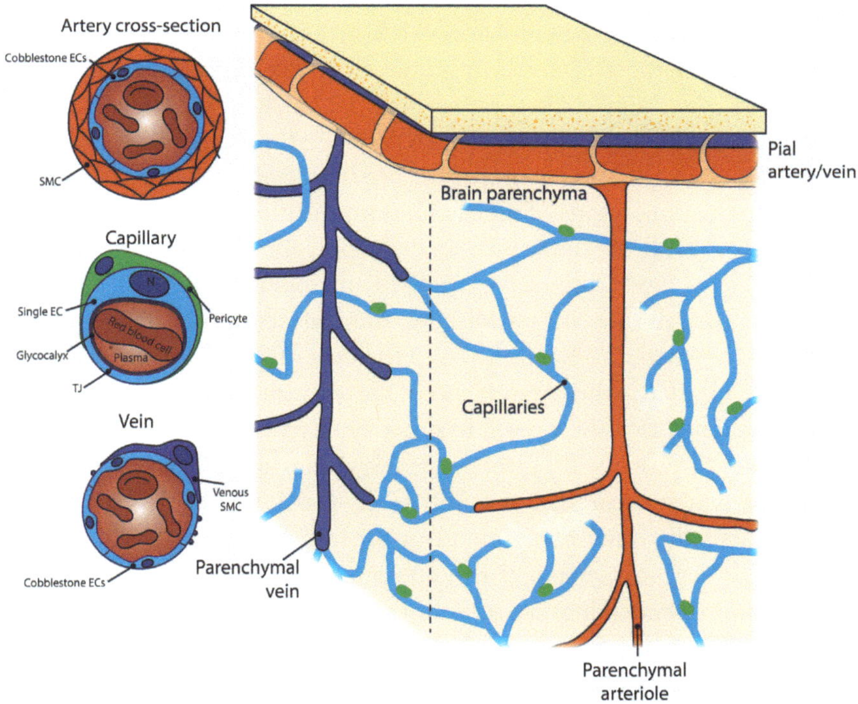

Fig. 1 Vascular anatomy of the brain. Pial arteries carry blood across the brain surface and penetrate orthogonally into the parenchyma. Typically, after diving for >100 μm, PAs branch out into the capillary bed. This transition is demarcated by continuity of the endothelium, but loss of elastin deposition and transition of mural cells from traditional SMCs to contractile cells with a distinct morphology. After $\sim$4 branches, these mural cells give way to capillary pericytes which extend long, thin-strand processes that cover the majority of the capillary bed. The capillaries are drained of blood by parenchymal veins, with disorganized non-contractile SMCs, which return blood to the brains surface pial veins.

2.2 Capillary endothelial cells

The exact details of the spatial organization of cECs within capillaries are not currently known. In cross section, it is clear that these small vessels consist of individual cECs that encircle the lumen of the vessel and make contact with their own distal processes. However, when considered in transverse or sagittal orientation, it is unclear how cECs organize with one another, and the degree of interdigitation between cells has not been studied in detail. High resolution imaging and 3-D reconstructions showing cell-cell borders are needed to answer this question, as this anatomical organization has important functional implications for how signaling along capillaries and between cECs and their environment is managed.

In the brain, these points of self-contact are the sites of tight junctions (TJs), forming the blood-brain barrier (BBB). Thus, energy substrates such as glucose need specialized transport to cross from blood to the brain, enabled by the high expression of the GLUT1 transporter in cECs (Vanlandewijck et al., 2018). Gases such as oxygen and carbon dioxide cross the BBB passively along their partial pressure gradients. In addition to preventing the transcytosis of molecules from the blood to the brain parenchyma, TJs also act as a molecular fence to prevent the lateral diffusion of parenchymally- and luminally-expressed membrane proteins, thereby playing a critical role in establishing luminal/abluminal cEC polarity (Zihni, Mills, Matter, & Balda, 2016). However, TJs are found exclusively in brain ECs, and thus the components of the fence that separates the luminal and parenchymal faces of the cell in other capillary beds is likely played by other proteins, possibly the proteinaceous strands of cadherin molecules, which are widely expressed in cECs (e.g., *Cdh5*; Vanlandewijck et al., 2018), and/or the other components of adherens junctions. Further work is needed to definitively answer this question.

The luminal and abluminal sides of the cell are covered in extracellular matrices. The glycocalyx is a carbohydrate-rich extracellular matrix that extends from the cEC surface into the vessel lumen by ~100 nm. It contains a number of proteins with diverse functions that ensure uninterrupted blood flow and contribute to EC homeostasis. Examples include anticoagulant factors that prevent clot formation in stalled vessels, proteins that control leukocyte adhesion, and anti-oxidant enzymes to quench free radicals and protect nitric oxide bioavailability (Reitsma, Slaaf, Vink, van Zandvoort, & oude Egbrink, 2007). Similarly, the parenchymal side of the cell is covered in the basement membrane, a dense matrix of proteins that provides physical support for the vasculature while also contributing to polarity, the trafficking and density of membrane proteins, and signaling between the vasculature and the parenchyma (Joutel, Haddad, Ratelade, & Nelson, 2015).

3. Ion channel expression in capillary ECs

Work in the 1990s made significant inroads toward understanding cEC ion channel expression and function (Bossu, Feltz, Rodeau, & Tanzi, 1989; Dittrich & Daut, 1999; Mederos y Schnitzler et al., 2000; Popp & Gögelein, 1992; Popp, Hoyer, Meyer, Galla, & Gögelein, 1992). These studies used primarily electrophysiological approaches to establish that cECs from a range of organ systems express an array of potassium (K^+), calcium (Ca^{2+}), and non-selective cation channels. At this time many channels had yet to be cloned and systematically studied to define their biophysical features. In particular, very little was known of the transient receptor potential (TRP) channel family, which is the likely source of undefined non-selective cation currents in this literature (Popp & Gögelein, 1992; Popp et al., 1992). Technological developments that have taken place over the past quarter-century now enable deep characterization of the gene expression profiles of both populations and of individual cells, which can be leveraged to accelerate explorations of ion channel expression and function. In particular, single-cell RNA-Seq allows the assessment of the expression of thousands of genes in individual cells. Although cell-to-cell expression will typically vary, expression between cells can be compared using bioinformatics approaches and clustering analyses to identify common expression patterns. From this, an aggregate picture of gene expression in a given population emerges. This approach was taken by Vanlandewijck and co-workers in a recent, seminal study that mapped the gene expression profiles of different classes of vascular cells in the mouse brain and lung (He et al., 2018; Vanlandewijck et al., 2018). These data, which are publicly available for further analysis, reveal that the endothelium varies dramatically in its expression profile as it transitions between arterioles and venules. Here, we mined this data set to guide our discussion of the landscape of ion channel expression and function in capillary ECs (Fig. 2). Note that to reasonably constrain the following review, we restrict our discussion to channels with a gene expression level exceeding 1 FPKM (fragment per kilobase of transcript per million mapped reads).

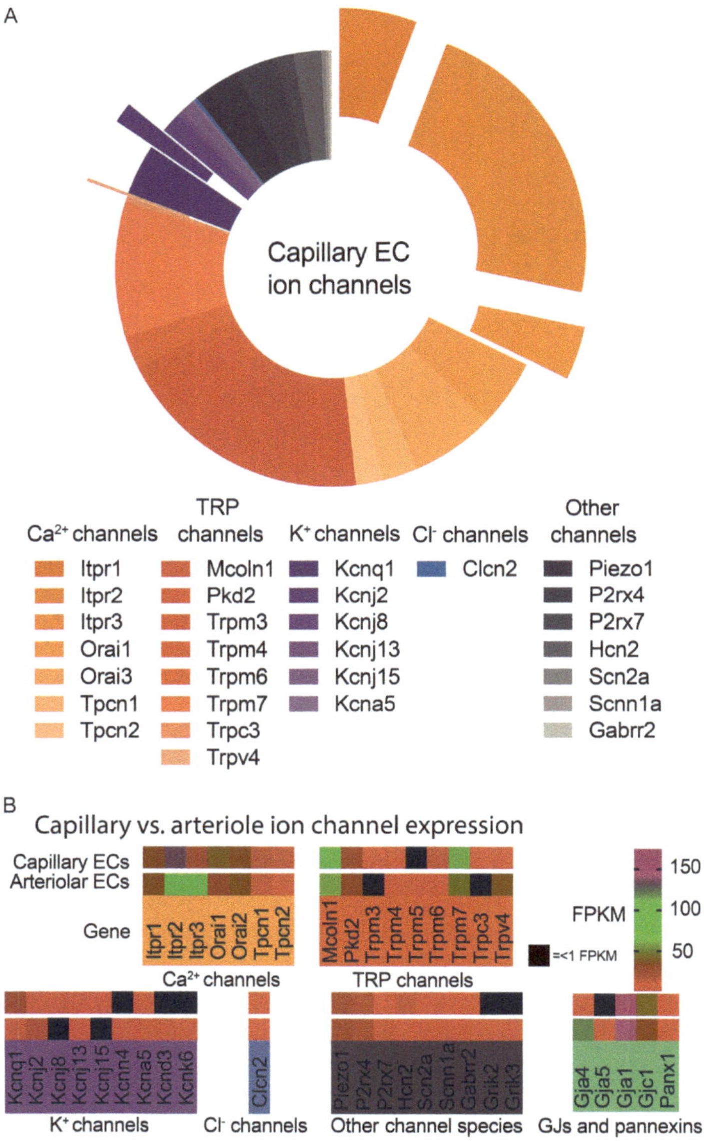

Fig. 2 Capillary EC ion channel expression. (A) RNA-Seq data mined from Vanlandewijck et al. (2018) and He et al. (2018) and reproduced with permission, showing the relative expression of ion channel genes*, represented by the size of the corresponding colored segment, in brain capillary ECs. Exploded segments correspond to ion channels with known functions that are a focus here (Kir2.1, TRPV4 and IP_3Rs). (B) Relative expression of ion channel genes in arteriolar vs capillary ECs. These molecular differences contribute to the differing functions of ECs in each compartment. Note that seemingly low expression of genes can still lead to important functional contributions of the encoded channel, as highlighted by Kir2.1 (*Kcnj2*), for example. *All known ion channel genes were analyzed, but only those with an expression level exceeding one fragment per kilobase of transcript per million mapped reads (FPKM) are included in this representation.

We also have not given further consideration to accessory subunits that are expressed in the absence of appreciable expression of their pore-forming α-subunits.

3.1 Overview

In surveying the ion channels expressed in brain cECs, we observed that Ca^{2+} and TRP channels are the most prominent classes, with each contributing about a third to overall ion channel expression. Thereafter, K^{+} channels contribute approximately a tenth of overall channel expression, and a handful of other channel subunits contribute the remainder (Fig. 2A). Notably, only a single chloride (Cl^{-}) channel was observed, expressed at a low level, suggesting the majority of Cl^{-} flux is instead carried by the more highly expressed proton (H^{+})/Cl^{-} exchangers CLC-3 (*Clcn3*) and CLC-4 (*Clcn4*) (He et al., 2018; Vanlandewijck et al., 2018). The channels encoded by these genes and their key properties are listed in Table 1, along with notation of whether they have been functionally confirmed.

4. Ca^{2+} channels and Ca^{2+} signaling in cECs

Until recently, the Ca^{2+} signaling properties of cECs were unexplored. In studies of the brain circulation, this can in part be attributed to difficulties in loading brain ECs with commonly used Ca^{2+} indicator dyes. Attempts to load these ECs with indicators typically result in poor staining and compartmentalization of the dye into puncta, likely reflecting their barrier function. Thus, thorough characterization of capillary EC Ca^{2+} signaling has had to await the development of mice expressing genetically encoded Ca^{2+} indicators (GECIs) in ECs. Initial mouse lines utilizing the connexin 40 (*Gja5*) promoter (Tallini et al., 2007) to drive transgene expression yielded excellent expression in aECs, but left cEC signaling invisible due to molecular differences between these two types of EC—connexin 40 is a major component of gap junctions in aECs, whereas cECs do not express this isoform and instead favor expression of connexins 43 and 45 (Fig. 2B). Recently, cadherin 5 (*Cdh5*; a pan-endothelial cell-cell adhesion glycoprotein), was instead used as a promoter for the expression of the high signal: noise Ca^{2+} indicator GCaMP8. This line, developed by the Kotlikoff laboratory (see: chromus.vet.cornell.edu, Jackson Laboratories strain 033342), exhibits GECI expression throughout the entire endothelium in a cell-specific manner, thus permitting the first characterization of Ca^{2+} signaling in cECs. We have used this mouse line to examine the Ca^{2+} signaling behaviors of cECs in vivo, and found stark contrasts compared to aEC Ca^{2+} signals, underscoring the fact that ECs are distinct from one another at different levels of the vascular tree (Longden et al., 2019).

4.1 IP_3 receptors are the predominant ion channel species expressed in capillary endothelium

According to expression data, the major Ca^{2+}-conducting ion channel species expressed in capillaries are those of the inositol 1,4,5-trisphosphate receptor (IP_3R) family (Vanlandewijck et al., 2018).

IP_3Rs are ubiquitous ion channels located in the membrane of the endoplasmic reticulum (ER). These are Ca^{2+}-activated Ca^{2+}-release channels with biphasic sensitivity to Ca^{2+}, with this ion and the phosphoinositide IP_3 acting as essential co-agonists (Foskett et al., 2007). At low intracellular Ca^{2+} concentrations, Ca^{2+} increases channel open probability, whereas at high concentrations it instead plays an inhibitory role (Bezprozvanny, Watras, & Ehrlich, 1991; Iino, 1990). When cytoplasmic IP_3 levels are very low, the channel is extremely sensitive to Ca^{2+} inhibition, which prevents substantial activity. In contrast, modest increases in IP_3 (to around 100 nM) dramatically reduce the sensitivity of the channel to Ca^{2+} inhibition, allowing for maximal activation of the IP_3R at sub-micromolar Ca^{2+} concentrations (Foskett et al., 2007). IP_3 binding occurs at a cytosolic region of the receptor distant from the pore, which produces a large conformational change over a region of more than 2000 amino acid residues in length, stretching from the IP_3 binding core to the Ca^{2+} channel (Hamada, Miyatake, Terauchi, & Mikoshiba, 2017). Although the precise mechanistic details remain to be determined, it is thought that this conformational change decreases the channels sensitivity to Ca^{2+} inhibition. As ER luminal Ca^{2+} is typically within the range of 100–800 μM (Burdakov, Petersen, & Verkhratsky, 2005), whereas cytosolic Ca^{2+} is maintained around 100 nM, opening of IP_3Rs leads to Ca^{2+} release into the cytoplasm.

IP_3 is produced by the phospholipase C (PLC)-mediated hydrolysis of the minor membrane phospholipid phosphatidylinositol 4,5-bisphosphate (PIP_2), which also yields the second messenger diacylglycerol (DAG; a key regulator of TRPC3 channel activity, see below). PLC, in turn, can be engaged by a number of upstream processes, including the activity of the $G_q\alpha$ subunit of G_q protein-coupled receptors (G_qPCRs) (Wu, Lee, Rhee, & Simon, 1992). Thus, G_qPCR activation typically leads to IP_3-mediated Ca^{2+} release. Depending on local conditions, release can occur from either a single IP_3R, a small cluster of channels, or a wave of

Ca^{2+} release can be evoked by Ca^{2+}-activation of neighboring IP_3Rs (Berridge, Lipp, & Bootman, 2000; a process termed Ca^{2+}-induced Ca^{2+} release; CICR), enabling a broad range of spatiotemporal profiles for Ca^{2+} signals. The IP_3- and Ca^{2+}-sensitivity of IP_3Rs also differs between subtypes (Miyakawa et al., 1999), which will further contribute to the diversity of Ca^{2+} signal patterning in different cell types. Capillary ECs express each of the three IP_3R genes (*Itpr1*, encoding IP_3R1; *Itpr2*, IP_3R2; *Itpr3*, IP_3R3), with the IP_3R2 subtype having the most robust expression of all channel-encoding genes in cECs (Vanlandewijck et al., 2018). Like aECs, no evidence for expression of the genes encoding ryanodine receptors is seen in cECs (Vanlandewijck et al., 2018).

4.2 IP_3R-mediated Ca^{2+} release in capillary and artery endothelium

G_qPCR-activating mediators such as PGE_2 or ATP may be released onto the endothelium during localized activity in a range of tissues, such as brain or muscle, to evoke IP_3-mediated Ca^{2+} increases (Horiuchi, Dietrich, Tsugane, & Dacey, 2001; Zonta et al., 2003). Alternatively, the increase in local CO_2/H^+ concentration accompanying increased metabolic activity could elicit activity of the recently characterized H^+-sensitive GPR4 G_qPCR to drive endothelial Ca^{2+} elevations (Wenzel et al., 2020). What effectors do such Ca^{2+} signals couple to in capillary and arterial ECs?

In aECs, basal Ca^{2+} signals are typically localized to fenestrations in the IEL, through which the endothelium extends filopodia-like myoendothelial projections that directly contact and form gap junctions with the overlying SMCs. These are major hubs of communication between these two cell types, and IP_3-mediated Ca^{2+} signals in this region—known as "pulsars"—activate small- and intermediate-conductance Ca^{2+}-activated K^+ (SK and IK) channels (Ledoux, Bonev, & Nelson, 2008; Ledoux, Taylor, et al., 2008), which generate hyperpolarization to influence SM membrane potential (Vm) and contractile state (Ledoux, Werner, Brayden, & Nelson, 2006).

Intriguingly, in capillaries functional SK and IK channels are absent (Longden et al., 2017), which removes a direct avenue for transduction of capillary Ca^{2+} signals to electrical activity of the membrane. Instead, Ca^{2+} signals here likely couple to other signaling effectors such as nitric oxide (NO), a classic EC-derived vasodilator and gasotransmitter (Ignarro, Cirino, Casini, & Napoli, 1999), and prostacyclin (PGI_2) derived

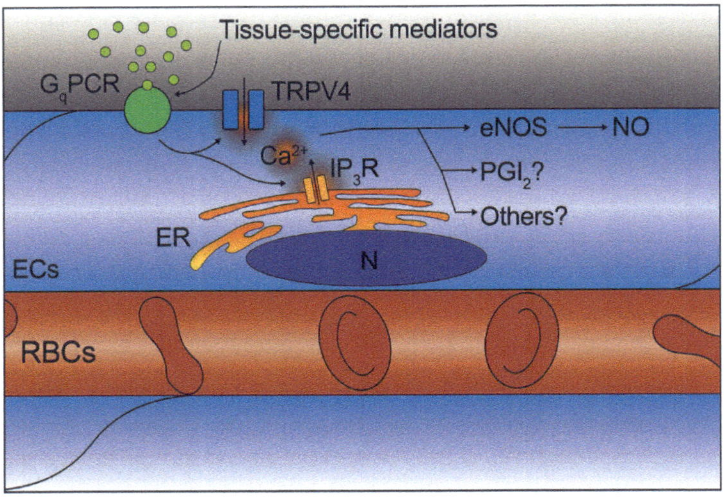

Fig. 3 A general mechanism for capillary Ca^{2+} signaling. G_qPCR mediators released onto capillaries can engage TRPV4 and IP_3Rs promote Ca^{2+} entry and release, respectively. Resultant cytosolic Ca^{2+} elevations are known to drive local NO production, and may stimulate other Ca^{2+}-dependent processes to generate further mediators.

from the action of cyclooxygenase enzymes on arachidonic acid (AA) formed by Ca^{2+}-dependent hydrolysis of membrane phospholipids by phospholipase A_2 (Bunting, Gryglewski, Moncada, & Vane, 1976; Moncada, Gryglewski, Bunting, & Vane, 1976) (Fig. 3). Further processing of AA by cytochrome P450 enzymes to epoxyeicosatrienoic acids (EETs) is another potential route for Ca^{2+}-dependent generation of signaling molecules in capillaries (Archer et al., 2003; Campbell & Falck, 2007; Campbell, Gebremedhin, Pratt, & Harder, 1996; Fleming & Busse, 2006). In the brain, it is unlikely that EC-secreted molecules travel far before meeting a receptor or enzyme upon which they act, which will limit their area of effect. Indeed, NO, which is unhindered by lipid membranes, has been estimated to only travel ~100 μm in any direction from its point source of release (Wood & Garthwaite, 1994). Therefore, given the considerable distance between capillaries and arterioles in many cases, it is unlikely that agents such as NO or PGI_2 released from cECs deep within the capillary bed diffuse to trigger dilation of upstream arterioles. Instead, electrical signaling within the capillary wall facilitates long-range communication between the capillary bed and upstream arterioles (Longden et al., 2017; see below). Thus, any Ca^{2+}-dependent release of signaling molecules from cECs is likely to be restricted to exerting local effects on nearby cells (e.g., pericytes, astrocytes, microglia or neurons).

Elegant recent work has demonstrated that distinct clusters of aECs in arteries are equipped to respond to different mediators, doing so with large, spreading, and oscillating Ca^{2+} signals. When multiple agonists are present, more complex behavior occurs, with cells generating emergent signals that are non-linear combinations of those generated in response to single mediators (Lee et al., 2018; McCarron et al., 2019). This feature of aEC Ca^{2+} signaling is thought to facilitate the ability of the endothelium to integrate the messages encoded by the panoply of blood-borne signals that they are exposed to and transduce these into changes in vascular tone. It is currently not known whether cECs exhibit similar heterogeneity in their receptor expression or their responses to single or multiple agonists, or comparable emergent behavior. Speculatively, inbuilt heterogeneity of receptor expression in cECs could contribute to physiological control by defining the spatial area of the network that will respond to locally released signaling molecules with Ca^{2+} elevations, for example, by limiting a response to a few adjacent cells around a capillary branch point. Alternatively, cEC heterogeneity could drive regionally-specific responses, for example—tailored toward the neurotransmitters and mediators specific to local neuroglial populations in the brain, or the heterogeneous parenchymal cells of other organs.

Further interesting possibilities for cEC IP_3R-mediated Ca^{2+} release await exploration. For example, given that the ER is continuous with the nuclear membrane, and that nuclear IP_3Rs can directly elevate nuclear Ca^{2+} (Bading, 2013), it is possible that capillary Ca^{2+} signals play a role in regulating gene expression. Alternatively, the components making up TJs of the BBB and endothelial cell-cell junctions in other tissues are also Ca^{2+}-sensitive (Bleich, Shan, & Himmerkus, 2012; Mehta et al., 2003), suggesting that cEC Ca^{2+} signals could also regulate permeability. Future experiments will undoubtedly reveal that these signals contribute to a broad and rich diversity of cEC functions.

4.3 Orai channel-mediated store-operated Ca^{2+} entry may fuel ongoing IP_3R Ca^{2+} signaling

In addition to IP_3Rs, molecular data indicate transcription of several other Ca^{2+}-permeable channel species, which await functional confirmation. These include the Orai genes *Orai1* and *Orai3* (Fig. 2A). These genes encode channels mediating store-operated calcium entry (SOCE). In this process, depletion of intracellular Ca^{2+} stores causes aggregation of ER STIM1 (stromal interaction molecule 1) proteins and drives their interaction with plasma membrane Orai channels to initiate Ca^{2+} entry to refill ER stores

(Prakriya & Lewis, 2015). Mechanistically, the ER-resident single transmembrane STIM1 and 2 proteins (both of which are expressed by cECs in addition to Orai1 and 3; Vanlandewijck et al., 2018) monitor ER luminal Ca^{2+} concentration through a low affinity Ca^{2+}-binding EF-hand motif (Stathopulos, Zheng, Li, Plevin, & Ikura, 2008; Zhang et al., 2005). Upon a fall in ER Ca^{2+}, the lack of Ca^{2+} bound to this site triggers oligomerization of STIM (Covington, Wu, & Lewis, 2010; Liou, Fivaz, Inoue, & Meyer, 2007), and evokes conformational changes: In the cell cytosol, an electrostatic interaction between acidic amino acids that tethers the CRAC activating domain/STIM1 Orai activating region (CAD/SOAR) to a coiled-coil region of STIM1 (which is in turn connected to the ER membrane region of the protein) is broken, allowing this domain to stretch toward the plasma membrane and interact with acidic residues in the C-terminal domain of Orai (Park et al., 2009; Wang, Deng, & Gill, 2010). This, in turn, evokes a conformational change in the Orai channel leading to its opening and Ca^{2+} influx across the plasma membrane. The incoming Ca^{2+} is then sequestered in the ER through the activity of the sarco-endoplasmic reticulum Ca^{2+} ATPase (SERCA). STIM and Orai expression strongly suggests that SOCE through these channels plays an important role in replenishing Ca^{2+} stores in cECs. This process is likely to be particularly important for Ca^{2+} homeostasis to enable ongoing IP_3R mediated Ca^{2+} events, as described above.

Also of note, cECs express the ubiquitous two-pore Ca^{2+} channels TPC1 and TPC2—Na^+- and Ca^{2+}-conducting channels with roles in endolysosomal trafficking and Ca^{2+} release from lysosomal stores. Similar roles are thus likely played by these channels in the cEC endolysosomal system.

5. TRP channels in capillaries

The TRP channel superfamily consists of a diverse group of 28 channels, which vary in form and function. All cells express at least one type of TRP channel, and in recent years their importance in the cardiovascular system has become apparent. The properties and roles of these channels in the vasculature has been extensively reviewed recently, and thus we direct readers to Earley and Brayden (2015) for a detailed discussion of this topic. Focusing on cECs, emerging roles for TRPV4 (Harraz, Longden, Hill-Eubanks, et al., 2018) and TRPA1 (Pires, Pritchard, & Earley, 2018) in mediating Ca^{2+} signals have been described. Moreover, a number of further TRP isoforms are transcribed by cECs, which may endow them with additional sensing and signal transduction capabilities.

5.1 TRPV4 channels in capillary ECs are opened by the same processes that underlie IP_3R activation

Despite relatively low expression, recent work has confirmed the presence of robust TRPV4 channel currents in cECs (Harraz, Longden, Hill-Eubanks, et al., 2018; Longden & Nelson, 2015b). TRPV4 channels are heat-, osmo-, and mechanosensitive and are permeable to Na^+, K^+, Ca^{2+} and magnesium (Mg^{2+}) and their large slope conductance (~100 pS at positive potentials; Watanabe et al., 2002) typically exerts a strong depolarizing influence on cellular Vm.

Harraz and co-workers recently demonstrated the tonic inhibition of TRPV4 in cECs by the membrane lipid PIP_2, which underpins the low open probability of this channel in the brain endothelium. However, G_qPCR activation increases TRPV4 channel activity several fold, suggesting a mechanism in which the engagement of $G_q\alpha$ activates PLC, hydrolyzing PIP_2 to IP_3 and DAG, with the resultant reduction of membrane phospholipid pools disinhibiting the channel (Harraz, Longden, Hill-Eubanks, et al., 2018). While the nature of PIP_2-TRPV4 interactions are not fully understood, PIP_2 is known to interact with an N-terminal stretch of positively-charged amino acids. Counterintuitively, this interaction also facilitates the activation of the channel by a range of physiological stimuli, such as changes in osmolarity, temperature, and EETs (Garcia-Elias et al., 2013). This latter interaction is particularly interesting, as EETs are known to contribute to functional hyperemia in the brain (Peng, Zhang, Alkayed, Harder, & Koehler, 2004) and to activate TRPV4 channels in astrocytic endfeet, closely opposed to the brain vasculature (Dunn, Hill-Eubanks, Liedtke, & Nelson, 2013). Thus, it is likely that any EETs released into the paravascular space between endfeet and cECs also activate tonically-inhibited cEC TRPV4 channels. This, in combination with the activation of G_qPCR signaling pathways as described above, could be a major regulator of TRPV4 channel activity in these cells.

The consequences of TRPV4 activation are Na^+ and Ca^{2+} influx (and K^+ efflux) and Vm depolarization. Therefore, activation of TRPV4 may be employed to repolarize the cEC membrane after hyperpolarizing electrical signaling (see K^+ channels, below), or to lock Vm in a depolarized state to prevent passage of electrical signals, thereby dampening blood flow through the capillary in question. Alternatively, Ca^{2+} influx through TRPV4 could couple to the activation of IP_3Rs, eNOS, or other Ca^{2+}-dependent signaling pathways.

5.2 Further TRP isoforms imbue capillaries with additional sensing capabilities

Gene expression for a range of other TRP channel isoforms has been observed in brain cECs, namely: the canonical TRPC3, the melastatin members TRPM3, TRPM4, TRPM6, and TRPM7, the mucolipin TRPML1, and the polycystin TRPP2. TRPM6, TRPM7 and TRPML1 are ubiquitously expressed, with the former pair having a major role in Mg^{2+} homeostasis and a range of basic cell processes such as proliferation, differentiation, migration and apoptosis. TRPML1, on the other hand, plays essential roles in controlling ion balance in lysosomes in all cells. Given the ubiquity of these channels, we refer readers to recent reviews (Cheng, Shen, Samie, & Xu, 2010; Schlingmann et al., 2007) and here focus on the potential vascular-specific roles of plasma membrane TRPC3, TRPM3, TRPM4 and TRPP2 channels in cECs.

TRPC3 is a non-selective cation channel with a modest selectivity for Ca^{2+} over Na^{+} (P_{Ca}:P_{Na}, 1.6) that is widely expressed (Flockerzi & Nilius, 2007). Roles for these channels have been described in mechanosensation and mechanotransduction (Quick et al., 2012; Sexton et al., 2015), and in SOCE when coexpressed with TRPC1 (Wu, Zagranichnaya, Gurda, Eves, & Villereal, 2004). In aECs, several reports have indicated the expression of TRPC3 channels (Yao & Garland, 2005; Yip et al., 2004). Intriguingly, the physiological activator of TRPC3 is DAG (Trebak et al., 2003), which is produced during the hydrolysis of PIP_2 by PLC in the same reaction that produces IP_3. Thus, in cECs the same G_qPCR signaling events that drive IP_3R-mediated Ca^{2+} signals and TRPV4 disinhibition might simultaneously engage TRPC3, which could amplify cytosolic Ca^{2+} entry or increase the duration of signals. Moreover, the mechanosensing properties of TRPC3 could be employed to sense local blood flow from the force imparted by passing red blood cells as they squeeze through the narrow capillary lumen, and transduce this into intracellular signals that feedback to modify blood flow or perform other functions such as the regulation of gene expression (Brenner & Dolmetsch, 2007).

TRPM3 is a non-selective channel that is permeable to Ca^{2+}. It has been described in vascular smooth muscle (Naylor et al., 2010), but evidence for its expression in aECs is sparse. This channel is responsive to changes in temperature, which have implicated it in pain sensation (Vriens et al., 2011), and it is activated by the neurosteroid pregnenolone sulfate (Wagner et al., 2008). This raises the interesting possibility of TRPM3 equipping the capillary network with sensitivity to locally released neurosteroids, which would presumably act through this channel to increase Ca^{2+} in the cEC. Neurosteroid-evoked Ca^{2+} signals could in turn engage a range of endothelial signaling pathways, as already outlined.

TRPM4 is a monovalent cation-permeable, Ca^{2+}-sensitive channel that is also modulated by cytosolic adenosine triphosphate (ATP) and PIP_2 (Guo et al., 2017). Its activity thus typically leads to membrane depolarization, and in the vasculature it is best characterized in SMCs, where TRPM4 is activated by Ca^{2+} released by IP_3Rs as a consequence of G_qPCR engagement. The resultant Na^+ influx and depolarization activates voltage-gated Ca^{2+} channels and promotes SMC contraction (Gonzales et al., 2014). Signaling by G_q-coupled P2Y4 and P2Y6 purinergic receptors contributes to TRPM4 activity in this context (Li, Baylie, Tavares, & Brayden, 2014) where P2Y activation results from either binding of locally released purines, or direct sensing of mechanical forces (Brayden, Li, & Tavares, 2012), eliciting an increase in cytosolic IP_3 levels to evoke Ca^{2+} release. In cECs, TRPM4 is uncharacterized, but it could be activated by IP_3-mediated Ca^{2+} signals, where it would contribute to membrane depolarization. If this is the case, TRPM4-activity would be expected to counteract propagating hyperpolarization (Longden et al., 2017) traveling through the cells in which it is active, thus helping to shape electrical signaling through the capillary network (see below).

TRPP2 channels have a large single-channel conductance and are also permeable to Ca^{2+}. This subunit has been implicated in mechanosensing processes when expressed with TRPP1, TRPC1, or TRPV4 (Du et al., 2016), and in intracellular Ca^{2+} homeostasis (Qian et al., 2003; Sammels et al., 2010; Wegierski et al., 2009). TRPP2s formation of heteromeric complexes with other TRP channel subunits, notably TRPV4 and TRPC3, which are also expressed in cECs, give rise to channels with unique biophysics and modes of activation (Du et al., 2016). TRPP2 has been described in SMCs (Qian et al., 2003) and aECs (Yao & Garland, 2005), and its gene expression in cECs provides yet another avenue for Ca^{2+} entry—likely adding further dimensions to the sensing capabilities of these cells and enabling them to respond to additional signaling molecules and modalities.

6. Capillary K^+ channels

K^+ channels play critical roles controlling the Vm of excitable and non-excitable cells. Under physiological conditions, the resting Vm of aECs in pressurized vessels is typically between −25 and −40 mV (Behringer & Segal, 2012; Welsh & Segal, 1998). In the brain, extracellular K^+ is maintained around 3 mM. Assuming 140 mM cytosolic K^+, this places the K^+ equilibrium potential (E_K) at approximately −100 mV, establishing an enormous outward driving force for K^+ ions. Accordingly, when K^+ channels open in cECs, K^+ outflow will drive profound membrane hyperpolarization.

6.1 Capillary Kir2.1 channels mediate long-distance electrical signaling to arterioles to regulate blood flow

A total of 15 subunit isoforms of Kir channels exist, comprising seven subfamilies (Kir1.1; Kir2.1-4; Kir3.1-4; Kir4.1-2; Kir5.1; Kir6.1-2; and Kir7.1) (Hibino et al., 2010). Brain cECs express an array of Kir channels genes including *kcnj2* (encoding Kir2.1), *kcnj15* (Kir4.2), *kcnj8* (Kir6.1), and *kcnj13* (Kir7.1) (Vanlandewijck et al., 2018; Fig. 2A). Recently, we confirmed that the expression of Kir2.1 in cECs translates into functional channels in the plasmalemma, which play a key role in transducing fluctuations in external K^+ into electrical activity of the cEC membrane (Longden et al., 2017).

Kir2 channels are composed of homo- or hetero-tetrameric assemblies of two-transmembrane α subunits. Their simple structure—each consisting of two membrane-spanning domains connected by a pore-forming loop that also contributes to the K^+-selectivity filter, with intracellular N- and C-terminal domains (Tao, Avalos, Chen, & MacKinnon, 2009)—belies an intriguing and unique biophysical profile. These channels conduct K^+ much more readily in the inward direction than outward. This characteristic conductance profile (originally termed "anomalous rectification" by Katz (1949), now termed "strong inward rectification") results from positively-charged polyamines and Mg^{2+} ions being driven into the pore at depolarized Vm (Lopatin, Makhina, & Nichols, 1994; Matsuda, 1988). Over ~20–30 mV of depolarization from E_K, this block is incomplete. This allows the outward passage of K^+, and gives rise to the "outward hump" sometimes seen in current-voltage relationships recorded from cells with a very high density of Kir channels (e.g., Dhamoon et al., 2004). This outward conduction is the basis for the physiological roles of Kir2 channels, as it permits them to rapidly hyperpolarize the membrane to E_K. Indeed, these channels serve to amplify membrane hyperpolarization through a feed-forward, self-unblocking process—an initial hyperpolarization will reduce the driving force for polyamines to occupy the pore and this will cause unblock of a small number of channels, permitting further hyperpolarizing K^+ efflux. This begets further relief of polyamine block, and further hyperpolarization. This process can be initiated by extracellular K^+ ions, upon which the degree of inward rectification and polyamine block depend (Hagiwara & Yoshii, 1979; Kubo, 1996; Lopatin & Nichols, 1996). Thus, in response to a small increase in K^+, Kir2 channels can promote a rapid and self-perpetuating hyperpolarization to E_K.

We recently showed that these properties position cEC Kir2.1 channels as major regulators of brain blood flow (Longden et al., 2017; Longden & Nelson, 2015a). During normal neuronal activity, dramatic increases in extracellular K^+ up to 10 mM have been measured (Ballanyi, Doutheil, & Brockhaus, 1996). Such increases activate Kir2.1-containing channels in cECs to drive rapid Vm hyperpolarization. This is in turn transferred to adjacent cells where it elicits further Kir channel activation by relieving local polyamine block, and in this fashion the signal propagates from cell-to-cell along the capillary at >2 mm/s (Longden et al., 2017). This mechanism is a major contributor to the process of functional hyperemia, and establishes the ability of capillaries to actively monitor their environment and regulate local blood flow on an as-needed basis (Longden et al., 2017). In addition to the brain circulation, similar mechanisms likely exist in other tissues in which external K^+ is raised during cellular activity, for example, cardiac and skeletal muscle, and bladder smooth muscle (Fig. 4).

The membrane lipid environment of Kir channels also regulates their function, and binding of one molecule of PIP_2 per subunit is essential for Kir2 channel gating, through an interaction with amino acids at the interface between the transmembrane and cytoplasmic domains of the channel subunit (Hansen, Tao, & Mackinnon, 2012). Depletion of membrane PIP_2 levels through G_qPCR activity is thus an important negative regulator of Kir channel activity. In the context of blood flow control, such modulation of PIP_2 levels sets the gain of capillary-initiated electrical signaling by adjusting Kir2.1 channel activity (Harraz, Longden, Dabertrand, et al., 2018). Indeed, depletion of PIP_2 through G_qPCR activation deactivates electrical capillary-to-arteriole signaling (Harraz, Longden, Dabertrand, et al., 2018). This in combination with the role of PIP_2 in actively inhibiting TRPV4 channel activity (see above) has prompted a model in which G_qPCR signaling acts as a molecular switch between Kir-mediated electrical and TRPV4-mediated Ca^{2+} signaling modes in cECs (Harraz, Longden, Hill-Eubanks, et al., 2018).

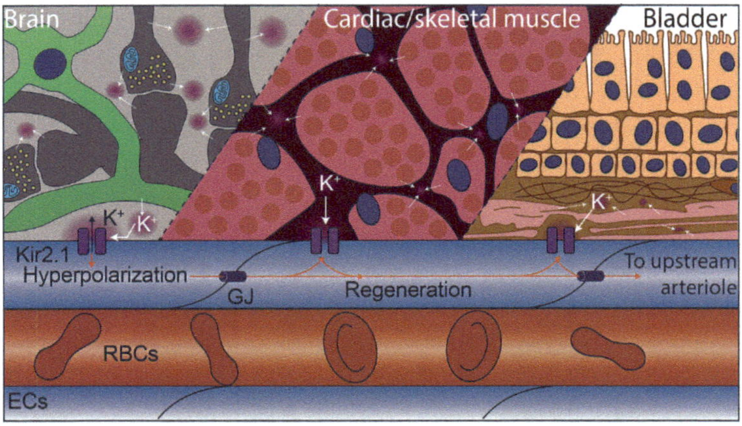

Fig. 4 Capillary electrical signaling is enabled by the extracellular K^+-sensing capability of Kir2.1-containing channels. Elevation of extracellular K^+ causes the release of K^+ from ECs via Kir channels, leading to membrane hyperpolarization. This is then transferred via gap junctions to adjacent cells where the signal is regenerated by engagement of resident Kir channels therein, by relief of voltage-dependent Mg^{2+} and polyamine block of the channel. In this way an electrical signal rapidly propagates along the capillary. Once it reaches upstream arterioles it is transferred to SMCs via gap junctions where it causes vasorelaxation to increase blood flow. This mechanism is known to operate in brain (left) and could hypothetically play a role in vasoregulation in other tissues in which K^+ is released during activity, such as muscle (middle) and bladder (right).

6.2 K_{ATP} channels may couple metabolism to cEC electrical activity

ATP-sensitive K^+ (K_{ATP}) channels couple cell metabolism to Vm and have a central role in insulin secretion, regulation of vascular tone, and excitability in muscle and neurons (Ashcroft, 2005). K_{ATP} channels are octamers composed of four α-subunits (Kir6.1 or Kir6.2) and four β-subunits (sulfonylurea receptors, SUR1 or SUR2) (Hibino et al., 2010). SURs are members of the ATP-binding cassette (ABC) family of proteins, which possess 17 transmembrane subunits grouped into three domains. Intracellular linkers connect each domain and each subunit has an extracellular N-terminus and an intracellular C-terminus (Hibino et al., 2010). The first two domains and the C-terminal tail possess nucleotide-binding domains that regulate open probability depending on the intracellular concentrations of ATP, adenosine diphosphate (ADP), and other nucleotides (Conti, Radeke, Shyng, & Vandenberg, 2001; Hibino et al., 2010; Linton & Higgins, 2006). Accordingly, decreases in intracellular ATP levels lead to increased channel activity, and vice versa. K_{ATP} channels are also regulated by a range of other factors: adenosine, calcitonin gene-related peptide, angiotensin II, pH, lactate, and the enzymes protein kinase (PK) A and PKC have all been found to modulate these channels (Mederos y Schnitzler et al., 2000). Protons (H^+) also directly activate K_{ATP} channels in vascular SMCs of mesenteric arteries (Wang et al., 2003).

The *kcnj8* gene, responsible for encoding the Kir6.1 channel, is expressed in cECs in the brain (Vanlandewijck et al., 2018), along with the SUR2 encoding gene *Abcc9*. Whether this translates into the presence of functional K_{ATP} channels here remains to be determined. If this is the case, these channels may have a role in sensing changes in the availability of local metabolites and converting this into electrical signaling to control blood flow. Notably, cardiac expression of both Kir6.1 and Kir6.2 subunits has been confirmed in cECs, and functional K_{ATP} channels have been recorded in freshly isolated capillaries (Mederos y Schnitzler et al., 2000).

6.3 Other Kir channels in capillaries

Modest expression of the intermediate rectifier Kir4.2 and weak rectifier Kir7.1 has also been detected in cECs (Vanlandewijck et al., 2018). Interestingly, Kir4.2 channels are also sensitive to K^+, but not on the rapid time scale of Kir2.1. Rather, external K^+ regulation of Kir4.2 takes tens of minutes and is thought to involve the engagement of silent channels in the membrane (Edvinsson et al., 2011). It is thus possible that these channels become active under pathophysiological conditions that involve chronic hyperkalemia, such as spreading depression. Kir7.1, on the other hand, does not exhibit an increase in outward current to increases in external K^+ (Krapivinsky et al., 1998). This channel has a very small conductance (Krapivinsky et al., 1998), which can be modulated by the activities of PKA and PKC (Zhang et al., 2008). Currents matching the electrophysiological fingerprint of either of these channels have yet to be observed in cECs.

6.4 K_v channels may be present in some capillary beds, but not others

Low expression of the voltage-gated K^+ (K_v) channel subunit $K_v1.5$ and more robust expression of the KCNQ channel subunit $K_v7.1$, encoded by *kcna5* and *kcnq1* genes, respectively, are also seen in cEC RNA-seq data (Vanlandewijck et al., 2018). K_v channels form a large group with 12 subfamilies (K_v1–K_v12), representing one of the most diverse families of K^+

channels (Gutman et al., 2005). These consist of four similar or identical pore-forming α-subunits that are often accompanied by auxiliary β-subunits that affect localization and/or channel function (Vacher, Mohapatra, & Trimmer, 2008). Each pore-forming subunit is composed of six transmembrane segments (S1–S6) with S4 possessing a variable number of positively-charged amino acids that form the voltage sensor, and S5–S6 constituting the pore domain (Yellen, 2002). The S4 voltage-sensing domain couples Vm depolarization to conformational changes that gate the pore (Mannuzzu, Moronne, & Isacoff, 1996; Swartz, 2008).

Despite this expression profile, the murine brain cortical cECs that have been tested to date show no evidence of functional K_V channels in electrophysiological recordings—depolarization as far as +100 mV has revealed no currents suggestive of the presence of functional K_V channels (Longden et al., 2017; Harraz, Longden, Dabertrand, et al., 2018; Harraz, Longden, Hill-Eubanks, et al., 2018). This disparity with RNA data could be due to regional differences in cells used for expression profiling and the cECs studied using electrophysiological approaches in the brain so far. Of note, however, K_v channel currents have been recorded in capillaries isolated from guinea pig heart (Dittrich & Daut, 1999), suggesting that at least some populations of cECs, or cECs from certain species, use these channels as a brake against depolarization.

7. Other ion channels in the capillary endothelium

In addition to the Ca^{2+}, TRP, and K^+ channels described above, molecular evidence hints at the potential expression of a range of other ion channels in cECs.

7.1 ClC-2 and HCN2 channels are candidates for membrane repolarization during electrical signaling

Cl^- channels are a structurally and functionally diverse group of anion-selective channels that regulate the excitability of many cell types (Nilius & Droogmans, 2003). Notably, the expression profile of cECs is almost devoid of Cl^--conducting channels, with only a low signal for a single Cl^- channel, ClC-2. In comparison, the expression of genes encoding H^+/Cl^- transporters is much more robust (Vanlandewijck et al., 2018). This expression profile suggests that H^+/Cl^- transporters play the predominant role in Cl^- handling in cECs, while ClC-2 may be expressed for a more specialized purpose.

ClC-2 is a plasma membrane protein of the voltage-dependent Cl^- channel (ClC) family, which contains 9 members (ClC-1 to ClC-7; ClC-Ka and ClC-Kb) (Jentsch, Stein, Weinreich, & Zdebik, 2002). These channels have a homodimeric double-barreled structure with each monomer containing 18 α helices, many of which are tilted and do not fully span the membrane, followed by a C-terminal region that contains cystathionine-β-synthase (CBS) domains, which play essential roles in the regulation of gating (Estévez, Pusch, Ferrer-Costa, Orozco, & Jentsch, 2004; Nilius & Droogmans, 2003). The ClC-2 channel possesses unique biophysical characteristics, molecular features, and pharmacological properties that distinguish it from other ClC family members—it inactivates very slowly, over seconds (Nilius & Droogmans, 2003), and it is activated by cell swelling, acidic pH, and PKA (Jordt & Jentsch, 1997). Intriguingly, a key biophysical characteristic of this channel is its activation by membrane hyperpolarization at potentials negative to ~ -40 mV (De Jesús-Pérez et al., 2016). This, in combination with its slow activation kinetics, makes it a reasonable candidate for repolarizing the membrane after Kir-mediated hyperpolarization. Indeed, its slow activation would allow plenty of time for the initiation of a propagating Kir-mediated hyperpolarization, following which slowly developing depolarization resulting from Cl^- efflux through ClC-2 would reset Vm to its resting level. (Here, we assume 50 mM intracellular Cl^-, similar to that recorded for SMCs (Aickin & Vermuë, 1983), and 130 mM extracellular Cl^-, as seen in cerebrospinal fluid (Longden et al., 2015), which places E_{Cl} at ~ -25 mV—close to resting endothelial Vm.)

Hyperpolarization-activated, cyclic nucleotide-gated (HCN) channels offer another possible mechanism for repolarization of the membrane following electrical signaling. These are Na^+- and K^+-permeable channels formed from tetrameric assemblies of six-transmembrane HCN1–4 subunits, with an S4 subunit with an opposite gating polarity to that of conventional K_V channels (Wahl-Schott & Biel, 2008). Thus, they are activated by hyperpolarization at voltages negative to ~ -50 mV, as are seen during capillary-to-arteriole electrical communication (Longden et al., 2017). The voltage-sensitivity of HCN2 and HCN4 channels can also be modulated by pathways that produce cyclic adenosine monophosphate, which shift it in a more positive direction and also accelerate the kinetics of activation (Wahl-Schott & Biel, 2008).

The *Hcn2* gene, encoding the HCN2 channel, is expressed in cECs (Vanlandewijck et al., 2018). The potential presence of functional HCN-2 channels thus raises the possibility that Na^+ influx resulting from its activation by hyperpolarization could also contribute to repolarization of Vm after an electrical signal has passed through the cEC in question.

7.2 Piezo1 may impart mechanical sensitivity to capillary endothelium

Piezo channels are large multipass transmembrane proteins that form mechanically activated ion channels important in touch and pain sensation. Piezos are Ca^{2+}-permeable, non-selective cation channels that form trimers with a central pore and propeller-like arms which are thought to sense mechanical forces, transducing these into channel activity (Coste et al., 2010; Wang et al., 2018; Wu, Lewis, & Grandl, 2017; Zhao et al., 2018).

Of the two Piezo isoforms that have been described (Piezo1 and Piezo2), Piezo1 is expressed robustly in cECs (Vanlandewijck et al., 2018), where it could mediate the sensation of mechanical forces associated with blood flow. In keeping with this, several studies have detected Piezo1 expression in ECs throughout the body, where it plays a crucial role in sensing blood-flow–associated shear stress to drive correct vascular development (Li, Hou, et al., 2014; Ranade et al., 2014). Indeed, loss of Piezo 1 impairs the ability of ECs to alter their alignment when subjected to directional shear stress (Ranade et al., 2014). Moreover, it has been proposed that Piezo1 directly regulates blood flow, through a mechanism involving NO formation in response to Piezo1 activation and subsequent vasodilation (Wang et al., 2016). Thus, in capillaries Piezo1 may play a similar role and convert mechanical forces into NO release.

7.3 Purinergic receptors may allow capillaries to sense local ATP release

Purinergic receptors are divided into three main classes: P1, P2Y, and P2X. P1 and P2Y are metabotropic receptors that bind to adenosine and ATP, respectively (von Kügelgen & Harden, 2011). P2X receptors are ionotropic, formed of homo- of heterotrimeric assembles of $P2X_1$ to $P2X_7$ subunits that function as ATP-gated cation channels and are permeable to Na^+, K^+ and Ca^{2+}, opening rapidly upon ATP binding (North, 2016; Surprenant & North, 2009). According to RNA-seq data, mRNAs for $P2X_4$ and $P2X_7$ receptors are robustly expressed in brain cECs (Vanlandewijck et al., 2018), which have similarly been detected in cultured venous ECs (Wilson et al., 2007). Notably, we have observed in preliminary work that the application of extracellular ATP to brain capillaries evokes robust Ca^{2+} elevations in these cells (Longden et al., 2019), although whether this occurs through activation of P2XRs remains to be determined. ATP also elicits a current with the pharmacological characteristics of $P2X_4$ in ECs of spiral ligament capillaries of the cochlea, causing a transient increase in Ca^{2+}, which contributes to the regulation of the cochlear blood flow (Wu et al., 2011). Furthermore, blocking P2X7Rs is neuroprotective against various neurological disorders, including stroke, traumatic brain injury and subarachnoid hemorrhage and this effect may be achieved through preservation of BBB integrity by reducing RhoA activity (Zhao et al., 2016). It is thus possible that, if functionally expressed in brain cECs, $P2X_4$ and $P2X_7$Rs could sense local ATP release from astrocytes, which glial cells employ to carry out intra- and intercellular communication (Guthrie et al., 1999).

Functional, amiloride-sensitive, epithelial Na^+ channels (ENaCs) have been described in aECs, where roles in sensing shear force, production of NO, and regulation of cell stiffness have been proposed (Tarjus et al., 2017; Wang et al., 2009). The *Scnn1a* gene, encoding the αENaC subunit, is expressed at a low level in brain cECs. It remains to be determined whether these channels are functional and contribute to cEC physiology.

The gene encoding GABA ρ subunits (*Gabrr2*) is also found at a low level in these cells. These can form homopentameric ligand-gated channels that respond to GABA with Cl^- influx. Thus, the expression of this gene could hint at a possible GABA-sensing capability of some cECs.

8. Capillary and arteriolar ECs express distinct ion channel complements

ECs form a continuous lining that permeates the entire body, separating blood from tissue. Given this macroscopic continuity, it is perhaps tempting to speculate that ECs are a homogenous cell type and that the proteins they express at disparate levels of the vascular tree are very similar. While superficially this may appear to be the case (capillary and arteriolar ECs express similar levels of key marker genes *Pecam1* and *Cdh5*, for example; Vanlandewijck et al., 2018), a closer examination reveals key differences in the molecular configurations of ECs at distinct levels of the vascular tree, which underscore differing functional roles.

Perhaps the most striking example among ion channels is the presence of SK and IK channels in arteriolar endothelium—for which a wealth of functional data exists (Burnham et al., 2002; Bychkov et al., 2002; Ledoux,

Bonev, et al., 2008; Ledoux, Taylor, et al., 2008)—contrasted with their complete absence in cECs (Longden et al., 2017). Indeed, we have found that Ca^{2+}-activated K^{+} currents could not be observed in cECs either in response to the selective SK and IK channel opener NS309, or in response to clamping cytosolic Ca^{2+} at 3 μM, well above the nM threshold required for channel activation (Longden et al., 2017). Thus, in cECs, Ca^{2+} signals lack this avenue to couple to Vm hyperpolarization, whereas in aECs coupling of Ca^{2+} to SK and IK channel activity plays a well-established and essential role in controlling smooth muscle Vm and contractile state.

Further differences in ion channel expression suggest additional degrees of distinction, and sensitivity to different modes of stimulation between arteriolar and capillary ECs—some ion channel genes that are expressed in cECs, such as *Trpc3*, *Trpm3*, *Kcnj15* and *Kcnj8* are absent in brain aECs. Similarly, brain aECs express a low levels of the ionotropic glutamate receptor genes *Grik1* and *Grik2*, whereas cECs do not (Fig. 2B). The relative expression of IP_3R genes is also an interesting point of difference between cECs and aECs, as cECs primarily express the IP_3R2 subtype, whereas aECs express a lower level of IP_3R2 which is approximately equal to their level of IP_3R3 expression, both of which are higher than IP_3R1. These molecular differences in particular, along with differences in cell morphology (cobblestone vs cylindrical orientation, the presence and absence of MEJs) and variation in expression and activity of other upstream receptors, channels and transporters, likely contributes to profound differences in EC Ca^{2+} signaling between the arteriolar and capillary compartments (for example, event frequencies, spatial spread, amplitudes, and durations). Future work is certain to reveal additional crucial differences between these vascular compartments, which will further underscore the divergent roles of ECs of capillaries and arterioles.

9. Summary and conclusion

It is clear that the ECs of capillaries are distinct from those of arteries and arterioles in terms of their ion channel expression and function. Modern "big data" approaches allow us to survey the transcriptional landscape of these cells in unprecedented, granular detail, which in turn promises to streamline functional explorations of ion channels and other molecules. The appraisal presented here makes it clear that we have only scratched the surface in terms of our understanding of ion channels in cECs and much work is needed before we can fully understand the function and roles of ion channels in capillaries throughout the body. Critical questions for future study include: (i) What are the spatiotemporal characteristics and roles of capillary Ca^{2+} signals, enabled by their expression of a range of Ca^{2+}-permeant ion channels, and how do these compare to Ca^{2+} signaling in aECs? (ii) Does the robust expression of a range of TRP channel isoforms and other channels impart currently unappreciated sensing capabilities to capillaries, such as the capability to sense mechanical forces, or substances released from parenchymal cells? (iii) What other K^+ channels are functional in cECs, and how do these contribute to control of Vm? (iv) How does cEC ion channel expression differ between organ systems—and do distinct ion channel expression patterns underlie differing roles and capabilities for these cells?

It is our hope that the information collated here can be used to guide further studies aiming to enrich our understanding of the ion channel expression and Ca^{2+} signaling properties of these fascinating and critically important vascular cells.

Acknowledgment

Support for this work was provided by the American Heart Association (17SDG33670237 and 19IPLOI34660108 to T.A.L.).

References

Aickin, C. C., & Vermuë, N. A. (1983). Microelectrode measurement of intracellular chloride activity in smooth muscle cells of guinea-pig ureter. *Pflügers Archiv*, *397*(1), 25–28.

Archer, S. L., Gragasin, F. S., Wu, X., Wang, S., McMurtry, S., & Kim, D. H. (2003). Endothelium-derived hyperpolarizing factor in human internal mammary artery is 11,12-epoxyeicosatrienoic acid and causes relaxation by activating smooth muscle BK_{Ca} channels. *Circulation*, *107*(5), 769–776.

Ashcroft, F. M. (2005). ATP-sensitive potassium channelopathies: Focus on insulin secretion. *Journal of Clinical Investigation*, *115*(8), 2047–2058.

Attwell, D., Mishra, A., Hall, C. N., O'Farrell, F. M., & Dalkara, T. (2016). What is a pericyte? *Journal of Cerebral Blood Flow and Metabolism*, *36*(2), 451–455.

Aydin, F., Rosenblum, W. I., & Povlishock, J. T. (1991). Myoendothelial junctions in human brain arterioles. *Stroke*, *22*(12), 1592–1597.

Bading, H. (2013). Nuclear calcium signalling in the regulation of brain function. *Nature Reviews Neuroscience*, *14*(9), 593–608.

Ballanyi, K., Doutheil, J., & Brockhaus, J. (1996). Membrane potentials and microenvironment of rat dorsal vagal cells in vitro during energy depletion. *The Journal of Physiology*, *495*(Pt. 3), 769–784.

Bayliss, W. M. (1902). On the local reactions of the arterial wall to changes of internal pressure. *The Journal of Physiology*, *28*(3), 220–231.

Behringer, E. J., & Segal, S. S. (2012). Tuning electrical conduction along endothelial tubes of resistance arteries through Ca^{2+}-activated K^+ channels. *Circulation Research*, *110*(10), 1311–1321.

Berridge, M. J., Lipp, P., & Bootman, M. D. (2000). The versatility and universality of calcium signalling. *Nature Reviews Molecular Cell Biology*, *1*(1), 11–21.

Bezprozvanny, L., Watras, J., & Ehrlich, B. E. (1991). Bell-shaped calcium-response curves of lns(1,4,5)P_3- and calcium-gated channels from endoplasmic reticulum of cerebellum. *Nature*, *351*(6329), 751–754.

Bleich, M., Shan, Q., & Himmerkus, N. (2012). Calcium regulation of tight junction permeability. *Annals of the New York Academy of Sciences*, *1258*(1), 93–99.

Blinder, P., Tsai, P. S., Kaufhold, J. P., Knutsen, P. M., Suhl, H., & Kleinfeld, D. (2013). The cortical angiome: An interconnected vascular network with noncolumnar patterns of blood flow. *Nature Neuroscience*, *16*(7), 889–897.

Bossu, J. L., Feltz, A., Rodeau, J. L., & Tanzi, F. (1989). Voltage-dependent transient calcium currents in freshly dissociated capillary endothelial cells. *FEBS Letters*, *255*(2), 377–380.

Brayden, J. E., Li, Y., & Tavares, M. J. (2012). Purinergic receptors regulate myogenic tone in cerebral parenchymal arterioles. *Journal of Cerebral Blood Flow & Metabolism*, *33*(2), 293–299.

Brenner, J. S., & Dolmetsch, R. E. (2007). TrpC3 regulates hypertrophy-associated gene expression without affecting myocyte beating or cell size. *PLoS One*, *2*(8), e802–e813.

Bunting, S., Gryglewski, R., Moncada, S., & Vane, J. R. (1976). Arterial walls generate from prostaglandin endoperoxides a substance (prostaglandin X) which relaxes strips of mesenteric and coeliac arteries and inhibits platelet aggregation. *Prostaglandins*, *12*(6), 897–913.

Burdakov, D., Petersen, O. H., & Verkhratsky, A. (2005). Intraluminal calcium as a primary regulator of endoplasmic reticulum function. *Cell Calcium*, *38*(3–4), 303–310.

Burnham, M. P., Bychkov, R., Feletou, M., Richards, G. R., Vanhoutte, P. M., Weston, A. H., et al. (2002). Characterization of an apamin-sensitive small-conductance Ca^{2+}-activated K^+ channel in porcine coronary artery endothelium: Relevance to EDHF. *British Journal of Pharmacology*, *135*(5), 1133–1143.

Bychkov, R., Burnham, M. P., Richards, G. R., Edwards, G., Weston, A. H., Félétou, M., et al. (2002). Characterization of a charybdotoxin-sensitive intermediate conductance Ca^{2+}-activated K^+ channel in porcine coronary endothelium: Relevance to EDHF. *British Journal of Pharmacology*, *137*(8), 1346–1354.

Campbell, W. B., & Falck, J. R. (2007). Arachidonic acid metabolites as endothelium-derived hyperpolarizing factors. *Hypertension*, *49*(3), 590–596.

Campbell, W. B., Gebremedhin, D., Pratt, P. F., & Harder, D. R. (1996). Identification of epoxyeicosatrienoic acids as endothelium-derived hyperpolarizing factors. *Circulation Research*, *78*(3), 415–423.

Cheng, X., Shen, D., Samie, M., & Xu, H. (2010). Mucolipins: Intracellular TRPML1-3 channels. *FEBS Letters*, *584*(10), 2013–2021.

Cipolla, M. J. (Ed.), (2009). Chapter 2, Anatomy and ultrastructure. In *The cerebral circulation*. San Rafael (CA): Morgan & Claypool Life Sciences.

Cipolla, M. J., Li, R., & Vitullo, L. (2004). Perivascular innervation of penetrating brain parenchymal arterioles. *Journal of Cardiovascular Pharmacology*, *44*(1), 1–8.

Conti, L. R., Radeke, C. M., Shyng, S. L., & Vandenberg, C. A. (2001). Transmembrane topology of the sulfonylurea receptor SUR1. *The Journal of Biological Chemistry*, *276*(44), 41270–41278.

Coste, B., Mathur, J., Schmidt, M., Earley, T. J., Ranade, S., Petrus, M. J., et al. (2010). Piezo1 and Piezo2 are essential components of distinct mechanically activated cation channels. *Science*, *330*(6000), 55–60.

Covington, E. D., Wu, M. M., & Lewis, R. S. (2010). Essential role for the CRAC activation domain in store-dependent oligomerization of STIM1. *Molecular Biology of the Cell*, *21*(11), 1897–1907.

De Jesús-Pérez, J. J., Castro-Chong, A., Shieh, R.-C., Hernández-Carballo, C. Y., De Santiago-Castillo, J. A., & Arreola, J. (2016). Gating the glutamate gate of CLC-2 chloride channel by pore occupancy. *The Journal of General Physiology*, *147*(1), 25–37.

Dhamoon, A. S., Pandit, S. V., Sarmast, F., Parisian, K. R., Guha, P., Li, Y., et al. (2004). Unique Kir2.x properties determine regional and species differences in the cardiac inward rectifier K^+ current. *Circulation Research*, *94*(10), 1332–1339.

Dittrich, M., & Daut, J. (1999). Voltage-dependent K^+ current in capillary endothelial cells isolated from guinea pig heart. *The American Journal of Physiology*, *277*(1), H119–H127.

Du, J., Fu, J., Xia, X.-M., & Shen, B. (2016). The functions of TRPP2 in the vascular system. *Acta Pharmacologica Sinica*, *37*(1), 13–18.

Dunn, K. M., Hill-Eubanks, D. C., Liedtke, W. B., & Nelson, M. T. (2013). TRPV4 channels stimulate Ca^{2+}-induced Ca^{2+} release in astrocytic endfeet and amplify neurovascular coupling responses. *Proceedings of the National Academy of Sciences of the United States of America*, *110*(15), 6157–6162.

Earley, S., & Brayden, J. E. (2015). Transient receptor potential channels in the vasculature. *Physiological Reviews*, *95*(2), 645–690.

Edvinsson, J. M., Shah, A. J., & Palmer, L. G. (2011). Potassium-dependent activation of Kir4.2 K^+ channels. *The Journal of Physiology*, *589*(24), 5949–5963.

Eng, C. L., Lawson, M., Zhu, Q., Dries, R., Koulena, N., Takei, Y., et al. (2019). Transcriptome-scale super-resolved imaging in tissues by RNA seqFISH. *Nature*, *568*(7751), 235–239.

Estévez, R., Pusch, M., Ferrer-Costa, C., Orozco, M., & Jentsch, T. J. (2004). Functional and structural conservation of CBS domains from CLC chloride channels. *The Journal of Physiology*, *557*(2), 363–378.

Fleming, I., & Busse, R. (2006). Endothelium-derived epoxyeicosatrienoic acids and vascular function. *Hypertension*, *47*(4), 629–633.

Flockerzi, V., & Nilius, B. (2007). *Transient receptor potential (TRP) channels*. Springer Science & Business Media (Part I).

Foskett, J. K., White, C., Cheung, K. H., & Mak, D. O. (2007). Inositol trisphosphate receptor Ca^{2+} release channels. *Physiological Reviews*, *87*(2), 593–658.

Garcia-Elias, A., Mrkonjic, S., Pardo-Pastor, C., Inada, H., Hellmich, U. A., & Rubio-Moscardó, F. (2013). Phosphatidylinositol-4, 5-biphosphate-dependent rearrangement of TRPV4 cytosolic tails enables channel activation by physiological stimuli. *Proceedings of the National Academy of Sciences of the United States of America*, *110*(23), 9553–9558.

Gonzales, A. L., Yang, Y., Sullivan, M. N., Sanders, L., Dabertrand, F., Hill-Eubanks, D. C., et al. (2014). A PLCγ1-dependent, force-sensitive signaling network in the myogenic constriction of cerebral arteries. *Science Signaling*, 7(327), ra49.

Gould, I. G., Tsai, P., Kleinfeld, D., & Linninger, A. (2016). The capillary bed offers the largest hemodynamic resistance to the cortical blood supply. *Journal of Cerebral Blood Flow & Metabolism*, *37*(1), 52–68.

Grant, R. I., Hartmann, D. A., Underly, R. G., Berthiaume, A. A., Bhat, N. R., & Shih, A. Y. (2017). Organizational hierarchy and structural diversity of microvascular pericytes in adult mouse cortex. *Journal of Cerebral Blood Flow & Metabolism*, *39*(3), 411–425.

Guo, J., She, J., Zeng, W., Chen, Q., Bai, X.-C., & Jiang, Y. (2017). Structures of the calcium-activated, non-selective cation channel TRPM4. *Nature*, *552*(7684), 205–209.

Guthrie, P. B., Knappenberger, J., Segal, M., Bennett, M. V., Charles, A. C., & Kater, S. B. (1999). ATP released from astrocytes mediates glial calcium waves. *Journal of Neuroscience*, *19*(2), 520–528.

Gutman, G. A., Chandy, K. G., Grissmer, S., Lazdunski, M., McKinnon, D., Pardo, L. A., et al. (2005). International Union of Pharmacology. LIII. Nomenclature and molecular relationships of voltage-gated potassium channels. *Pharmacological Reviews*, *57*(4), 473–508.

Haas, T. L., & Duling, B. R. (1997). Morphology favors an endothelial cell pathway for longitudinal conduction within arterioles. *Microvascular Research*, *53*(2), 113–120.

Hagiwara, S., & Yoshii, M. (1979). Effects of internal potassium and sodium on the anomalous rectification of the starfish egg as examined by internal perfusion. *Journal of Physiology*, *292*, 251–265.

Hamada, K., Miyatake, H., Terauchi, A., & Mikoshiba, K. (2017). IP3-mediated gating mechanism of the IP3 receptor revealed by mutagenesis and X-ray crystallography. *Proceedings of the National Academy of Sciences of the United States of America*, *114*(18), 4661–4666.

Hansen, S. B., Tao, X., & Mackinnon, R. (2012). Structural basis of PIP_2 activation of the classical inward rectifier K^+ channel Kir2.2. *Nature*, *477*(7365), 495–498.

Harraz, O. F., Longden, T. A., Dabertrand, F., Hill-Eubanks, D., & Nelson, M. T. (2018). Endothelial G_qPCR activity controls capillary electrical signaling and brain blood flow through PIP_2 depletion. *Proceedings of the National Academy of Sciences of the United States of America*, *115*(15), E3569–E3577.

Harraz, O. F., Longden, T. A., Hill-Eubanks, D., & Nelson, M. T. (2018). PIP_2 depletion promotes TRPV4 channel activity in mouse brain capillary endothelial cells. *eLife*, *7*, e38689.

Hartmann, D. A., Underly, R. G., Grant, R. I., Watson, A. N., Lindner, V., & Shih, A. Y. (2015). Pericyte structure and distribution in the cerebral cortex revealed by high-resolution imaging of transgenic mice. *Neurophotonics*, *2*(4), 041402.

He, L., Vanlandewijck, M., Mäe, M. A., Andrae, J., Ando, K., Del Gaudio, F., et al. (2018). Single-cell RNA sequencing of mouse brain and lung vascular and vessel-associated cell types. *Scientific Data*, *5*, 180160. See. http://betsholtzlab.org/VascularSingleCells/database.html.

Hibino, H., Inanobe, A., Furutani, K., Murakami, S., Findlay, I., & Kurachi, Y. (2010). Inwardly rectifying potassium channels: Their structure, function, and physiological roles. *Physiological Reviews*, *90*(1), 291–366.

Hill, R. A., Tong, L., Yuan, P., Murikinati, S., Gupta, S., & Grutzendler, J. (2015). Regional blood flow in the normal and ischemic brain is controlled by arteriolar smooth muscle cell contractility and not by capillary pericytes. *Neuron*, *87*(1), 95–110.

Horiuchi, T., Dietrich, H. H., Tsugane, S., & Dacey, R. G., Jr. (2001). Analysis of purine- and pyrimidine-induced vascular responses in the isolated rat cerebral arteriole. *American Journal of Physiology—Heart and Circulatory Physiology*, *280*(2), H767–H776.

Ignarro, L. J., Cirino, G., Casini, A., & Napoli, C. (1999). Nitric oxide as a signaling molecule in the vascular system: An overview. *Journal of Cardiovascular Pharmacology*, *34*(6), 879–886.

Iino, M. (1990). Biphasic Ca^{2+} dependence of inositol 1,4,5-trisphosphate-induced Ca release in smooth muscle cells of the guinea pig taenia caeci. *The Journal of General Physiology*, *95*(6), 1103–1122.

Jentsch, T. J., Stein, V., Weinreich, F., & Zdebik, A. A. (2002). Molecular structure and physiological function of chloride channels. *Physiological Reviews*, *82*(2), 503–568.

Jones, P. A., Scott-Burden, T., & Gevers, W. (1979). Glycoprotein, elastin, and collagen secretion by rat smooth muscle cells. *Proceedings of the National Academy of Sciences of the United States of America*, *76*(1), 353–357.

Jordt, S. E., & Jentsch, T. J. (1997). Molecular dissection of gating in the ClC-2 chloride channel. *The EMBO Journal*, *16*(7), 1582–1592.

Joutel, A., Haddad, I., Ratelade, J., & Nelson, M. T. (2015). Perturbations of the cerebrovascular matrisome: A convergent mechanism in small vessel disease of the brain? *Journal of Cerebral Blood Flow & Metabolism*, *36*(1), 143–157.

Katz, B. (1949). Les constantes electriques de la membrane du muscle. *Archives des Sciences Physiologiques*, *2*, 285–299.

Krapivinsky, G., Medina, I., Eng, L., Krapivinsky, L., Yang, Y., & Clapham, D. E. (1998). A novel inward rectifier K^+ channel with unique pore properties. *Neuron*, *20*(5), 995–1005.

Kubo, Y. (1996). Effects of extracellular cations and mutations in the pore region on the inward rectifier K^+ channel IRK1. *Receptors & Channels*, *4*(2), 73–83.

Ledoux, J., Bonev, A. D., & Nelson, M. T. (2008). Ca^{2+}-activated K^+ channels in murine endothelial cells: Block by intracellular calcium and magnesium. *The Journal of General Physiology*, *131*(2), 125–135.

Ledoux, J., Taylor, M. S., Bonev, A. D., Hannah, R. M., Solodushko, V., Shui, B., et al. (2008). Functional architecture of inositol 1, 4, 5-trisphosphate signaling in restricted spaces of myoendothelial projections. *Proceedings of the National Academy of Sciences of the United States of America*, *105*(28), 9627–9632.

Ledoux, J., Werner, M. E., Brayden, J. E., & Nelson, M. T. (2006). Calcium-activated potassium channels and the regulation of vascular tone. *Physiology*, *21*(1), 69–78.

Lee, R. M. (1995). Morphology of cerebral arteries. *Pharmacology and Therapeutics*, *66*(1), 149–173.

Lee, M. D., Wilson, C., Saunter, C. D., Kennedy, C., Girkin, J. M., & McCarron, J. G. (2018). Spatially structured cell populations process multiple sensory signals in parallel in intact vascular endothelium. *Science Signaling*, *11*(561), eaar4411.

Li, Y., Baylie, R. L., Tavares, M. J., & Brayden, J. E. (2014). TRPM4 channels couple purinergic receptor mechanoactivation and myogenic tone development in cerebral parenchymal arterioles. *Journal of Cerebral Blood Flow & Metabolism*, *34*(10), 1706–1714.

Li, J., Hou, B., Tumova, S., Muraki, K., Bruns, A., Ludlow, M. J., et al. (2014). Piezo1 integration of vascular architecture with physiological force. *Nature*, *515*(7526), 279–282.

Linton, K. J., & Higgins, C. F. (2006). Structure and function of ABC transporters: The ATP switch provides flexible control. *Pflügers Archiv*, *453*(5), 555–567.

Liou, J., Fivaz, M., Inoue, T., & Meyer, T. (2007). Live-cell imaging reveals sequential oligomerization and local plasma membrane targeting of stromal interaction molecule 1 after Ca^{2+} store depletion. *Proceedings of the National Academy of Sciences of the United States of America*, *104*(22), 9301–9306.

Longden, T. A., Dabertrand, F., Koide, M., Gonzales, A. L., Tykocki, N. R., Brayden, J. E., et al. (2017). Capillary K^+-sensing initiates retrograde hyperpolarization to increase local cerebral blood flow. *Nature Neuroscience*, *20*(5), 717–726.

Longden, T. A., Harraz, O. F., Hennig, G., Shui, B., Lee, F., Lee, J., et al. (2019). Neural activity drives dynamic Ca^{2+} signals in capillary endothelial cells that shape local brain blood flow. *FASEB Journal*, *33*, 688.8. (abstract).

Longden, T. A., Hill-Eubanks, D. C., & Nelson, M. T. (2015). Ion channel networks in the control of cerebral blood flow. *Journal of Cerebral Blood Flow & Metabolism*, *36*(3), 492–512.

Longden, T. A., & Nelson, M. T. (2015a). Vascular inward rectifier K^+ channels as external K^+ sensors in the control of cerebral blood flow. *Microcirculation*, *22*(3), 183–196.

Longden, T., & Nelson, M. (2015b). Unique ion channel properties of brain capillary endothelial cells. *The FASEB Journal*, *29*, 832.9. (abstract).

Lopatin, A. N., Makhina, E. N., & Nichols, C. G. (1994). Potassium channel block by cytoplasmic polyamines as the mechanism of intrinsic rectification. *Nature*, *372*(6504), 366–369.

Lopatin, A. N., & Nichols, C. G. (1996). [K^+] dependence of open-channel conductance in cloned inward rectifier potassium channels (IRK1, Kir2.1). *Biophysical Journal*, *71*(2), 682–694.

Mann, G. E., Yudilevich, D. L., & Sobrevia, L. (2003). Regulation of amino acid and glucose transporters in endothelial and smooth muscle cells. *Physiological Reviews*, *83*(1), 183–252.

Mannuzzu, L. M., Moronne, M. M., & Isacoff, E. Y. (1996). Direct physical measure of conformational rearrangement underlying potassium channel gating. *Science*, *271*(5246), 213–216.

Matsuda, H. (1988). Open-state substructure of inwardly rectifying potassium channels revealed by magnesium block in guinea-pig heart cells. *The Journal of Physiology*, *397*, 237–258.

McCarron, J. G., Wilson, C., Heathcote, H. R., Zhang, X., Buckley, C., & Lee, M. D. (2019). Heterogeneity and emergent behaviour in the vascular endothelium. *Current Opinion in Pharmacology*, *45*, 23–32.

Mederos y Schnitzler, M., Derst, C., Daut, J., & Preisig-Müller, R. (2000). ATP-sensitive potassium channels in capillaries isolated from guinea-pig heart. *The Journal of Physiology*, *525*(2), 307–317.

Mehta, D., Ahmmed, G. U., Paria, B. C., Holinstat, M., Voyno-Yasenetskaya, T., Tiruppathi, C., et al. (2003). RhoA interaction with inositol 1,4,5-trisphosphate receptor and transient receptor potential channel-1 regulates Ca^{2+} entry. *The Journal of Biological Chemistry*, *278*(35), 33492–33500.

Miyakawa, T., Maeda, A., Yamazawa, T., Hirose, K., Kurosaki, T., & Iino, M. (1999). Encoding of Ca^{2+} signals by differential expression of IP3 receptor subtypes. *The EMBO Journal*, *18*(5), 1303–1308.

Moncada, S., Gryglewski, R., Bunting, S., & Vane, J. R. (1976). An enzyme isolated from arteries transforms prostaglandin endoperoxides to an unstable substance that inhibits platelet aggregation. *Nature*, *263*(5579), 663–665.

Naffaa, M. M., Hung, S., Chebib, M., Johnston, G. A. R., & Hanrahan, J. R. (2017). GABA-ρ receptors: Distinctive functions and molecular pharmacology. *British Journal of Pharmacology*, *174*(13), 1881–1894.

Naylor, J., Li, J., Milligan, C. J., Zeng, F., Sukumar, P., Hou, B., et al. (2010). Pregnenolone sulphate- and cholesterol-regulated TRPM3 channels coupled to vascular smooth muscle secretion and contraction. *Circulation Research*, *106*(9), 1507–1515.

Nilius, B., & Droogmans, G. (2003). Amazing chloride channels: An overview. *Acta Physiologica Scandinavica*, *177*(2), 119–147.

Nishimura, N., Schaffer, C. B., Friedman, B., Lyden, P. D., & Kleinfeld, D. (2007). Penetrating arterioles are a bottleneck in the perfusion of neocortex. *Proceedings of the National Academy of Sciences of the United States of America*, *104*(1), 365–370.

North, R. A. (2016). P2X receptors. *Philosophical Transactions of the Royal Society B*, *371*(1700), 20150427.

O'Nejll, J. F. (1947). The effects on venous endothelium of alterations in blood flow through the vessels in vein walls, and the possible relation to thrombosis. *Annals of Surgery*, *126*(3), 270–288.

Park, C. Y., Hoover, P. J., Mullins, F. M., Bachhawat, P., Covington, E. D., Raunser, S., et al. (2009). STIM1 clusters and activates CRAC channels via direct binding of a cytosolic domain to Orai1. *Cell*, *136*(5), 876–890.

Peng, X., Zhang, C., Alkayed, N. J., Harder, D. R., & Koehler, R. C. (2004). Dependency of cortical functional hyperemia to forepaw stimulation on epoxygenase and nitric oxide synthase activities in rats. *Journal of Cerebral Blood Flow & Metabolism*, *24*(5), 509–517.

Pires, P. W., Pritchard, H. A. T., & Earley, S. (2018). Cerebral capillary TRPA1 channels mediate functional hyperemia via retrograde conducted vasodilation. *FASEB Journal*, *32*, 843.7 (abstract).

Popp, R., & Gögelein, H. (1992). A calcium and ATP sensitive nonselective cation channel in the antiluminal membrane of rat cerebral capillary endothelial cells. *Biochimica et Biophysica Acta, 1108*(1), 59–66.
Popp, R., Hoyer, J., Meyer, J., Galla, H. J., & Gögelein, H. (1992). Stretch-activated non-selective cation channels in the antiluminal membrane of porcine cerebral capillaries. *Journal of Physiology, 454*(1), 435–449.
Prakriya, M., & Lewis, R. S. (2015). Store-operated calcium channels. *Physiological Reviews, 95*(4), 1383–1436.
Purves, D., Augustine, G. J., Fitzpatrick, D., et al. (Eds.), (2001). Chapter 1. The blood supply of the brain and spinal cord. In *Neuroscience* (2nd ed.). Sunderland (MA): Sinauer Associates.
Qian, Q., Hunter, L. W., Li, M., Marin-Padilla, M., Prakash, Y. S., Somlo, S., et al. (2003). Pkd2 haploinsufficiency alters intracellular calcium regulation in vascular smooth muscle cells. *Human Molecular Genetics, 12*(15), 1875–1880.
Quick, K., Zhao, J., Eijkelkamp, N., Linley, J. E., Rugiero, F., Cox, J. J., et al. (2012). TRPC3 and TRPC6 are essential for normal mechanotransduction in subsets of sensory neurons and cochlear hair cells. *Open Biology, 2*(5), 120068.
Ranade, S. S., Qiu, Z., Woo, S. H., Hur, S. S., Murthy, S. E., Cahalan, S. M., et al. (2014). Piezo1, a mechanically activated ion channel, is required for vascular development in mice. *Proceedings of the National Academy of Sciences of the United States of America, 111*(28), 10347–10352.
Reitsma, S., Slaaf, D. W., Vink, H., van Zandvoort, M. A., & oude Egbrink, M. G. (2007). The endothelial glycocalyx: Composition, functions, and visualization. *Pflügers Archiv, 454*(3), 345–359.
Rodrigues, S. F., & Granger, D. N. (2015). Blood cells and endothelial barrier function. *Tissue Barriers, 3*(1–2), e978720.
Ruas, M., Chuang, K. T., Davis, L. C., Al-Douri, A., Tynan, P. W., Tunn, R., et al. (2014). TPC1 has two variant isoforms, and their removal has different effects on endo-lysosomal functions compared to loss of TPC2. *Molecular and Cellular Biology, 34*(21), 3981–3982.
Sammels, E., Devogelaere, B., Mekahli, D., Bultynck, G., Missiaen, L., Parys, J. B., et al. (2010). Polycystin-2 activation by inositol 1,4,5-trisphosphate-induced Ca^{2+} release requires its direct association with the inositol 1,4,5-trisphosphate receptor in a signaling microdomain. *The Journal of Biological Chemistry, 285*(24), 18794–18805.
Schaffer, C. B., Friedman, B., Nishimura, N., Schroeder, L. F., Tsai, P. S., Ebner, F. F., et al. (2006). Two-photon imaging of cortical surface microvessels reveals a robust redistribution in blood flow after vascular occlusion. *PLoS Biology, 4*(2), 258–270.
Schlingmann, K. P., Waldegger, S., Konrad, M., Chubanov, V., & Gudermann, T. (2007). TRPM6 and TRPM7—Gatekeepers of human magnesium metabolism. *Biochimica et Biophysica Acta, 1772*(8), 813–821.
Schwartz, E., Adamany, A. M., & Blumenfeld, O. O. (1981). Isolation and characterization of the internal elastic lamina from calf thoracic aorta. *Experimental and Molecular Pathology, 34*(3), 299–306.
Sender, R., Fuchs, S., & Milo, R. (2016). Revised estimates for the number of human and bacteria cells in the body. *PLoS Biology, 14*(8), e1002533.
Sexton, J. E., Desmonds, T., Quick, K., Taylor, R., Abramowitz, J., Forge, A., et al. (2015). The contribution of TRPC1, TRPC3, TRPC5 and TRPC6 to touch and hearing. *Neuroscience Letters, 610*, 36–42.
Shih, A. Y., Blinder, P., Tsai, P. S., Friedman, B., Stanley, G., Lyden, P. D., et al. (2013). The smallest stroke: Occlusion of one penetrating vessel leads to infarction and a cognitive deficit. *Nature Neuroscience, 16*(1), 55–63.
Stathopulos, P. B., Zheng, L., Li, G.-Y., Plevin, M. J., & Ikura, M. (2008). Structural and mechanistic insights into STIM1-mediated initiation of store-operated calcium entry. *Cell, 135*(1), 110–122.

Straub, A. C., Zeigler, A. C., & Isakson, B. E. (2014). The myoendothelial junction: Connections that deliver the message. *Physiology*, *29*(4), 242–249.
Surprenant, A., & North, R. A. (2009). Signaling at purinergic P2X receptors. *Annual Review of Physiology*, *71*(1), 333–359.
Swartz, K. J. (2008). Sensing voltage across lipid membranes. *Nature*, *456*(7224), 891–897.
Tallini, Y. N., Brekke, J. F., Shui, B., Doran, R., Hwang, S. M., Nakai, J., et al. (2007). Propagated endothelial Ca^{2+} waves and arteriolar dilation in vivo: Measurements in Cx40BAC GCaMP2 transgenic mice. *Circulation Research*, *101*(12), 1300–1309.
Tao, X., Avalos, J. L., Chen, J., & MacKinnon, R. (2009). Crystal structure of the eukaryotic strong inward-rectifier K^+ channel Kir2. 2 at 3.1 Å resolution. *Science*, *326*(5960), 1668–1674.
Tarjus, A., Maase, M., Jeggle, P., Martinez-Martinez, E., Fassot, C., Loufrani, L., et al. (2017). The endothelial αENaC contributes to vascular endothelial function in vivo. *PLoS One*, *12*(9) e0185319.
Trebak, M., Bird, G. S. J., McKay, R. R., Birnbaumer, L., & Putney, J. W., Jr. (2003). Signaling mechanism for receptor-activated canonical transient receptor potential 3 (TRPC3) channels. *The Journal of Biological Chemistry*, *278*(18), 16244–16252.
Vacher, H., Mohapatra, D. P., & Trimmer, J. S. (2008). Localization and targeting of voltage-dependent ion channels in mammalian central neurons. *Physiological Reviews*, *88*(4), 1407–1447.
Vanlandewijck, M., He, L., Mäe, M. A., Andrae, J., Ando, K., Del Gaudio, F., et al. (2018). A molecular atlas of cell types and zonation in the brain vasculature. *Nature*, *554*(7693), 475–480. Seehttp://betsholtzlab.org/VascularSingleCells/database.html.
Venkatachalam, K., Wong, C.-O., & Zhu, M. X. (2015). The role of TRPMLs in endolysosomal trafficking and function. *Cell Calcium*, *58*(1), 48–56.
von Kügelgen, I., & Harden, T. K. (2011). Molecular pharmacology, physiology, and structure of the P2Y receptors. *Advances in Pharmacology*, *61*, 373–415.
Vriens, J., Owsianik, G., Hofmann, T., Philipp, S. E., Stab, J., Chen, X., et al. (2011). TRPM3 is a nociceptor channel involved in the detection of noxious heat. *Neuron*, *70*(3), 482–494.
Wagner, T. F., Loch, S., Lambert, S., Straub, I., Mannebach, S., Mathar, I., et al. (2008). Transient receptor potential M3 channels are ionotropic steroid receptors in pancreatic β cells. *Nature Cell Biology*, *10*(12), 1421–1430.
Wahl-Schott, C., & Biel, M. (2008). HCN channels: Structure, cellular regulation and physiological function. *Cellular and Molecular Life Sciences*, *66*(3), 470–494.
Wang, S., Chennupati, R., Kaur, H., Iring, A., Wettschureck, N., & Offermanns, S. (2016). Endothelial cation channel PIEZO1 controls blood pressure by mediating flow-induced ATP release. *Journal of Clinical Investigation*, *126*(12), 4527–4536.
Wang, Y., Chi, S., Guo, H., Li, G., Wang, L., Zhao, Q., et al. (2018). A lever-like transduction pathway for long-distance chemical- and mechano-gating of the mechanosensitive Piezo1 channel. *Nature Communications*, *9*(1), 1300.
Wang, Y., Deng, X., & Gill, D. L. (2010). Calcium signaling by STIM and Orai: Intimate coupling details revealed. *Science Signaling*, *3*(148), pe42.
Wang, S., Meng, F., Mohan, S., Champaneri, B., & Gu, Y. (2009). Functional ENaC channels expressed in endothelial cells: A new candidate for mediating shear force. *Microcirculation*, *16*(3), 276–287.
Wang, X., Wu, J., Li, L., Chen, F., Wang, R., & Jiang, C. (2003). Hypercapnic acidosis activates K_{ATP} channels in vascular smooth muscles. *Circulation Research*, *92*(11), 1225–1232.
Watanabe, H., Vriens, J., Suh, S. H., Benham, C. D., Droogmans, G., & Nilius, B. (2002). Heat-evoked activation of TRPV4 channels in a HEK293 cell expression system and in native mouse aorta endothelial cells. *The Journal of Biological Chemistry*, *277*(49), 47044–47051.

Wegierski, T., Steffl, D., Kopp, C., Tauber, R., Buchholz, B., Nitschke, R., et al. (2009). TRPP2 channels regulate apoptosis through the Ca^{2+} concentration in the endoplasmic reticulum. *The EMBO Journal, 28*(5), 490–499.

Welsh, D. G., & Segal, S. S. (1998). Endothelial and smooth muscle cell conduction in arterioles controlling blood flow. *The American Journal of Physiology, 274*(1), H178–H186.

Wenzel, J., Hansen, C. E., Bettoni, C., Vogt, M. A., Lembrich, B., Natsagdorj, R., et al. (2020). Impaired endothelium-mediated cerebrovascular reactivity promotes anxiety and respiration disorders in mice. *Proceedings of the National Academy of Sciences of the United States of America, 117*, 1753–1761, pii: 201907467.

Wilson, H. L., Varcoe, R. W., Stokes, L., Holland, K. L., Francis, S. E., Dower, S. K., et al. (2007). P2X receptor characterization and IL-1/IL-1Ra release from human endothelial cells. *British Journal of Pharmacology, 151*(1), 115–127.

Wood, J., & Garthwaite, J. (1994). Models of the diffusional spread of nitric oxide: Implications for neural nitric oxide signalling and its pharmacological properties. *Neuropharmacology, 33*(11), 1235–1244.

Wu, T., Dai, M., Shi, X. R., Jiang, Z. G., & Nuttall, A. L. (2011). Functional expression of P2X4 receptor in capillary endothelial cells of the cochlear spiral ligament and its role in regulating the capillary diameter. *American Journal of Physiology Heart and Circulatory Physiology, 301*(1), H69–H78.

Wu, D. Q., Lee, C. H., Rhee, S. G., & Simon, M. I. (1992). Activation of phospholipase C by the alpha subunits of the Gq and G11 proteins in transfected Cos-7 cells. *The Journal of Biological Chemistry, 267*(3), 1811–1817.

Wu, J., Lewis, A. H., & Grandl, J. (2017). Touch, tension, and transduction—The function and regulation of piezo ion channels. *Trends in Biochemical Sciences, 42*(1), 57–71.

Wu, X., Zagranichnaya, T. K., Gurda, G. T., Eves, E. M., & Villereal, M. L. (2004). A TRPC1/TRPC3-mediated increase in store-operated calcium entry is required for differentiation of H19-7 hippocampal neuronal cells. *The Journal of Biological Chemistry, 279*(42), 43392–43402.

Yao, X., & Garland, C. J. (2005). Recent developments in vascular endothelial cell transient receptor potential channels. *Circulation Research, 97*(9), 853–863.

Yellen, G. (2002). The voltage-gated potassium channels and their relatives. *Nature, 419*(6902), 35–42.

Yip, H., Chan, W.-Y., Leung, P.-C., Kwan, H.-Y., Liu, C., Huang, Y., et al. (2004). Expression of TRPC homologs in endothelial cells and smooth muscle layers of human arteries. *Histochemistry and Cell Biology, 122*(6), 553–561.

Yu, F. H., & Catterall, W. A. (2003). Overview of the voltage-gated sodium channel family. *Genome Biology, 4*(3), 207.

Zhang, H., Prabhakar, P., Sealock, R., & Faber, J. E. (2010). Wide genetic variation in the native pial collateral circulation is a major determinant of variation in severity of stroke. *Journal of Cerebral Blood Flow and Metabolism, 30*(5), 923–934.

Zhang, S. L., Yu, Y., Roos, J., Kozak, J. A., Deerinck, T. J., Ellisman, M. H., et al. (2005). STIM1 is a Ca^{2+} sensor that activates CRAC channels and migrates from the Ca^{2+} store to the plasma membrane. *Nature, 437*(7060), 902–905.

Zhang, W., Zitron, E., Bloehs, R., Müller-Krebs, S., Scholz, E., Zeier, M., et al. (2008). Dual regulation of renal Kir7.1 potassium channels by protein Kinase A and protein Kinase C. *Biochemical and Biophysical Research Communications, 377*(3), 981–986.

Zhao, H., Zhang, X., Dai, Z., Feng, Y., Li, Q., Zhang, J. H., et al. (2016). P2X7 receptor suppression preserves blood-brain barrier through inhibiting RhoA activation after experimental intracerebral hemorrhage in rats. *Scientific Reports, 6*, 23286.

Zhao, Q., Zhou, H., Chi, S., Wang, Y., Wang, J., Geng, J., et al. (2018). Structure and mechanogating mechanism of the Piezo1 channel. *Nature, 554*(7693), 487–492.

Zihni, C., Mills, C., Matter, K., & Balda, M. S. (2016). Tight junctions: From simple barriers to multifunctional molecular gates. *Nature Reviews Molecular Cell Biology, 17*(9), 564–580.

Zlokovic, B. V. (2008). The blood-brain barrier in health and chronic neurodegenerative disorders. *Neuron*, *57*(2), 178–201.

Zong, X., Schieder, M., Cuny, H., Fenske, S., Gruner, C., Rötzer, K., et al. (2009). The two-pore channel TPCN2 mediates NAADP-dependent Ca^{2+}-release from lysosomal stores. *Pflügers Archiv*, *458*(5), 891–899.

Zonta, M., Angulo, M. C., Gobbo, S., Rosengarten, B., Hossmann, K. A., Pozzan, T., et al. (2003). Neuron-to-astrocyte signaling is central to the dynamic control of brain microcirculation. *Nature Neuroscience*, *6*(1), 43–50.

CHAPTER TEN

Excitability and contractility in arterioles and venules from the urinary bladder

Nathan R. Tykocki[b,*], **Frederick C. Monson**[a]
[a]West Chester University, West Chester, PA, United States
[b]Michigan State University, East Lansing, MI, United States
*Corresponding author: e-mail address: tykockin@msu.edu

Contents

Abstract

The urinary bladder performs two key physiological functions: (1) to store urine, and (2) void urine at an appropriate time. While these two functions seem simple, both processes exert prolonged stretch and compressive forces on the urinary bladder vasculature that are greater than those seen by vessels in any other hollow organ. To compensate for these forces, the urinary bladder vasculature has adapted several key features that maintain blood flow during bladder filling and prevent damaging pressure fluctuations during emptying. This chapter first describes key anatomical features of the urinary bladder vasculature and how these features aid in maintaining blood flow in the milieu of the functioning bladder. Next, we investigate the mechanisms regulating excitability of urinary bladder arterioles with emphasis on the development and regulation of myogenic tone. We then discuss the physiological significance and excitability of urinary bladder capillaries and venules, and their important roles in maintaining tissue perfusion. Finally, the functionality of the urinary bladder vasculature will be explored in terms of bladder dysfunction, to understand if lower urinary tract symptoms associated with disease can be considered vascular in nature. Also included are perspectives on the urinary bladder itself as a model for understanding ischemia/reperfusion injury and the possibility that the urinary bladder holds a key to mitigating deleterious effects that result when blood flow is occluded and rapidly restored to other organs.

Current Topics in Membranes, Volume 85
ISSN 1063-5823
https://doi.org/10.1016/bs.ctm.2020.01.003

1. Anatomy of the bladder vasculature

Normal urinary bladder function is a balancing act between urine storage and urine excretion. It is because of the time spent in these two phases that the urinary bladder vasculature finds itself subjected to forces not commonly seen in other hollow organs. When compared to the gall bladder, lung, or stomach, the urinary bladder has the added capability of continuously storing large volumes of noxious fluid for long periods of time. Additionally, the bladder can change its shape to accommodate these fluid volumes while maintaining very low pressures inside it, even when faced with the stressors placed upon the bladder in the abdominal cavity during sitting, standing, and movement (Robertson, 1999). Thus, every cell in the urinary bladder is designed to adapt to the specific cycles of expansion and contraction associated with normal bladder function (Fig. 1).

Detrusor smooth muscle remains relatively quiescent as the bladder fills, allowing the bladder wall to distend as stored urine volume increases (Andersson & Arner, 2004). This distention can be quite profound. In mice, bladder surface area increases from about 60 mm^2 when empty to over

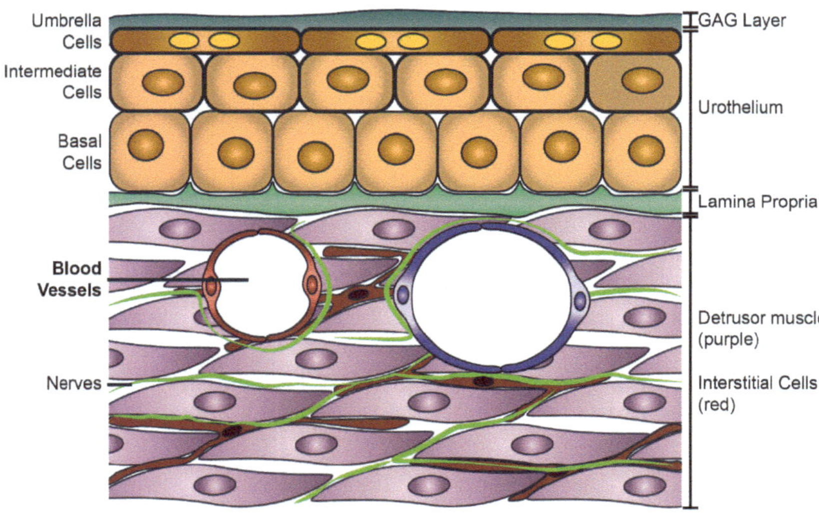

Fig. 1 Cartoon depicting the different layers of the murine urinary bladder wall. GAG, glycosaminoglycan. *Adapted from Neuhaus, J., & Schwalenberg, T. (2012). Intravesical treatments of bladder pain syndrome/interstitial cystitis.* Nature Reviews. Urology, *9(12), 707–720. doi: 10.1038/nrurol.2012.217.*

220 mm^2 immediately prior to micturition, all with minimal increases in intravesical pressure (Dorr, 1992; Fabiyi & Brading, 2006; Jost, 1985). As expected, during filling the bladder wall thickness also changes from around 400 μm when empty (Qiao et al., 2014) to less than 100 μm when full. This change is so profound that a mouse bladder is thin enough to appear translucent immediately prior to voiding. Thus, maintaining adequate blood flow during these changes is a physiological challenge. In this chapter, we will examine the unique anatomy and physiology of the urinary bladder, and how the contractile and biophysical properties of these vessels are essential for urinary bladder tissue perfusion and function. Special emphasis will be placed on urinary bladder arteriole and venule excitability, since it is in this aspect that bladder vessels markedly differ from other similar-sized microcirculation beds.

The urinary bladder is capable of adapting to such expansion because it is highly vascularized with a specialized network of arterioles, venules, and collateral capillaries that maintain perfusion during distention (Fig. 2A) (Hossler & Kao, 2007; Hossler, Kao, & Monson, 2005; Hossler & Monson, 1998a, 1998b). To further compensate for the longitudinal and axial stretch bladder vessels experience during bladder filling, the arterioles are also coiled within a specialized elastic sheath that allows them to elongate freely when stretched (Fig. 2B) (Hossler & Monson, 1998a; Kowalewska, Burrows, & Fox-Robichaud, 2011; Miodonski & Litwin, 1999). This serves to mitigate occlusion as the bladder wall thins. The arteriolar circulation feeds into a vascular plexus sandwiched between the detrusor muscle and the urothelium in the lamina propria, where it branches into the capillary bed and ultimately reconvenes into the venous circulation (Hossler & Kao, 2007). The dense capillary bed contains an extraordinary amount of collateral and arcade capillaries that form into a web of vascularization (Fig. 2C). The capillary bed is also covered with a large quantity of pericytes, present both along the length of the capillary and at capillary junctions (Hashitani & Lang, 2016). The role of junctional pericytes in the bladder is not known, but work from other vascular beds suggests they may serve to fine-tune blood flow through the distal capillary networks (Hall et al., 2014).

The convergence of bladder capillaries into the venular circulation is not at all like one might picture: a series of streams converging, two by two, into ever-wider creeks and eventually rivers. Instead, the capillaries form a hydra-like structure at the proximal venule where multiple capillaries converge to empty into a single vessel (Fig. 2D) (Hossler & Kao, 2007).

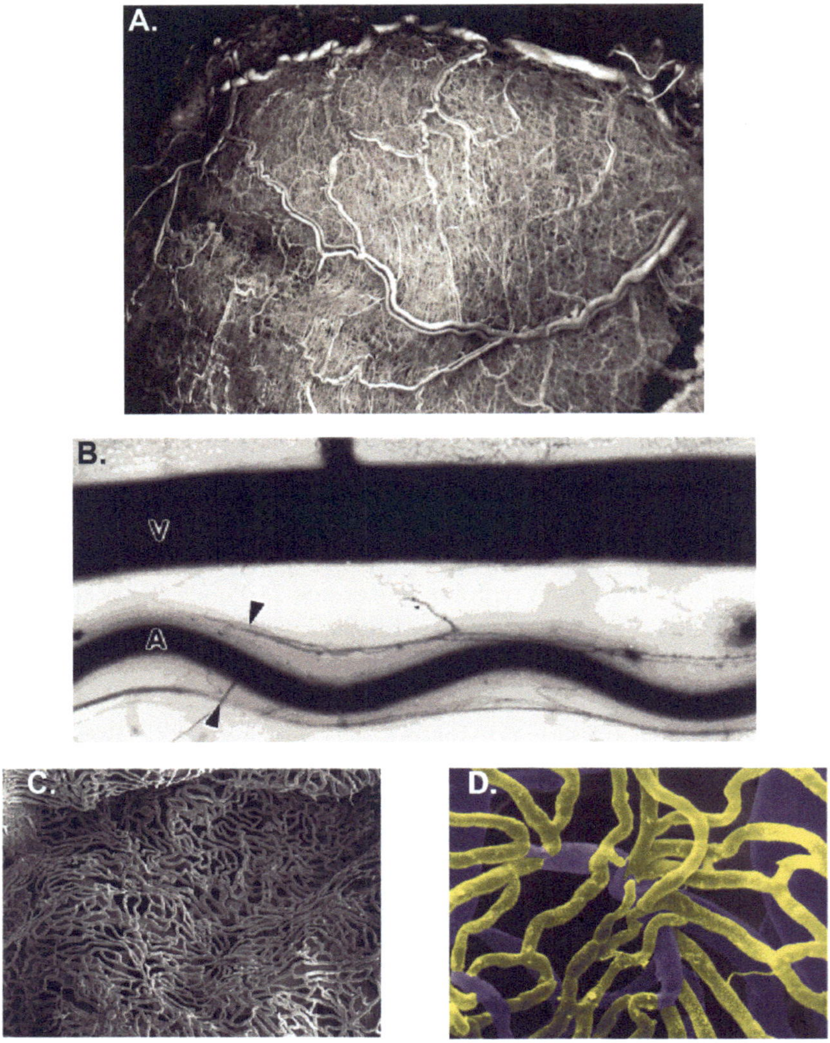

Fig. 2 (A) Corrosion casting of dog urinary bladder. (B) Bladder venule (V) and matching arteriole (A), showing elastic sheath surrounding the vessel.(C) Dense capillary bed in the dog urinary bladder. Capillaries are arranged in a net-like fashion and contain multiple arcade vessels and interconnections. (D) The capillaries connect to venules in a hydra-like manner, where multiple capillaries enter a single venule along the length of the vessel image is pseudo-colored to show capillaries in yellow and venules in blue. *Panel (A): From Hossler, F. E., & Kao, R. L. (2007). Microvasculature of the urinary bladder of the dog: A study using vascular corrosion casting.* Microscopy and Microanalysis, 13*(3), 220–227. doi: 10.1017/S1431927607070249. Panel (B): From Hossler, F. E., & Monson, F. C. (1998a). Evidence for a unique elastic sheath surrounding the vesicular arteries of the rabbit urinary bladder—Studies of the microvasculature with microscopy and vascular corrosion casting.* Anatomical Record, *252(3), 472–476. Panel (D): From Hossler, F. E., & Kao, R. L. (2007). Microvasculature of the urinary bladder of the dog: A study using vascular corrosion casting.* Microscopy and Microanalysis, 13*(3), 220–227. doi: 10.1017/S1431927607070249.*

As seen in other microcirculatory beds, the venules also contain multiple valve-like structures to prevent backflow of blood into the capillaries (Hossler & Kao, 2007). The purpose of a "hydra-like" arrangement is unclear but seems to be a result of the anatomical associations between the muscular and urothelial layers in an empty bladder. While the muscle is able to maintain its shape when full or empty, the urothelial layer has frequent deep invaginations due to its less-compliant nature and the lack of robust attachments to the underlying muscle (Chai, Russo, Yu, & Lu, 2016). Thus, as the bladder fills the urothelium expands and flattens, spreading the capillary bed outward without necessarily affecting the penetrating arterioles and venules needed to feed and maintain the vascular circuit. This arrangement further mitigates forces being transferred to the larger feed vessels as the bladder begins to fill, and still allows for adequate capillary perfusion of the mucosal layers of the bladder. The capillary web is so well interconnected that blood flow to the bladder wall is rapidly restored even after partial cystectomy with nothing more than a transient ischemic response, if at all (Hossler & Kao, 2007). This suggests that the arrangement of vessels within the wall is able to adapt to the ever-changing shape and size of the urinary bladder and circumvent any small area of disruption to the vascular network through adjacent collateral connections.

2. Physiology of the bladder vasculature

Yet even with all of the anatomical adaptations highlighted above, bladder blood flow does not remain constant. Bladder blood flow progressively decreases each and every time the bladder fills and leads to ischemia and hypoxia across the bladder wall (Pontari & Ruggieri, 1999). Amazingly, non-diseased urinary bladder smooth muscle is still able to generate voiding contractions in the face of prolonged states of ischemia normally associated with end-organ damage (Booth, Obeid, & Lucchesi, 2005; Wendt & Paul, 2003). The reasons appear to be twofold: (1) urinary bladder smooth muscle is able to generate sufficient ATP to maintain contractile functionality through glycolysis; and (2) hypoxia decreases the rate of cross-bridge cycling, thus increasing the economy of force maintenance and preserving contractile function during hypoxia (Nagai et al., 2016; Wendt & Paul, 2003). It appears that both the bladder vasculature and the bladder smooth muscle itself have adapted to the unique anatomical changes present during bladder function to ensure proper voiding can occur during the less-than-ideal circumstances (with regard to blood flow) of having a really full bladder. This change in blood flow as the bladder fills is also a clue that there is more—*and in some ways less*—to the regulation of bladder vascular tone than meets the eye.

2.1 Internal influences on bladder vascular tone

As referenced in previous chapters and discussed in many extensive reviews (Earley, 2013; Faraci & Sobey, 1998; Knot & Nelson, 1995), a central tenet of microvascular physiology is that small arteries and arterioles generate "myogenic tone." This means that when intraluminal pressure increases, smooth muscle cells contract to maintain constant blood flow through the capillary bed and to prevent large shifts in pressure from causing vascular damage. Myogenic tone is evident in the brain, for example, where cerebral blood flow remains constant even in the face of relatively large changes in blood pressure (Santos et al., 2016). Myogenic tone also provides the peripheral vascular resistance necessary for blood pressure homeostasis (Mulvany & Halpern, 1977). Generally speaking, the amount of myogenic tone a blood vessel is capable of producing is inversely proportional to its diameter. Small resistance arteries in the mesenteric circulation generate more tone than much larger-diameter cerebral vessels, and large conduit arteries produce little or no tone at all in the absence of endothelial cell damage. The arterioles of the urinary bladder, however, are an exception to this rule. In the presence of increasing intraluminal pressure, urinary bladder arterioles do not constrict (Tykocki, Bonev, Longden, Heppner, & Nelson, 2017). As shown in Fig. 3A, bladder arterioles lack any myogenic response. While increasing pressures do cause a smaller and smaller increase in intraluminal diameter, it is not indicative of myogenic tone. Rather this is caused by the passive stiffness of the vessel wall's extracellular matrix and muscle (VanBavel, Siersma, & Spaan, 2003). Furthermore, the lack of the myogenic response could not be attributed to traditional mechanisms by which endothelial cells regulate vascular tone, as block of nitric oxide synthase, small-conductance Ca^{2+}-activated K^+ (SK) channels, or intermediate-conductance Ca^{2+}-activated K^+ (IK) channels has no effect on vessel diameter (Tykocki, Bonev, et al., 2017). Thus, the lack of tone in these vessels was not simply due to shear- and flow-mediated mechanisms generated by the endothelium but rather something completely different.

The hallmark of myogenic tone is pressure-induced smooth muscle cell depolarization, which drives the opening of voltage-gated Ca^{2+} channels, calcium influx, and ultimately smooth muscle contraction.

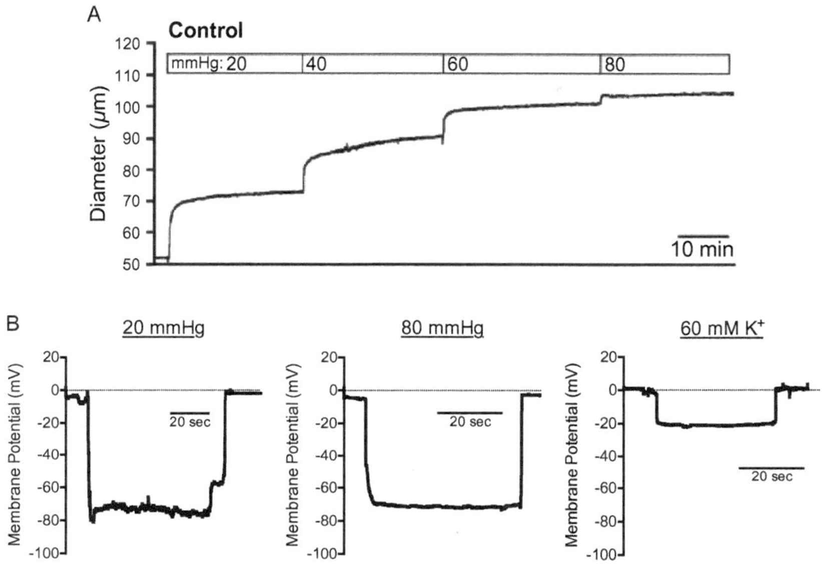

Fig. 3 (A) Diameter response of bladder arterioles to increasing intraluminal pressures, showing the absence of myogenic tone. (B) Sharp-electrode membrane potential recordings of intact bladder arterioles at 20 and 80 mmHg. Membrane potential is unchanged as pressure increases; the addition of 60 mM K^+ drives the membrane potential to −22 mV and causes a significant constriction (not shown). *From Tykocki, N. R., Bonev, A. D., Longden, T. A., Heppner, T. J., & Nelson, M. T. (2017). Inhibition of vascular smooth muscle inward-rectifier K(+) channels restores myogenic tone in mouse urinary bladder arterioles.* American Journal of Physiology. Renal Physiology, 312*(5), F836–F847. doi: 10.1152/ajprenal.00682.2016.*

In cerebral arteries, smooth muscle cell membrane potential will be around −65 mV at 20 mmHg. As pressure is increased to 80 mmHg, membrane potential depolarizes to nearly −40 mV and the vessel constricts (Knot & Nelson, 1998). Not so in the bladder vasculature: smooth muscle membrane potential stays roughly at −72 mV regardless of the intraluminal pressure applied (Fig. 3B). The cells are not incapable of depolarizing; when exposed to 60 mM extracellular K^+, membrane potential rapidly depolarizes to −22 mV (the reversal potential for K^+ (E_K)) and a robust constriction occurs in the vessel. Thus, the problem is not a lack of reactivity or inability to contract; bladder vessels simply appear to be insensitive to pressure-induced depolarization. But why? The responsible party turns out to be an inward-rectifying K^+ (K_{IR}) channel, particularly $K_{IR}2.1$ (Tykocki, Bonev, et al., 2017). In order to understand why, we first must understand how these peculiar ion channels work.

The biophysical properties of inward-rectifier K^+ channels are described in-depth in a previous review (Longden & Nelson, 2015); but, in short, $K_{IR}2$ channels are exquisitely sensitive to both extracellular K^+ concentration and membrane potential. Small increases in extracellular K^+ concentration (e.g., from 5 to 10 mM, $E_K = -68$ mV) cause nearly a 60-fold increase in K^+ conductance through the channel and lead to a robust hyperpolarization that approaches the reversal potential for K^+ (E_K). The result is a feed-forward effect, as hyperpolarization also activates the channel. Conversely, K_{IR} channels are inhibited by depolarization. If external K^+ concentration was increased more substantially (e.g., from 5 to 60 mM, $E_K = -22$ mV), membrane potential would depolarize, inhibit K_{IR} channels thus prevent K^+ efflux, and lead to smooth muscle contraction (Longden & Nelson, 2015). While it may seem confusing, the general idea is this: small changes in extracellular K^+ concentration that result in an E_K negative of smooth muscle cell resting membrane potential (~ -50 mV) activate K_{IR} channels and hyperpolarize the cell. Larger changes in extracellular K^+ concentration that cause E_K to be positive of resting membrane potential prevent K_{IR} channel opening and lead to depolarization. As such, the $K_{IR}2$ channel family is well-suited to respond to small changes in extracellular K^+ and membrane potential to maintain hyperpolarization in excitable cells.

Bladder blood vessel smooth muscle cells exhibit robust K_{IR} channel currents, typified by their increasing inward current at more negative membrane potentials and their sensitivity to Ba^{2+}, which blocks the K_{IR} channel pore (Alagem, Dvir, & Reuveny, 2001; Hsieh, Kuo, & Huang, 2015; Quayle, McCarron, Brayden, & Nelson, 1993) (Fig. 4A). In intact bladder arterioles, bladder smooth muscle cell membrane potential in the presence of 100 μM Ba^{2+} and at 20 mmHg is roughly −50 mV (Fig. 4B) (Tykocki, Bonev, et al., 2017). This value is similar to many other resistance arterioles that exhibit myogenic tone, at low intraluminal pressure where tone does not yet generate (Knot & Nelson, 1995; Kotecha & Hill, 2005). With K_{IR} channels blocked, increasing intraluminal pressure to 80 mmHg causes a significant depolarization (Fig. 4B) and a robust contraction. These findings suggest that K_{IR} channels are present in bladder arteriolar smooth muscle cells and are responsible for the maintained hyperpolarization and insensitivity to pressure. This was confirmed through the use of smooth muscle specific inducible $K_{IR}2.1$ knockout mice, which showed significant myogenic constriction as intraluminal pressure increased (Tykocki, Bonev, et al., 2017) (Fig. 4C). Thus, it is not that bladder arterioles are incapable

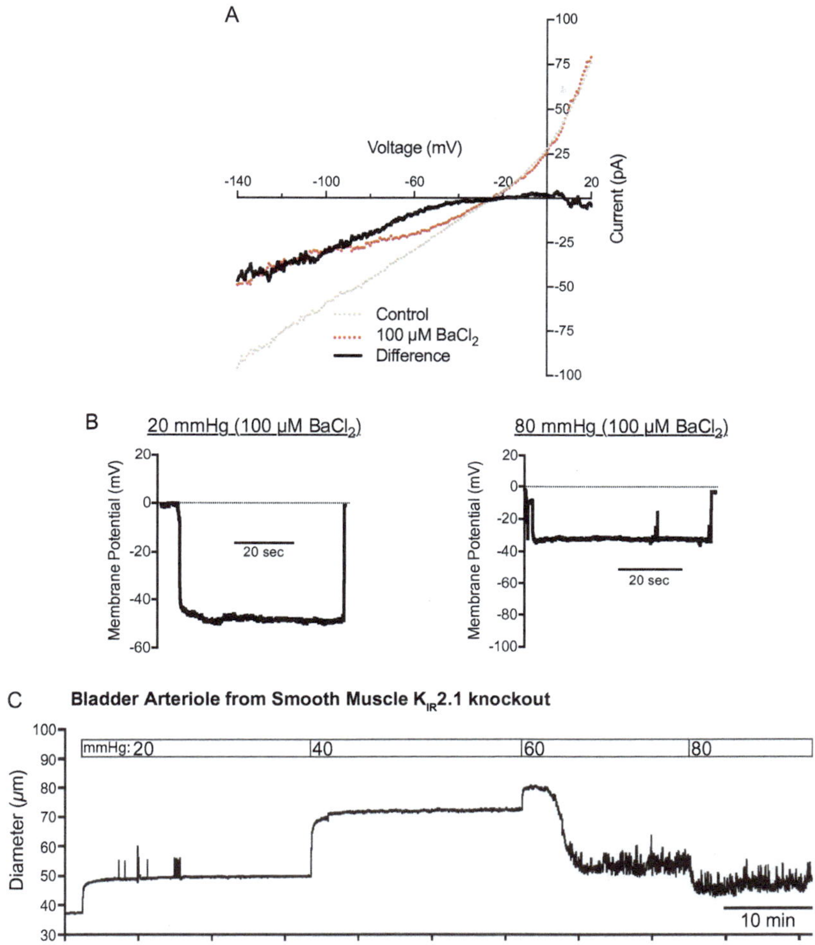

Fig. 4 (A) Barium-sensitive whole cell currents from isolated bladder arteriole smooth muscle cells. The inward-rectifying current is typical of members of the inward-rectifier K^+ (K_{IR}) channel family. (B) Knockout of $K_{IR}2.1$ channels from smooth muscle cells results in bladder arterioles that generate significant myogenic tone. (C) In the presence of $BaCl_2$, which blocks K_{IR} channels, bladder arteriolar smooth muscle cells depolarize in a pressure-dependent manner. *From Tykocki, N. R., Bonev, A. D., Longden, T. A., Heppner, T. J., & Nelson, M. T. (2017). Inhibition of vascular smooth muscle inward-rectifier K(+) channels restores myogenic tone in mouse urinary bladder arterioles.* American Journal of Physiology. Renal Physiology, 312*(5), F836–F847. doi: 10.1152/ajprenal.00682.2016.*

of myogenic tone; rather, the pressure-induced depolarizing conductances are incapable of overcoming the overwhelming hyperpolarizing influence of the $K_{IR}2.1$ channel and the K^+ conductance out of smooth muscle cells.

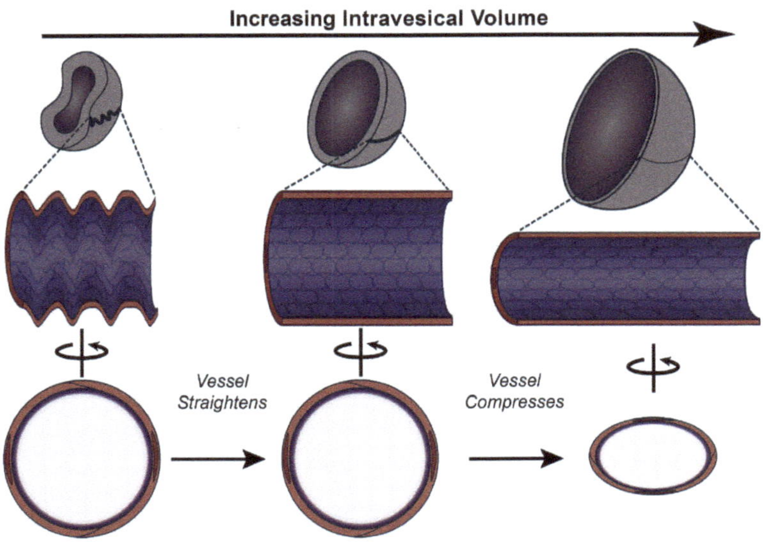

Fig. 5 Cartoon depicting the effects of bladder filling on bladder vascular shape and size. When empty, the bladder arterioles are coiled in an elastic sheath to allow free movement during filling. As the bladder begins to fill, these vessels extend and straighten without affecting intraluminal diameter. As the bladder continues filling, longitudinal stretch elongates the vessel, which in turn reduces intraluminal diameter. Simultaneously, thinning of the bladder wall compresses blood vessel shape, making the vessel less cylindrical and altering the characteristics of blood flow. *Adapted from Kozlowski, R., Siroky, M. B., Krane, R. J., & Azadzoi, K. M. (2002). Regulation of blood flow and microcirculation resistance in rabbit bladder.* The Journal of Urology, 168*(4 Pt. 1), 1608–1614. doi: 10.1097/01.ju.0000028482.95914.ba.*

When considering, from a physiological perspective, the tissue environment in which the bladder vasculature resides, having such a mechanism to prevent pressure-induced constriction makes logical sense. The cartoon in Fig. 5 represents the effects of bladder filling on small arterioles within the bladder wall. As the bladder wall stretches to accommodate urine, the vasculature is being pulled longitudinally while simultaneously being flattened as the bladder wall thins. The coiling of the vessels in their elastic sheath allows for some degree of elongation without affecting lumen diameter, but eventually the stretch is significant enough to cause the vessel diameter to decrease and thus reducing blood flow. The reason is illustrated below, using a simple version of Poiseuille's Law: flow (F) is inversely proportional to the viscosity (η) and tube length (L), and directly proportional to the change in pressure (ΔP) and radius (r) to the 4th power:

$$F \propto \frac{r^4 \Delta P}{\eta L}$$

Here is the same idea in plainer language: As the bladder fills, the vessel length increases and the vessel radius decreases, which markedly reduces blood flow. Since in a full bladder the vessel can no longer shorten or increase in radius, the only way to return blood flow to normal is to: (a) decrease the viscosity of blood, or (b) increase the pressure. Changing blood viscosity would have deleterious effects to all the other organs in the body, so increasing pressure is the only option. Thus, these vessels are specialized to stay as maximally dilated as possible during bladder filling and as pressure increases, to maintain perfusion over the prolonged time of urine storage. If bladder arterioles exhibited myogenic tone, *any* increase in pressure would feed forward to rapidly reduce blood flow even more and increase the potential for prolonged tissue ischemia and damaging hypoxia. Even so, the lack of tone is not enough to compensate and bladder tissue perfusion still decreases during bladder filling, resulting in tissue ischemia (Pontari, Hanno, & Ruggieri, 1999; Pontari & Ruggieri, 1999).

To summarize: bladder vessels experience significant mechanical forces that limit their ability to adequately perfuse the bladder tissue, simply through the normal expansion of the bladder as it serves to store urine. These vessels lack traditional pressure-induced constriction, or "myogenic tone," as an adaptation to maintain tissue perfusion for as long as possible. But how, then, is blood flow regulated in an empty bladder? The next section will discuss external influences on bladder vascular tone and propose potential mechanisms to prevent capillary damage when the bladder is empty.

2.2 External influences on bladder vascular tone

2.2.1 Neuronal influences: Sympathetic nerves

While we know some of the intrinsic factors (or lack thereof) that regulate bladder arteriolar tone, very little is known about the contractile properties of the bladder vasculature with regard to their response to external vasoconstrictors and/or vasodilators. Vessel diameter regulation does appear to be attributable, at least in part, to perivascular nerve innervation. Isolated bladder arterioles constrict significantly to the adrenergic agonist phenylephrine (Tykocki, Bonev, et al., 2017) and electrical field stimulation-induced constriction of submucosal arterioles is blocked by the α_1 antagonist prazosin. Together with immunochemical staining of perivascular nerves, these findings indicate sympathetic innervation of bladder blood vessels (Hashitani, Takano, Fujita, Mitsui, & Suzuki, 2011). However, other nerves—including nitrergic, cholinergic and sensory nerves—also innervate these vessels, suggesting that neural control of bladder blood flow is heterogeneous and complex (Mitsui & Hashitani, 2013).

2.2.2 Neuronal influences: Sensory nerves

While exploring the roles of sensory nerves in regulating bladder vascular tone, it was discovered that the transient receptor potential vanilloid type 1 (TRPV1) channel, often considered the hallmark indicator of bladder C-fiber afferents, was sporadically expressed in bladder arteriolar smooth muscle cells (Phan, Ton, Chen, Basha, & Ahern, 2016) (see Tykocki, Boerman, and Jackson (2017) for more information on the properties of TRPV channels). As seen in Fig. 6, we have also recapitulated these findings and show that TRPV1-expressing sensory nerves are in close apposition to the bladder vasculature, with TRPV1 expression also being present sparsely in vascular smooth muscle. Our findings are not

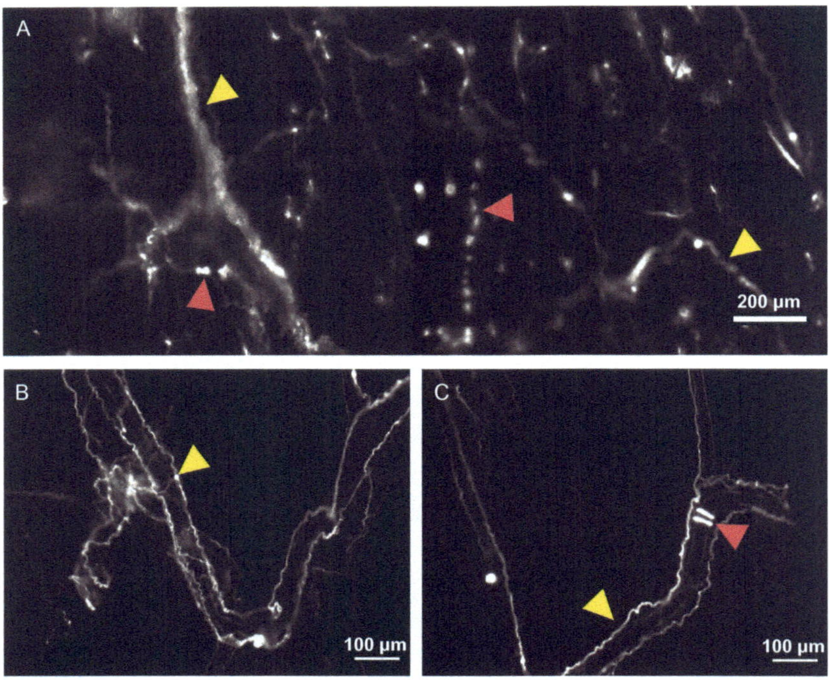

Fig. 6 Expression of transient receptor potential vanilloid type 1 (TRPV1) channels in urinary bladder nerves and arteries. (A) Wide view of a flat mounted urinary bladder, showing tdTomato expression driven by the *Trpv1* promotor. (B) Close-up of sensory nerves traveling in close apposition to blood vessels, presumably venules dues to the lack of smooth muscle cell expression. (C) Close-up of sensory nerves along a bladder arteriole, with two smooth muscle cells also expressing TRPV1, showing the intermittent nature of TRPV1 expression in bladder vessels. Yellow arrows: nerves. Pink arrows: vascular smooth muscle cells.

just a fate map of cells previously expressing TRPV1 channels, as Phan et al. demonstrated that TRPV1$^+$ smooth muscle cells individually constricted in response to the prototypical TRPV1 agonist capsaicin (Phan et al., 2016). Thus, factors that affect the sensitization of sensory nerves in close proximity to the bladder vasculature (e.g., mast cell degranulation) may also indirectly influence contractility of the vasculature itself, independent of release of any neurotransmitter. Further research is needed to determine if changes in vascular tone that were previously attributed to sensory nerve activity are instead direct actions on the vascular smooth muscle cells themselves.

2.2.3 Neuronal influences: Nitrergic nerves

While bladder blood flow does significantly reduce during filling, it also significantly increases during micturition (Pontari & Ruggieri, 1999). This is not due solely to alleviation of mechanical stress of the vessels, as inhibition of nitric oxide synthase (and thus NO production) blunted the increase in blood flow. Since inhibiting NO production did not increase tone in isolated bladder arterioles (see above), the NO sensitivity of blood flow increases during voiding are likely through endothelial NO-independent mechanisms. For example, detrusor smooth muscle appears to be insensitive to NO, yet nitrergic nerves are found throughout the bladder wall (Ehren, Iversen, Jansson, Adolfsson, & Wiklund, 1994; Gillespie, 2004; Longhurst, Briscoe, Rosenberg, & Leggett, 1997). These nerves may serve to release NO directly on vessels to oppose other vasoconstrictor elements, or to act on the poorly-understood network of interstitial cells that crisscross the bladder wall and interact with the vasculature, nerves, and urinary bladder smooth muscle (Persson, Alm, Uvelius, & Andersson, 1998). The latter is not outside the realm of possibility, as nitrergic and cholinergic nerve interactions are documented in Interstitial Cells of Cajal in the nearby colon (Lies et al., 2015).

2.2.4 The influence of non-neuronal vasoactive substances

It is also possible that bladder blood flow is regulated locally through the release of vasoactive substances from within the bladder wall from non-nerve sources, specifically the bladder urothelium and vascular endothelium. NO, acetylcholine, endothelin-1 (ET-1), and ATP are all released from the urothelium as the bladder fills (Ferguson, Kennedy, & Burton, 1997; Hanna-Mitchell et al., 2007; Khan et al., 2003; Mumtaz et al., 1999). Also, changes in NO, ET-1, and ATP release from the urothelium correlate with urinary bladder dysfunction (Khan et al., 2003; Munoz, Smith, Boone, & Somogyi, 2011). All of these substances have profound effects on vascular tone and could alter blood flow significantly during filling. This suggests that the bladder vasculature is regulated locally through the release of vasoactive substances from the urothelium, and that the resultant changes in blood flow regulate bladder function. In all, these findings suggest that the decrease in blood flow during filling is not solely the result of mechanical deformation, and that active regulation of bladder blood flow occurs before, during, and after micturition through neuronal and non-neuronal means.

2.3 Bladder venules: More than just capacitive vessels

In nearly every organ, the roles of venules are overlooked. Often thought of as nothing other than a catch-all of capillary outflow, the venous circulation has not had nearly the research attention given to the arteriolar circulation. Given that our circulatory system is a closed loop, it seems short-sighted to ignore a segment of the vasculature that can have profound effects on tissue perfusion and blood pressure regulation. In fact, because roughly 70% of total circulating blood is held in the venous system, a relatively small shift of blood from the venous to arterial circulation would profoundly affect blood pressure and systemic hemodynamics (Fink, 2009). Also important to consider are the roles of veins in maintaining capillary perfusion pressure and flow since these vessels serve to allow blood to leave the capillary bed in a timely and controlled fashion (Segal, 2005). This second role becomes immensely important in the bladder, where blood could easily become trapped as flow reduces during filling. But for the venous circulation to affect change in any hemodynamic parameters, they must be able to constrict and dilate. So, can they? Do veins have smooth muscle, or are they just covered in pericytes? And if they are vasoactive, do veins have a role in regulating vascular resistance in the bladder?

The answer depends on how you define each cell type. If smooth muscle α-actin defines a smooth muscle cell, then both vessels contain smooth muscle. The contractile cells in both bladder arterioles and venules contains this protein, which is considered requisite for smooth muscle contractility (Mitsui & Hashitani, 2013). Additionally, if you define a pericyte as being positive for markers like c-Kit, Ano-1, or PDGFRβ, (all common to capillary pericytes) then the case for veins containing smooth muscle is even stronger (Hashitani, Mitsui, Shimizu, Higashi, & Nakamura, 2012).

However, morphological, contractile, and electrophysiological characteristics of the contractile cells found in bladder venules tell a very different (and important) story. Contractile cells in bladder arterioles have the "finger-like" appearance typical of the classically-defined smooth muscle cells from other microvascular beds (Billaud et al., 2011) (Fig. 7). By contrast, the α-SMA-positive cells on venules have stellate processes that

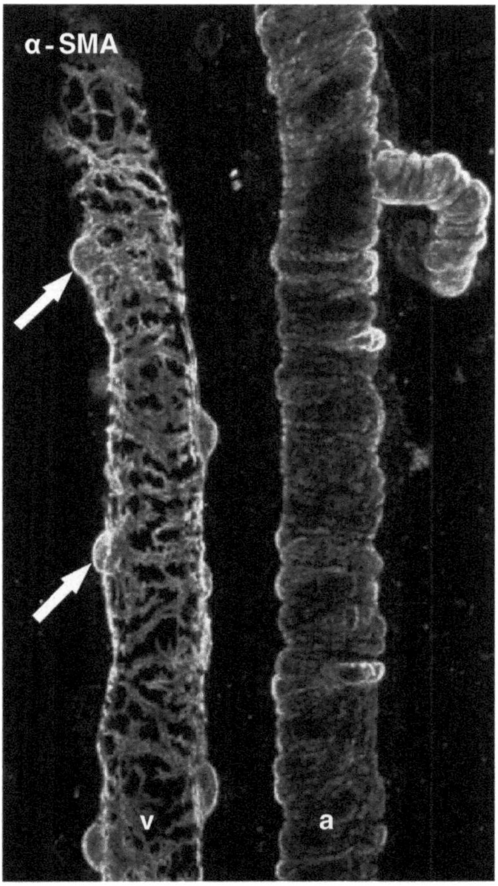

Fig. 7 Expression of smooth muscle α-actin (α-SMA) in a bladder arteriole (A) and paired venule (V). Cells expressing α-SMA in the arteriole have the "finger-like" appearance typical of classically defined smooth muscle cells from other microvascular beds. By contrast, α-SMA-positive cells on venules have stellate processes that enwrap the vessel and "bump on a log" nuclei traditionally associated with capillary pericytes. *From Mitsui, R., & Hashitani, H. (2013). Immunohistochemical characteristics of suburothelial microvasculature in the mouse bladder.* Histochemistry and Cell Biology, 140*(2), 189–200. doi: 10.1007/s00418-012-1074-5.*

enwrap the vessel and "bump on a log" nuclei, both of which are traditionally associated with pericytes (Sakagami, Wu, & Puro, 1999). Furthermore, unlike their arteriolar counterparts, bladder venules exhibit phasic contractions (around 5 per minute) that propagate along the length of the vessel. These contractions are insensitive to tetrodotoxin but are blocked by nicardipine and cyclopiazonic acid, suggesting that their activity depends on calcium influx through voltage-gated Ca^{2+} channels and intracellular calcium stores release through IP_3 receptors, and not through nerve-mediated mechanisms (Hashitani et al., 2011). Interestingly, the phasic contractions are insensitive to ryanodine, meaning that contraction of bladder venules is uncoupled from RyR-mediated calcium release (another fact differentiating venous pericytes from arteriolar smooth muscle). This relationship is also noted in large veins from the rat, suggesting that it may be common throughout the venous circulation (Tykocki, Thompson, Jackson, & Watts, 2013). The amplitude and frequency of bladder venule phasic contractions are also altered by vasoactive substances in several peculiar ways. The prototypical endothelium-dependent vasodilator acetylcholine *increased* the frequency of venular contractions, whereas ATP transiently *constricted* then *dilated* venules in an NO-dependent manner. Venules were also sensitive to substance-P and calcitonin gene-related peptide (CGRP), both of which are released from sensory nerves closely apposed to the vessel (Shimizu, Mochizuki, Mitsui, & Hashitani, 2014).

The importance of these findings with regard to urinary bladder physiology is enormous. It is long-known that acetylcholine is the primary neurotransmitter driving bladder muscle contraction, but the role of ATP released from the urothelium has remained somewhat contentious. The findings by Shimizu et al. suggest that both ACh and ATP, originally thought to be solely affecting detrusor contractility, instead have significant roles in altering venous tone/function and thus bladder circulation. Thus, it appears that it is arterioles—not veins and venules—which serve as the passive conduit for blood, and blood flow regulation through neurohumoral cross-talk and urothelium-dependent feedback occurs in the venous circulation of the bladder.

3. Changing bladder blood flow changes bladder function

An interesting dichotomy exists in bladder function with regard to the role of the vasculature. On one hand, the mechanisms outlined above suggest that the bladder muscle has a significant ability to generate forceful contractions even in the absence of adequate blood flow, and thus the role of the vasculature in bladder function may be inconsequential. On the other hand, many of the same findings suggest that changes in bladder blood flow are integral in the sensation of bladder fullness, such that sensory information to the brain is influenced by tissue perfusion/ischemia and can directly result in (or prevent) micturition. This second point has been woefully under-explored, even though it supports the idea that bladder dysfunction could be, in part, a vascular disease.

One of the most striking (and under-investigated) links between bladder blood flow and function is the fact that chronic bladder ischemia alone causes symptoms of overactive bladder (Andersson, Boedtkjer, & Forman, 2017; Azadzoi, Tarcan, Kozlowski, Krane, & Siroky, 1999). Further prolonging the ischemia, however, results in tissue damage and a loss of bladder function (i.e., underactive bladder) (Azadzoi et al., 1999). Exaggerated hypoxia, caused by inflammation, outlet obstruction or vesical artery occlusion, chronically reduces bladder blood flow and causes bladder dysfunction that progresses from overactivity to underactivity and ultimately to bladder necrosis if blood flow is not restored (Azadzoi et al., 1999; Pontari et al., 1999). Vesical artery occlusion is also associated with significant remodeling of the bladder wall, including collagen deposition and urothelial barrier disruption, and production of pro-angiogenic factors and reactive oxygen species (Azadzoi et al., 1999; Ghafar et al., 2002; Lin & Juan, 2012). This reinforces and provides direct evidence for several key ideas about bladder blood flow as it relates to function which were suggested previously in this chapter: (1) that the bladder is sensitive to changes in blood flow, (2) that even a mild reduction in blood flow leads to hypoxia during bladder filling, and (3) that prolonged changes in blood flow can cause bladder overactivity, remodeling, and progressive loss of bladder function if untreated.

3.1 Is the bladder vasculature protected from ischemia/reperfusion injury?

One of the most interesting things about the bladder vasculature, and the bladder itself, is their resilience in the face of acute ischemic/hypoxic conditions. As stated above, blood flow markedly drops during bladder filling before massively rebounding while the bladder is emptying (Pontari & Ruggieri, 1999). Thus, the urinary bladder is undergoing constant cycles of ischemia and reperfusion (I/R) during our entire lifetime with minimal deleterious effects, assuming normal bladder function and the absence of outlet obstruction. To cause significant I/R-mediated bladder dysfunction in a rabbit bladder requires complete occlusion of both vesical artery branches for 2h (Juan et al., 2009). For the sake of comparison, significant myocardial infarction and tissue injury in the heart occurs after just 30 min of occlusion of the left anterior descending carotid artery (Booth et al., 2005). Even after 2h, no infarct or overt tissue damage is visible in the bladder, suggesting that changes in bladder function instead require ischemia-driven remodeling to occur (Nomiya et al., 2012). Why is it that acute ischemia/reperfusion injury damages other organs and other vascular beds so extensively, yet the bladder appears to be protected from it?

The difference, in part, may come from the varying metabolic demands of the various layers of the bladder wall. The metabolic demand of detrusor smooth muscle is nearly three times lower than that of the mucosal layers, suggesting that simply fewer nutrients and oxygen are required to maintain muscle cell viability compared to urothelial cells (Hypolite et al., 1993). Chronic hypoxia/ischemia led to a loss of urothelium-derived prostaglandins and leukotrienes, which directly altered detrusor smooth muscle contractility in isolated muscle strips (Azadzoi, Heim, Tarcan, & Siroky, 2004). Yet in nearly all these studies, detrusor smooth muscle contractility is maintained even in the face of prolonged occlusion of blood flow, suggesting the tissues are alive and functional (Azadzoi et al., 2004; Bajory, Hutter, Krombach, & Messmer, 2002; Nomiya et al., 2012). The reasons for such a metabolic discrepancy are not clearly understood, and represent an area of investigation that could offer significant insights into how the different cells within the bladder wall reciprocally interact to maintain bladder tissue health and function during varying states of normoxia and hypoxia.

Ischemic damage is only part of the issue, however, abrupt reperfusion leads to inflammation, generation of multiple reactive oxygen species, and tissue remodeling (Tano & Gollasch, 2014). Indeed, the bladder vasculature is not immune to such effects either: reperfusion for 30 min after 60 min of bilateral occlusion resulted in increased inflammation, decreased blood flow, and macromolecular damage in an I/R-induced cystitis rat model (Bajory et al., 2002). Increased protein oxidation and mitochondrial dysfunction is also evident in outlet obstruction/overdistension models that coincidentally decrease bladder blood flow (Lin & Juan, 2012), but it is unclear if such damage is related to changes in bladder blood flow alone. This is because many of the studies analyzing the effects of I/R injury in the bladder posit altered voiding function from altered sensitivity of isolated muscle strips to different contractile agonists, without providing cystometric or in vivo voiding behavior data (Miyata et al., 2019). The studies that do show differences in voiding function note that the changes occur well after the I/R injury, suggesting an important role for tissue remodeling in response to chronic oxidative stress, inflammation, and hypoxia (Andersson et al., 2017; Nomiya et al., 2012).

So what does it all mean? Overall, the vasculature of the bladder is designed with resilience in mind—both in terms of adaptation to mechanical stress and resistance to ischemic conditions. But, as with all body systems, this only holds true to a point. Eventually, prolonged ischemia and extreme reperfusion result in inflammation and tissue remodeling that fundamentally alter the functionality of the urinary bladder. Further research is needed to understand how the bladder vasculature has adapted to buffer the ischemia/reperfusion injuries associated with normal bladder filling, in hopes that we can learn new approaches for preserving vascular function and tissue viability in other organs that are less capable of coping with such changes.

3.2 Is bladder dysfunction a vascular dysfunction?

As discussed in previous chapters, multiple pathological conditions, including hypertension, obesity, and aging, are associated with profound vascular dysfunction. Many of these same diseases correlate with significant lower urinary tract symptoms; but unfortunately, very little evidence exists directly linking the vascular dysfunction to the bladder dysfunction in nearly all of these pathologies. However, one disease where vascular dysfunction and bladder dysfunction coincide is type II diabetes mellitus.

With nearly 50% of diabetic patients reporting one or more lower urinary tract symptoms, bladder dysfunction is a substantial contributor to reduced life quality in this population (Abrams, Chapple, Junemann, & Sharpe, 2012; Brown et al., 2005; Liu et al., 2011; Palleschi et al., 2014). Even with the high prevalence of lower urinary tract dysfunction in type II diabetes, little is known about the pathophysiological changes that lead to diabetes-related bladder dysfunction (Brown et al., 2005). While diabetes is a disease defined by the inability of the body to regulate blood glucose, it has numerous deleterious effects on the vascular system. An important hallmark of diabetes is vascular pathologies, including endothelial and smooth muscle cell dysfunction and hyperreactivity (Fleischhacker et al., 1999; Zimmermann, Knot, Stevenson, & Nelson, 1997). In large arteries, diabetes causes atherosclerosis and medial calcification; in the microvasculature, dysfunction leads to blindness and renal failure (Creager, Luscher, Cosentino, & Beckman, 2003). Several molecular mechanisms responsible for many of these complications have been identified, and include decreased NO production (Williams et al., 1998), increased prostanoid and ET-1 synthesis

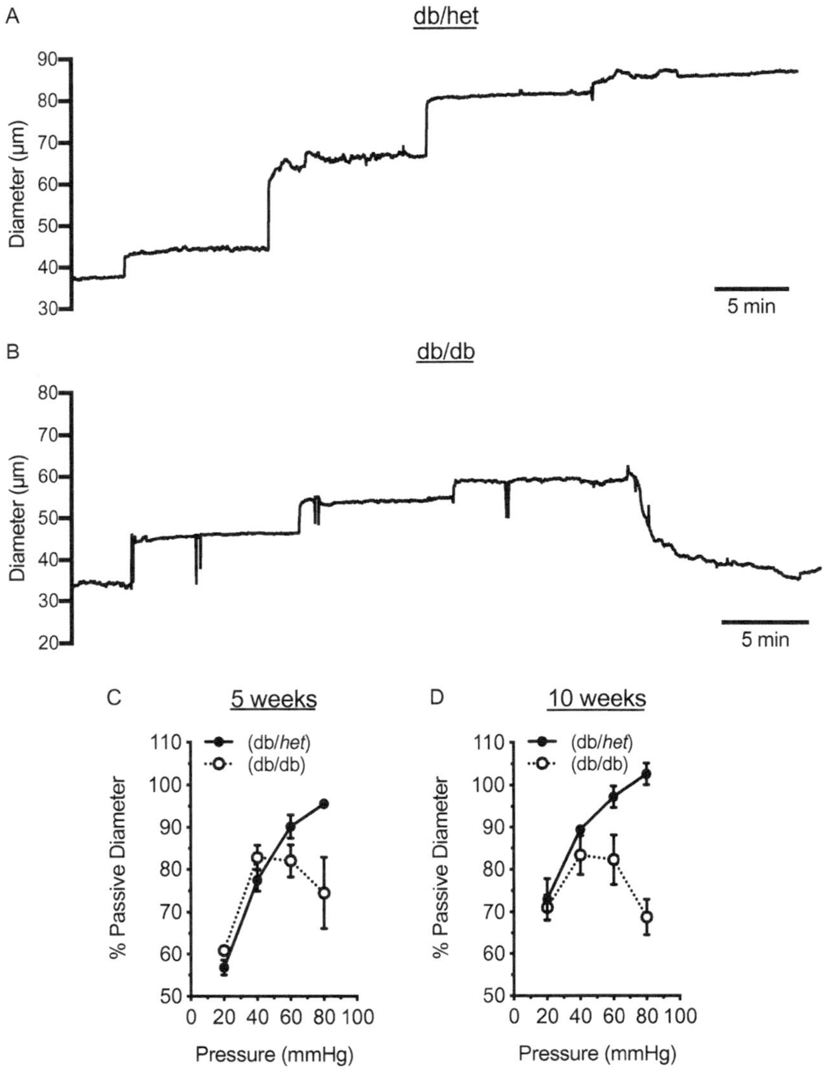

Fig. 8 Representative traces of pressurized bladder arteriolar diameter in response to increasing pressure (20–80 mmHg) from *db/het* (A) and *db/db* (B) mice. Arterioles from 5-week (C) and 10-week (D) old diabetic mice both developed myogenic tone, with more substantial tone developing after 10 weeks of age. *N* = 3.

(De Vriese, Verbeuren, Van de Voorde, Lameire, & Vanhoutte, 2000; Luft, 2002), and increased smooth muscle protein kinase C activity (Tesfamariam, Brown, & Cohen, 1991). The traditional cause of diabetic lower urinary tract symptoms was solely considered to be autonomic neuropathy (Frimodt-Moller, 1980), but most investigators now agree that the urological complications of diabetes are multifactorial and include pathophysiological changes to detrusor, urethra, urothelium, and/or autonomic nerves (Daneshgari, Liu, Birder, Hanna-Mitchell, & Chacko, 2009). While the link between impaired bladder blood flow and urological complications of diabetes has not yet been established in humans, we discovered that bladder arterioles from type II diabetic (db/db) mice developed significant myogenic tone after the onset of hyperglycemia, and even prior to onset of metabolic dysfunction/obesity (Fig. 8).

The relationship between diabetes, vascular dysfunction, and bladder disorders suggests that vascular impairments also indeed play a role in the progression of diabetic lower urinary tract symptoms and should be investigated further. These studies should be recapitulated in models of obesity, hypertension, peripheral artery disease, and aging to further uncover how small changes to vascular function can affect the lower urinary tract, since it is very unlikely that type II diabetes is the only disease that affects the bladder vasculature in such a way as to alter voiding function.

4. Conclusions

The urinary bladder vasculature has uniquely adapted to the environment in which it resides. Bladder blood vessels have developed a special arrangement of arterioles, capillaries, and venules that maintain tissue perfusion to best of their ability in the face of huge changes in shape and size of the bladder during filling. To meet these challenges, the arterioles have adapted to lack myogenic tone in order to maintain blood flow during increasing resistance and pressure caused by bladder wall distention. Likewise, the capillary architecture has become a vast web of collateral vessels and hydra-like connections to the venous circulation that compensate for any small occlusions or obstructions that can (and do) occur. Finally, the venules exhibit phasic waves of neuronally-independent contractions, which help maintain perfusion pressure and drive blood out of the bladder to minimize stagnation and maximize oxygen delivery to the urothelium. These properties of the bladder microvasculature highlight the heterogeneity of the microcirculation and its ability to adapt to the different organs in which it resides.

References

Abrams, P., Chapple, C. R., Junemann, K. P., & Sharpe, S. (2012). Urinary urgency: A review of its assessment as the key symptom of the overactive bladder syndrome. *World Journal of Urology*, *30*(3), 385–392. https://doi.org/10.1007/s00345-011-0742-8.

Alagem, N., Dvir, M., & Reuveny, E. (2001). Mechanism of Ba(2+) block of a mouse inwardly rectifying K+ channel: Differential contribution by two discrete residues. *The Journal of Physiology*, *534*(Pt. 2), 381–393. https://doi.org/10.1111/j.1469-7793.2001.00381.x.

Andersson, K. E., & Arner, A. (2004). Urinary bladder contraction and relaxation: Physiology and pathophysiology. *Physiological Reviews*, *84*(3), 935–986. https://doi.org/10.1152/physrev.00038.2003.

Andersson, K. E., Boedtkjer, D. B., & Forman, A. (2017). The link between vascular dysfunction, bladder ischemia, and aging bladder dysfunction. *Therapeutic Advances in Urology*, *9*(1), 11–27. https://doi.org/10.1177/1756287216675778.

Azadzoi, K. M., Heim, V. K., Tarcan, T., & Siroky, M. B. (2004). Alteration of urothelial-mediated tone in the ischemic bladder: Role of eicosanoids. *Neurourology and Urodynamics*, *23*(3), 258–264. https://doi.org/10.1002/nau.20029.

Azadzoi, K. M., Tarcan, T., Kozlowski, R., Krane, R. J., & Siroky, M. B. (1999). Overactivity and structural changes in the chronically ischemic bladder. *The Journal of Urology*, *162*(5), 1768–1778.

Bajory, Z., Hutter, J., Krombach, F., & Messmer, K. (2002). Microcirculation of the urinary bladder in a rat model of ischemia-reperfusion-induced cystitis. *Urology*, *60*(6), 1136–1140. https://doi.org/10.1016/s0090-4295(02)01952-0.

Billaud, M., Lohman, A. W., Straub, A. C., Looft-Wilson, R., Johnstone, S. R., Araj, C. A., et al. (2011). Pannexin1 regulates alpha1-adrenergic receptor-mediated vasoconstriction. *Circulation Research*, *109*(1), 80–85. https://doi.org/10.1161/CIRCRESAHA.110.237594.

Booth, E. A., Obeid, N. R., & Lucchesi, B. R. (2005). Activation of estrogen receptor-alpha protects the in vivo rabbit heart from ischemia-reperfusion injury. *American Journal of Physiology. Heart and Circulatory Physiology*, *289*(5), H2039–H2047. https://doi.org/10.1152/ajpheart.00479.2005.

Brown, J. S., Wessells, H., Chancellor, M. B., Howards, S. S., Stamm, W. E., Stapleton, A. E., et al. (2005). Urologic complications of diabetes. *Diabetes Care*, *28*(1), 177–185.

Chai, T. C., Russo, A., Yu, S., & Lu, M. (2016). Mucosal signaling in the bladder. *Autonomic Neuroscience*, *200*, 49–56. https://doi.org/10.1016/j.autneu.2015.08.009.

Creager, M. A., Luscher, T. F., Cosentino, F., & Beckman, J. A. (2003). Diabetes and vascular disease: Pathophysiology, clinical consequences, and medical therapy: Part I. *Circulation*, *108*(12), 1527–1532. https://doi.org/10.1161/01.CIR.0000091257.27563.32.

Daneshgari, F., Liu, G., Birder, L., Hanna-Mitchell, A. T., & Chacko, S. (2009). Diabetic bladder dysfunction: Current translational knowledge. *Journal of Urology*, *182*(6 Suppl), S18–S26. https://doi.org/10.1016/j.juro.2009.08.070.

De Vriese, A. S., Verbeuren, T. J., Van de Voorde, J., Lameire, N. H., & Vanhoutte, P. M. (2000). Endothelial dysfunction in diabetes. *British Journal of Pharmacology*, *130*(5), 963–974. https://doi.org/10.1038/sj.bjp.0703393.

Dorr, W. (1992). Cystometry in mice—Influence of bladder filling rate and circadian variations in bladder compliance. *The Journal of Urology*, *148*(1), 183–187. https://doi.org/10.1016/s0022-5347(17)36549-7.

Earley, S. (2013). TRPM4 channels in smooth muscle function. *Pflügers Archiv*, *465*(9), 1223–1231. https://doi.org/10.1007/s00424-013-1250-z.

Ehren, I., Iversen, H., Jansson, O., Adolfsson, J., & Wiklund, N. P. (1994). Localization of nitric oxide synthase activity in the human lower urinary tract and its correlation with neuroeffector responses. *Urology*, *44*(5), 683–687.

Fabiyi, A. C., & Brading, A. F. (2006). The use of the isolated mouse whole bladder for investigating bladder overactivity. *Journal of Pharmacology and Experimental Therapeutics*, *319*(3), 1386–1394. https://doi.org/10.1124/jpet.106.108902.

Faraci, F. M., & Sobey, C. G. (1998). Role of potassium channels in regulation of cerebral vascular tone. *Journal of Cerebral Blood Flow and Metabolism*, *18*(10), 1047–1063. https://doi.org/10.1097/00004647-199810000-00001.

Ferguson, D. R., Kennedy, I., & Burton, T. J. (1997). ATP is released from rabbit urinary bladder epithelial cells by hydrostatic pressure changes—A possible sensory mechanism? *Journal of Physiology*, *505*(Pt. 2), 503–511.

Fink, G. D. (2009). Arthur C. Corcoran Memorial Lecture. Sympathetic activity, vascular capacitance, and long-term regulation of arterial pressure. *Hypertension*, *53*(2), 307–312. https://doi.org/10.1161/HYPERTENSIONAHA.108.119990.

Fleischhacker, E., Esenabhalu, V. E., Spitaler, M., Holzmann, S., Skrabal, F., Koidl, B., et al. (1999). Human diabetes is associated with hyperreactivity of vascular smooth muscle cells due to altered subcellular Ca^{2+} distribution. *Diabetes*, *48*(6), 1323–1330.

Frimodt-Moller, C. (1980). Diabetic cystopathy: Epidemiology and related disorders. *Annals of Internal Medicine*, *92*(2 Pt. 2), 318–321.

Ghafar, M. A., Anastasiadis, A. G., Olsson, L. E., Chichester, P., Kaplan, S. A., Buttyan, R., et al. (2002). Hypoxia and an angiogenic response in the partially obstructed rat bladder. *Laboratory Investigation*, *82*(7), 903–909.

Gillespie, J. I. (2004). Phosphodiesterase-linked inhibition of nonmicturition activity in the isolated bladder. *BJU International*, *93*(9), 1325–1332. https://doi.org/10.1111/j.1464-410X.2004.04840.x.

Hall, C. N., Reynell, C., Gesslein, B., Hamilton, N. B., Mishra, A., Sutherland, B. A., et al. (2014). Capillary pericytes regulate cerebral blood flow in health and disease. *Nature*, *508*(7494), 55–60. https://doi.org/10.1038/nature13165.

Hanna-Mitchell, A. T., Beckel, J. M., Barbadora, S., Kanai, A. J., de Groat, W. C., & Birder, L. A. (2007). Non-neuronal acetylcholine and urinary bladder urothelium. *Life Sciences*, *80*(24–25), 2298–2302. https://doi.org/10.1016/j.lfs.2007.02.010.

Hashitani, H., & Lang, R. J. (2016). Spontaneous activity in the microvasculature of visceral organs: Role of pericytes and voltage-dependent Ca(2+) channels. *The Journal of Physiology*, *594*(3), 555–565. https://doi.org/10.1113/JP271438.

Hashitani, H., Mitsui, R., Shimizu, Y., Higashi, R., & Nakamura, K. (2012). Functional and morphological properties of pericytes in suburothelial venules of the mouse bladder. *British Journal of Pharmacology*, *167*(8), 1723–1736. https://doi.org/10.1111/j.1476-5381.2012.02125.x.

Hashitani, H., Takano, H., Fujita, K., Mitsui, R., & Suzuki, H. (2011). Functional properties of suburothelial microvessels in the rat bladder. *The Journal of Urology*, *185*(6), 2382–2391. https://doi.org/10.1016/j.juro.2011.02.046.

Hossler, F. E., & Kao, R. L. (2007). Microvasculature of the urinary bladder of the dog: A study using vascular corrosion casting. *Microscopy and Microanalysis*, *13*(3), 220–227. https://doi.org/10.1017/S1431927607070249.

Hossler, F., Kao, R., & Monson, F. (2005). Microvasculature of the mammalian urinary bladder—A preliminary comparative study using corrosion casting. *Microscopy and Microanalysis*, *11*(S02), 1204–1205. https://doi.org/10.1017/s1431927605502356.

Hossler, F. E., & Monson, F. C. (1998a). Evidence for a unique elastic sheath surrounding the vesicular arteries of the rabbit urinary bladder—Studies of the microvasculature with microscopy and vascular corrosion casting. *Anatomical Record*, *252*(3), 472–476.

Hossler, F. E., & Monson, F. C. (1998b). Structure and blood supply of intrinsic lymph nodes in the wall of the rabbit urinary bladder—Studies with light microscopy, electron microscopy, and vascular corrosion casting. *Anatomical Record*, *252*(3), 477–484.

Hsieh, C. P., Kuo, C. C., & Huang, C. W. (2015). Driving force-dependent block by internal Ba(2+) on the Kir2.1 channel: Mechanistic insight into inward rectification. *Biophysical Chemistry, 202*, 40–57. https://doi.org/10.1016/j.bpc.2015.04.003.

Hypolite, J. A., Longhurst, P. A., Gong, C., Briscoe, J., Wein, A. J., & Levin, R. M. (1993). Metabolic studies on rabbit bladder smooth muscle and mucosa. *Molecular and Cellular Biochemistry, 125*(1), 35–42. https://doi.org/10.1007/bf00926832.

Jost, S. P. (1985). Postnatal growth of the mouse bladder. *Journal of Anatomy, 143*, 39–43.

Juan, Y. S., Li, S., Levin, R. M., Kogan, B. A., Schuler, C., Leggett, R. E., et al. (2009). The effect of ischemia/reperfusion on rabbit bladder—Role of Rho-kinase and smooth muscle regulatory proteins. *Urology, 73*(5), 1126–1130. https://doi.org/10.1016/j.urology.2008.02.063.

Khan, M. A., Thompson, C. S., Dashwood, M. R., Mumtaz, F. H., Morgan, R. J., & Mikhailidis, D. P. (2003). Endothelin-1 and nitric oxide in the pathogenesis of urinary tract disorders secondary to bladder outlet obstruction. *Current Vascular Pharmacology, 1*(1), 27–31.

Knot, H. J., & Nelson, M. T. (1995). Regulation of membrane potential and diameter by voltage-dependent K+ channels in rabbit myogenic cerebral arteries. *The American Journal of Physiology, 269*(1 Pt. 2), H348–H355. https://doi.org/10.1152/ajpheart.1995.269.1.H348.

Knot, H. J., & Nelson, M. T. (1998). Regulation of arterial diameter and wall [Ca^{2+}] in cerebral arteries of rat by membrane potential and intravascular pressure. *The Journal of Physiology, 508*(Pt. 1), 199–209. https://doi.org/10.1111/j.1469-7793.1998.199br.x.

Kotecha, N., & Hill, M. A. (2005). Myogenic contraction in rat skeletal muscle arterioles: Smooth muscle membrane potential and Ca(2+) signaling. *American Journal of Physiology. Heart and Circulatory Physiology, 289*(4), H1326–H1334. https://doi.org/10.1152/ajpheart.00323.2005.

Kowalewska, P. M., Burrows, L. L., & Fox-Robichaud, A. E. (2011). Intravital microscopy of the murine urinary bladder microcirculation. *Microcirculation, 18*(8), 613–622. https://doi.org/10.1111/j.1549-8719.2011.00123.x.

Lies, B., Beck, K., Keppler, J., Saur, D., Groneberg, D., & Friebe, A. (2015). Nitrergic signalling via interstitial cells of Cajal regulates motor activity in murine colon. *The Journal of Physiology, 593*(20), 4589–4601. https://doi.org/10.1113/JP270511.

Lin, A. T., & Juan, Y. S. (2012). Ischemia, hypoxia and oxidative stress in bladder outlet obstruction and bladder overdistention injury. *Lower Urinary Tract Symptoms, 4*(Suppl. 1), 27–31. https://doi.org/10.1111/j.1757-5672.2011.00134.x.

Liu, R. T., Chung, M. S., Lee, W. C., Chang, S. W., Huang, S. T., Yang, K. D., et al. (2011). Prevalence of overactive bladder and associated risk factors in 1359 patients with type 2 diabetes. *Urology, 78*(5), 1040–1045. https://doi.org/10.1016/j.urology.2011.05.017.

Longden, T. A., & Nelson, M. T. (2015). Vascular inward rectifier K+ channels as external K+ sensors in the control of cerebral blood flow. *Microcirculation, 22*(3), 183–196. https://doi.org/10.1111/micc.12190.

Longhurst, P. A., Briscoe, J. A., Rosenberg, D. J., & Leggett, R. E. (1997). The role of cyclic nucleotides in guinea-pig bladder contractility. *British Journal of Pharmacology, 121*(8), 1665–1672. https://doi.org/10.1038/sj.bjp.0701328.

Luft, F. C. (2002). Proinflammatory effects of angiotensin II and endothelin: Targets for progression of cardiovascular and renal diseases. *Current Opinion in Nephrology and Hypertension, 11*(1), 59–66.

Miodonski, A. J., & Litwin, J. A. (1999). Microvascular architecture of the human urinary bladder wall: A corrosion casting study. *Anatomical Record, 254*(3), 375–381.

Mitsui, R., & Hashitani, H. (2013). Immunohistochemical characteristics of suburothelial microvasculature in the mouse bladder. *Histochemistry and Cell Biology, 140*(2), 189–200. https://doi.org/10.1007/s00418-012-1074-5.

Miyata, Y., Matsuo, T., Mitsunari, K., Asai, A., Ohba, K., & Sakai, H. (2019). A review of oxidative stress and urinary dysfunction caused by bladder outlet obstruction and treatments using antioxidants. *Antioxidants (Basel)*, *8*(5), 132. https://doi.org/10.3390/antiox8050132.

Mulvany, M. J., & Halpern, W. (1977). Contractile properties of small arterial resistance vessels in spontaneously hypertensive and normotensive rats. *Circulation Research*, *41*(1), 19–26. https://doi.org/10.1161/01.res.41.1.19.

Mumtaz, F. H., Khan, M. A., Sullivan, M. E., Thompson, C. S., Mikhailidis, D. P., Morgan, R. J., et al. (1999). Potential role of endothelin and nitric oxide in physiology and pathophysiology of the lower urinary tract. *Endothelium*, 7(1), 1–9.

Munoz, A., Smith, C. P., Boone, T. B., & Somogyi, G. T. (2011). Overactive and underactive bladder dysfunction is reflected by alterations in urothelial ATP and NO release. *Neurochemistry International*, *58*(3), 295–300. https://doi.org/10.1016/j.neuint.2010.12.002.

Nagai, Y., Kaneda, T., Miyamoto, Y., Nuruki, T., Kanda, H., Urakawa, N., et al. (2016). Effects of hypoxia and glucose-removal condition on muscle contraction of the smooth muscles of porcine urinary bladder. *The Journal of Veterinary Medical Science*, *78*(1), 55–59. https://doi.org/10.1292/jvms.15-0269.

Nomiya, M., Yamaguchi, O., Andersson, K. E., Sagawa, K., Aikawa, K., Shishido, K., et al. (2012). The effect of atherosclerosis-induced chronic bladder ischemia on bladder function in the rat. *Neurourology and Urodynamics*, *31*(1), 195–200. https://doi.org/10.1002/nau.21073.

Palleschi, G., Pastore, A. L., Maggioni, C., Fuschi, A., Pacini, L., Petrozza, V., et al. (2014). Overactive bladder in diabetes mellitus patients: A questionnaire-based observational investigation. *World Journal of Urology*, *32*, 1021–1025. https://doi.org/10.1007/s00345-013-1175-3.

Persson, K., Alm, P., Uvelius, B., & Andersson, K. E. (1998). Nitrergic and cholinergic innervation of the rat lower urinary tract after pelvic ganglionectomy. *The American Journal of Physiology*, *274*(2), R389–R397. https://doi.org/10.1152/ajpregu.1998.274.2.R389.

Phan, T. X., Ton, H. T., Chen, Y., Basha, M. E., & Ahern, G. P. (2016). Sex-dependent expression of TRPV1 in bladder arterioles. *American Journal of Physiology. Renal Physiology*, *311*(5), F1063–F1073. https://doi.org/10.1152/ajprenal.00234.2016.

Pontari, M. A., Hanno, P. M., & Ruggieri, M. R. (1999). Comparison of bladder blood flow in patients with and without interstitial cystitis. *Journal of Urology*, *162*(2), 330–334.

Pontari, M. A., & Ruggieri, M. R. (1999). Sex differences and role of nitric oxide in blood flow of canine urinary bladder. *American Journal of Physiology*, *276*(2 Pt. 2), R407–R413.

Qiao, Z., Xia, C., Shen, S., Corwin, F. D., Liu, M., Guan, R., et al. (2014). Suppression of the PI3K pathway in vivo reduces cystitis-induced bladder hypertrophy and restores bladder capacity examined by magnetic resonance imaging. *PLoS One*, *9*(12), e114536. https://doi.org/10.1371/journal.pone.0114536.

Quayle, J. M., McCarron, J. G., Brayden, J. E., & Nelson, M. T. (1993). Inward rectifier K+ currents in smooth muscle cells from rat resistance-sized cerebral arteries. *The American Journal of Physiology*, *265*(5 Pt. 1), C1363–C1370. https://doi.org/10.1152/ajpcell.1993.265.5.C1363.

Robertson, A. S. (1999). Behaviour of the human bladder during natural filling: The Newcastle experience of ambulatory monitoring and conventional artificial filling cystometry. *Scandinavian Journal of Urology and Nephrology. Supplementum*, *201*, 19–24.

Sakagami, K., Wu, D. M., & Puro, D. G. (1999). Physiology of rat retinal pericytes: Modulation of ion channel activity by serum-derived molecules. *The Journal of Physiology*, *521*(Pt. 3), 637–650. https://doi.org/10.1111/j.1469-7793.1999.00637.x.

Santos, G. A., Petersen, N., Zamani, A. A., Du, R., LaRose, S., Monk, A., et al. (2016). Pathophysiologic differences in cerebral autoregulation after subarachnoid hemorrhage. *Neurology*, *86*(21), 1950–1956. https://doi.org/10.1212/WNL.0000000000002696.
Segal, S. S. (2005). Regulation of blood flow in the microcirculation. *Microcirculation*, *12*(1), 33–45. https://doi.org/10.1080/10739680590895028.
Shimizu, Y., Mochizuki, S., Mitsui, R., & Hashitani, H. (2014). Neurohumoral regulation of spontaneous constrictions in suburothelial venules of the rat urinary bladder. *Vascular Pharmacology*, *60*(2), 84–94. https://doi.org/10.1016/j.vph.2014.01.002.
Tano, J. Y., & Gollasch, M. (2014). Hypoxia and ischemia-reperfusion: A BiK contribution? *American Journal of Physiology. Heart and Circulatory Physiology*, *307*(6), H811–H817. https://doi.org/10.1152/ajpheart.00319.2014.
Tesfamariam, B., Brown, M. L., & Cohen, R. A. (1991). Elevated glucose impairs endothelium-dependent relaxation by activating protein kinase C. *Journal of Clinical Investigation*, *87*(5), 1643–1648. https://doi.org/10.1172/JCI115179.
Tykocki, N. R., Boerman, E. M., & Jackson, W. F. (2017). Smooth muscle ion channels and regulation of vascular tone in resistance arteries and arterioles. *Comprehensive Physiology*, 7(2), 485–581. https://doi.org/10.1002/cphy.c160011.
Tykocki, N. R., Bonev, A. D., Longden, T. A., Heppner, T. J., & Nelson, M. T. (2017). Inhibition of vascular smooth muscle inward-rectifier K(+) channels restores myogenic tone in mouse urinary bladder arterioles. *American Journal of Physiology. Renal Physiology*, *312*(5), F836–F847. https://doi.org/10.1152/ajprenal.00682.2016.
Tykocki, N. R., Thompson, J. M., Jackson, W. F., & Watts, S. W. (2013). Ryanodine receptors are uncoupled from contraction in rat vena cava. *Cell Calcium*, *53*(2), 112–119. https://doi.org/10.1016/j.ceca.2012.10.006.
VanBavel, E., Siersma, P., & Spaan, J. A. (2003). Elasticity of passive blood vessels: A new concept. *American Journal of Physiology. Heart and Circulatory Physiology*, *285*(5), H1986–H2000. https://doi.org/10.1152/ajpheart.00248.2003.
Wendt, I. R., & Paul, R. J. (2003). Energy cost of contraction in rat urinary bladder smooth muscle during anoxia. *Clinical and Experimental Pharmacology & Physiology*, *30*(8), 565–569. https://doi.org/10.1046/j.1440-1681.2003.03882.x.
Williams, S. B., Goldfine, A. B., Timimi, F. K., Ting, H. H., Roddy, M. A., Simonson, D. C., et al. (1998). Acute hyperglycemia attenuates endothelium-dependent vasodilation in humans in vivo. *Circulation*, *97*(17), 1695–1701.
Zimmermann, P. A., Knot, H. J., Stevenson, A. S., & Nelson, M. T. (1997). Increased myogenic tone and diminished responsiveness to ATP-sensitive K+ channel openers in cerebral arteries from diabetic rats. *Circulation Research*, *81*(6), 996–1004.

CHAPTER ELEVEN

Intrinsic regulation of microvascular tone by myoendothelial feedback circuits

Hamish A.L. Lemmey, Christopher J. Garland, Kim A. Dora*
Department of Pharmacology, University of Oxford, Oxford, United Kingdom
*Corresponding author: e-mail address: kim.dora@pharm.ox.ac.uk

Contents

Abstract

The endothelium is an important regulator of arterial vascular tone, acting to release nitric oxide (NO) and open Ca^{2+}-activated K^+ (K_{Ca}) channels to relax vascular smooth muscle cells (VSMCs). While agonists acting at endothelial cell (EC) receptors are widely used to assess the ability of the endothelium to reduce vascular tone, the intrinsic

Current Topics in Membranes, Volume 85
ISSN 1063-5823
https://doi.org/10.1016/bs.ctm.2020.01.004

EC-dependent mechanisms are less well characterized. In small resistance arteries and arterioles, the presence of heterocellular gap junctions termed myoendothelial gap junctions (MEGJs) allows the passage of not only current, but small molecules including Ca^{2+} and inositol trisphosphate (IP_3). When stimulated to contract, the increase in VSM Ca^{2+} and IP_3 can therefore potentially pass through MEGJs to activate adjacent ECs. This activation releases NO and opens K_{Ca} channels, which act to limit contraction. This myoendothelial feedback (MEF) is amplified by EC Ca^{2+} influx and release pathways, and is dynamically modulated by processes regulating gap junction conductance. There is a remarkable localization of key signaling and regulatory proteins within the EC projection toward VSM, and the intrinsic EC-dependent signaling pathways occurring with this highly specialized microdomain are reviewed.

1. Introduction

Endothelium-dependent vasodilation in resistance arteries is a fundamental mechanism for regulating the pressure and flow of blood, with endothelial dysfunction recognized as a ubiquitous hallmark of cardiovascular disease. The central, essential role of endothelial cells (ECs) was discovered by Furchgott and Zawadzki, when acetylcholine (ACh) was found to stimulate vasodilation via ECs in rabbit isolated thoracic aorta (Furchgott & Zawadzki, 1980). The majority of subsequent investigations have adopted a similar approach to probe endothelium-dependent vasodilation, by activating ECs directly with pharmacological agents or changes in intraluminal flow. Not long after a crucial role for the endothelium was described, several groups reported that mechanical destruction of the endothelium enhanced adrenergic vasoconstriction in arteries from multiple species and different vascular beds (Amerini, Mantelli, & Ledda, 1995; Carrier & White, 1985; Eglème, Godfraind, & Miller, 1984; Greenberg, Diecke, Peevy, & Tanaka, 1989; Jackson, 1988; Martin, Furchgott, Villani, & Jothianandan, 1986; Tesfamariam, Halpern, & Osol, 1985; Tesfamariam, Weisbrod, & Cohen, 1987; Urabe, Kawasaki, & Takasaki, 1991; Verrecchia, Sercombe, & Seylaz, 1985; Vinet, Brieva, Pinardi, & Penna, 1991). These observations were initially attributed to loss of constitutive EC nitric oxide (NO) release or release stimulated by EC α_2-adrenoceptors (α_2-ARs). However, an alternative possibility was introduced when depolarization and contraction of arteries with KCl was also found to be suppressed by the endothelium. EC depolarization per se would not be expected to stimulate vasodilation, because the cells lack voltage-gated Ca^{2+} channels (VGCCs), so the endothelium might be activated via the vascular smooth muscle (VSM) (Amerini et al., 1995; Jackson, 1988). This is now recognized to be the case, with considerable evidence to indicate the endothelium is activated during vasoconstriction by VSM-derived signals that indirectly facilitate endothelium-dependent dilator pathways. This negative feedback circuit has been termed myoendothelial feedback (MEF). By limiting vasoconstriction, MEF may represent the primary physiological mechanism for activating NO synthesis and endothelium-derived hyperpolarization (EDH) in resistance arteries and arterioles in vivo. Interest in understanding how arterial vasoconstriction is regulated by the endothelium has not waned since these initial observations, with recent publications utilizing advanced fluorescence imaging and molecular approaches in order to probe the complex heterocellular signaling responsible for MEF.

The mechanisms suggested to mediate myoendothelial feedback and their significance in setting vasoreactivity vary between studies, particularly with regard to the identification of a possible "principal" VSM signal to the endothelium and whether endothelial activation attenuates the potency and/or efficacy of vasoconstrictor agents. A key aim of this chapter is to highlight these discrepancies and attempt to provide a unifying framework for future investigations. We will summarize the key experimental evidence for heterocellular signaling from VSM to ECs, the proposed mechanisms for the induction of myoendothelial feedback, the physiological implications of the negative feedback loop within the vascular wall, and identify some further avenues for exploration.

2. Endothelial suppression of vasoconstriction: Basal or stimulated NO synthesis?

Following the work of Furchgott and Zawadzki it was soon recognized that the endothelium can suppress the action of vasoconstrictor agonists, and this influence was explained by a constitutive release of NO (Martin et al., 1986). The evidence for this assertion came from studies with rabbit isolated aorta using oxyhemoglobin to bind and remove NO, which significantly decreased basal cGMP levels in the VSM. The possibility NO release might be stimulated to evoke MEF was thought unlikely, as direct stimulation of aortic VSM with the α_1-adrenoreceptor (α_1-AR) agonist phenylephrine (PE) did not alter cGMP levels, while direct endothelial stimulation with ACh did (Martin et al., 1986). Using a bioassay setup, the perfusate from flow through an endothelium-intact donor rabbit isolated aorta was also shown to stimulate ~25% vasodilation in a denuded recipient artery (Rubanyi, Romero, & Vanhoutte, 1986). However, the possibility shear stress might be responsible for releasing NO (over and above basal release) to evoke vasodilation cannot be ruled out.

To draw more robust conclusions, measurement of NO release at rest, without a vasoconstrictor agonist present, are desirable. Use of NO electrodes (Simonsen, Wadsworth, Buus, & Mulvany, 1999; Stankevicius et al., 2006) and fluorescent indicators (Itoh et al., 2000; Kojima et al., 1998; Rodriguez, Specian, Maloney, Jourd'heuil, & Feelisch, 2005) rely on accurate calibration and inference as to the concentrations of NO available within VSM to stimulate relaxation or vasodilation. Nevertheless it does appear that rat mesenteric arteries can generate a basal release of ~6 nM NO, which is sufficient to suppress tension-associated myogenic contraction (Simonsen et al., 1999). Data showing the effect of inhibitors of nitric oxide synthase (NOS) such as ADMA (Stankevicius et al., 2006) against basal NO release in intact, tensioned arteries has yet to be established. Another approach is to quantify tissue content of *S*- and *N*-nitroso species using gas phase chemiluminescence, alongside nitrate and nitrate levels using high performance liquid chromatography (Bryan et al., 2004). This approach is somewhat confounded by the presence of blood plasma and cells in homogenized whole tissues, but may be ideal for ex vivo measurements in luminally cleared pressurized arteries studied at rest and in the presence of vasoconstrictors. Taken together these studies suggest there may be basal release of NO sufficient to modulate, at least in part, the action of vasoconstrictor agonists.

While basal NO synthesis may help suppress vasoconstriction in some arteries, direct stimulation of the endothelium is also a possibility to consider. In large coronary arteries from dog and pig, both noradrenaline and 5-HT stimulate endothelium-dependent vasorelaxation, the former via EC α_2-ARs (Angus, Cocks, & Satoh, 1986a; Cocks & Angus, 1983). Interestingly, α_2-AR mediated vasorelaxation was also observed in a number of other large canine arteries, although maximum relaxation was much less than in the coronary artery at 70% of the potential maximum (Angus et al., 1986a). A common feature of studies demonstrating basal NO synthesis and endothelial activation by vasoconstrictor agents was the use of large conduit arteries as the bioassay tissue. As MEF depends on the presence of myoendothelial gap junctions (MEGJs), and as the incidence of these heterocellular connections is inversely related to artery size (Straub, Zeigler, & Isakson, 2014), it might be predicted large arteries would not display highly developed MEF. Consistent with this suggestion, in smaller arteries where myoendothelial gap junctions are widespread, heterocellular Ca^{2+} signaling appears to be an important mechanism.

Therefore, before considering MEF may be a primary signaling mechanism in specific arteries these characteristics should be considered:

(1) Is there evidence to suggest basal NO production or activation of EDH, before the development of myogenic tone or adding a vasoconstrictor agent?

(2) Where are the receptors for a vasoconstrictor, and can direct stimulation of ECs be excluded as significant?

(3) Are there VGCCs in native ECs?

(4) Is there evidence for MEGJs in the vascular bed to be studied? i.e., do gap junction uncouplers prevent MEF?

3. Defining MEF according to function alone

While it is clear the endothelium can modulate vasoconstrictor-mediated tone, the temporal characteristics of vasoconstriction can vary between agonists and vascular beds. Therefore, to improve MEF nomenclature we highlight various forms of MEF which may occur in response to artery pressurization or tension (myogenic) and/or in the presence of vasoconstrictors, resulting in an attenuation in contraction over time (fade) and often with superimposed cycles of contraction (vasomotion) (Fig. 1). Note that fade and vasomotion can also occur against myogenic tone.

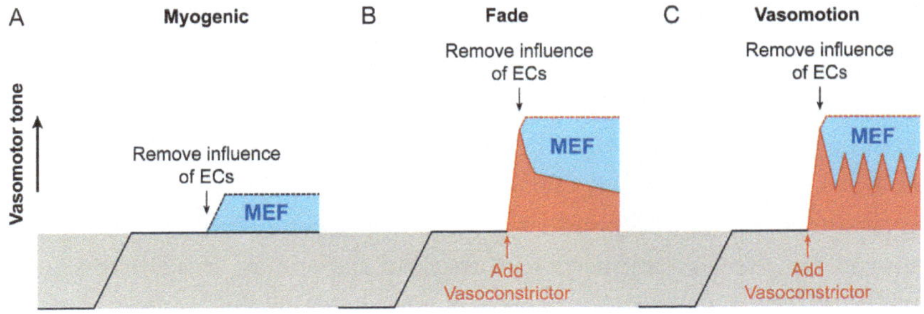

Fig. 1 Measurement of arterial diameter or tension enables a functional readout of MEF. (A) In myogenic arteries and arterioles, removing the influence of the endothelium can augment or unmask myogenic tone. In either the presence or absence of myogenic tone, vasoconstrictor agents such as α_1-AR agonists or KCl can activate MEF apparent as (B) fade in contraction or (C) the development of vasomotion; cycles of vasoconstriction and vasorelaxation. Each is compared to the vasomotor response in the absence of an influence of the endothelium (dashed lines). An understanding of the electrical and chemical signaling processes that underlie MEF will provide a framework for establishing the endothelium-dependent aspects of the intrinsic regulation of arterial diameter, which act in parallel to the intrinsic smooth muscle-dependent feedback mechanisms considered in Chapter 2.

Consideration of these temporal aspects during vasoconstrictor responses during experiments is key, and reinforces the importance of presenting representative, original traces within publications. As the mechanisms responsible for MEF unfold, much can be learnt by revisiting the literature. For example, most of the original papers using large arteries in the 1980s and 1990s to assess the role of the endothelium against vasoconstrictors showed original time courses for responses. The 1988 paper by Jackson using hamster aortae under isometric tension (Jackson, 1988) shows vasomotion in response to PE, and once the endothelium was damaged, vasomotion was abolished and contraction augmented ~10-fold. Methylene blue and hemoglobin (to prevent NO signaling and used prior to the development of NOS inhibitors) replicated this loss of the endothelium. Furthermore, once Ca^{2+} influx pathways were blocked by removing extracellular Ca^{2+} or by using a L-type VGCC blocker, EC-feedback and vasomotion were also abolished. An interesting question is whether these data may be explained by heterocellular coupling, or by another signaling pathway that perhaps involves the release of a diffusible factor such as ATP from activated VSMs? If a diffusible factor from the smooth muscle is involved, it might also be defined as MEF, introducing the concept of a myoendothelial feedback factor (MEFF).

4. Direct observation of VSM to EC Ca^{2+} signaling

Since both the production of NO and activation of Ca^{2+}-activated K^+ (K_{Ca}) channels occurs secondary to an increase in EC Ca^{2+}, one method of establishing whether tensioning an artery and then adding vasoconstrictor agents could activate the endothelium is by EC Ca^{2+} measurement before and during vasoconstriction. Then by defining the mechanism for this EC activation, the MEF-dependent and MEF-independent processes responsible for the activation of endothelium-dependent dilation might be established. This then opens the same avenue of research when defining EDH vs EDHF, the latter also defined as a diffusible factor.

The first clear observation showing a vasoconstrictor could indirectly stimulate the endothelium and initiate MEF as a result came from isolated and pressurized hamster cheek pouch arterioles (Table 1). In myogenically active arteries, basal EC $[Ca^{2+}]_i$ was increased during constriction to either PE or raised extracellular potassium (elevated $[K^+]_o$) (Dora et al., 1997). Furthermore, vasoconstriction was enhanced in each case by block of NOS and taking these data together it was hypothesized $[Ca^{2+}]_i$ and/or a

Table 1 Chronological summary of published investigations into VSMC to EC signaling and myoendothelial feedback.

	Model/vascular bed	Vasoreactivity methodology	Ca^{2+} imaging methodology	VSM-dependent activator of ECs/ feedback vasodilation	Proposed VSM-derived signal	Proposed mechanism of feedback vasodilation	Inhibitors of VSM-dependent EC activation/ feedback vasodilation
Dora, Doyle, and Duling (1997)	Hamster cheek pouch	PM	ECs or VSMCs (pressurized arteries)	PE, elevated $[K^+]_o$	Ca^{2+}	NO	Nifedipine
Dora, Hinton, Walker, and Garland (2000)	Rat mesentery	WM	Freshly isolated ECs	PE	–	NO, EDH (SK_{Ca}, IK_{Ca})	–
Yashiro and Duling (2000)	Hamster cremaster muscle	PM	ECs or VSMCs (pressurized arteries)	PE	Ca^{2+}	EDH (SK_{Ca}, IK_{Ca})	BAPTA
Oishi et al. (2001)	Rat mesentery	–	ECs and VSMCs simultaneously (*en face* arteries)	Elevated $[K^+]_o$, PE	Ca^{2+}, IP_3	–	–
Schuster, Hirotaka, Beny, Stergiopulos, and Meister (2001)	Rat mesentery	PM	ECs or VSMCs (pressurized arteries)	PE	–	–	Palmitoleic acid
Budel, Schuster, Stergiopoulos, Meister, and Beny (2001)	Pig heart	Arterial strips	ECs or VSMCs (*en face* arteries), cultured primary ECs	Elevated $[K^+]_o$, ACh	Ca^{2+}, IP_3	–	Palmitoleic acid, halothane

Continued

Table 1 Chronological summary of published investigations into VSMC to EC signaling and myoendothelial feedback.—cont'd

	Model/vascular bed	Vasoreactivity methodology	Ca^{2+} imaging methodology	VSM-dependent activator of ECs/ feedback vasodilation	Proposed VSM-derived signal	Proposed mechanism of feedback vasodilation	Inhibitors of VSM-dependent EC activation/ feedback vasodilation
Tuttle and Falcone (2001)	Rat cremaster muscle	Arterial strips, PM	ECs or VSMCs (pressurized arteries)	NE, PE	–	NO	–
Lamboley et al. (2005)	Rat mesentery	PM	ECs and VSMCs simultaneously (*en face* arteries)	Elevated $[K^+]_o$, PE	IP_3	–	Palmitoleic acid, U 73122, 2-APB
Isakson, Ramos, and Duling (2007)	Mouse aorta	VSMC/EC co-culture	Cultured ECs and VSMCs simultaneously	PE	Ca^{2+}, IP_3	–	BAPTA, 18α-glycyrrhetinic acid, thapsigargin
Kansui, Garland, and Dora (2008)	Rat mesentery	PM	ECs and VSMCs simultaneously (pressurized arteries)	Intraluminal pressure, PE, elevated $[K^+]_o$	–	–	Nifedipine, U 73122, 18α-glycyrrhetinic acid, carbenoxolone, reduced $[Ca^{2+}]_o$, SKF 96365, $NiCl_2$, KB-R 7943
Jackson et al. (2008)	Hamster cremaster muscle	PM	Freshly isolated ECs, ECs or VSMCs (pressurized arteries)	PE, NE	–	–	–
Nausch et al. (2012)	Mouse mesentery	PM	ECs (*en face* arteries)	EFS	IP_3	EDH (IK_{Ca})	U 73122

Tran et al. (2012)	Hamster retractor muscle	PM, sharp microelectrode	ECs (pressurized and *en face* arteries)	PE	IP_3	EDH (SK_{Ca}, IK_{Ca})	BAPTA, 2-APB, thapsigargin, xestospongin C
Bagher et al. (2012)	Rat cremaster muscle	PM	ECs and VSMCs simultaneously (pressurized arteries)	Low intraluminal pressure	–	EDH (IK_{Ca})	U 73122, reduced $[Ca^{2+}]_o$, xestospongin C, RN 1734
Kerr et al. (2015)	Rat brain and mesentery	PM, WM, sharp microelectrode	–	5-HT, PE	–	NO, EDH (IK_{Ca})	18α-Glycyrrhetinic acid, xestospongin C, Pyr3
Garland et al. (2017)	Rat cremaster muscle	PM	ECs (pressurized arteries and EC tubes)	PE, elevated $[K^+]_o$, Bay-K 8644, 4-AP	Ca^{2+}, IP_3	EDH (IK_{Ca})	Nifedipine, heparin
Looft-Wilson et al. (2017)	Mouse mesentery	PM	–	PE	–	NO	–
Hong et al. (2018)	Mouse mesentery	PM	ECs (*en face* arteries)	PE, U 46619	IP_3	NO, EDH (SK_{Ca}, IK_{Ca})	U 73122, GSK 2193874, 18β-glycyrrhetinic acid
Wei et al. (2018)	Rat mesentery	PM, WM, perfused mesenteric bed	–	NE, U 46619, EFS	IP_3	NO, EDH (SK_{Ca}, IK_{Ca})	–
Mishra et al. (2018)	Rat cremaster muscle	PM, sharp microelectrode	–	PE	Ca^{2+}	NO, EDH	Nifedipine
Wilson et al. (2019)	Rat mesentery	Pinned arterial strips	ECs (*en face* arteries)	PE	IP_3	–	BAPTA

PM, pressure myography; WM, wire myography.

Ca^{2+} signal passed from the VSM to the ECs through myoendothelial gap junctions (Dora et al., 1997). An increase in EC $[Ca^{2+}]_i$ to these vasoconstrictors was confirmed soon after in rat mesenteric arteries (Kansui et al., 2008; Oishi et al., 2001). Importantly, it was established that VGCCs and α_1-ARs are not expressed in the endothelium (Angus, Cocks, & Satoh, 1986b; Nilius, Viana, & Droogmans, 1997), a point since corroborated in freshly isolated ECs and EC tubes using both molecular and functional approaches (Dora et al., 2000; Garland et al., 2017; Jackson, Boerman, Lange, Lundback, & Cohen, 2008; Lamboley et al., 2005), strongly suggesting VSMC Ca^{2+} influx was obligatory for an increase in EC $[Ca^{2+}]_i$. Potentiated vasoconstriction to PE or to elevated $[K^+]_o$ during inhibition of NOS was consistent with previous studies with isolated arteries in the absence of intraluminal flow (Aldasoro, Martinez, Vila, Medina, & Lluch, 1996; Bennett, Watt, & Thurston, 1992; Martínez et al., 1995; Shinozuka, Kobayashi, Shimoura, & Hattori, 1992; Toda, Yoshida, & Okamura, 1991). Soon after the initial observation of increasing EC $[Ca^{2+}]_i$, activation of α_1-ARs or opening of VGCCs (in VSM) was also shown to elevate EC Ca^{2+} in cremaster arterioles (Tuttle & Falcone, 2001; Yashiro & Duling, 2000), mesenteric (Oishi et al., 2001; Schuster et al., 2001) and coronary arteries (Budel et al., 2001), indicating that MEF may be a ubiquitous signaling mechanism for resistance arteries. Notably, Schuster et al. (2001) showed that increases in EC $[Ca^{2+}]_i$ were blocked by a gap junction uncoupler, palmitoleic acid, without altering vasoconstrictor $[Ca^{2+}]_i$ increases in the VSM. This observation was significant, as it supported the suggestion of a signal passing through myoendothelial gap junctions (MEGJs) from the VSM to activate the endothelium. If the response was due to activation of receptors on the surface of the endothelium, the inhibitor would not be expected to alter EC $[Ca^{2+}]_i$. However, it is also important to acknowledge gap junction inhibitors such as palmitoleic acid, halothane, carbenoxolone and 18α-glycyrrhetinic acid may also have off-target effects on ECs (Budel et al., 2001). That being said, these original observations have been confirmed in pressurized arteries with 18α-glycyrrhetinic acid and with carbenoxolone (Kansui et al., 2008) and further support for the essential involvement of MEGJs in myoendothelial feedback has now been demonstrated with non-pharmacological approaches, which will be discussed later. The basic principle of signaling to stimulate MEF is summarized in Fig. 2.

With regard to the functional outcome to follow an increase in EC Ca^{2+}, the majority of studies initially focused on the involvement of eNOS activity

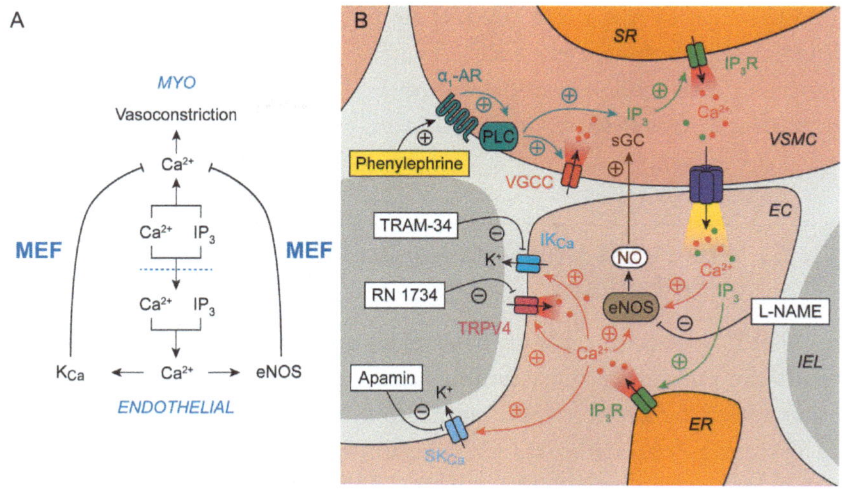

Fig. 2 Schematic summarizing the activation of MEF during vasoconstriction to PE in (A) skeletal form and (B) with signaling pathways and commonly used drugs.

and NO generation. However, Dora et al. (2000) showed that while PE-induced vasoconstriction in rat mesenteric arteries was enhanced by block of NO synthase with L-NAME, it was also enhanced by block of endothelium-dependent hyperpolarization (EDH) using charybdotoxin and apamin to target the EC K_{Ca} channels. Consistent with these data, the same combination of K_{Ca} channel blockers was found to prolong PE-induced vasoconstriction in hamster isolated cremasteric arterioles (Yashiro & Duling, 2000). EDH follows K^+ efflux through endothelial small- and intermediate-conductance K_{Ca} channels (SK_{Ca} and IK_{Ca}; Edwards, Dora, Gardener, Garland, & Weston, 1998; Plane, Holland, Waldron, Garland, & Boyle, 1997), and these were the first instances where a role for their activation was detected without direct EC stimulation with agonists such as ACh. A confounding influence was the fact that as well as blocking IK_{Ca}, charybdotoxin will also block large-conductance (BK_{Ca}) channels found in VSM but not ECs. VSM BK_{Ca} channels are activated basally by Ca^{2+}-sparks and also serve to suppress vasoreactivity (Nelson et al., 1995). However, subsequent studies using the more selective IK_{Ca} channel blocking agent TRAM-34 have confirmed the role of EDH in MEF (Table 1) (Garland et al., 2017; Kerr et al., 2015; Tran et al., 2012). Most pertinently, overall these studies identified a role for EDH in modulating vasoconstriction (alongside NO), and this mechanism seems to be a primary driver for MEF vasodilation in small resistance arteries and arterioles.

In the span of just a few years, these publications and others (Table 1) provided compelling evidence for EC activation and subsequent endothelium-dependent vasodilation due to the transmission of a signal/s originating in the VSM, giving rise to the notion of an intrinsic regulatory feedback mechanism within the vascular wall: myoendothelial feedback.

5. What VSM signal passes to increase calcium in the endothelium?

MEGJs are permeable to ions and second messengers of <1 KDa, and as such can potentially allow bidirectional electrical and chemical signaling within resistance arteries. Two candidate molecules implicated in EC activation during contraction are Ca^{2+} and inositol 1,4,5-trisphosphate (IP_3). Both these signaling molecules mediate VSM contraction, and in ECs IP_3 receptors (IP_3Rs) are responsible for Ca^{2+} release. In contrast to VSM, ECs do not contain ryanodine receptors. Our understanding of signaling via MEGJs was greatly facilitated by the development of a mouse aortic co-culture model in which the VSM and ECs form MEGJs and with each cell type accessible for selective targeting with pharmacological agents to probe heterocellular transmission (Isakson, 2008; Isakson et al., 2007). Using this model, stimulation of VSM α_1-ARs was shown to evoke an increase in EC Ca^{2+} that persisted even when the IP_3R antagonist xestospongin C was loaded into the ECs, but not when loaded into the VSM. Furthermore, VSMC to EC signaling was not altered by loading the Ca^{2+} chelator BAPTA into the former, or by depleting EC Ca^{2+} stores with the sarco/endoplasmic reticulum Ca^{2+}-ATPase (SERCA) inhibitor thapsigargin. Taken together, these data show that simulating the action of a vasoconstrictor in cultured VSM can lead to the passage of both Ca^{2+} and IP_3 into cultured ECs, informing studies with intact arteries. As an aside, it appears that Ca^{2+} signaling from ECs to VSM can also occur, with only Ca^{2+} but not IP_3 detected in VSM in the co-culture model (Isakson et al., 2007).

So the question becomes, in intact arteries do both Ca^{2+} and IP_3 contribute to the activation of EC feedback via NO and EDH or does one signal predominate? This is a question that is far more difficult to answer than you might think, for a number of reasons. First, available values for diffusion coefficients and effective ranges for Ca^{2+} and IP_3 in extracted cytoplasm suggest IP_3 moves faster and is effective at a distance about five times greater than Ca^{2+} (Allbritton, Meyer, & Stryer, 1992). However, these values were derived under quite contrived conditions, using cytosolic extracts from Xenopus oocytes in the presence of inhibitors to prevent Ca^{2+} sequestration and IP_3 degradation. The diffusion of Ca^{2+} was limited by rapid binding to cytosolic targets, such as calmodulin, as increasing the cytosolic $[Ca^{2+}]$ also increased its diffusion coefficient. In contrast, increasing IP_3 concentration did not alter the rate of diffusion. The diffusion constants derived from this study, $\sim 38\,\mu m^2/s$ for Ca^{2+} and $\sim 280\,\mu m^2/s$ for IP_3 have been widely accepted ever since. However, more recent data suggest the diffusion of IP_3 is actually much slower (Dickinson, Ellefsen, Dawson, Pearson, & Parker, 2016). In a neuroblastoma cell line, SH-SY5Y cells, the release of a caged but poorly metabolizable form of IP_3, generated a diffusion coefficient around $10\,\mu m^2/s$, assessed by visualizing local IP_3-evoked Ca^{2+} puffs. This value is approximately 30 times slower than previously thought and attributed to binding to immobile IP_3Rs, which are more concentrated and uniformly distributed in mammalian cells than in cytosolic extracts from Xenopus oocytes. Thus, in reality, the effective range for Ca^{2+} and IP_3 may actually be very similar. This suggestion is supported by studies using photolysis of caged IP_3 in the dendrites of cerebellar Purkinje cells, showing Ca^{2+} signals with a radius of just a few microns (Finch & Augustine, 1998). Second, the EC signaling pathways underpinning MEF occur within a short distance of the adjacent VSMs. Many key proteins, and therefore binding sites, for both Ca^{2+} and IP_3 cluster within the restricted space of the myoendothelial projections that link VSM and ECs. Principal sites are IK_{Ca} channels, ER IP_3Rs and eNOS (Bagher et al., 2012; Ledoux et al., 2008; Straub et al., 2011), and as these underpin the activation of EDH and the generation of NO responsible for MEF (Table 1), attempting to define and distinguish separate signals due to IP_3 and Ca^{2+} in the EC bulk cytoplasm may be of little functional relevance. A third consideration is that both Ca^{2+} and IP_3 are necessary for IP_3-mediated Ca^{2+} release from the ER, and this interdependence will complicate any attempt to separate these signals. Finally, regardless of whether IP_3 and/or Ca^{2+} passes to ECs, activation of EC IP_3Rs or Ca^{2+} influx pathways such as transient receptor potential (TRP) channels will then be augmented by adjacent receptors and channels which may themselves not be the primary target of the heterocellular signal.

That being said, evidence to support a role for Ca^{2+} ion as a transferable signal has come from experiments involving direct activation of VSM VGCCs. Thus, VSM depolarization and vasoconstriction to elevated $[K^+]_o$, which does not rely on the generation of IP_3, leads to a parallel increase in EC $[Ca^{2+}]_i$ (Dora et al., 1997; Lamboley et al., 2005; Oishi et al., 2001) with vasoconstriction potentiated by endothelial denudation (Budel et al., 2001). Direct pharmacological manipulation of L-type VGCCs also supports the notion VSM Ca^{2+} influx initiates myoendothelial feedback. Selective activation of L-type VGCCs with Bay-K 8644 increased local and propagating EC Ca^{2+} events in pressurized rat cremasteric arterioles under basal myogenic tone, and these events were not recapitulated in EC tubes isolated from the same vessels and devoid of VSM (Garland et al., 2017). Furthermore, block of L-type VGCCs with nifedipine markedly attenuated both spontaneous and PE-induced EC Ca^{2+} activity in pressurized rat cremasteric arteries, again supporting a central role for VSM VGCC Ca^{2+} influx (Dora et al., 1997; Garland et al., 2017; Kansui et al., 2008). Consistent with these observations is the demonstration that EDH-associated vasodilation, which is due to activation of EC SK_{Ca} and IK_{Ca} channels, was enhanced in myogenically-active rat cremasteric arterioles constricted with PE. The enhanced vasodilation was abolished in the presence of nifedipine (Mishra et al., 2018). However, if VSM is activated with PE rather than by simply opening VGCCs, nifedipine only partially inhibits the appearance of EC Ca^{2+} events (Dora et al., 1997; Garland et al., 2017; Kansui et al., 2008; Mishra et al., 2018). So not surprisingly, it seems Ca^{2+} is unlikely to be the only signal transmitted to the endothelium.

Support for IP_3 as the primary signal passing to the endothelium to activate MEF comes from experiments activating perivascular sympathetic nerves with electrical field stimulation (EFS) in *en face* mouse mesenteric arteries (Nausch et al., 2012) (Table 1). EFS initiated small EC Ca^{2+} events, defined as pulsars, restricted close to holes in the internal elastic lamina (IEL) that represent sites of MEGJs. Recruitment of Ca^{2+} pulsars was attenuated by the voltage-gated Na^+ channel blocker tetrodotoxin, the selective α_1-AR antagonist prazosin, and the non-selective inhibitor of phospholipase C (PLC) U 73122, yet were unaffected by block of L-type VGCCs with nifedipine, arguing for a central role for IP_3 (Table 1). Additionally, in hamster retractor muscle arterioles studied *en face*, PE stimulated EC Ca^{2+} transients, which were prevented with either the non-selective IP_3R antagonist 2-aminoethoxydiphenylborate (2-APB) or the more selective IP_3R antagonist xestospongin C (Tran et al., 2012) (Table 1). Most recently, activation of L-type VGCCs with Bay-K 8644 was reported not to evoke any increase in EC $[Ca^{2+}]_i$ in cut open *en face* mouse mesenteric arteries exposed to the SERCA inhibitor cyclopiazonic acid (CPA) to deplete stored Ca^{2+} (Hong et al., 2018). This report is consistent with the observations of

Lamboley et al. (2005) using rat mesenteric arteries, where the authors used both PE and KCl in the presence of 500 nM U 73122 and 100 μM 2-APB and found full block of the usually robust vasoconstrictor-evoked EC Ca^{2+} signals (Table 1). These studies strongly support a role for IP_3 rather than Ca^{2+} as the heterocellular signal in this vascular bed.

Perhaps the most direct evidence for detectable and effective amounts of Ca^{2+} passing from VSM to ECs during vasoconstriction comes from the use of heparin to block IP_3Rs in ECs. In these experiments responses to PE, which could rely at least partly on IP_3 diffusion from VSM, were compared to responses to the activators of L-type VGCCs KCl and Bay-K 8644, which should be independent of IP_3 diffusion (Garland et al., 2017) (Table 1). Responses to all vasoconstrictors were markedly blocked, suggesting that the Ca^{2+} passing from VSM to ECs is amplified by further release by EC IP_3Rs. The selective EC loading of fluorescent heparin (heparin-Cy5, a pan-IP_3R blocker) in pressurized cremasteric arteries closely aligned with IP_3R1 and simultaneous imaging of EC Ca^{2+} activity and elastin revealed discrete heparin-insensitive EC Ca^{2+} events that rapidly disappeared both temporally (duration, ~1 to 2 s) and radially (total width, <15 μm, often 5–10 μm) such that residual events remained and were localized to the region of myoendothelial projections in holes through the IEL. These localized events were termed VECTors, VGCC-dependent endothelial calcium transients. Compared to untreated cells, heparin loading halved the frequency of EC Ca^{2+} events stimulated by 30 nM Bay-K 8644, did not affect the number of active cells, yet halved the incidence of Ca^{2+} waves. Despite the loss of Ca^{2+} signaling, the transient receptor potential cation channel subfamily V member 4 (TRPV4) agonist GSK 1016790A was able to increase endothelial cell Ca^{2+} by direct stimulation of the cells. Therefore, Bay-K 8644-induced VECTors reflect Ca^{2+} passing to the ECs via MEGJs where they are normally amplified by endothelial cell IP_3Rs, forming Ca^{2+} puffs evolving into propagating cell-wide Ca^{2+} waves.

Overall, there is considerable evidence to suggest both Ca^{2+} and IP_3 pass through MEGJs from VSM to ECs to activate MEF, which then may use both EDH and NO to suppress vasoconstriction.

6. Confounding factors to consider

6.1 Does depolarization to KCl activate PLC to generate IP_3 in VSMCs?

One note of caution with experiments that elevate VSM Ca^{2+} by direct, receptor-independent activation of VGCCs, is that the increase in cytoplasmic Ca^{2+} may be sufficient to generate IP_3 simply by stimulating the Ca^{2+}-sensitive isozyme PLC-δ. It is important to define whether block of PLC with U 73122 or IP_3Rs with 2-APB alter VSM Ca^{2+} increases. One report suggests they do not (Lamboley et al., 2005), but this should be tested in each vascular bed to avoid off-target effects. In freshly isolated guinea pig colonic myocytes, 1 μM U 73122 decreased Ca^{2+} transients that followed the release of caged IP_3, which was attributed to an inhibition of SERCA and depletion of the intracellular Ca^{2+} store (Macmillan & McCarron, 2010). In rat cremasteric arterioles, exposure to 3 μM U 73122 inhibited $[K^+]_o$-evoked vasoconstriction and compromised the integrity of the endothelial cell barrier. Off target effects of 2-APB include block of connexins at the MEGJ (e.g., connexins 40 and 43) (Bai, del Corsso, Srinivas, & Spray, 2006), and when used at 100 μM, 2-APB itself caused asynchronous, slow cycles of vasoconstriction and vasodilation, and full block of vasodilation to the allosteric EC K_{Ca} channel modulator, NS 309 (to activate EDH), consistent with block of MEGJs (Garland et al., 2017). Indeed, 2-APB might be considered a useful gap junction uncoupler if it were not for the fact it also affects Ca^{2+} channels.

6.2 Do functional vascular responses match EC Ca^{2+} signaling?

Due to the technical difficulty in observing EC Ca^{2+} signals during vasoconstriction in pressurized or tensioned arteries, the majority of studies have monitored Ca^{2+} signaling in unpressurized arteries (Table 1). Although both function and EC Ca^{2+} events can be assessed in pinned arteries, usually cut open (Lamboley et al., 2005; Wilson et al., 2019) the functional readout cannot accurately be placed in a physiological context, especially if VSM length-tension has not been optimized. This would not only confound measurements of contraction but could also alter voltage-dependent processes, such as activation of VGCCs in arteries where stretch-mediated depolarization processes have not been effectively activated.

7. The long and short of EC Ca^{2+} activity during MEF vasodilation

This section provides a chronological description of EC Ca^{2+} events driving MEF, their nomenclature, and spatial and temporal characteristics. These Ca^{2+} signals can be visualized by loading the endothelium with Ca^{2+}-sensitive indicators or utilizing transgenic mice expressing a genetically encoded Ca^{2+} indicator exclusively in the endothelium.

7.1 MEF in response to bolus delivery of vasoconstrictors

When first described in pressurized hamster cheek pouch arterioles, short picospritz pulses were used to deliver agents as a bolus, so the effective concentration was not known (Dora et al., 1997). In these experiments, activation of VSM α_1-ARs with 10 μM PE or depolarization following 250 mM $[K^+]_o$ caused a rapid increase in EC fluorescence measured with photomultiplier tubes, allowing temporal resolution of a calibrated signal at 5 Hz, but without any spatial resolution. The increase in fluorescence signal was equivalent to a change in $[Ca^{2+}]_i$ of ~150 nM and of a similar magnitude to responses elicited by picospritzing 100 μM ACh directly to activate the endothelium (Dora et al., 1997). Crucially, the peak Ca^{2+} response in VSM and ECs occurred prior to the onset of vasoconstriction, suggesting the contraction was not obligatory for heterocellular communication. In hamster cremasteric arterioles, picospritzing 1 mM PE caused a similar magnitude of EC Ca^{2+} increase, but due to lower temporal resolution (1 Hz) the peak change in EC Ca^{2+} coincided more closely with the peak change in diameter (Yashiro & Duling, 2000). This problem was circumvented using a ratiometric Ca^{2+} indicator, fura-2, and a camera-based system to closely monitor signal integrity during motion. In other experiments, bath-applied α_1-AR agonists (<10 μM) also caused transient global EC Ca^{2+} events of variable magnitude in rat mesenteric (Oishi et al., 2001; Schuster et al., 2001), rat cremasteric (Tuttle & Falcone, 2001) and hamster cremasteric vessels (Jackson et al., 2008) (Table 1). In each case, when fluorescence was measured simultaneously with diameter change, EC $[Ca^{2+}]_i$ increases also preceded or coincided with vasoconstriction, rather than after the fact. Any cell-wide responses were attributed to the mobilization of Ca^{2+} from the endoplasmic reticulum (ER).

7.2 Rapid image acquisition is necessary for temporal resolution of EC Ca^{2+} events

Subcellular EC Ca^{2+} activity during stimulation with submaximal concentrations of PE and extracellular K^+ was first described in pressurized rat mesenteric arteries (Kansui et al., 2008), aided by the use of high sensitivity, temporal and spatial resolution imaging with the Nipkow spinning-disk confocal microscope. As opposed to eliciting a global increase in EC fluorescence, the vasoconstriction was accompanied by an increase in the frequency of localized EC Ca^{2+} signals appearing to originate from the opening of IP_3Rs in the ER. These discrete events were postulated to be capable of activating neighboring IP_3Rs to initiate a Ca^{2+} "wavelet" to open adjacent SK_{Ca} and IK_{Ca} channels in the plasma membrane to hyperpolarize ECs (and thereby VSM) to modulate vascular tone, similar to the way in which Ca^{2+} release through ryanodine receptors in cerebral artery myocytes limits contraction by activating BK_{Ca} channels (Nelson et al., 1995). Notably, discreet wavelets propagated within ECs to form "cell-wide waves" (CWWs). The formation of CWWs requires the presence of both Ca^{2+} and IP_3 to activate calcium-induced Ca^{2+} release from IP_3Rs. These events may also be amplified by Ca^{2+} influx channels sensitive to Ca^{2+} and/or IP_3, which were blocked non-selectively with SKF-96365, although TRP channel blockers were not considered at the time. Gap junction uncouplers used to prevent the movement of signals from VSM to ECs did block EC Ca^{2+} events to PE and KCl, and importantly did not block EC Ca^{2+} responses to ACh in control experiments.

7.3 Not all EC Ca^{2+} events evolve into CWWs

At around the same time, Ledoux et al. (2008) described local EC Ca^{2+} events in *en face* mouse mesenteric arteries restricted to holes in the IEL, which are synonymous with myoendothelial projections, referring to them as Ca^{2+} "pulsars." Direct activation of ECs with ACh increased the frequency of these fixed, spontaneous pulsars, as well as global $[Ca^{2+}]_i$. Furthermore, Ca^{2+} pulsars were completely abrogated using the SERCA inhibitor CPA to interfere with Ca^{2+} release from the ER/SR, and by either block of VSM and EC PLC with U 73122 or IP_3Rs with xestospongin C. The authors concluded that Ca^{2+} pulsars represent a fundamental component of ACh-induced endothelium-dependent vasodilation, providing a means for Ca^{2+} release via IP_3Rs localized to myoendothelial projections to activate closely associated K^+ channels in the EC membrane. Another type of subcellular Ca^{2+} event was described in the context of myoendothelial feedback using a spinning-disk imaging setup in arterioles isolated from the hamster retractor muscle (Tran et al., 2012). These local EC Ca^{2+} events were recorded during adrenergic stimulation. They appeared to be distinct from pulsars as they were of longer duration, larger area and greater frequency, and as such designated Ca^{2+} wavelets. Wavelets occurred spontaneously or following exposure to PE, and were sensitive to block of IP_3Rs with 2-APB or xestospongin C, not dissimilar to findings by Kansui et al. (2008). It is now clear that subcellular EC Ca^{2+} events such as pulsars or wavelets may occur spontaneously and their origin can be attributed to either Ca^{2+} influx and/or release from stores. This can be assessed by removing extracellular Ca^{2+}, using selective Ca^{2+} influx blockers (e.g., blockers of TRP channels) or blockers of IP_3Rs, and may be linked to MEF; an aspect not yet fully explored in all vascular beds, and not reviewed here. Superimposed on any such defined spontaneous events the responses to vasoconstrictors can then be established and attributed to Ca^{2+} and/or IP_3 movement through MEGJs. Importantly, both spontaneous and vasoconstrictor-mediated EC Ca^{2+} responses appear to be restricted to the vicinity of sites of physical contact between ECs and VSMCs, with a dependence on Ca^{2+} release through EC IP_3Rs to initiate and/or amplify EC Ca^{2+} events. Clearly there is no set universal source of EC Ca^{2+} events linked to MEF when considering the multiple vascular beds and species studied (again, yet to be fully explored, Table 1 provides some detail), and the forms of EC Ca^{2+} activity during vasoconstriction are also heterogeneous.

7.4 Amplification of EC Ca^{2+} events by TRP channels

Influx of extracellular Ca^{2+} through TRP channels has also been suggested to play a role in endothelium-dependent vasodilation and myoendothelial feedback in resistance arteries (Sonkusare et al., 2012). EC Ca^{2+} events in *en face* mouse mesenteric arteries followed the activation of TRPV4s with GSK 1016790A. These signals, termed "sparklets," persisted during block of SERCA with CPA and PLC with U 73122, suggesting TRPV4 Ca^{2+} influx is not amplified by intracellular Ca^{2+} store release. Furthermore, and unlike pulsars, sparklets were not restricted to the region close to holes in the IEL, and exhibited a very short, but detectable, plateau phase of discrete amplitudes, indicative of single-channel gating. In pressurized arteries, GSK 1016790A evoked endothelium-dependent vasodilation insensitive to block of NOS or SERCA, but was sensitive to the blockers of K_{Ca} channels charybdotoxin and apamin, suggesting an association of endothelial TRPV4 channels and K_{Ca} channels. Additionally, artery relaxation following direct stimulation of endothelial receptors was attenuated by TRPV4 block. The authors concluded that TRPV4 activation was obligatory for endothelium-dependent vasodilation. With respect to feedback, a role for sparklets was also demonstrated in *en face* mouse mesenteric arteries, with PE or the thromboxane A_2 receptor agonist U 46619 stimulating EC Ca^{2+} sparklets sensitive to the TRPV4 blocker GSK 2193874 or the PLC inhibitor U 73122 (Hong et al., 2018). Compellingly, GSK 2193874 or genetic ablation of TRPV4 attenuated the feedback vasodilation observed following PE-induced vasoconstriction in pressurized arteries. That PLC inhibition prevented PE-induced sparklet activation led to the suggestion VSM IP_3 generated during vasoconstriction passes through MEGJs to directly activate TRPV4 channels, rather than requiring amplification of the signal by binding to IP_3Rs in the ECs. This aspect warrants further consideration. Involvement of other TRP channels localized to myoendothelial junctions (MEJs) has also been suggested, such as transient receptor potential cation channel subfamily C member 3 (TRPC3) in basilar arteries (Kerr et al., 2015).

7.5 Vascular-bed specific thresholds for developing MEF

Another important concept to consider when measuring MEF is the possibility of a threshold for activation of EC Ca^{2+} signaling. In myogenically-active rat cremaster muscle arterioles, myogenic tone per se is associated with a low frequency of EC Ca^{2+} events, and the tone is not augmented following block of NOS or EC SK_{Ca} or IK_{Ca} channels (Bagher et al., 2012; Garland et al., 2017). Furthermore, the functional response to PE is a peak response followed by fade in contraction, sometimes observed as vasomotion (Fig. 1). When EC Ca^{2+} events were measured during the onset and progression of constriction it was clear that substantial contraction occurred before the ECs became active: first as localized wavelets, and as contraction developed, cycles of propagating CWWs became evident, not unlike responses to direct activation of the ECs with ACh. Remarkably, this was true also for responses to openers of L-type VGCCs, KCl and Bay-K 8644. This delay in onset of activation of MEF beyond a "threshold" level of depolarization or VSM Ca^{2+} signaling was proposed to enable the maintenance of myogenic tone at rest in muscle, serving to help limit unnecessary blood flow to inactive muscle (Garland et al., 2017). The as yet unexplored inference is that tissues such as the heart and brain may not have the same level of intrinsic regulation of EC-dependent MEF. It is already known that, for example, the cerebral and cremasteric microcirculation have different mechanisms for feedback regulation of myogenic tone intrinsic to the VSM (see chapter "Ion channels and the regulation of myogenic tone in peripheral arterioles" by Jackson).

Since endothelial Ca^{2+} activity was first described during vasoconstriction, focus has shifted toward the importance of local Ca^{2+} events being responsible for the activation of nearby K^+ channels that facilitate feedback dilation (Fig. 3). However, how the various types of Ca^{2+} events evolve and integrate during contraction in vivo has yet to be determined.

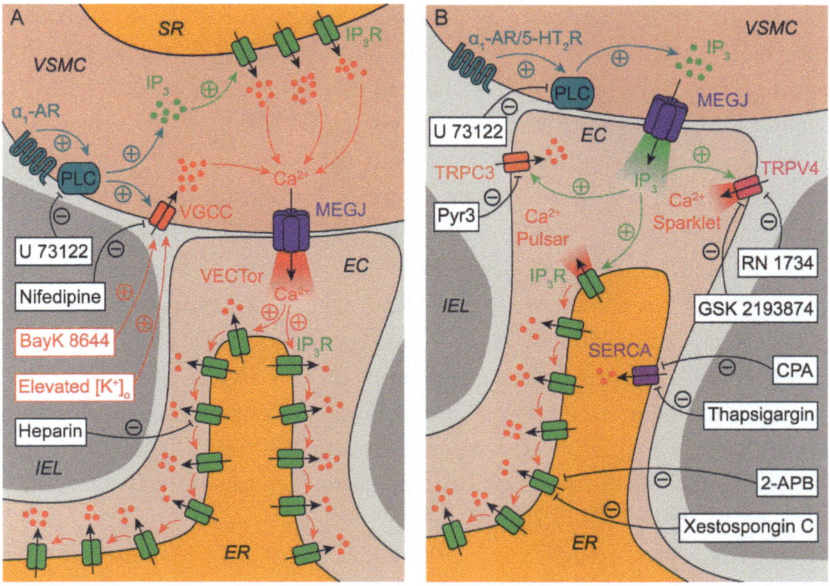

Fig. 3 Schematics of the proposed mechanisms of Ca^{2+} (A) and IP_3 (B) transmission through MEGJs during myoendothelial feedback, the resulting Ca^{2+} mobilization pathways, and the pharmacological agents used to probe them.

8. Regulation of MEF

Following activation of ECs by a Ca^{2+} signal, the resulting generation of vasodilators may vary between vascular beds. There are multiple possible reasons for response heterogeneity and these include (1) the expression of proteins producing the dilator pathways, such as NOS and K_{Ca} channels; (2) the expression of proteins responsible for amplifying responses, such as Ca^{2+} release and Ca^{2+} influx channels; (3) anchoring of proteins within the microdomain associated with the MEJ; (4) regulation of the MEGJs themselves, and (5) buffering of the released signal before reaching the VSM.

8.1 Establishing the localization of key signaling intermediates to MEJs

Expression profiles can be probed by immunolabeling whole mount arteries, either cut and pinned *en face* so the EC surface is exposed, or by pumping antibodies into the lumen of cannulated and pressurized arteries and imaging while pressurized. These approaches have established that IK_{Ca} channels in particular are highly localized to the MEJs, within ECs, in multiple vascular beds and species (Bagher et al., 2012; Dora, Gallagher, McNeish, & Garland, 2008; Kerr et al., 2015; Ledoux et al., 2008; Sandow, Neylon, Chen, & Garland, 2006). This enables 3D imaging through the artery wall showing, with reasonable resolution, the relative spatial arrangement of proteins within ECs, VSMCs and their alignment (or not) with holes in the IEL. Very high spatial resolution can be obtained by immunogold labeling and transmission electron microscopy in slices through the MEJ (Straub et al., 2010, 2011, 2012; Yeh, Rothery, Dupont, Coppen, & Severs, 1998) but rarely are images reported in 3D. The advantage of this approach is it defines MEGJs, or at least membrane appositions, as well as localizing the proteins of interest. Finally, tissue fractionation into ECs, VSMCs and MEJs using co-culture has been used successfully not only to detect proteins of interest but also allowing screening for proteins enriched in the MEJ fraction, often combined with the visual localization approaches (Isakson & Duling, 2005; Straub et al., 2010, 2011, 2012).

Similarly, as the complex involvement of TRP channels emerges, their localization to the MEJ is also being established. To date, both TRPV4 (Bagher et al., 2012) and TRPC3 (Kerr et al., 2015; Senadheera et al., 2012) channels have been shown aligned with holes in the IEL and alongside EC IK_{Ca} channels. Both these channels are known to be activated by diacylglycerol, and while Ca^{2+} or IP_3 can activate TRPV4 channels, actions at TRPC3 channels are less clear. How these channels are targeted and anchored within the MEJ is not known, but the PKC-anchoring protein AKAP150 appears to align PKC with TRPV4 channels in EC projections and, again due to the proximity to IK_{Ca} channels, can facilitate MEF (Sonkusare et al., 2014).

8.2 Regulation of MEGJs during MEF

Connexins forming MEGJs are also subject to regulation (Billaud et al., 2014). In terms of MEF, one key observation has shown PE can modulate connexin opening. Phosphorylation of connexin 43 at serine 368 by a PKC-dependent mechanism (here on the VSM-side of the junction) reduced Ca^{2+} and dye movement through the channel pore, which acts to reduce MEF (Straub et al., 2010). From the EC-side of the junction, released NO appears to nitrosylate connexin 43 at serine 271, increasing conductance through the channel, potentially facilitating MEF. In the presence of PE the regulation of connexin 43 would be biphasic or perhaps oscillatory and certainly complex. The time course of *S*-nitrosylation shows a transient de-nitrosylation of connexin 43 with an associated reduction in Ca^{2+}

signaling to the ECs (following uncaging of IP_3 in VSM) (Straub et al., 2011). Based on these observations alone, PE-initiated heterocellular Ca^{2+} signaling to ECs will generate NO, which may then serve to counter any PKC-dependent inhibition of connexin 43 opening. Then the EC Ca^{2+} (or other) signal acts via *S*-nitrosoglutathione reductase (GSNOR), de-nitrosylating connexin 43 and reducing Ca^{2+} signals through the MEGJ. This circuit then reduces GSNOR activation enabling NO to re-nitrosylate the connexins. It will clearly be important to establish the temporal dynamics of these processes and other possible mechanisms that may modulate, in particular, the opening of connexin 43 and connexin 40.

8.3 A role for hemoglobin α in buffering NO

Another consideration regarding the regulation of MEF is the potential for EC projections to buffer NO. Hemoglobin α (Hbα) expression is suggested to be enriched in EC projections (MEJs), and depending on the oxidation state of iron may bind NO and prevent movement to the adjacent VSM, thereby reducing MEF both due to reduced actions of NO in VSM and at connexin 43 in the MEGJ. In the Fe^{3+} state Hbα permits NO signaling by releasing NO, but following reduction by endothelial cytochrome b5 reductase 3 (CytB5R3) retains NO. Putting these concepts together (Fig. 4), the regulation of CytB5R3 activity could be fundamental in establishing whether NO either acts as a diffusible vasodilator in VSM, and/or modulates the ability of EDH and MEF to stimulate vasodilation/feedback dilation. How these processes might help unravel any interaction between NO and EDH signaling and interpret non-additive data, remains to be established.

8.4 Does Ca^{2+}-buffering paradoxically enable MEF?

The complexity of the signaling involved in MEF is further evident by the possibility of Ca^{2+} buffering external to the ER within the MEJ (Biwer et al., 2018). This highly novel and, on the face of it, confounding method of regulation relies on a signal derived during contraction to PE (e.g., Ca^{2+} or IP_3) causing the accumulation of free Ca^{2+} in the MEJ only when calreticulin, a Ca^{2+}-binding protein, is expressed at the MEJ. In EC-specific calreticulin knockout mice, PE-mediated EC Ca^{2+} events were reduced and contraction to PE was augmented ex vivo (Biwer et al., 2018). This suggests EC calreticulin either normally enhances the ability of IP_3 to release Ca^{2+} associated with MEF, directly allows Ca^{2+} itself to activate MEF, and/or other as yet undefined signaling intermediates are involved.

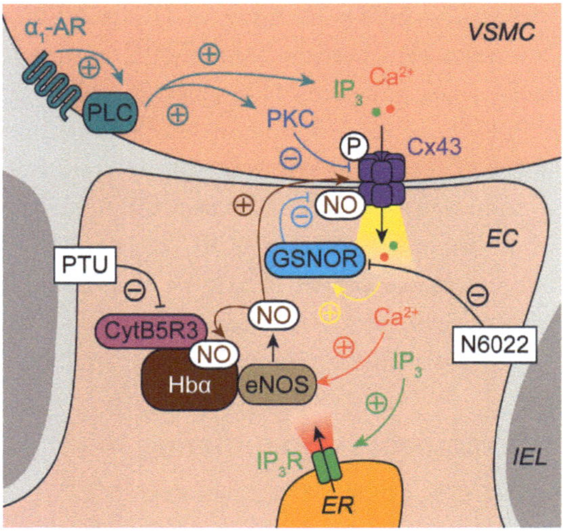

Fig. 4 Schematic of the proposed mechanisms of regulation of MEGJs during MEF. PKC-dependent phosphorylation of connexin 43 (Cx43) inhibits heterocellular Ca^{2+} signaling, while *S*-nitrosylation of Cx43 acts to increase channel conductance. The oxidation state of NO and hemoglobin α (Hbα) affects the bioavailability of NO for actions within VSM alongside those at Cx43. PTU, propylthiouracil, inhibiter of CytB5R3 (Straub et al., 2012); N6022, inhibitor of GSNOR (Green et al., 2012); P, phosphorylated Cx43.

9. Influence of MEF on blood pressure in vivo

An ability to study and define the contribution of MEF to blood pressure in vivo is not simple. Blocking NO or EDH in vivo would also include block of hemodynamic and EC-agonist activated pathways, which cannot be interrogated independently. Isakson's group has instead used the novel approach of infusing PE in vivo. While this could still be interpreted in different ways, it does support the idea that PE contraction in vivo is usually modulated. TRPV4 channel knockout mice had comparable blood pressures to wild-type controls, yet acute bolus infusion of PE significantly increased blood pressure (peak change almost double compared to control) in the TRPV4 channel knockout mice (Hong et al., 2018). The observation that VSM TRPV4 channels are linked to feedback via VSM BK_{Ca} channels in other vascular beds (Earley, Heppner, Nelson, & Brayden, 2005) was not assessed. Nevertheless, both EC- and VSM-intrinsic pathways may operate in vivo. Another approach demonstrating a possible role for MEF in vivo involved knockout of calreticulin. Biwer et al. (2018) showed that this Ca^{2+} buffering protein which is usually thought confined to the ER is also expressed in high density in the MEJ and appears external to the ER at this location. When calreticulin is knocked out from ECs, the blood pressure of mice is elevated (Biwer et al., 2018). This again is supportive of MEF usually limiting blood pressure, however, the subtlety of establishing whether compromising calreticulin usually found in the ER does affect hemodynamic and agonist-mediated EC responses in vivo has yet to be shown. Despite these possible limitations it is compelling to suggest that this area of research is vital to our understanding of blood pressure control.

10. Future prospects and unanswered questions

While it is clear the endothelium can be activated by vasoconstrictor agents, much has still to be defined. Accurate and reliable measurement of NO release by ECs at rest and during over the entire time course of vasoconstriction would help draw together the established pathways for promoting MEF, both NO dependent and -independent. Dissecting out the mechanisms responsible for generating MEF offers the potential to design ways to enhance this pathway in situations where blood flow is compromised within the microcirculation.

References

Aldasoro, M., Martinez, C., Vila, J. M., Medina, P., & Lluch, S. (1996). Influence of endothelial nitric oxide on adrenergic contractile responses of human cerebral arteries. *Journal of Cerebral Blood Flow and Metabolism*, *16*, 623–628.

Allbritton, N. L., Meyer, T., & Stryer, L. (1992). Range of messenger action of calcium ion and inositol 1,4,5-trisphosphate. *Science*, *258*, 1812–1815.

Amerini, S., Mantelli, L., & Ledda, F. (1995). Enhancement of the vasoconstrictor response to KCl by nitric oxide synthesis inhibition: A comparison with noradrenaline. *Pharmacological Research*, *31*, 175–181.

Angus, J. A., Cocks, T. M., & Satoh, K. (1986a). Alpha 2-adrenoceptors and endothelium-dependent relaxation in canine large arteries. *British Journal of Pharmacology*, *88*, 767–777.

Angus, J. A., Cocks, T. M., & Satoh, K. (1986b). The alpha adrenoceptors on endothelial cells. *Federation Proceedings*, *45*, 2355–2359.

Bagher, P., Beleznai, T., Kansui, Y., Mitchell, R., Garland, C. J., & Dora, K. A. (2012). Low intravascular pressure activates endothelial cell TRPV4 channels, local Ca^{2+} events, and IK_{Ca} channels, reducing arteriolar tone. *Proceedings of the National Academy of Sciences of the United States of America*, *109*, 18174–18179.

Bai, D., del Corsso, C., Srinivas, M., & Spray, D. C. (2006). Block of specific gap junction channel subtypes by 2-aminoethoxydiphenyl borate (2-APB). *The Journal of Pharmacology and Experimental Therapeutics*, *319*, 1452–1458.

Bennett, M. A., Watt, P. A., & Thurston, H. (1992). Endothelium-dependent modulation of resistance vessel contraction: Studies with NG-nitro-L-arginine methyl ester and NG-nitro-L-arginine. *British Journal of Pharmacology*, *107*, 616–621.

Billaud, M., Lohman, A. W., Johnstone, S. R., Biwer, L. A., Mutchler, S., & Isakson, B. E. (2014). Regulation of cellular communication by signaling microdomains in the blood vessel wall. *Pharmacological Reviews*, *66*, 513–569.
Biwer, L. A., Good, M. E., Hong, K., Patel, R. K., Agrawal, N., Looft-Wilson, R., et al. (2018). Non-endoplasmic reticulum-based Calr (Calreticulin) can coordinate heterocellular calcium signaling and vascular function. *Arteriosclerosis, Thrombosis, and Vascular Biology*, *38*, 120–130.
Bryan, N. S., Rassaf, T., Maloney, R. E., Rodriguez, C. M., Saijo, F., Rodriguez, J. R., et al. (2004). Cellular targets and mechanisms of nitros(yl)ation: An insight into their nature and kinetics in vivo. *Proceedings of the National Academy of Sciences of the United States of America*, *101*, 4308–4313.
Budel, S., Schuster, A., Stergiopoulos, N., Meister, J.-J., & Beny, J.-L. (2001). Role of smooth muscle cells on endothelial cell cytosolic free calcium in porcine coronary arteries. *The American Journal of Physiology*, *281*, H1156–H1162.
Carrier, G. O., & White, R. E. (1985). Enhancement of alpha-1 and alpha-2 adrenergic agonist-induced vasoconstriction by removal of endothelium in rat aorta. *The Journal of Pharmacology and Experimental Therapeutics*, *232*, 682–687.
Cocks, T. M., & Angus, J. A. (1983). Endothelium-dependent relaxation of coronary arteries by noradrenaline and serotonin. *Nature*, *305*, 627–630.
Dickinson, G. D., Ellefsen, K. L., Dawson, S. P., Pearson, J. E., & Parker, I. (2016). Hindered cytoplasmic diffusion of inositol trisphosphate restricts its cellular range of action. *Science Signaling*, *9*, ra108.
Dora, K. A., Doyle, M. P., & Duling, B. R. (1997). Elevation of intracellular calcium in smooth muscle causes endothelial cell generation of NO in arterioles. *Proceedings of the National Academy of Sciences of the United States of America*, *94*, 6529–6534.
Dora, K. A., Gallagher, N. T., McNeish, A., & Garland, C. J. (2008). Modulation of endothelial cell $K_{Ca}3.1$ channels during endothelium-derived hyperpolarizing factor signaling in mesenteric resistance arteries. *Circulation Research*, *102*, 1247–1255.
Dora, K. A., Hinton, J. M., Walker, S. D., & Garland, C. J. (2000). An indirect influence of phenylephrine on the release of endothelium-derived vasodilators in rat small mesenteric artery. *British Journal of Pharmacology*, *129*, 381–387.
Earley, S., Heppner, T. J., Nelson, M. T., & Brayden, J. E. (2005). TRPV4 forms a novel Ca^{2+} signaling complex with ryanodine receptors and BK_{Ca} channels. *Circulation Research*, *97*, 1270–1279.
Edwards, G., Dora, K. A., Gardener, M. J., Garland, C. J., & Weston, A. H. (1998). K^+ is an endothelium-derived hyperpolarizing factor in rat arteries. *Nature*, *396*, 269–272.
Eglème, C., Godfraind, T., & Miller, R. C. (1984). Enhanced responsiveness of rat isolated aorta to clonidine after removal of the endothelial cells. *British Journal of Pharmacology*, *81*, 16–18.
Finch, E. A., & Augustine, G. J. (1998). Local calcium signalling by inositol-1,4,5-trisphosphate in Purkinje cell dendrites. *Nature*, *396*, 753–756.
Furchgott, R. F., & Zawadzki, J. V. (1980). The obligatory role of endothelial cells in the relaxation of arterial smooth muscle by acetylcholine. *Nature*, *288*, 373–376.
Garland, C. J., Bagher, P., Powell, C., Ye, X., Lemmey, H. A. L., Borysova, L., et al. (2017). Voltage-dependent Ca^{2+} entry into smooth muscle during contraction promotes endothelium-mediated feedback vasodilation in arterioles. *Science Signaling*, *10*, 1–14.
Green, L. S., Chun, L. E., Patton, A. K., Sun, X., Rosenthal, G. J., & Richards, J. P. (2012). Mechanism of inhibition for N6022, a first-in-class drug targeting S-nitrosoglutathione reductase. *Biochemistry*, *51*, 2157–2168.
Greenberg, S., Diecke, F. P., Peevy, K., & Tanaka, T. P. (1989). The endothelium modulates adrenergic neurotransmission to canine pulmonary arteries and veins. *European Journal of Pharmacology*, *162*, 67–80.

Hong, K., Cope, E. L., DeLalio, L. J., Marziano, C., Isakson, B. E., & Sonkusare, S. K. (2018). TRPV4 (Transient Receptor Potential Vanilloid 4) channel-dependent negative feedback mechanism regulates Gq protein-coupled receptor-induced vasoconstriction. *Arteriosclerosis, Thrombosis, and Vascular Biology*, *38*, 542–554.

Isakson, B. E. (2008). Localized expression of an Ins(1,4,5)P_3 receptor at the myoendothelial junction selectively regulates heterocellular Ca^{2+} communication. *Journal of Cell Science*, *121*, 3664–3673.

Isakson, B. E., & Duling, B. R. (2005). Heterocellular contact at the myoendothelial junction influences gap junction organization. *Circulation Research*, *97*, 44–51.

Isakson, B. E., Ramos, S. I., & Duling, B. R. (2007). Ca^{2+} and inositol 1,4,5-trisphosphate-mediated signaling across the myoendothelial junction. *Circulation Research*, *100*, 246–254.

Itoh, Y., Ma, F. H., Hoshi, H., Oka, M., Noda, K., Ukai, Y., et al. (2000). Determination and bioimaging method for nitric oxide in biological specimens by diaminofluorescein fluorometry. *Analytical Biochemistry*, *287*, 203–209.

Jackson, W. F. (1988). Oscillations in active tension in hamsters aortas: Role of the endothelium. *Blood Vessels*, *25*, 144–156.

Jackson, W. F., Boerman, E. M., Lange, E. J., Lundback, S. S., & Cohen, K. D. (2008). Smooth muscle $alpha_{1D}$-adrenoceptors mediate phenylephrine-induced vasoconstriction and increases in endothelial cell Ca^{2+} in hamster cremaster arterioles. *British Journal of Pharmacology*, *155*, 514–524.

Kansui, Y., Garland, C. J., & Dora, K. A. (2008). Enhanced spontaneous Ca^{2+} events in endothelial cells reflects signalling through myoendothelial gap junctions in pressurized mesenteric arteries. *Cell Calcium*, *44*, 135–146.

Kerr, P. M., Wei, R., Tam, R., Sandow, S. L., Murphy, T. V., Ondrusova, K., et al. (2015). Activation of endothelial IK_{Ca} channels underlies NO-dependent myoendothelial feedback. *Vascular Pharmacology*, *74*, 130–138.

Kojima, H., Nakatsubo, N., Kikuchi, K., Kawahara, S., Kirino, Y., Nagoshi, H., et al. (1998). Detection and imaging of nitric oxide with novel fluorescent indicators: Diaminofluoresceins. *Analytical Chemistry*, *70*, 2446–2453.

Lamboley, M., Pittet, P., Koenigsberger, M., Sauser, R., Beny, J. L., & Meister, J. J. (2005). Evidence for signaling via gap junctions from smooth muscle to endothelial cells in rat mesenteric arteries: Possible implication of a second messenger. *Cell Calcium*, *37*, 311–320.

Ledoux, J., Taylor, M. S., Bonev, A. D., Hannah, R. M., Solodushko, V., Shui, B., et al. (2008). Functional architecture of inositol 1,4,5-trisphosphate signaling in restricted spaces of myoendothelial projections. *Proceedings of the National Academy of Sciences of the United States of America*, *105*, 9627–9632.

Looft-Wilson, R. C., Goodell, C. R., Mutch, C. A., Mutchler, S. M., Miller, K. L., & Guraya, M. (2017). Increased myoendothelial feedback is associated with increased connexin37 and IK1 channel expression in mesenteric arteries of diet-induced hyperhomocysteinemic mice. *Microcirculation*, *24*, 643–653.

Macmillan, D., & McCarron, J. G. (2010). The phospholipase C inhibitor U-73122 inhibits Ca^{2+} release from the intracellular sarcoplasmic reticulum Ca^{2+} store by inhibiting Ca^{2+} pumps in smooth muscle. *British Journal of Pharmacology*, *160*, 1295–1301.

Martin, W., Furchgott, R. F., Villani, G. M., & Jothianandan, D. (1986). Depression of contractile responses in rat aorta by spontaneously released endothelium-derived relaxing factor. *The Journal of Pharmacology and Experimental Therapeutics*, *237*, 529–538.

Martínez, C., Cases, E., Vila, J. M., Aldasoro, M., Medina, P., Marco, V., et al. (1995). Influence of endothelial nitric oxide on neurogenic contraction of human pulmonary arteries. *The European Respiratory Journal*, *8*, 1328–1332.

Mishra, R. C., Rahman, M. M., Davis, M. J., Wulff, H., Hill, M. A., & Braun, A. P. (2018). Alpha$_1$-adrenergic stimulation selectively enhances endothelium-mediated vasodilation in rat cremaster arteries. *Physiological Reports*, *6*, e13703.

Nausch, L. W., Bonev, A. D., Heppner, T. J., Tallini, Y., Kotlikoff, M. I., & Nelson, M. T. (2012). Sympathetic nerve stimulation induces local endothelial Ca^{2+} signals to oppose vasoconstriction of mouse mesenteric arteries. *American Journal of Physiology Heart and Circulatory Physiology*, *302*, H594–H602.

Nelson, M. T., Cheng, H., Rubart, M., Santana, L. F., Bonev, A. D., Knot, H. J., et al. (1995). Relaxation of arterial smooth muscle by calcium sparks. *Science*, *270*, 633–637.

Nilius, B., Viana, F., & Droogmans, G. (1997). Ion channels in vascular endothelium. *Annual Review of Physiology*, *59*, 145–170.

Oishi, H., Budel, S., Schuster, A., Stergiopulos, N., Meister, J., & Beny, J. (2001). Cytosolic-free calcium in smooth-muscle and endothelial cells in an intact arterial wall from rat mesenteric artery in vitro. *Cell Calcium*, *30*, 261–267.

Plane, F., Holland, M., Waldron, G. J., Garland, C. J., & Boyle, J. P. (1997). Evidence that anandamide and EDHF act via different mechanisms in rat isolated mesenteric arteries. *British Journal of Pharmacology*, *121*, 1509–1511.

Rodriguez, J., Specian, V., Maloney, R., Jourd'heuil, D., & Feelisch, M. (2005). Performance of diamino fluorophores for the localization of sources and targets of nitric oxide. *Free Radical Biology & Medicine*, *38*, 356–368.

Rubanyi, G. M., Romero, J. C., & Vanhoutte, P. M. (1986). Flow-induced release of endothelium-derived relaxing factor. *The American Journal of Physiology*, *250*, H1145–H1149.

Sandow, S. L., Neylon, C. B., Chen, M. X., & Garland, C. J. (2006). Spatial separation of endothelial small- and intermediate-conductance calcium-activated potassium channels (K_{Ca}) and connexins: Possible relationship to vasodilator function? *Journal of Anatomy*, *209*, 689–698.

Schuster, A., Hirotaka, O., Beny, J. L., Stergiopulos, N., & Meister, J. J. (2001). Simultaneous arterial calcium dynamics and diameter measurements: Application to myoendothelial communication. *American Journal of Physiology Heart and Circulatory Physiology*, *280*, H1088–H1096.

Senadheera, S., Kim, Y., Grayson, T. H., Toemoe, S., Kochukov, M. Y., Abramowitz, J., et al. (2012). Transient receptor potential canonical type 3 channels facilitate endothelium-derived hyperpolarization-mediated resistance artery vasodilator activity. *Cardiovascular Research*, *95*, 439–447.

Shinozuka, K., Kobayashi, Y., Shimoura, K., & Hattori, K. (1992). Role of nitric oxide from the endothelium on the neurogenic contractile responses of rabbit pulmonary artery. *European Journal of Pharmacology*, *222*, 113–120.

Simonsen, U., Wadsworth, R. M., Buus, N. H., & Mulvany, M. J. (1999). In vitro simultaneous measurements of relaxation and nitric oxide concentration in rat superior mesenteric artery. *The Journal of Physiology*, *516*, 271–282.

Sonkusare, S. K., Bonev, A. D., Ledoux, J., Liedtke, W., Kotlikoff, M. I., Heppner, T. J., et al. (2012). Elementary Ca^{2+} signals through endothelial TRPV4 channels regulate vascular function. *Science*, *336*, 597–601.

Sonkusare, S. K., Dalsgaard, T., Bonev, A. D., Hill-Eubanks, D. C., Kotlikoff, M. I., Scott, J. D., et al. (2014). AKAP150-dependent cooperative TRPV4 channel gating is central to endothelium-dependent vasodilation and is disrupted in hypertension. *Science Signaling*, 7, ra66.

Stankevicius, E., Lopez-Valverde, V., Rivera, L., Hughes, A. D., Mulvany, M. J., & Simonsen, U. (2006). Combination of Ca^{2+}-activated K^+ channel blockers inhibits acetylcholine-evoked nitric oxide release in rat superior mesenteric artery. *British Journal of Pharmacology*, *149*, 560–572.

Straub, A. C., Billaud, M., Johnstone, S. R., Best, A. K., Yemen, S., Dwyer, S. T., et al. (2011). Compartmentalized connexin 43 *S*-nitrosylation/denitrosylation regulates heterocellular communication in the vessel wall. *Arteriosclerosis, Thrombosis, and Vascular Biology, 31*, 399–407.
Straub, A. C., Johnstone, S. R., Heberlein, K. R., Rizzo, M. J., Best, A. K., Boitano, S., et al. (2010). Site-specific connexin phosphorylation is associated with reduced heterocellular communication between smooth muscle and endothelium. *Journal of Vascular Research, 47*, 277–286.
Straub, A. C., Lohman, A. W., Billaud, M., Johnstone, S. R., Dwyer, S. T., Lee, M. Y., et al. (2012). Endothelial cell expression of haemoglobin alpha regulates nitric oxide signalling. *Nature, 491*, 473–477.
Straub, A. C., Zeigler, A. C., & Isakson, B. E. (2014). The myoendothelial junction: Connections that deliver the message. *Physiology (Bethesda), 29*, 242–249.
Tesfamariam, B., Halpern, W., & Osol, G. (1985). Effects of perfusion and endothelium on the reactivity of isolated resistance arteries. *Blood Vessels, 22*, 301–305.
Tesfamariam, B., Weisbrod, R. M., & Cohen, R. A. (1987). Endothelium inhibits responses of rabbit carotid artery to adrenergic nerve stimulation. *The American Journal of Physiology, 253*, H792–H798.
Toda, N., Yoshida, K., & Okamura, T. (1991). Analysis of the potentiating action of N(G)-nitro-L-arginine on the contraction of the dog temporal artery elicited by transmural stimulation of noradrenergic nerves. *Naunyn-Schmiedeberg's Archives of Pharmacology, 343*, 221–224.
Tran, C. H., Taylor, M. S., Plane, F., Nagaraja, S., Tsoukias, N. M., Solodushko, V., et al. (2012). Endothelial Ca^{2+} wavelets and the induction of myoendothelial feedback. *American Journal of Physiology Cell Physiology, 302*, C1226–C1242.
Tuttle, J. L., & Falcone, J. C. (2001). Nitric oxide release during α_1-adrenoceptor-mediated constriction of arterioles. *The American Journal of Physiology, 281*, H873–H881.
Urabe, M., Kawasaki, H., & Takasaki, K. (1991). Effect of endothelium removal on the vasoconstrictor response to neuronally released 5-hydroxytryptamine and noradrenaline in the rat isolated mesenteric and femoral arteries. *British Journal of Pharmacology, 102*, 85–90.
Verrecchia, C., Sercombe, R., & Seylaz, J. (1985). Influence of endothelium on noradrenaline-induced vasoconstriction in rabbit central ear artery. *Clinical and Experimental Pharmacology & Physiology, 12*, 169–179.
Vinet, R., Brieva, C., Pinardi, G., & Penna, M. (1991). Modulation of alpha-adrenergic-induced contractions by endothelium-derived relaxing factor in rat aorta. *General Pharmacology, 22*, 137–142.
Wei, R., Lunn, S. E., Tam, R., Gust, S. L., Classen, B., Kerr, P. M., et al. (2018). Vasoconstrictor stimulus determines the functional contribution of myoendothelial feedback to mesenteric arterial tone. *The Journal of Physiology, 596*, 1181–1197.
Wilson, C., Zhang, X., Buckley, C., Heathcote, H. R., Lee, M. D., & McCarron, J. G. (2019). Increased vascular contractility in hypertension results from impaired endothelial calcium signaling. *Hypertension, 74*, 1200–1214.
Yashiro, Y., & Duling, B. R. (2000). Integrated Ca^{2+} signaling between smooth muscle and endothelium of resistance vessels. *Circulation Research, 87*, 1048–1054.
Yeh, H. I., Rothery, S., Dupont, E., Coppen, S. R., & Severs, N. J. (1998). Individual gap junction plaques contain multiple connexins in arterial endothelium. *Circulation Research, 83*, 1248–1263.

FSC
www.fsc.org
MIX
Papier | Fördert
gute Waldnutzung
FSC® C083411

Kolibri 360 GmbH
Ferdinand-Jühlke-Straße 7
99095 Erfurt, Deutschland
produktsicherheit@kolibri360.de